Flight Center
logy 1957–2009

| 1984 | 1985 | 1986 | 1987 | 1988 | 1989 | 1990 | 1991 | 1992 | 1993 | 1994 | 1995 | 1996 | 1997 | 1998 | 1999 | 2000 | 2001 | 2002 | 2003 | 2004 | 2005 | 2006 | 2007 | 2008 | 2009 |

- Challenger Accident
- MSFC Spacelab Mission Control Opens
- First Tests for Super Light-weight External Tank
- Vision for Space Exploration Announced
- Ares 1-X Seperation Test
- President Reagan Approves Space Station Development Work Begins on Payload Operations Control Center
- Shuttle Returns to Flight
- First United States Microgravity Laboratory–1 Flight 
- Chandra Launched
- Columbia Accident
- Solid Rocket Motor Redesign
- First U.S./MIR Rendezvous
- First Exp. 1 Crew on International Space Station
- LEED Certified Engineering Facility
- Hubble Launched
- United States Microgravity Laboratory–2 Flight
- Ares I & V Launch Vehicle Announced
- Ares 1-X Stacking Begins
- Shuttle Pathfinder Display Alabama Space and Rocket Center
- Gamma-Ray Observatory
- Completed Ares I Preliminary Design Review
- First Spacelab Mission (STS-9)
- First Tethered Satellite System (STS-46)
- Gravity Probe B Launched
- Astro-1
- International Space Station Assembly Begins
- Apollo 11 Anniversary
- Burst and Transient Source Experiment
- MSFC Named to Manage Inertial Upper Stage for Interplanetary Probes
- Major Ares I Element Contracts Awarded
- ISS Water Recovery System Launched
- Program Support Communication Network Becomes Operational
- MSFC Opens Space Optics Manufacturing Technology Center
- MAF R&D Administration Building Groundbreaking
- Transient Pressure Test Article for Return to Flight
- ISS Oxygen Generation System Launched
- Plans Announced for Advanced Solid Rocket Motor 
- Fastrac Engine Tested
- Deep Space Impact Launched
- CRRES (Combined Release Radiation Effects)
- Lightfoot Named Director of Marshall
- MSFC Engineer Jan Davis Selected for Astronaut Corps
- STS–121 Verifies External Tank Redesign
- X-Ray Calibration Facility
- Shuttle Upgrades
- Building 4600
- Heavy Lift Launch Vehicle Studies
- Building 4203
- Redesigned Solid Rocket Motor
- Space Launch Initiative Authority to Proceed Established
- NASA 50th Anniversary
- Dr. J. Wayne Littles Center Director
- Fermi Gamma-Ray Space Telescope Launched
- James R. Thompson Center Director
- T.J. Lee Center Director
- G. Porter Bridwell Center Director
- Arthur G. Stephenson Center Director
- David A. King Center Director

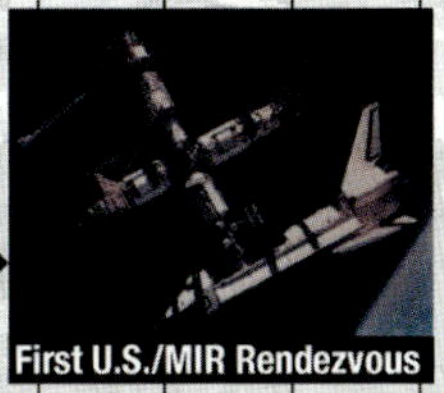

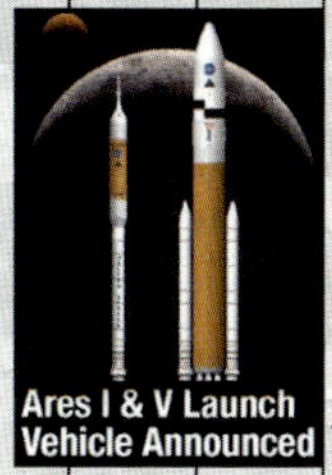

onology

- Joe Davis Stadium Opened: Huntsville Stars Baseball Team Established
- Alabama Supercomputer Network Headquarters in Huntsville
- I-565 Completed
- EarlyWorks Museum Opened
- Columbia High School Opened
- International Intermodel Center Opened at Airport
- First International Cargo Shipped From Huntsville
- Von Braun Research Hall at UAH Dedicated
- Davidson Center for Space Exploration Opened
- Northeast Alabama Regional Small Business Development Center Combines with Chamber of Commerce
- National Space Science and Technology Center Opened
- oreign Trade Zone stablished
- Huntsville Botanical Garden Established

| 984 | 1985 | 1986 | 1987 | 1988 | 1989 | 1990 | 1991 | 1992 | 1993 | 1994 | 1995 | 1996 | 1997 | 1998 | 1999 | 2000 | 2001 | 2002 | 2003 | 2004 | 2005 | 2006 | 2007 | 2008 | 2009 |

50 Years of Rockets & Spacecraft
NASA Marshall Space Flight Center

Acclaim Press
MORLEY, MISSOURI

Acclaim Press
—*Your Next Great Book*—

P.O. Box 238
Morley, MO 63767
(573) 472-9800
www.acclaimpress.com

Designer: M. Frene Melton
Cover Design: M. Frene Melton

Copyright © 2009
The Marshall Retiree Association
All Rights Reserved.

No part of this book shall be reproduced or transmitted in any form or by any means, electronic or mechanical, including photocopying, recording or by an information or retrieval system, except in the case of brief quotations embodied in articles and reviews without the prior written consent of the publisher. The scanning, uploading, and distribution of this book via the Internet or via any other means without permission of the publisher is illegal and punishable by law.

Library of Congress Cataloging-in-Publication Data

Buckbee, Ed.
  50 years of rockets & spacecraft : NASA-Marshall Space Flight Center commemorative history / by Ed Buckbee.
    p. cm.
  Includes bibliographical references and index.
  ISBN-10: 1-935001-17-5 (alk. paper)
  ISBN-13: 978-1-935001-17-1 (alk. paper)
 1. George C. Marshall Space Flight Center--History. 2. Astronautics--United States--History. I. Title. II. Title: Fifty years of rockets and spacecraft.
  TL862.G4B83 2009
  629.40973--dc22

                                    2009027376

All photographs, unless otherwise noted, appear courtesy of NASA, U.S. Army, U.S. Space & Rocket Center, or the University of Alabama-Huntsville.

First Printing: 2009
Printed in the United States of America
10 9 8 7 6 5 4 3 2 1

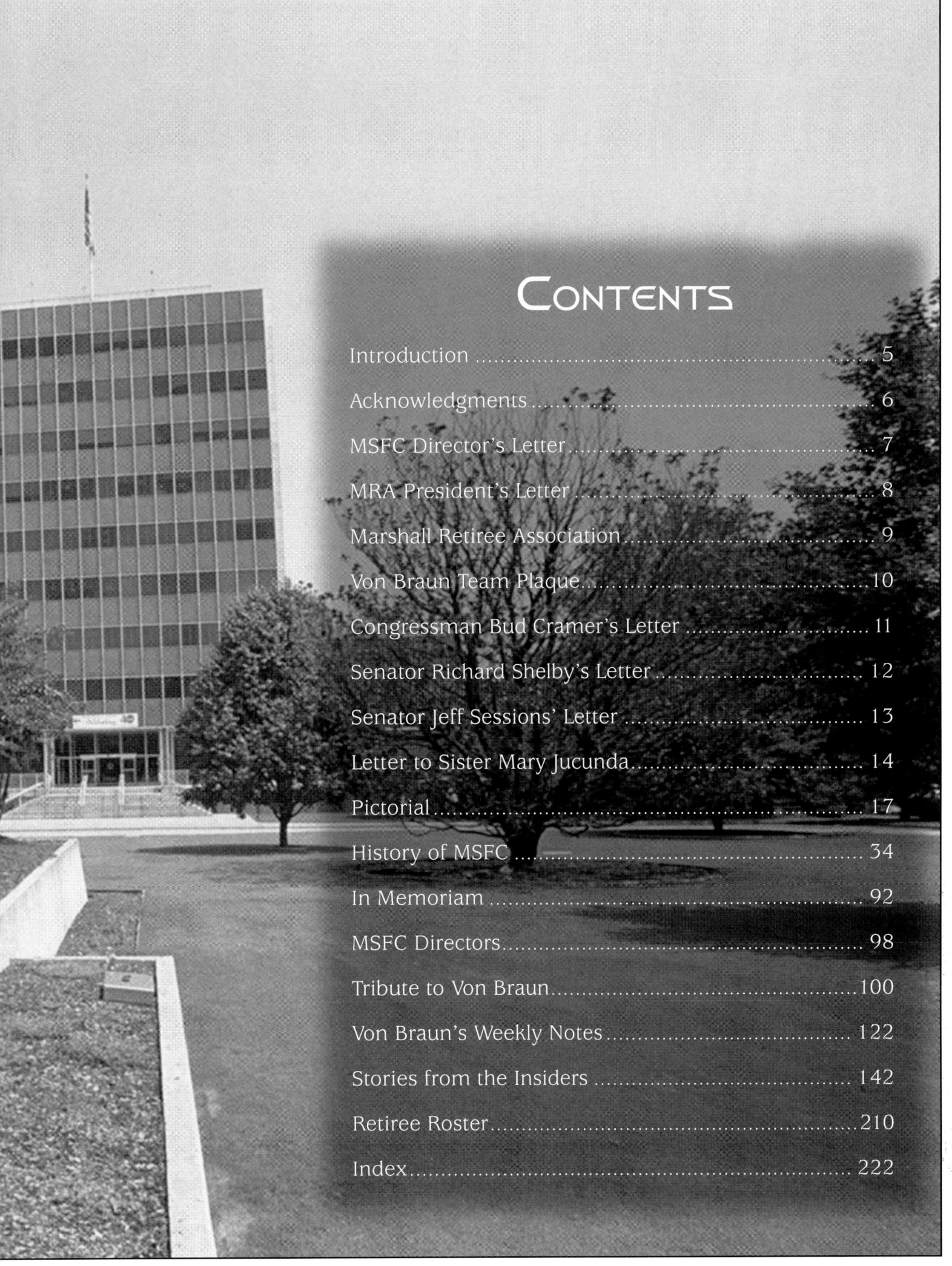

Contents

Introduction 5

Acknowledgments 6

MSFC Director's Letter 7

MRA President's Letter 8

Marshall Retiree Association 9

Von Braun Team Plaque 10

Congressman Bud Cramer's Letter 11

Senator Richard Shelby's Letter 12

Senator Jeff Sessions' Letter 13

Letter to Sister Mary Jucunda 14

Pictorial 17

History of MSFC 34

In Memoriam 92

MSFC Directors 98

Tribute to Von Braun 100

Von Braun's Weekly Notes 122

Stories from the Insiders 142

Retiree Roster 210

Index 222

50 YEARS
AND COUNTING

INTRODUCTION

In 1961, President John F. Kennedy sought to overtake the Russians in the race to the moon and issued a challenge to his fellow Americans to land man on the moon before the end of the decade. Eight years later, on July 20, 1969, Apollo 11 achieved that goal. It's been 40 years since U.S. astronauts Buzz Aldrin, Neil Armstrong and Michael Collins first landed on the moon and returned safely to Earth. Looking back at that time in our history, there remains a sense of wonder at what was accomplished. Perhaps Sir Arthur C. Clarke, author of *2001: A Space Odyssey* captured the full magnificence of the feat when he noted, "Apollo will be the only achievement this generation will be remembered for a thousand years from now."

No other country has done it and no one but America's Saturn rocket riders heard the call, "You are go for TLI" [trans lunar insertion] which means you are going to the moon at escape velocity––25,000 mph," heralded Houston mission control.

This book is a narrative history commemorating 50 years of rockets and spacecraft. This history begins with Redstone, Explorer I and Jupiter, and continues with Saturn, Skylab, Space Shuttle, Spacelab, Hubble, Chandra, the International Space Station and the current Ares program. The early programs, while devoted to developing military vehicles, provided the infrastructure for space flight––the facilities, equipment, processes and most importantly, the working relationships between Wernher von Braun's rocket team and the American engineers. The early years also served to define the roles and interfaces between the myriad technical, scientific, administrative and management disciplines required for the complexities of manned space flight.

In the beginning, the rocket team was uniquely capable to conceive, design, develop, fabricate, test and launch an entirely new rocket system. The heart of the organization was found in the laboratories where research, design, fabrication, testing and quality control took place. Joining the Germans were hundreds of American-born engineers who were fast tracking Rocketry 101. Most laboratory directors were from von Braun's original rocket team, but some of this new breed of American rocketeers filled deputy and senior positions. The Americans learned quickly from their German mentors and together they worked as a dedicated and passionate team driven by Kennedy's challenge.

Over time and the passage of programs, the original members of the von Braun team retired and the American-born engineers began assuming positions held by their German colleagues. This seasoned group led the Marshall Center through several decades of exciting space exploration projects.

In this book, members of the Marshall team tell the "insider" stories of memorable events, successes and failures. These stories range from humorous anecdotes to gut-wrenching decisions required to safely fly humans into space. Many stories come from the very first engineers who had to contemplate the variables and processes required to sustain life in space. Starting with Alan Shepard's flight aboard a Mercury-Redstone rocket in 1961, there are people who assumed responsibilities for certifying the launch vehicle as reliable and safe to fly with humans aboard. That final decision to launch was made by a handful of people who depended upon data and information furnished by colleagues. This special breed of rocketeer had to be disciplined and skilled in the system. Each accepted the challenge of working in a highly sophisticated and technical field that often required fast, critical, calculated and intuitive decisions. The astronauts might have been the most obvious risk-takers, but the people on the ground shared in that risk. Even today, human space flight continues to be a high-risk business.

The famous von Braun Weekly Notes of the 1960s have been found and what a wealth of knowledge they have produced! The notes offer a case study of von Braun's management style, together with an inside view of the hot topics of the day. Most of the people named among the pages were laboratory directors and program office mangers. However, it's important to note that the content of these notes often came from down in the organization––in the trenches––where critical decisions and issues were being dealt with by the hour. These are the people, many in this association, who contributed to critical decisions made through the channel of Weekly Notes.

Working for von Braun was the gift that kept on giving. His limitless ideas and passion sparked a corresponding chord across the organization. Von Braun was the inventor of rockets and we witnessed the incredible birth of the moon rocket in his laboratories. It was a privilege to join him in extending the human presence to the moon and beyond. He truly was a legend in his own time.

Yes, that sense of wonder still endures, but there is something else that is equally important that we carry in our collective memory. America is at its best when answering a challenge. Space flight, especially human space flight, is not easy. It *is* rocket science and the laws of physic are unchanged. It is no less complex getting into low Earth orbit now than it was when the Marshall team launched pounds instead of tons. The men and women of today's NASA Marshall Space Flight Center possess the will, intellect and imagination to continue mankind's space exploration through new programs and missions. Furthermore, their creation and implementation of new space technologies will continue to afford untold advancements to environments and operations right here on Earth.

It is an awesome task to cover over 50 years of dramatic events that have taken place at the world's premier rocket center, but I know that with the support of the association, it's just another challenge to be met.

Ed Buckbee, Editor
Member of the Marshall Retiree Association

Acknowledgments

This book is dedicated to the men and women of Marshall Space Flight Center who made it possible for humans to walk on the moon, to study our home planet Earth, to explore the universe and to future generations who will use the technology of space for the benefit of all mankind.

We offer special thanks to the following for their many contributions in remembering significant events, acknowledging people who made important contributions and spending countless hours correcting and editing the manuscript:

J.A. "Woody" Bethay
James Downey
J.N. "Jay" Foster
Nancy Guire
Bonnie Holmes
Jack Lee
Robert Lindstrom
Brooks Moore
Robert Naumann
Fred Ordway
Axel Roth
Ruth von Saurma

And a thanks for reviewing, researching materials and providing photography to:

Mike Baker
Kaylene Hughes
Mike Wright
Bob Jaques
Chieko Inman

General George C. Marshall. The Marshall Space Flight Center was established in 1960 and named in honor of General Marshall, the Army Chief of Staff during World War II, Secretary of State, and Nobel Prize Winner for his world-renowned Marshall Plan.

MSFC Director's Letter

National Aeronautics and Space Administration
George C. Marshall Space Flight Center
Marshall Space Flight Center, AL 35812

Reply to Attn of: DA01

TO: MSFC Retiree Association

FROM: DA01/Robert M. Lightfood

SUBJECT: Letter of Appreciation

"50 Years of Rockets and Spacecraft" is a first-person account of America's compelling journey into world leadership in space. It is a story of successes, failures, hopes, and dreams told by the individuals who lived it, and that makes it invaluable.

We are indebted to the NASA-MSFC Retiree Association for their hard work and dedication in producing this account and to the individuals who took their time to add their recollections of historical events.

We are justly proud of our history and excited about the future. We stand on the threshold of another great age of space exploration, just as our predecessors did 50 years ago. Our nation, once again, has a plan for exploring space with clear goals and great challenges. Soon a new generation of rockets will rise over Huntsville, ready to carry new crews of explorers to the Moon, a first step toward voyages to Mars and beyond. Once again, Marshall and Huntsville will play a key role in this historic endeavor.

Thank you for this commemoration for the continuing support of the men and women of the NASA-MSFC Retiree Association.

Robert M. Lightfoot
Director

MRA President's Letter

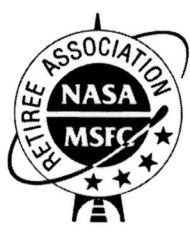

The Marshall Retiree Association

The Marshall Retiree Association (MRA) provides a forum where the Marshall Space Flight Center retirees can gather in a social setting to enjoy the reminiscing of collective contributions in the development of capability for space exploration.

While daily enjoying the technological advancements from their performance on past programs, Marshall retirees continue to build upon those evolving advancements for even greater benefits for future space exploits and an ever-improving life for all mankind.

Many changes have occurred in the space program since the publication of "50 Years of Rockets and Spacecraft in the Rocket City," in 2003. A reflection of just a few of those changes include:

- An International Space Station that continues to grow in size, complexity and capability for the evolvement of space science through experiments conducted in zero gravity. This is a critical milestone for our space faring longevity.
- An enormous expansion of the U.S. Space & Rocket Center, with the addition of the Davidson Center for Space Exploration. This museum is the largest in the world devoted exclusively to the exhibition of space artifacts, including the only Saturn V designated as a National Historic Landmark. This Saturn V was completely renovated and is impressively displayed horizontally overhead in the Davidson Center, permitting social, civic, community, and technical conference events to be held in a unique setting. The MRA played a major role in the fund-raising for this marvelous museum. Similarly, fund-raising activities are already underway for a similar renovation and display of Skylab hardware, America's first space station.
- The MRA sponsored a recognition plaque for the 121-member German Rocket Team who developed and launched America's first satellite and first manned space flight. A memorable celebration was held to recognize both German and American Teammates, with a headcount exceeding 375, to commemorate their mutual respect and collaborative efforts in achieving incredible goals. This plaque is permanently and appropriately mounted at the Davidson Center for Space Exploration.
- With the pending retirement of Space Shuttle, NASA-Marshall Space Flight Center is currently developing an exploration vehicle system, called Constellation, to expand the space program capabilities for manned exploration of Moon, Mars and beyond. Ares I vehicle will provide the capabilities for crew launch; Ares V vehicle will provide heavy cargo lift capability. Additionally, robotic missions will gather significant data for both current and future explorations.
- NASA-Marshall Space Flight Center innovative research and scientific pursuits continue to foster and advance unique capabilities from theoretical into tangible technologies. These efforts include telescopes, satellites, space sensors, scientific spacecraft, and complex instruments studying Earth science, solar science, lunar science and science of other planets.
- The members of the MRA continue involvement in many aspects of the development and demonstration of flight-worthy hardware. They not only contribute their time and talents, but their financial support as well.

The MRA members continue to contribute and shape the current and future programs by volunteering their knowledge and experiences from yesterday's exciting and rewarding space achievements.

Jim Splawn, President
Marshall Retiree Association

Marshall Retiree Association

*T*he Marshall Retiree Association was established in 1987 as a chapter of the NASA Alumni Association based in Washington, D.C. In 2000 the chapter voted to reorganize as a local organization, with its overall objective to be more responsive to the Marshall Space Flight Center and the Huntsville community. Our objectives are to provide regular fellowship and social opportunities for members, their spouses and guests, to promote community interest in current and planned NASA and Marshall activities and to provide a source of volunteers for historic, technical and general information requirements.

In our efforts to preserve the Marshall Team legacy and share the vision of human exploration of space with the next generation of space explorers, the association has sponsored several projects:

- **FIRESIDE CHATS**—Reminiscence of Space Exploration History. Over twenty members of our association participated in a four part series discussing, "Huntsville Before Rockets; The Army Years; The Lunar Program; and Space Shuttle and Beyond."
- **ANNIVERSARIES AND CELEBRATIONS**—The association has traditionally supported the Marshall Center Open House and joined the Center and the Huntsville-Madison County Chamber of Commerce in celebrating Marshall's 40th birthday and anniversaries of Apollo 11, 17, Skylab and NASA and Marshall's 50th celebrations. Others are being planned for the future.
- **UNIVERSITY OF ALABAMA-HUNTSVILLE SCHOLARSHIP**—The association has pledged $10,000 to an endowment that will create a scholarship for a freshman pursing a degree in engineering or physical sciences. Our first scholarship was awarded in 2002 to Ryan T. Bowerman of Moulton, Alabama.
- **ARCHIVES AND SPACE COLLECTION**—The association is collaborating with the University of Alabama in Huntsville Louis Salmon library to assist in the collection of papers, books, documents and artifacts.
- **NEW BOOK**—The association has published a new book commemorating the 50 years of space exploration and the 40th anniversary of Apollo 11.

The Marshall Retiree Association is an organization rich in tradition and pride in its accomplishment. We strive to preserve the Marshall Team's legacy.

Past Presidents:

Jim Murphy	1987-1988
George Hardy	1989-1991
Brooks Moore	1992-1996
Jay Foster	1997-1998
Ed Buckbee	1999-2002
Harold "Bill" Hallisey	2003-2004
Bob G. Noblitt	2004-2005

Officers:

Jim Splawn	President, 2006-09
Gary Wicks	Vice-President, 2006-09
Linda Posey	Events & Communications Chairman
Mary Spaulding	Membership Chairman
Harold "Bill" Hallisey	Program Chairman
Sue Light	Newsletter Editor
Batrice Glen	Treasurer

Association Members:

Adair, Billy M.
Akridge, Charles M.
Aldrup, Alice A.
Askew, William C.
Atkins, Harry L.
Atherton, James R.
Austin, Robert E.
Bechtel, R.T.
Bethay, J.A.
Bledsoe, Ronald B.
Bledsoe, Theodore T.
Bradford, William C.
Bradley, Jesse
Brown, Richard L.
Buckbee, Edward O.
Buhmann, Eugene J.
Calhoun, James D.
Clark, Floyd M.
Coates, Keith D.
Compton, Eugene T.
Connor, Edward J.
Counts, Parker V.
Courtenay Jr., St. John
Cox Sr., Charles E.
Cozelos, Charles L.
Craft Jr., Harry G.
Daniels, Jane
Darwin, Charles R.
Daussman, Grover F.
Davies, Robert J.
Deaton, Emsley T.
Delillo, John
Digesu, Fred E.
Doane III, George B.
Dorland, Wade D.
Downey, James A.
Eoff, William L.
Emanuel, William A.
Fee, John S.
Field, Elmer L.
Foster, J.N.
Fountain, James A.
Freedman, Jack A.
Fuchtman, Charles M.
Garriott, Dr. Owen K.
Gillespie, Charles S.
Glen, Batrice T.
Golden, Harvey
Graff, Charles B.
Greenwood, Dr. Lawrence R.
Gresham, Albert E.
Grimes, Victor S.
Griner, Carolyn S.
Guire, Nancy M.
Guynes, Barry V.
Hallisey, Harold W.
Hamieter, Leon C.
Hardy, George B.
Harrell, William B.
Helmick, Richard A.
Henke, Charles E.
Henritze, Richard M.
Hester, Howard B.
Hildreth, Edward D.
Hight, Hermon H.
Holmes, Bonnie G.
Hooper, William H.
Hueter, Uwe
Humphries, Dr. William R.
Igou, James M.
Ise, Rein
Jobe, Grady S.
Johns, Stanley A.
Johnson, Charles W.
Johnson, William H.
Johnstone, Harry M.
Jones III, Charlie T.
Jones, Jack A.
Jones, Lee W.
Kastanakis, John A.
Key, C.F.
Kingsbury, James E.
Kinney, Linda H.
Klan, Richard L.
Kurtz, Herman F.
Laue, Jay H.
Lechner, Larry E.
Lee, Thomas J.
LeMay, Jacqueline R.
Lester, Howard B.
Lide, George M.
Lifer, Charles E.
Light, Susan K.
Lindstrom, Robert E.
Littles, Dr. J.W.
Lucas, Dr. William R.
Lundquist, Dr. Charles A.
Machen, Jyles
Malkmus, Bernard R.
Mann Jr., William H.
Marshall, Ronald C.
Mauldin, Charles R.
May, Ellery B.
McAnnally, Robert C.
McBrayer, Robert O.
McCullar, B.R.
McMillion, James M.
Middleton, Robert L.
Mobley, David H.
Monk, Jan C.
Moody, Jewell W.
Moore, F.B.
Morea, Saverio F.
Morris, John K.
Murphy, Ralph W.
Neece, Ronald L.
Neighbors, Dr. Joyce K.
Nevins, Clyde D.
Newby, David H.
Noblitt, Bob G.
Nurre, Dr. Gerald S.
Odom, James B.
Olsen, Ronald K.
Osborn, Leroy
Owen, Clark M.
Pace Jr., Robert E.
Page, Rabon P.
Parmer, Patsy C.
Partain, Mary E.
Payne, Molly H.
Pearson, James C.
Peasae, Robert E.
Pessin, Myron A.
Porter, Doris J.
Posey, Linda B.
Pryor, Donald
Ralls, Gail A.
Randolph, Joanne W.
Reaves, John H.
Reynolds, William R.
Riehl, Wilbur A.
Ritter, Glen D.
Robinson, James H.
Rohr, Stephen V.
Rosenthal, Max E.
Rutland, Cary H.
Ryan, Robert S.
Sacheim, Norman L.
Schwinghamer, Robert J.
Self, Deemer O.
Sells, Harold R.
Settle, Gray L.
Shelton, Harvey L.
Sheppard, Robert G.
Sisson, James M.
Smith, Gerald W.
Smith, Richard G.
Smith, Thomas H.
Sneed, Bill H.
Snoddy, William C.
Southerland, Thomas H.
Spaulding, Mary D.
Speer, Dr. Fridtjof A.
Splawn, James L.
Stapler, Vachel
Staples, Evelyn M.
Stewart III, Frank M.
Stocks, Charles D.
Struck, Heinz G.
Susko, Dr. Michael
Tanner, E.R.
Thomas Jr., James W.
Thomas, Joe H.
Thomas, Sylvia B.
Thompson Jr., James R.
Thompson, Richard L.
Thompson, Zack
Tomlin, Donald D.
Torreulla, Antonio R.
Turner, Tom W.
Vann, James W.
Vardaman, William K.
Vaughan, Dr. William W.
Vaughan, Otha H.
Vedane, Charles R.
Vick, Howell G.
Velvet, Camille S.
Von Saurma, Ruth
Waite Sr., Jack H.
Walls, Samuel L.
Webb, Horton L.
Wallace, Gabriel R.
Westrope Jr., Dewitt T.
Whitaker, Ann F.
Wilhold, Gilbert A.
Wicks, Thomas G.
Williams, John R.
Williams, Frank
Wittenstein, Gerald
Wojtalik, Fred S.
Zoller, Lowell K.

Von Braun Team Plaque

U.S. Space & Rocket Center
January 29, 2009

On January 29, 2009, this plaque of recognition, proudly sponsored by the Marshall Space Flight Center Retiree Association, was presented to the German team who has served our Huntsville community team, our NASA - Marshall Space Flight Center team, and our United States team, in achieving space superiority. The first step in this space superiority was the successful launch of Explorer I on January 31, 1958 atop a Jupiter-C rocket, a product of the U.S. Army Ballistic Missile Agency (ABMA). The first American satellite was on orbit!

On May 5, 1961, just forty months after the launch of Explorer I, Alan B. Shepard, Jr. became the first U.S. astronaut in space. Flying Freedom 7, the Mercury capsule atop the Redstone rocket, Shepard rocketed 115 miles for his suborbital flight.

Eight years and two months later, Apollo 11 landed on the moon's surface with Neil Armstrong and Buzz Aldrin at the controls.

Eleven years after the first satellite was launched, the U.S. Space Program had placed man on the surface of the Moon and returned him safely to Earth. What a marvelous achievement!

The purpose of the plaque is to record permanent recognition of so many that gave so much in leadership for our space program. This plaque honors the German team for their intellect, their technical capabilities, and their persistence to excellence, their professionalism, and their total contributions to all mankind.

Through the years, there have been numerous recognitions of German individuals and small team efforts, but never a specific recognition of the entire German group as a total team. Our Marshall Retiree Association felt it worthy to correct this void and, therefore, launched a major effort to identify all of the team members. As a result, a total of 121 names, notably and nobly led by Dr. Wernher von Braun, adorn this plaque.

And the German team, possessing the dream, the vision, the conviction, the determination, the commitment, and the persistent leadership, coupled with the duplicate traits of their American teammates, established the world-wide reputation that the United States has the greatest space program in the world.

Attending this ceremony were seven original German team members, 156 German family members and relatives, and 250 American teammates.

Jim Splawn, President
Marshall Retiree Association

Congressman's Letter

BUD CRAMER
5TH DISTRICT
ALABAMA

COMMITTEE ON APPROPRIATIONS
SUBCOMMITTEES
DEFENSE
TRANSPORTATION, HOUSING AND URBAN DEVELOPMENT, AND RELATED AGENCIES
FINANCIAL SERVICES AND GENERAL GOVERNMENT

SELECT INTELLIGENCE OVERSIGHT PANEL

PERMANENT SELECT COMMITTEE ON INTELLIGENCE
SUBCOMMITTEES
OVERSIGHT AND INVESTIGATIONS
TECHNICAL AND TACTICAL INTELLIGENCE

Congress of the United States
House of Representatives
Washington, D.C. 20515

OFFICES:
2184 RAYBURN HOUSE OFFICE BUILDING
WASHINGTON, DC 20515
(202) 225-4801

200 PRATT AVENUE NORTHEAST
SUITE A
HUNTSVILLE, AL 35801
(256) 551-0190

MORGAN COUNTY COURTHOUSE
BOX 668
DECATUR, AL 35602
(256) 355-9400

THE BEVILL CENTER
1011 GEORGE WALLACE BOULEVARD
TUSCUMBIA, AL 35674
(256) 381-3450

E-mail: budmail@mail.house.gov
Web page: http://cramer.house.gov

Marshall Space Flight Center
Huntsville, Alabama 35812

To the Marshall Team:

I want to send my sincere congratulations to the Marshall Space Flight Center Team on the 50th anniversary of NASA. This is truly a time of celebration.

Since its inception 50 years ago, NASA has inspired generations of space scientists and engineers and has remained the world leader in aeronautics and space exploration. It has been a great privilege to represent in Congress the thousands of NASA employees and contractors who work at the Marshall Space Flight Center. I commend the agency, and particularly the space community at Marshall Space Flight Center and across North Alabama, for their hard work during the past 50 years.

Sincerely,

Bud

Bud Cramer
Member of Congress

This mailing was prepared, published, and mailed at taxpayer expense.
Printed on recycled paper with soy ink.

Senator's Letter

UNITED STATES SENATE
WASHINGTON, D.C. 20510

RICHARD SHELBY
ALABAMA

To the Marshall Team,

Please accept my heartfelt congratulations to you all as you celebrate your 50th Anniversary. Since the dawn of civilization, humans have desired to travel to and explore the cosmos. The humble beginnings of the National Aeronautics and Space Administration in 1958 have since led to remarkable feats of space exploration. From capturing stunning views of the far reaches of the Universe to watching our astronauts walk on the moon, each major milestone on the path to space exploration has run straight through Huntsville, Alabama.

With the arrival of Dr. Wehrner von Braun to Redstone Arsenal, the people of Huntsville were called upon to usher our nation into space. Fifty years ago, a U.S. Army team led by Dr. von Braun lofted a modified Redstone rocket to successfully launch the first American satellite into the Earth's orbit. This launch marked the United States' entry into the space race and solidified the role Alabama would play in our nation's space program for the next half century.

Over the next fifty years, Marshall would be intimately involved in some of the greatest achievements of our space program. On May 5, 1961, Alan Shepard became the first American launched into space – on a Redstone rocket. The development of the Saturn V that lofted our astronauts to the moon in 1969 was achieved at Marshall. The Center's activities broadened to an era of scientific research in the 1970s – confirming scientific theories and developing technology to ease our everyday lives.

Today, Marshall is on the forefront of returning man to the moon. Receiving over $2.7 billion a year in funding from NASA and employing 7,000 engineers, scientists and support personnel, Marshall continues as the center of the rocket development world with the fundamental role of designing our nation's next manned rocket and lunar robotic exploration missions.

The Apollo moon landings; observations of the sun, our solar system, and Earth; robotic explorations of the planets; and the inauguration of the International Space Station are achievements that have revolutionized our view of the universe and inspired people around the world. None of this would have been realized if not for the people and work accomplished in Huntsville.

Now, in the new millennium, the time has come to answer the question of expanding humankind's future in space. With the return of man to the moon, Marshall will remain the critical centerpiece that delivers our astronauts to space, while growing its new role in robotic exploration. As these goals are realized, the missions currently under development at Marshall will provide the groundwork and inspiration for the next generation of scientists, engineers and explorers. The historic role Marshall plays in our space program will continue for years to come.

Please accept my warmest regards as you celebrate this important milestone. Your previous 50 years have been stellar and I know your next 50 will be just as exceptional.

Sincerely,

Richard Shelby

Senator's Letter

JEFF SESSIONS
ALABAMA

COMMITTEES:
ARMED SERVICES
JUDICIARY
HEALTH, EDUCATION, LABOR, AND PENSIONS
JOINT ECONOMIC

United States Senate
WASHINGTON, DC 20510–0104

Marshall Space Flight Center
Huntsville, Alabama

Dear Friends:

    I want to take a moment and express my admiration and gratitude for all that has been accomplished in our nation's space program as a result of the work at Marshall Space Flight Center. The next great adventure for the United States is the exploration and development of space. Throughout my term as United States Senator, I have witnessed excellent work performed at MSFC and know of the quality of individuals who strive on a daily basis to enhance our nation's space exploration capabilities. We are a nation of explorers and inventors–proud, hardworking and brave. Our legacy as a nation is one of unmatched proportion. We must do our part to continue to build upon the past for the benefit of future generations. I am pleased that this book of space history has been dedicated to all the remarkable accomplishments of MSFC. The region, our state, and the nation are immensely proud of all you do.

Very truly yours,

Jeff Sessions
United States Senator

JS:jwd

LETTER TO SISTER MARY JUCUNDA

by Dr. Ernst Stuhlinger

Editorial Introduction
by Charles Lundquist

On May 6, 1970, in the year following the first human lunar landing, Dr. Ernst Stuhlinger, then Associate Director for Science at the Marshall Center, answered a letter from Sister Mary Jucunda, a nun striving to save starving children in Zambia. In her letter, Sister Mary asked Dr. Stuhlinger how he could suggest the expenditure of billions of dollars for a voyage to Mars at a time when many children on this Earth were starving to death. Ernst's thoughtful and comprehensive reply has been widely distributed and quoted.

On November 18, 2008, a copy of the reply was included in a souvenir document prepared for the Dr. Ernst Stuhlinger Recognition Symposium sponsored by the University of Alabama Huntsville. The Symposium discussed four principal areas of research productively pursued by Ernst: Space Science, Electric Vehicles, Electrical Propulsion and Advanced Space Transportation. The results of his work on these topics are historic, but the discussions of them fail to capture to his humble humanity. The reply to Sister Mary does convey the overriding, caring character of Ernst Stuhlinger. It is reproduced below.

The inclusion of the reply in this 2009 volume is timely and appropriate. It is timely because the United States is again contemplating a return to the Moon and the human exploration of Mars. It is appropriate because it speaks eloquently for all of Ernst's colleagues who have contributed to this volume.

Ernst Stuhlinger was born on December 19, 1913 in Germany and died May 25, 2008 in Huntsville, Alabama.

Dear Sister Mary Jucunda,

Your letter was one of many which are reaching me every day, but it has touched me more deeply than all the others because it came so much from the depths of a searching mind and a compassionate heart.

I will try to answer your question as best as I possibly can.

First, however, I would like to say what great admiration I have for you, and for all your many brave sisters, because you are dedicating your lives to the noblest cause of man: help for his fellowmen who are in need.

You asked in your letter how I can suggest the expenditure of billions of dollars for a voyage to Mars, at a time when many children on this earth are starving to death.

I know that you do not expect an answer such as "Oh, I did not know that there are children dying from hunger, but from now on I will desist from any kind of space research until mankind has solved that problem!"

In fact, I have known of famined children long before I knew that a voyage to the planet Mars is technically feasible; however, I believe, like many of my friends, that traveling to the Moon and eventually to Mars and to other planets is a venture which we should undertake now and I even believe that this project, in the long run, will contribute more to the solution of these grave problems we are facing here on Earth than many other potential projects of help which are debated and discussed year after year, and which are so extremely slow in yielding tangible results.

Before trying to describe in more detail how our space program is contributing to the solution of our earthly problems, I would like to relate briefly a supposedly true story which may help support the argument.

About 400 years ago, there lived a count in a small town in Germany. He was one of the benign counts and he gave a large part of his income to the poor in his town. This was much appreciated because poverty was abundant during medieval times, and there were epidemics of the plague which ravaged the country frequently.

One day, the count met a strange man. He had a workbench and little laboratory in his house, and he labored hard during the daytime so that he could afford a few hours every evening to work in his laboratory.

He ground small lenses from pieces of glass; he mounted the lenses in tubes; and he used these gadgets to look at very small objects. The count was particularly fascinated by the tiny creatures that could be observed with the strong magnification and which he had never seen before.

He invited the man to move with his laboratory to the castle, to become a member of the count's household, and to devote henceforth all his time to the development and perfection of his optical gadgets as a special employee of the count.

The townspeople, however, became angry when they realized that the count was wasting his money, as they thought, on a stunt without purpose. "We are suffering from this plague," they said, "while he is paying that man for a useless hobby!"

But the count remained firm. "I give you as much as I can afford," he said, "but I also support this man and his work, because I know that someday something will come out of it!"

Indeed, something very good came out of this work, and also out of similar work done by others at other places: the microscope. It is well known that the microscope has contributed more than any other invention to the progress of medicine, and that the elimination of the plague and many other contagious diseases from most parts of the world is largely a result of studies which the microscope made possible.

The count, by retaining some of his spending money for research and discovery, contributed far more to the relief of human suffering than he could have contributed by giving all he could possibly spare to his plague-ridden community.

The situation which we are facing today is similar in many respects. The president of the United States is spending about $200 billion in his yearly budget. This money goes to health, education, welfare, urban renewal, highways, transportation, foreign aid, defense, conservation, science, agriculture and many installations inside and outside the country.

About 1.6 per cent of this national budget was allocated to space exploration this year. The space program includes Project Apollo, and many other smaller projects in space physics, space astronomy, space biology, planetary projects, earth resources projects, and space engineering.

To make this expenditure for the space program possible, the average American taxpayer with $10,000 income per year is paying about 30 tax dollars for space.

The rest of his income, $9,970, remains for his subsistence, his recreation, his savings, his taxes and all his other expenditures.

You will probably ask now: "Why don't you take 5 or 3 or 1 dollar out of the 30 space dollars which the average American taxpayer is paying and send these dollars to the hungry children?"

To answer this question, I have to explain briefly how the economy of this country works. The situation is very similar in other countries.

The government consists of a number of departments (Interior; Justice; Health, Education and Welfare; Transportation; Defense; and others), and of bureaus (National Science Foundation; National Aeronautics and Space Administration; and others).

All of them prepare their yearly budgets according to their assigned missions, and each of them must defend its budget against extremely severe screening by congressional committees, and against heavy pressure for economy from the Bureau of the Budget and the President. When the funds are finally appropriated by Congress, they can be spent only for the line items specified and approved in the budget.

The budget of the National Aeronautics and Space Administration, naturally, can contain only items directly related to aeronautics and space. If this budget were not approved by Congress, the funds proposed for it would not be available for something else; they would simply not be levied from the taxpayer, unless one of the other budgets had obtained approval for a specific increase which would then absorb the funds not spent for space.

You may realize from this brief discourse that support for hungry children, or rather a support in addition to what the United States is already contributing to this very worthy cause in the form of foreign aid, can be obtained only if the appropriate department submits a budget line item for this purpose and if this line item is then approved by Congress.

You may ask now whether I personally would be in favor of such a move by our government. My answer is an emphatic yes. Indeed, I would not mind it at all if my annual taxes were increased by a number of dollars for the purpose of feeding hungry children wherever they may live.

I know that all of my friends feel the same way; however, we could not bring such a program to life merely by desisting from making plans for voyages to Mars. On the contrary, I even believe that by working for the space program I can make some contribution to the relief and eventual solution of such grave problems as poverty and hunger on earth.

Basic to the hunger problem are two functions: the production of food and distribution of food. Food production by agriculture, cattle ranching, ocean fishing and other large scale operations is efficient in some parts of the world, but drastically deficient in many others.

For example, large areas of land could be utilized far better if efficient methods of watershed control, fertilizer use, weather forecasting, fertility assessment, plantation programming, field selection, planting habits, timing of cultivation, crop survey and harvest planning were applied.

The best tool for the improvement of all these functions, undoubtedly, is the artificial earth satellite. Circling the globe at a high altitude, it can screen wide areas of land within a short time; it can observe and measure a large variety of factors indicating the status and conditions of crops, soil, droughts, rainfall, snow cover, etc., and it can radio this information to ground stations for appropriate use.

It has been estimated that even a modest system of earth satellites equipped with earth resources sensors, working within a program for worldwide agricultural improvement, will increase the yearly crops by an equivalent of many billions of dollars.

The distribution of the food to the needy is a completely different problem. The question is not so much one of shipping volume; it is one of international cooperation.

The ruler of a small nation may feel very uneasy about the prospects of having large quantities of food shipped into his country by a large nation, simply because he fears that along with the food there may also be an import of influence and foreign power.

Efficient relief from hunger, I am afraid, will not come before the boundaries between nations have become less divisive than they are today.

I do not believe that space flight will accomplish this miracle overnight; however, the space program is certainly among the most promising and powerful agents working in this direction.

Let me only remind you of the recent near-tragedy of Apollo 13. When the time of the crucial entry of the astronauts approached, the Soviet Union discontinued all Russian radio transmissions in the frequency bands used by the Apollo Project in order to avoid any possible interference, and Russian ships stationed themselves in the Pacific and the Atlantic oceans in case an emergency rescue would become necessary.

Had the astronaut capsule touched down near a Russian ship, the Russians would undoubtedly have expended as much care and effort in their rescue as if Russian cosmonauts had returned from a space trip.

If Russian space travelers should ever be in a similar emergency situation, Americans would do the same, without any doubt.

Higher food production, through survey and assessment from orbit, and better food distribution through improved international relations, are only two examples of how profoundly the space program will impact life on earth.

I would like to quote two other examples: stimulation

of technological development and generation of scientific knowledge.

The requirements for high precision and for extreme reliability which must be imposed upon the components of a moon-traveling spacecraft are entirely unprecedented in the history of engineering.

The development of systems which meet these severe requirements has provided us a unique opportunity to find new materials and methods, to invent better technical systems, to improve manufacturing procedures, to lengthen the lifetimes of instruments and even to discover new laws of nature.

All this newly acquired technical knowledge is also available for applications to earth-bound technologies. Every year, about a thousand technical innovations generated in the space program find their ways into our earthly technology where they lead to better kitchen appliances and farm equipment, better sewing machines and radios, better ships and airplanes, better weather forecasting and storm warning, better communications, better medical instruments, better utensils and tools for everyday life.

Presumably, you will ask now why we must develop first a life support system for our moon-traveling astronauts, before we can build a remote-reading sensor system for heart patients.

The answer is simply: significant progress in the solution of technical problems is frequently made not by a direct approach, but by first setting a goal of high challenge which offers a strong motivation for innovative work, which fires the imagination and spurs men to expend their best efforts, and which acts as a catalyst by including chains of other reactions.

Space flight, without any doubt, is playing exactly this role. The voyage to Mars will certainly not be a direct source of food for the hungry; however, it will lead to so many new technologies and capabilities that the spinoffs from this project alone will be worth many times the cost of its implementation.

Besides the need for new technologies, there is a continuing great need for new basic knowledge in the sciences if we wish to improve the conditions of human life on earth.

We need more knowledge in physics and chemistry, in biology and physiology, and very particularly in medicine to cope with all these problems which threaten man's life: hunger, disease, contamination of food and water, pollution of the environment.

We need more young men and women who choose science as a career, and we need better support for those scientists who have the talent and the determination to engage in fruitful research work.

Challenging research objectives must be available, and sufficient support for research projects must be provided. Again, the space program with its wonderful opportunities to engage in truly magnificent research studies of the Moon and planets, of physics and astronomy, of biology and medicine, is an almost ideal catalyst which induces the reaction between the motivation for scientific work, opportunities to observe exciting phenomena of nature, and material support needed to carry out the research effort.

Among all the activities which are directed, controlled and funded by the American government, the space program is certainly the most visible, and probably the most debated activity, although it consumes only 1.6 per cent of the total national budget and less than one-third of 1 per cent of the gross national product.

As a stimulant and catalyst for the development of new technologies, and for research in the basic sciences, it is unparalleled by any other activity. In this respect, we may even say that the space program is taking over a function which for three or four thousand years has been the sad prerogative of wars.

How much human suffering can be avoided if nations, instead of competing with their bomb-dropping fleets of airplanes and rockets, compete with their moon-traveling space ships! This competition is full of promise for brilliant victories, but it leaves no room for the bitter fate of the vanquished which breeds nothing but revenge and new wars.

Although our space program seems to lead us away from our Earth and out toward the Moon, the Sun, the planets and the stars, I believe that none of these celestial objects will find as much attention and study by space scientists as our Earth.

It will become a better Earth, not only because of all the new technological and scientific knowledge which we will apply to the betterment of life, but also because we are developing a far deeper appreciation of our Earth, of life, and of man.

The photograph which I enclose with this letter shows a view of our Earth as seen from Apollo 8 when it orbited the moon at Christmas, 1968.

Of all the many wonderful results of the space program so far, this picture may be the most important one.

It opened our eyes to the fact that our Earth is a beautiful and most precious island in an unlimited void, and that there is no other place for us to live but the thin surface layer of our planet, bordered by the bleak nothingness of space.

Never before did so many people recognize how limited our Earth really is, and how perilous it would be to tamper with its ecological balance.

Ever since this picture was first published, voices have become louder and louder, warning of the grave problems that confront man in our times: pollution, hunger, poverty, urban living, food production, water control, overpopulation.

It is certainly not by accident that we begin to see the tremendous tasks waiting for us at a time when the young space age has provided us the first good look at our own planet.

Very fortunately, though, the space age not only holds out a mirror in which we can see ourselves; it also provides us with the technologies, the challenge, the motivation, and even with the optimism to attack these tasks with confidence.

What we learn in our space program, I believe, is fully supporting what Albert Schweitzer had in mind when he said:

"I am looking at the future with concern, but with good hope."

My very best wishes will always be with you and with your children.

Very Sincerely Yours,

Ernst Stuhlinger

This aerial view depicts the campus of the U.S. Space and Rocket Center, which serves as the official visior center for Marshall Space Flight Center. Conceived by Wernher von Braun, the center opened in 1970 and today is the home of the U.S. Space Camp programs.

Right: Test firing of a Redstone missile at Redstone Test Stand in the early 1950's.

Below: This was the test stand where the modified Redstone missile that launched into space the first American, Alan Shepard, was static tested as the last step before the flight occurred.

The Ares I rocket's first-stage development motor, or DM-1, completed its first successful test firing Sept. 10. The flame exits the motor at Mach 3 and the test firing lasted for 122 seconds. DM-1 is managed by the Ares Projects at NASA's Marshall Space Flight Center in Huntsville, Alabama. The stationary test firing was conducted by ATK Space Systems, a division of Alliant Techsystems of Brigham City, Utah, the prime contractor for the Ares I first stage. (Credit: ATK)

The Space Shuttle External Tank 120, slated for launch on the Orbiter Discovery, is shown here during transfer in NASA's Michoud Assembly Facility in New Orleans.

The launch of the Mercury-Redstone (MR-3), Freedom 7, MR-3 placed the first American astronaut, Alan Shepard, in suborbit on May 5, 1961.

Billows of smoke and steam rise above Launch Pad 39A at NASA's Kennedy Space Center in Florida alongside Space Shuttle Discovery as it races toward space on the STS-128 mission, August 28, 2009. Photo by NASA/Sandra Joseph and Kevin O'Connell.

Saturn V S-IC flight stages being assembled in the horizontal assembly area at the Michoud Assembly Facility.

Saturn I (SA-I), Block-1, on the launch pad ready for liftoff, carrying a dummy upperstage, October 27, 1961.

Repair of the Hubble Space Telescope (1993) - orbiting earth at an altitude of 325 nautical miles. Perched atop a foot restraint on shuttle Endeavour's remote manipulator system arm, astronauts Story Musgrave and Jeffrey Hoffman wrap up the final of five space walks.

The Shuttle Orbiter Enterprise is removed from Marshall Space Flight Center's Dynamic Test Stand following its first Mated Vertical Ground Vibration test. The tests marked the first time ever that the entire shuttle complement (including Orbiter, External Tank, and Solid Rocket Boosters) were mated vertically. The Test Stand will be used with the Ares Program.

The Shuttle Orbiter Enterprise atop a 747 landing at Redstone Arsenal Airfield.

On the 25th Anniversary of the Apollo-11 Moon landing, Marshall celebrated with a test firing of the Space Shuttle Main Engine at the Technology Test Bed (July 20, 1994). This drew a large crowd who stood in the fields around the test site and watched as plumes of white smoke verified ignition.

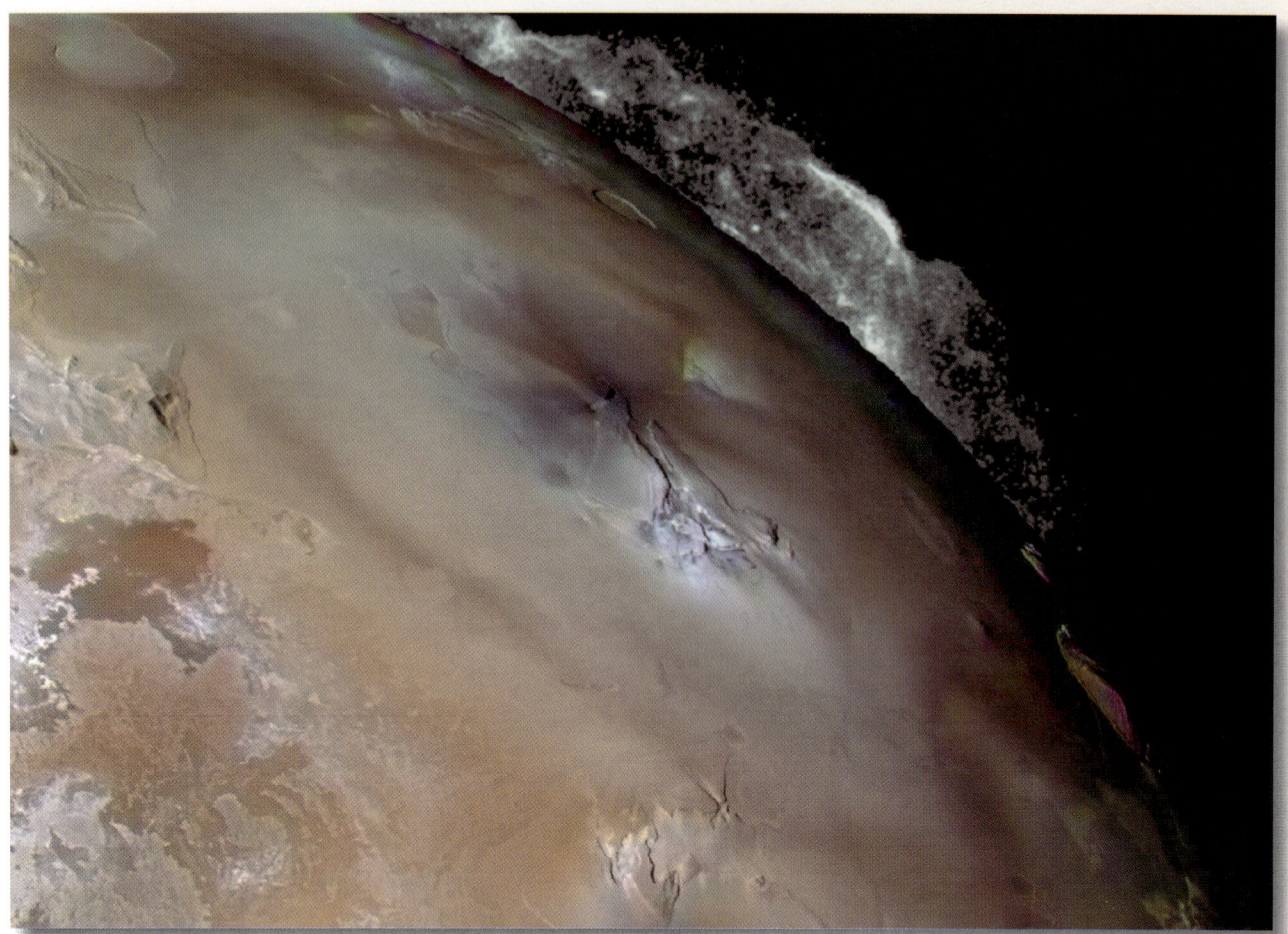

This image of Jupiter's moon, Io, was taken by the Chandra X-Ray Observatory. Shown here is the most extreme example of the effect of tidal forces as Io is being pulled by massive Jupiter on one side and by the outer moons Europa, Callisto, and Ganymede on the other. The opposing tidal forces alternately squeeze and stretch its interior, causing the solid surface to rise and fall by about 100 meters. The enormous amount of heat and pressure generated by the resulting friction creates colossal volcanoes and fractures on the surface of this moon.

The International Space Station is seen from Space Shuttle Discovery as the two spacecraft begin their relative separation on March 25, 2009.

The Saturn V launch vehicle (SA-506) for the Apollo 11 mission liftoff at 8:32 am CDT, July 16, 1969, from launch complex 39A at the Kennedy Space Center.

Space Shuttle Discovery and its seven-member crew launched at 2:38 p.m. (EDT) on a journey to the International Space Station on the historic Return to Flight STS-121 mission, being the first human-occupying spacecraft to launch on Independence Day.

SA-210 Apollo-Soyuz Test Project (ASTP) awaits the launch scheduled on July 15, 1975 on the launch pad at the Kennedy Space Center, the ASTP mission with astronauts Thomas Stafford, Vance Brand, and Donald "Deke" Slayton.

Nighttime at the U.S. Space and Rocket Center.

The Wide Field Camera 3 (WFC3), a new camera aboard NASA's Hubble Space Telescope, snapped this image of the planetary nebula, catalogued as NGC 6302, but more popularly called the Bug Nebula or the Butterfly Nebula. WFC3 was installed by NASA astronauts in May 2009, during the servicing mission to upgrade and repair the 19-year-old Hubble telescope.

Tethered to the end of the remote manipulator system arm, which was controlled from inside Atlantis' crew cabin, STS-125 astronaut Andrew Feustel navigates near the Hubble Space Telescope, during the mission's third spacewalk on May 16, 2009. Astronaut John Grunsfeld signals to his crewmate from just a few feet away. Astronauts Feustel and Grunsfeld were continuing servicing work on the giant observatory, which was locked down in the cargo bay of shuttle Atlantis.

Standing tall at its fully assembled height of 327 feet, the Ares I-X is one of the largest rockets ever processed in the Vehicle Assembly Building's High Bay 3, Super Stack 5 at the Kennedy Space Center. Ares I-X rivals the height of the Apollo Program's 364-foot-tall Saturn V.

Fireworks brighten the sky above the Davidson Center for Space Esploration, the new front door for the U.S. Space and Rocket Complex. A Saturn V rocket towers above the center.

MSFC Organizational Structures

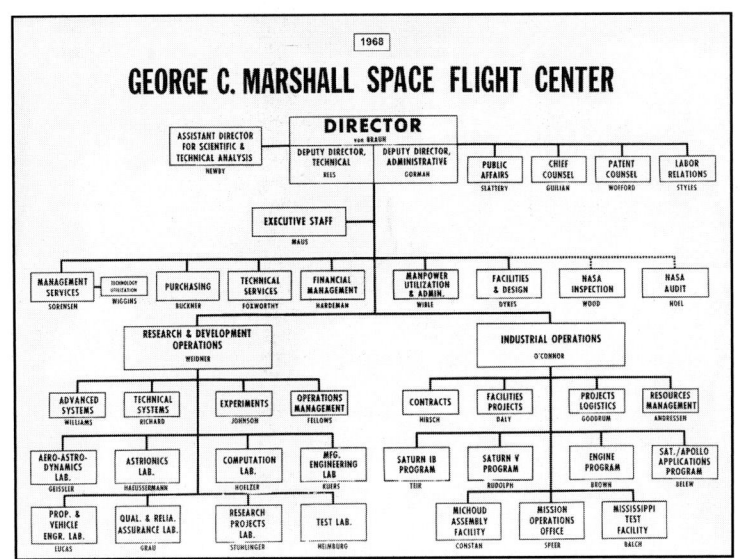

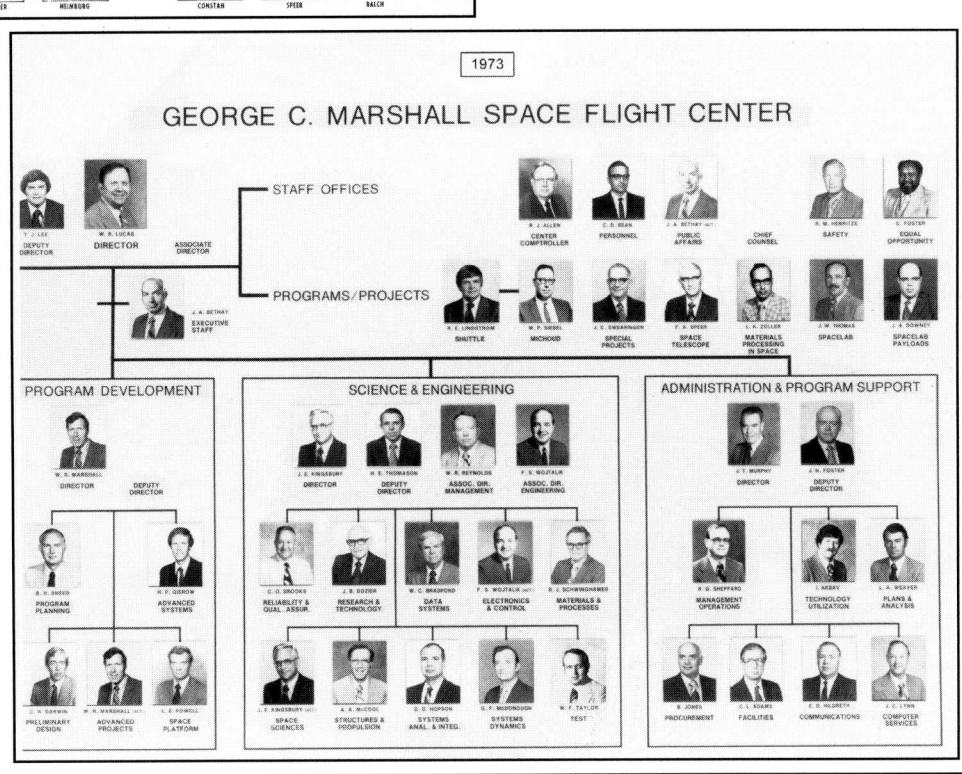

The top 27 officials at MSFC, 1960.

GEORGE C. MARSHALL
INITIAL MANAGEM

Bart J. Slattery, Jr.
Public Information Office

Jerry C. McCall
Assistant to the Director

Erich W. Neubert
Associate Deputy Director,
Research & Development

Eberhard F.M. Rees
Deputy Director,
Research & Development

Oswald H. Lange
Saturn Systems Office

Hans Hueter
Agena & Centaur
Systems Office

Werner G. Tiller
Weapons Systems Office

Heinz H. Koelle
Future Projects Office

Ernst D. Geissler
Aeroballistics Division

Helmut Hoelzer
Computation Division

Hans H. Maus
Fabrication & Assembly
Engineering Division

Walter Haeusserman
Guidance & Control
Division

History

SPACE FLIGHT CENTER
NT STRUCTURE 1960

Braun

Delmar M. Morris
Deputy Director,
Administration

Harry H. Gorman
Associate Deputy Director,
Administration

Chauncey W. Huth
Management Analysis
Office

William E. Guilian
General Counsel

nstan
gram
Office

David H. Newby
Office of
Technical Services

Victor C. Sorenson
Office of
Management Services

Claude E. Stockton
Office of
Financial Management

Wilber S. Davis
Office of Procurement
& Contracts

ous
tions
e

Ernst Stuhlinger
Research Projects
Division

William A. Mrazek
Structures & Mechanics
Division

Dieter Grau
Systems Analysis
& Reliability Division

Karl L. Heimburg
Test Division

Chapter 1

The Early Years

America entered mid-20th century as the undisputed leader of the free world. Victorious in World War II, its only real rival for power was the Soviet Union. Almost a footnote of both nation's victories was their acquisition of Germany's best rocket scientists and engineers, those gifted experts who had given Germany the means to unleash explosive power on targets hundreds of miles away. Yet, history would show their work in Germany was only the beginning.

For Wernher von Braun and his team of about 120 rocket engineers, scientists, technicians and a staff of U.S. Army personnel, mid-century began in Fort Bliss, Texas, south of the U.S. Army's White Sands Proving Ground. The charismatic and dynamic von Braun and his team had developed Germany's rocket capabilities at their research complex at Peenemunde, a small village on the Baltic coast. By early 1945, the stark reality of Germany's eminent defeat forced a decision: with the Red Army less than 75 miles away, von Braun and his team decided surrender to American forces was preferable to captivity by the Soviets. America, they agreed, had the resources to allow them to pursue their dream of rocket development for space travel. Little did they suspect that space travel was also the dream of many in Russia's Communist hierarchy, and that they would be competing with many of their colleagues who had fallen into Russian hands.

1950 saw the beginning of a string of events that put America solidly in the space race, with the commitment by a popular American president just a decade later to put a man on the Moon. That moon landing would be the fulfillment of the dream von Braun had nurtured since his youth. It would call for the most monumental commitment of scientific and engineering resources since the Manhattan Project that developed the atomic bomb. And it would completely change the face of a small southern town, Huntsville, Alabama.

Rocketry Comes to Huntsville

In postwar Huntsville, Alabama, civic leaders pondered how to market a huge bloc of government surplus real estate. The U.S. Army had come to Madison County in 1941, with war clouds gathering, and purchased a 34,000-acre tract south and west of town as the site of a huge chemical installation. Within a year the Army had developed both a chemical manufacturing operation, dubbed the Huntsville Arsenal, and an Ordnance shell loading and assembly plant which it called the Redstone Arsenal after the red dirt and rock found in the area. At peak production, the facilities turned out hundreds of thousands of shells, grenades and smoke bombs daily; employment reached more than 10,000, nearly half of them women. But the end of the war saw the end of production for the complex, and by late 1945 employment dropped to fewer than 600. Though the Army decided to keep the smaller Redstone Arsenal, the larger Huntsville Arsenal property was declared war surplus.

Salvation arrived in stages. In 1948 the newly formed U.S. Air Force showed a brief interest in the property. That interest soon evaporated, and the Army put a large "For Sale" sign up at the Arsenal property.

Civic leaders were determined to find a new tenant for the Army-owned property. Huntsville soon learned of plans by the Air Force to build a major research center and wind tunnel somewhere in the South for the development and testing of new aircraft designs. Air Force General Henry Harley "Hap" Arnold soon narrowed his search to Huntsville Arsenal's property and a site near Tullahoma, Tennessee. The final site selection decision became a tug of war between two powerful U.S. Senators, Alabama's John Sparkman and Tennessee's Estes Kefauver. When Kefauver finally prevailed, Tullahoma became the site of the new Arnold Engineering and Development Center.

But Sparkman soon found out about the Army's search for a larger site for rocket and missile development. Teaming up with Alabama's Congressman Bob Jones, Sparkman helped persuade the head of Army missile development, Colonel Holger N. Toftoy, to take a look at the Army installation in Huntsville.

Since World War II, Toftoy had been a key figure in the Army's missile program. As a major in the Ordnance Corps in Germany at the close of the war, he personally accepted the surrender of von Braun and his Peenemunde rocket team. Toftoy launched Operation Paperclip, arranging for the group's relocation to the U.S. Also shipped were components for nearly 100 V-2s, snatched almost out from under the Soviet's Red Army, as well as with extensive documentation of Germany's World War II rocket development efforts.

This German rocket team would form the nucleus of America's space program through the 1960s and well into the 1970s. Toftoy had the foresight to keep the group intact: von Braun had led the group to stellar achievements, so he remained as their natural leader. The team settled into quarters at the Army's Fort Bliss, near El Paso, Texas, 80 miles south of the Army's White Sands Proving Ground in New Mexico. Toftoy directed von Braun and his team to assemble and test fire the V-2s, using the launches to support the development of an American military rocket program, as well as supporting the nation's budding interest in upper atmospheric and space research. The team achieved impressive results: a V-2 launch in February 1949 reached an altitude of 244 miles, surpassing the 109 miles of the best Peenemunde launch in 1944.

Missile was a new word to post-war Americans, but the nation quickly became aware of their defensive possibilities. With the Army's growing interest in missile development, Toftoy knew he needed a larger, more permanent research and development facility. By 1948 he realized that the management of the Ordnance Corp's rocket activities needed to be decentralized into a field organization. After learning that Fort Bliss could not accommodate the new facilities needed for the Army's growing guided missile mission, Toftoy became the catalyst for relocating the von Braun team to the facilities located at the Huntsville Arsenal.

Toftoy often told the story of being sent by the Chief of Ordnance to meet with General Matthew Ridgeway, the Army Vice Chief of Staff, to try to convince him to permit the move of the Fort Bliss mission to the old Huntsville Arsenal area. Toftoy recalled that his map was so large that he had to spread it across the floor. He soon found that his pointer was not long enough, so he had to crawl around on the map as he pled his case. Finally, Toftoy found himself on his hands and knees right next to General Ridgeway. He looked up at the General, gave him a great smile, and said, "I'm really on my hands and knees, literally and figuratively, begging for this place." Upon Toftoy's recommendation, the Secretary of the Army in October 1949 gave his stamp of approval to the consolidation of rocket and missile research at the merged Huntsville and Redstone Arsenal complex.

Huntsvillians observed with curiosity the Spring of 1950 arrival of the first of the more than 100 German engineers and scientists. For many, the first reaction was suspicion: after all, these men and their families had been the enemy a scant five years earlier. But these feelings soon gave way to wholehearted acceptance and friendship as the Germans embraced their new home, and Huntsville embraced them. Few could imagine at the time how much the revival of Redstone Arsenal and arrival of the Germans would transform Huntsville and change its identity over the next decade.

Dr. Ernst Stuhlinger, an original member of the von Braun team who later directed the Marshall Space Science Laboratory, remembered the elation the team members felt upon arriving in Huntsville. "It was green and hilly, much more like Germany than Ft. Bliss was. For the first time since the war, we could buy homes and live where we chose." By 1956, most of the German team had become American citizens.

The Germans along with 500 U. S. military personnel, 180 General Electric contractors and 120 civil service employees joined a contingent of American rocket developers already assembled at Redstone Arsenal. Rohm and Haas and the Thiokol Corporation, a leader in solid-fuel propulsion, had moved some engineers to Huntsville in 1949. Employment at the Army's Guided Missile Development Division grew quickly as more military and civilian engineers and technicians were recruited to join the missile development effort. The number of people working for Redstone Arsenal surpassed 5,000 by June 1951. In 1952 the Army changed the name to Ordnance Missile Lab, with Toftoy, who had been promoted to Brigadier General, as its first Director. In 1954, General Toftoy became the commander of Redstone Arsenal.

The mid-1950s buildup at Redstone was an exciting time, one that saw the arrival of many of the Americans who would assume positions of authority by the early to mid 70s. The Army's 9330th Technical Services Unit stands out by the number of American rocketry leaders it produced. Formed at the end of the War, the 9330th was the Army's first guided missile organization, and its original mission was to support the von Braun team at Fort Bliss. The unit took part in the first U.S. V-2 launch on April 16, 1946, and followed the Germans to Huntsville in 1950.

With rapid growth in the missile development program, the Army lacked sufficient technical staff to support the rapidly growing enterprise. The Germans suggested a solution that had worked for them in the early 1940s: scour the Army ranks for technically trained people. Throughout the Army, swelling again due to manpower needs of the Korean War, the search began for recruits with scientific and engineering backgrounds.

Among those pulled from the ranks was a young Penn State professor of engineering named Charles Lundquist. Arriving in 1954, he was assigned to the Technical Feasibility Study Office under Dr. Adolf Thiel. When his group became part of the Army Ballistic Missile Agency in 1956, Lundquist found himself in the thick of the frenetic pace of missile development.

"Everyone was pressed to get things done," he recalled. Development of the Jupiter was well underway, which Lundquist said "had the Army's highest priority."

Another draftee was Robert Lindstrom. With a degree in ceramics engineering from the University of Illinois, the Army felt his knowledge could be put to use solving the problem of the intensive heat a missile experienced during reentry. In 1951, he found himself in Huntsville, working for the von Braun team.

"There were a bunch of us that worked for the Germans," he recalled. "It was Army life. We lived in the barracks, even did KP duty."

Lindstrom went on to larger responsibilities, becoming a project engineer for the Jupiter-C program in 1957, shuttling between Huntsville and the Jet Propulsion Laboratory in Pasadena, California. Soon afterward, he began work on the "big booster" that eventually became the Saturn program. Lindstrom went on in 1973 to head the Shuttle Projects Office.

James Kingsbury was another early arrival, sent to the Arsenal in 1951 when the rest of his unit went to Korea. He progressed through the engineering ranks on the Saturn and Skylab programs, functioning as Chief Engineer for the latter. Kingsbury became Director of Science and Engineering in 1975.

Alex McCool arrived in Huntsville in 1954, after earning a masters degree in Mechanical Engineering and beginning his career at the Army Corps of Engineers in Vicksburg, Mississippi. "I got this call from a friend, he was telling me what they were doing here in Huntsville. So I came over and got a job." He began

In the years following World War II, the Army directed that the Huntsville Arsenal be advertised for sale. The decision was reversed because the Army found it needed this land for the new missile work that would occur at Redstone Arsenal. January 1, 1940.

a long career in propulsion, arriving in time to work on improvements to the Redstone's propellant tanks. "We were trying to figure out how best to resize the tanks to get more performance." By the mid-70s, McCool was directing Marshall's Structures and Propulsion Lab. He was still a Marshall employee in 2002, heading the Shuttle Projects Office.

Alabama itself boasted several who played prominent roles in the space program. Bill Snoddy arrived in 1958 for a summer internship, on his way to a physics degree from the University of Alabama. He remembers the lack of space caused by the ABMA's rapid growth in the 50s. "There were several of us from the University of Alabama. They didn't have enough office space for us, so several of us just sat in the hall." Snoddy soon found himself in the Space Sciences Lab under Ernst Stuhlinger, analyzing temperature data from the Explorer satellites, and later worked with Stuhlinger on the Operation Highwater experiments and the Pegasus programs. He later went on to become the Deputy Director of the Program Development Office from 1983 to 1993.

T. J. "Jack" Lee came to Huntsville from the tiny town of Wedowee, Alabama. By the mid-70s he was working closely with the European Space Agency, NASA's partner in the Spacelab program as the manager of the project. Deputy Director of the Marshall Center in the 80s, he became Center Director in July 1989.

The Redstone Project

By 1950 the U.S. clearly recognized the threat posed by the Soviet Union, which by then possessed atomic weapons. The von Braun team found itself at the center of frenetic efforts to develop a missile that could deliver a nuclear warhead to distant targets. Their efforts paid off in 1952 when the Army Ordnance group perfected its first heavy, liquid-fueled ballistic missile, the Redstone, a versatile weapon that had evolved from and took advantage of German V-2 technology.

The first major rocket developed for the U.S. Army, the Redstone, was a marked improvement over its German predecessor. It utilized a more powerful engine than the V-2, with improved turbopumps and propellant delivery systems. While von Braun's V-2 design enclosed separate fuel and oxidizer tanks inside a steel body, the Redstone saved weight by joining together two aluminum fuel and oxidizer tanks without an external skin. And, unlike the V-2, which was designed to hit the target as a unit, the Redstone used a detachable warhead, which continued on to the target after engine cutoff. Perhaps most importantly, Redstone's advanced guidance system made it much more accurate over its 200 mile range.

Despite Cold War military spending, there were fiscal limits to U.S. rocket research in the days after World War II and later Korea, a situation that forced the von Braun team to economize whenever possible. When Redstone development required a test stand for engine static firing tests, the Army was dismayed to get contractor estimates exceeding $75,000, a large sum in those days. Undaunted, Karl Heimburg, head of the Test Laboratory, scrounged scrap steel and other materials, and provided the labor to build one for less than $1,000. That test stand, now a half-century old, still stands in the Marshall test area, a designated national historical site.

The first Redstones were actually assembled at the Arsenal, since the Army found it could convert the industrial works left over from World War II days to manufacture the missiles more economically in house than private contractors using other sites. But the Arsenal assembled the rocket from a number of sources. Engines came from the Rocketdyne division of North American Aviation, with several other contractors providing avionics and many other systems. Later, the Army contracted

The first Redstone Rocket launch.

with Chrysler Corporation to produce additional missiles using the Michigan Ordnance (later Army) Missile Plant near Detroit.

The "can-do" attitude that developed resulted from the arsenal system, under which the research center designed, tested and produced the missiles. When outside contractors were used to produce various components, they did so under the strict supervision of the Army's missile team. Where the Air Force quickly adopted a system of contractor reliance, the Army preferred this "hands-on" approach. "We had a core of technical people to oversee the contractors," recalled J.A. "Woody" Bethay, who later became Marshall's Associate Director. "We had the capability to actually do the work." Lindstrom reflected on the control this arrangement provided. "If we did not like a contractor, we got rid of them," although he admitted that sometime the contractors "were better than the civil service people."

These men and hundreds of others would complete an apprenticeship, of sorts, under the tutelage of von Braun and his German team. The Americans voice their respect and admiration for their boss in hundreds of stories, jokes, and reminisces of their years spent working for him. "He could make you feel good about yourself," reflected Jay Foster, who became an assistant to von Braun and spent many hours with him. Lindstrom recalled that, "von Braun wanted everyone in a place where he could contribute."

Spaceflight: First Satellites...

By 1956, the Ordnance Missile Laboratory at Redstone had grown to nearly 1,600 people, and soon that number would triple. In February of that year, the Army Ballistic Missile Agency came into being, charged with the continued development of the Redstone as well as the newer, 1,500-mile range Jupiter intermediate range ballistic missile. Von Braun became Director of ABMA's Development Operations Division, reporting to ABMA commander Major General John Bruce Medaris. Von Braun and

Medaris became supportive associates; though a military man, Medaris shared the vision of von Braun of manned space flight for exploration and discovery.

The ABMA team also developed the Jupiter-C nosecone reentry test vehicle, consisting of a Redstone first stage and two upper stages. It was developed to simulate the reentry of the still-under development Jupiter IRBM's nosecone, which explained the "Jupiter-C" (for composite) designation. A Jupiter-C launch in September 1956 traveled nearly 3,400 miles and reached an altitude of more than 680 miles. Von Braun and Medaris were so confident of the Jupiter-C's capabilities that they lobbied the Department of Defense to use it as America's prime satellite launch vehicle. It would achieve that purpose with the addition of an upper stage powered by a cluster of solid propellant motors, enabling the rocket to put a payload in orbit.

The Soviet Union stunned the world with the October 1957 launch of Sputnik, the world's first satellite. Americans reacted with shock and the realization that if the Soviets could orbit a satellite; they might be ahead in the missile race. In Huntsville, Medaris and von Braun confidently asserted that a Jupiter-C could launch a satellite within 90 days. But when the Soviets followed up their initial Sputnik triumph one month later with the launch of the much larger Sputnik II with a live dog on board, the Defense Department hurried plans for the launch of America's first satellite.

The nation held its breath on December 6, 1957 as the countdown for the launch of the Navy's Vanguard missile proceeded. At ignition, the Vanguard rose only a few inches from the launch pad before it exploded. Its dramatic destruction shifted attention to the Army's proposed launch vehicle, and the von Braun team finally had its chance. On January 31, 1958, the Army launched Explorer I into orbit on the Jupiter-C.

From ARMY to NASA

There was celebration in the streets of Huntsville with the successful Explorer launch, and von Braun was carried around the courthouse square on his colleague's shoulders like a winning football coach. But in Washington and around the nation, questions begged answers: how would America's space program develop? What level of resources should be devoted to it? Should there be a manned spaceflight element? And perhaps most importantly, should the military continue to control the space program, or should it become a civilian function?

President Eisenhower saw space exploration largely in scientific terms. His Presidential Science Advisory Commission (SAC) proposed a new agency to handle space exploration and national research on astronautics. Aeronautics had been the domain of the National Advisory Council for Aeronautics, formed during World War I to conduct research into flight, a technology then just over a decade old. The SAC proposal would transfer many of these functions to a new agency, the National Aeronautics and Space Administration. The President told Congress in April 1958 that space exploration should be a civilian function, separate from possible military uses of space.

Back at Redstone Arsenal, Medaris and von Braun continued to advance spaceflight endeavors, some of which were funded by the Defense Department's Advanced Research Projects Agency (ARPA). By October 1958, the von Braun group had launched several more satellites in the Explorer series.

Already, von Braun envisioned a much larger launch vehicle. He proposed that ABMA develop a huge booster, a multi-stage, liquid-fueled rocket with a clustered first stage powerful enough to produce 1.5 million pounds of thrust. Rocketdyne, which was already doing research on advanced engine concepts, was also at work developing engines for such a booster. This project, which

Major General John B. Medaris, (left) who was a Commander of the Army Ballistic Missile Agency (ABMA) in Redstone Arsenal, Alabama, during 1955 to 1958, shakes hands with Major General Holger Toftoy (right), who consolidated U.S. missile and rocketry development.

Jet Propulsion Laboratory Director Dr. James Pickering, Dr. James van Allen of the University of Iowa, and Army Ballistic Missile Agency Technical Director Dr. Wernher von Braun triumphantly display a model of the Explorer I, America's first satellite, shortly after the satellite's launch on January 31, 1958.

50 Years of Rockets and Spacecraft

Juno I, a slightly modified Jupiter-C launch vehicle, shortly before the January 31, 1958 lauch of America's first satellite, Explorer I.

would later be called the Saturn I, became the defining mission of von Braun's Development Operations Division both before and after its inclusion into NASA.

Urged on by the President, Congress passed the National Aeronautics and Space Act creating the National Aeronautics and Space Administration. The agency's first official day was October 1, 1958. Eisenhower named Dr. T. Keith Glennan as its first administrator. It absorbed the 8,000 personnel and all projects of the old NACA, the Jet Propulsion Laboratory, which had been operated by the California Institute of Technology under contract to the Army, the Navy's Vanguard, and several other projects.

But it did not, initially, include von Braun's German team and the Army employees of ABMA's Development Operations Division. Medaris and von Braun opposed Glennan's efforts to absorb the group into NASA, fearing a possible splitting up of the prized development team and loss of program control to the civilian agency. When Glennan would not relent, an uneasy compromise was reached under which the von Braun team remained under Army control, but would be "continually responsive to NASA requirements," making it, in effect, a NASA contractor.

For more than a year, NASA and ABMA tried to make the arrangement work. The agreement called for ABMA to provide Redstone rockets as boosters for the early Project Mercury suborbital flights. But by this time, development of the huge Saturn I was well under way. And it posed a dilemma: its huge booster capability was much larger than anything the Army needed to accomplish its missions, and it drained resources the Army needed for other programs. On the other hand, only the military seemed capable of providing the necessary funding for its development. Von Braun himself had experienced that in 1932, when only the German Army had the resources to support rocket development efforts.

The pressures came to a head in late 1959 when the Defense Department and NASA finally agreed to the transfer of ABMA's Development Operations Division to NASA. Glennan suggested that NASA's new Huntsville operations be named after General George C. Marshall, Chief of Staff during World War II. Some saw irony in naming a civilian agency after a military general, but as Secretary of State, Marshall had initiated the Marshall Plan for European recovery and had later won the Nobel Peace prize. It was a proud day for Eisenhower to come to Huntsville and name the Marshall Space Flight Center after his old World War II boss.

The transfer became effective on July 1, 1960. Von Braun became the first Center Director, a post he retained until 1970. More than 4,700 former Army employees transferred to NASA, along with 1,800 acres of Redstone Arsenal property and facilities valued at more than $100 million, including the launch facilities at Cape Canaveral.

To some, the change meant little in their day-to-day work. "You kept the same desk," recalled Jim Downey, who later became Deputy Director of the Program Development Office. Jay Foster recalled that "there were fewer regulations than there were with the Army," although he did note the changes in management with von Braun now fully in charge.

But for others, the incorporation into NASA was an exciting time of change and challenge. "We had to build every process, every procedure," said Woody Bethay, who transferred over from the Army shortly after Marshall was formed. "We didn't even have a budgetary process, or many of the things needed by a major agency. We had to develop everything. We wrote the book as we went along."

The final transfer of ABMA's Development Operations Division to NASA set the stage for what is acknowledged as the most exciting, historic chapter of Marshall's history. President Kennedy's announcement of the manned lunar mission and the development of the huge Saturn boosters for that program would set the tone for Marshall Space Flight Center during the 1960s. Along with it, Huntsville, Alabama would explode in population and national prominence through its leadership role in America's space program, becoming truly "the Rocket City."

Jupiter-C missile number 27 assembly at Redstone Arsenal, 1957.

Chapter 2
The Saturn Program

The 1957 launch of Sputnik propelled the growing contest between the U.S. and the Soviet Union into a vast new theater, outer space. But exploring space was a quantum leap upward from developing ICBMs. Space exploration signaled larger payloads, and larger payloads required larger launch vehicles. NASA recognized that a major part of the talent for development of those larger launch vehicles resided with the U.S. Army's Army Ballistic Missile Agency and its development team led by Dr. Wernher von Braun at Redstone Arsenal. Yet, with the establishment of NASA in 1958, money for Saturn development began to be funneled through NASA instead of the ABMA. Reluctantly, Von Braun and his boss General Medaris both realized the ultimate necessity of their becoming a part of NASA if the Saturn program were to continue under von Braun's control.

The Saturn program made Marshall Space Flight Center familiar to millions of Americans. Throughout the 1960s, it was MSFC's major program, consuming the major portion of its resources, employing the vast majority of its people, and accounting for the lion's share of its budget. Few who were not a part of the program understand the magnitude of the changes Saturn brought to the Marshall Center and to Huntsville, Alabama.

When MSFC was established in July 1960, Huntsville's population had grown to 72,365–more than quadruple the 16,500 residents of 1950. But by 1966, the tremendous impact of the Saturn program accelerated city growth to more than 143,000 residents. Huntsville took on a different social character as thousands of educated engineers, scientists and technicians flooded into what had long been a sleepy agricultural and textile town. The city took on a different character from others in Alabama and the rest of the Southeast. The explosive growth presented a staggering challenge to city leaders to provide roads, schools and infrastructure.

The thousands of new civil service and contract workers put a strain on the resources of Marshall Space Flight Center. A new headquarters complex was completed in 1963, crowned by a 10-story building with a commanding view of the development surroundings. Construction of new static and dynamic test stands, assembly buildings, and office space easily surpassed any of the other NASA centers, and ultimately drew dozens of aerospace firms from across the United States.

Cold War Realities

America had egg on its face after Sputnik. "The Russians suprised us," recalled a former MSFC manager. "We thought they were just tractor drivers." Not only did the Soviet Union put the first satellite in orbit, they added insult to injury less than one month later by orbiting a satellite weighing more than half a ton, with the dog Laika as passenger. America's response the following January with Jupiter-C and Explorer I helped save American pride, but across the U.S., it was clear Russia had gotten the jump on the United States in the space race.

Cold War threats, both perceived and real, compounded the national malaise. Many felt the Soviet Union had taken the lead in the missile race. Although U-2 spy plane photos showed no "missile gap" existed, the Eisenhower administration refused to compromise intelligence by revealing the source of the information. Consequently, the late 50s witnessed a major national debate as to how America could best catch up.

Wernher von Braun also sensed America's growing interest in space, and used his considerable public relations skills to stimulate it. A series of articles in *Colliers* magazine in the early 50s presented a vision of space exploration not too different from the late 60s reality. When Walt Disney approached him in 1956 for help in scripting a series of programs for "Tomorrowland," a Disney TV series, von Braun jumped at the chance.

In May 1959 the team launched monkeys Able and Baker aboard a Jupiter rocket on a 10,000-mile sub-orbital flight that proved animals could withstand the rigors of space flight and helped prepare the Marshall team for manned space flight. Baker, who became known as the "first lady of space" and somewhat of a celebrity, lived her final years at the U.S. Space & Rocket Center, where she is memorialized.

The first flight configuration of the giant Saturn I rocket is seen in the Fabrication and Assembly Engineering Division. Dwarfed by the 162-foot Saturn I are a Juno II (left rear) and a Mercury-Redstone (front foreground).

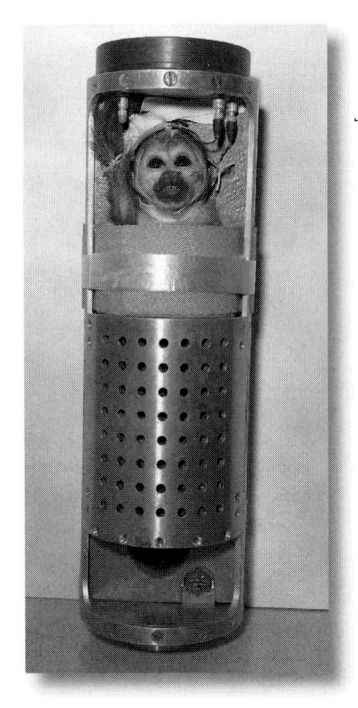

Miss Baker in a bio-pack couch being readied for Jupiter (AM-18 flight) which also carried an American-born rhesus monkey, Able into suborbit. The flight was successful and both monkeys were recovered in good condition. AM-18 was launched on May 28, 1959.

John F. Kennedy assumed the Presidency in January 1961, and before long the specter of foreign policy concerns began to consume more of the administration's attention. The Cuban missile crisis of 1962 demonstrated clearly that the Soviet threat was real, and could even be transported to a nation just 90 miles from American shores. When Cuban expatriates attempted an invasion of Castro's Cuba at the Bay of Pigs, the result was a diplomatic as well as military failure. But the big blow to American pride, and a shock to American security, came early in the Administration when the Russians, in April 1961, put their first cosmonaut, Yuri Gagarin, into orbit in a capsule weighing over 10,000 pounds.

By 1961, with America's interest in space exploration growing by the month, the new president realized the need for a dramatic program that would establish American supremacy in the space race. So in his first State of the Union speech in May 1961, he put forth the goal of "landing a man on the moon, and returning him safely to Earth, before the end of this decade." Though Kennedy knew that technical data showed the goal to be possible, his decision was clearly based more on political considerations. And perhaps faith. The famous speech came just three weeks after Alan Shepard completed his early May 1961 sub-orbital flight that lasted 15 minutes. A Huntsville-made Redstone rocket was the launch vehicle.

Kennedy's electrifying announcement gave rise to the Apollo-Saturn program. By the time of Kennedy's speech, Marshall Space Flight Center had been in operation less than a year. But the next five years would see phenomenal growth at MSFC in terms of budget, personnel and national prestige. More than anything else, the Apollo-Saturn program put Marshall and Huntsville on the map.

The Early Saturn Program

Long before Kennedy's speech, however, the Army had begun planning for a much larger booster, a sort of "super Jupiter," to be used solely for space science and exploration. In 1957, von Braun's group began preliminary development work for the huge booster, but serious funds were not forthcoming until the following year when the project won the support of the Defense Department's newly created Advanced Research Projects Agency. ARPA was born in 1958 when DoD realized the need for a group that could independently and critically examine competing rocket and missile programs from the services, and make decisions as to which should be funded. ARPA backed the ABMA proposal for a multi-stage, liquid-fueled rocket that could boost 40,000 pound payloads into Earth orbit, or 12,000 pound payloads to escape velocity. They urged the use of existing engine hardware, believing this would allow more rapid and less costly development. In August, ARPA directed von Braun's Huntsville group to develop the new booster within a two-year time frame. The initial directive only called for a successful static firing to occur by late 1960. Authorization of the new program energized the von Braun/ABMA team, now finally working on their dream since the early 1940s: vehicles for space exploration.

Continuing with the Juno name designation, the Army referred to the booster as Juno V, although the Army soon adopted the Saturn designations and the first was renamed Saturn I. Essentially hybrid vehicles, they employed modified versions of existing rocket engines in a clustered configuration around a central oxidizer tank.

The engine would be the H-1 engine already under development at Rocketdyne's facility at Canoga Park, California. There would be eight H-1 engines in the S-1 first stage, fueled by an RP-1 kerosene distillate and liquid oxygen (LOX) mixture, these engines each produced nearly 200,000 pounds of thrust, for a total of 1,500,000 pounds. One ABMA report referred to the Saturn as "the first real space vehicle, as the Douglas DC-3 was the first real airliner."

But the American space program was still in its infancy during these early years of Saturn. ABMA operated under acute budget constraints, with much of the meager allocation for Saturn going to Rocketdyne to finance upgrading of the Jupiter H-1 engine for Saturn use. Most of what remained had to be allocated for test stand construction, tank configuration studies and development, guidance systems, tooling, and other necessities. ABMA settled on an arrangement of eight 70 inch diameter Redstone fuel tanks surrounding a single 102 inch diameter Jupiter tank. While not exactly an engineering masterpiece, it worked, and it proved reliable. Saturn I rockets propelled Project Highwater, a water dispersion experiment, and Pegasus, a meteoroid detection satellite into low earth orbit, as well as unmanned Apollo spacecraft. But the decision to base Saturn largely on existing hardware, while saving money and time, created other problems. Many engineers doubted whether eight engines could be made to fire at exactly the same time. Engineers called the clustered S-I first stage "Clusters Last Stand." The use of multiple, existing tanks created a plumber's nightmare. Mr. Willie Mrazek, Director of the Structures and Mechanics Lab, shuddered at the thought of having to dismantle the huge first stage for shipping, a problem averted when it was decided to ship completed first stages from Huntsville by barge.

The year 1960 was one of rapid progress in the Saturn I program. In March, the executive order transferring the Saturn program from ABMA to NASA became effective. That month also saw the first successful test of the Saturn I first stage. In April, NASA awarded a contract to Douglas Aircraft for development of the Saturn 2nd stage, which would be powered by six Pratt & Whitney RL-10 liquid oxygen/liquid hydrogen engines of 15,000 pounds thrust each. NASA awarded Rocketdyne a contract the following month to proceed with work on the high-thrust J-2 engine, which would power the second stage of the more advanced Saturn IB.

The Saturn 1B: Improvement of an Existing Design

The Saturn 1 was planned primarily for test purposes, one of which was to see if eight engines could be made to fire simultaneously. Actual orbital missions would be launched by the more powerful Saturn IB, using the clustered first stage, but which featured an upper stage, the S-IVB, powered by the more advanced J-2 engine. The J-2 would derive its high, 200,000-pound thrust by the use of liquid hydrogen as a fuel. Liquid hydrogen was a more efficient fuel than RP-1 kerosene, but its lower density required larger tanks. To be practical, the fuel had to be carried in a liquefied state at extremely low temperatures with special tanks and pumping systems, made possible by advances in cryogenic technology by the late 60s. The Saturn IB was pivotal to the space program: it was the first American booster capable of launching a three-man Apollo crew to earth orbit.

The Rendezvous Issue

The decision to develop an even larger launch vehicle to support Kennedy's end of the 60s goal had already been made, but the size and configuration of the vehicle depended largely on settling the mode of mission issue. The key question: should the moon launch utilize earth orbital rendezvous, lunar orbital rendezvous or direct flights using Nova, a rocket more powerful than Saturn V. Earth orbital rendezvous (EOR) would require two launches to boost the Apollo spacecraft and the lunar lander into earth orbit, where they would link up before making the 250,000-mile journey to the moon. The lunar orbital rendezvous (LOR), however, would launch everything needed for the mission at one time but required rendezvous and docking in lunar orbit. The Nova-class vehicle, nearly twice the power of Saturn V, would permit a direct flight to the moon. Von Braun and his team had early favored the EOR approach, because EOR would require infrastructure that would be more readily adaptable to future projects, particularly the Space Station.

The Space Technology Group, centered at NASA's Langley Research Center, favored the LOR approach. They calculated that a single Saturn V launch vehicle could propel the entire lunar payload to escape velocity, thus reaching the moon. But Von Braun and the Marshall team had serious reservations about the complex docking procedures that would be required on the dark side of the moon, away from Earth contact. The issue was settled after IBM developed a computer in 1962 that could handle the complex calculations. With the issue settled, work on Saturn V moved into high gear. The decision to go with LOR stopped the Nova project from continuing.

Saturn V

Perhaps the greatest engineering challenge in American history was the development of the mammoth Saturn V launch vehicle. Of all American research and development projects, only the Manhattan Project to develop the world's first atomic weapon came close in terms of complexity, expense and time required.

View of Saturn V boosters being assembled at the Michoud Assembly Facility, New Orleans, LA.

The gigantic Saturn V was conceived in the minds of NASA planners as a three stage, liquid–fueled rocket. When launched, the S-IC first stage, powered by a cluster of five F-1 engines, would develop 7.5 million pounds of thrust and propel the launch vehicle to a speed of nearly 6,000 miles per hour. When it separated from the vehicle at an altitude of 38 miles, the S-II second stage ignited its five J-2 engines, further accelerating the vehicle to the edge of space, an altitude of 110 miles, and a speed of 17,000 miles per hour before being jettisoned. Finally, the S-IVB third stage took over, its single J-2 engine, now largely free of earth gravity, propelling the lunar payload into an elliptical orbit, and achieving a speed increase of another 2,000 miles per hour. Following three earth orbits to check systems, the S-IVB would re-ignite at the peak of the orbit, sending the lunar payload on its way to the moon at more than 24,000 miles per hour.

At Marshall, work began in 1960 on what would become the S-IC. In keeping with the arsenal concept, so well proven on the Redstone and Jupiter programs, Marshall performed all of the initial development work on the S-IC first stage, as well as further testing on the huge F-1 engine. Marshall built in-house several test articles including the S-IC-S structural test article, S-IC-F facilities test article, S-IC-T test article and the first two flight boosters. The S-IC-T, the "shop queen" or "T-Bird", was used for numerous ground test firings held at Marshall over the next several years. At 140 feet in height, the S-IC was almost as tall as the combined first and second stages of the Saturn I. When fully fueled with more than 200,000 gallons of RP-1 kerosene, it weighed more than 2 million pounds. It captured the attention of the national press and decision-makers in Washington when it was rolled out of the Fab Lab for the first time on its way to the test stand. Even though missing many of its critical parts that had to be added in the test stand, it was the first assembled piece of the Saturn V moon rocket to be seen by the public, thus previewing the immense size of the machine that would carry Americans to the moon.

In 1961 the Boeing Company was selected as the S-IC contractor. Boeing quickly converted an abandoned textile mill in Huntsville to serve as its engineering center for the project. Boeing joined other Saturn-Apollo contractors, Chrysler and IBM, who established a Huntsville presence. Later, Boeing and Chrysler would produce flight articles at Marshall's Michoud Assembly Facility in New Orleans.

A Myriad of Engineering Challenges

The F-1 engine, fueled by kerosene instead of the more exotic liquid hydrogen, lacked the sophistication of the smaller J-2, which powered the second stage. However, numerous problems had to be solved in its development, many of them brought about by the sheer size of the engine. For every second of the F-1's approximately two-minute burn, roughly two tons of RP-1 and one ton of LOX had to be pumped to the engine's

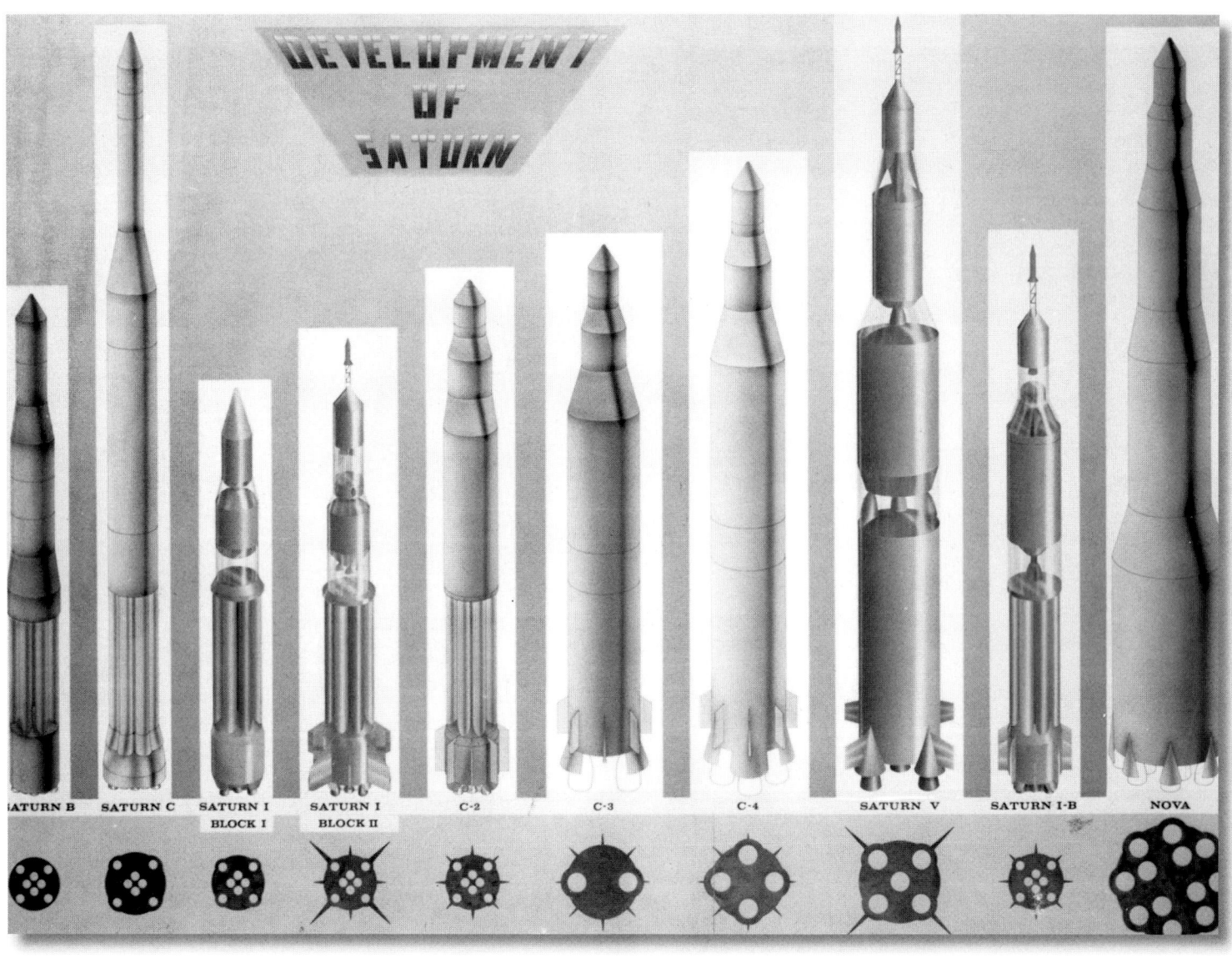

The Saturn V booster (S-IC) being prepared for shipping at the Manufacturing Engineering Laboratory at MSFC.

combustion chamber. Temperatures in the massive fuel pumps increased to 1,200 degrees, while combustion chamber temperatures reached over 5,000 degrees. The RP-1 fuel was circulated through tubes in the wall of the engine chamber, cooling the chamber and pre-heating the fuel before entering the combustion chamber. Dealing with these problems required advances in metallurgy and brazing techniques that were uncalled for in earlier engines.

A particularly sticky problem was combustion instability. Caused in part by slight design imperfections in the fuel injector, it caused vibrations and stresses that could destroy the engine within seconds. The dimensions of the problem were such that both Marshall and Rocketdyne established special committees in 1961 to address the issue. Marshall made tremendous efforts to solve the problem; indeed, some of those efforts bordered on the bizarre, such as prompting instability for test purposes by setting off small bombs in the exhaust blast. Von Braun even urged universities to ask their engineering doctoral students to make research on combustion instability the subject of their doctoral dissertations. Finally, in 1964, Rocketdyne perfected a baffled injector design that could be mission-rated. But solving the problem was expensive, "about $30 million," recalled Alex McCool, who at that time worked in the Propulsion Laboratory under William Mrazek.

Design of the turbopump posed major challenges. It had the task of pumping more than 15,000 gallons of RP-1 fuel and 24,000 gallons of oxidizer per minute to the F-1's combustion chamber. To enhance reliability, design was kept simple. Rocketdyne engineers devised a method to use the RP-1 fuel, with additives, to lubricate the bearings, thus eliminating a separate oil supply. Despite its simplicity, design of the turbopump required more time and effort than did any other engine component. "We had a lot of those pumps to blow up in development," McCool recalled.

Meanwhile, development proceeded on the S-II and S-IVB stages. Both stages would be powered by the J-2 engine, also developed by Rocketdyne. At NASA, Abe Silverstein directed the Office of Space Flight Development, which in conjunction with von Braun, Manned Space Center Director Robert Gilruth, and others had concluded that the second stage would need more efficient propulsion. Plans for the J-2 envisioned a self-contained engine that could start in flight, shut down, and then restart. J-2 design borrowed heavily from the smaller RL-10 liquid hydrogen/LOX engine.

Marshall's issuance of a Request for Proposals for the S-II second stage in the spring of 1961 might be considered the formal start of that program. Competition for the contract was intense, with North American Aviation finally besting Aerojet, Convair and Douglas Aircraft for the project. Though the contract, let in September, actually preceded the S-IC contract, the S-II encountered numerous problems over its five-year development period that caused it to become the de facto pacing item for the entire Saturn program. Some of these resulted from the uniqueness of the new engines, efforts to trim weight, and manufacturing techniques never before attempted. The S-II was the final Saturn stage to become man-rated: designed and fabricated to transport humans into space.

In contrast to the problem-ridden S-II program, development of the third stage S-IVB proceeded much more smoothly. This resulted partly because it had a head start, developed initially for the Saturn I and IB rockets. Douglas Aircraft was awarded the contract in June 1960, and remained the contractor through the various S-IV configurations. Reflecting the tried and true philosophy of the Saturn I program, initial S-IV stages relied on a cluster of six RL-10 engines of 15,000 pounds thrust. But as plans developed for larger payloads in manned lunar missions, the S-IVB configuration changed to a larger, 22 foot diameter stage powered by a single advanced J-2 engine of 225,000 pounds thrust.

Mounted atop the S-IVB, between the spacecraft and the stage, was a three-foot slice called the Instrument Unit or IU. The Instrument Unit, which was considered the "brain" or the "nerve center" of the Saturn IB and V, wasn't as glamorous when compared to the fiery show put on by the rocket powered stages, but it contained an assortment of highly sophisticated equipment required to perform the many complex functions that were vital to the success of the launch vehicle's mission. It contained the measuring and telemetry system to monitor the performance and status of all elements of the launch vehicle, the radio frequency communication system to transmit information to the ground and to receive command signals from the ground, and the guidance and control equipment for the vehicle which gave commands for engine gimballing, in-flight sequencing of engines, staging operations and all primary timing signals. The Instrument Unit also contained the electrical power and distribution system for the launch vehicle and the environmental control system to provide an acceptable thermal environment for the sensitive electronic equipment.

The various functional elements contained in the IU consisted of adaptations of existing technologies and in some cases advancements in the technological state-of-the-art. The sensing system of the Saturn inertial guidance system was the ST-124 gyro-stabilized platform. It used an ingenious air-bearing system, which was an outgrowth of stabilized platforms used on

50 Years of Rockets and Spacecraft

earlier missile systems. The other major element of the guidance and control system, the on-board digital computer and its peripheral equipment, utilized what was then advanced state-of-the-art microminiature unit logic devices. To significantly enhance reliability in the very complex digital system, a completely new concept of "triple modular redundancy" was incorporated in seven separate functional stages within the computer, which would assure continued proper operation of the computer even in the presence of multiple internal component failures.

The initial design of the IU and a number of its critical components was done in-house at Marshall. Several companies were then brought on board to develop the major functional elements (e.g., development and assembly of the stabilized platform was contracted to Bendix Corporation). International Business Machines (IBM) was selected as the prime contractor for the design of the digital computer system and for development, integration and fabrication of the overall IU. IBM built a facility in Cummings Research Park in Huntsville, where they developed and built the IU under the watchful eye of Marshall. The IU was unique in the regard that the structure was built and the functional elements were integrated in Huntsville, Alabama, whereas other major elements of the vehicle were developed and manufactured in other parts of the country.

Although admitting that his views might be somewhat parochial, Brooks Moore, Director of the Astrionics Laboratory which had the responsibility within MSFC for the technical oversight of the IU development, offered this assessment: "The Instrument Unit was a major functional element of the Saturn V system. Without the IU, the Saturn V was a 'Big Dumb Booster.' The IU added the intelligence to make it a living vehicle able to accomplish a remarkable and unprecedented mission. In spite of its complexity, the IU functioned reliably and successfully on every Saturn V mission."

Test Procedures

A key aspect of the arsenal system was intense testing: every component, every system, every stage was tested and retested in the effort to achieve "man rating" for space flight. Chuck Vedane, a Test Conductor in the Quality and Reliability Lab during those years, recalled how detailed the testing program was. "You'd have specialists in Guidance, Instrumentation, Electrical networks and RF systems," he said. "It was a building block test program." Marshall engineers developed automated test systems, he said, that were later adopted by the contractors to perform their own test procedures.

The culmination of testing was static firing of the engines and stages. Much larger test stands than those used to test the 188,000 pound thrust H-1 engines would be required to test the massive F-1 engine and S-IC first stage. Since the F-1 began as an Air Force project, Rocketdyne had built a massive test stand at Edwards Air Force Base in the Mojave Desert. Research and development testing took place at Edwards, while production engines individually and in the final cluster arrangement were tested in Marshall after the new stand was built.

The T-stand, on the west side of the Marshall test area, was built in the early 50s for the Redstone program, and modified in 1961 to withstand the 1.6 million pounds of thrust produced by the complete Saturn I first stage. Marshall built a massive stand in 1964 to accommodate the entire five-engine S-IC first stage. This stand was designed to test the Nova booster that had been envisioned as a mammoth rocket to replace the Saturn V, producing 12 million pounds of thrust. Tests of the mighty F-1 engines had the effect of small earthquakes. When Marshall officials learned that low-lying clouds increased the rocket's thunderous roar and damage to surrounding homes and businesses from the tremendous acoustical energy, they restricted testing to clear days.

S-IC static test firings were exciting events that attracted the media from far and wide. The *New York Times*, *Washington Post*, and other major papers would send reporters, who would stand, at a safe distance, and gaze in awe at the massive S-IC Test Stand. When the five F-1 engines ignited, the resulting flash was brighter than the sun. A couple of seconds later, spectators would feel a tremendous shock wave that seemed to pummel the chest. What they could not see were the 320,000 gallons of water per second being fed by gigantic pumps to cool the flame bucket, which took the full force and heat from the engines. At Building 4200, von Braun would interrupt important meetings so those present could witness the display from the top floor.

The tremendous forces of acceleration and the sloshing of fuels would subject the giant rocket to stresses and vibration during flight. So the largest structure at Marshall was, and still is, the gigantic dynamic test stand built on the north side of the main test area. For years it was the tallest structure in the state of Alabama. Standing 360 feet tall, topped by a 64-foot derrick crane of 200 tons capacity, the structure referred to

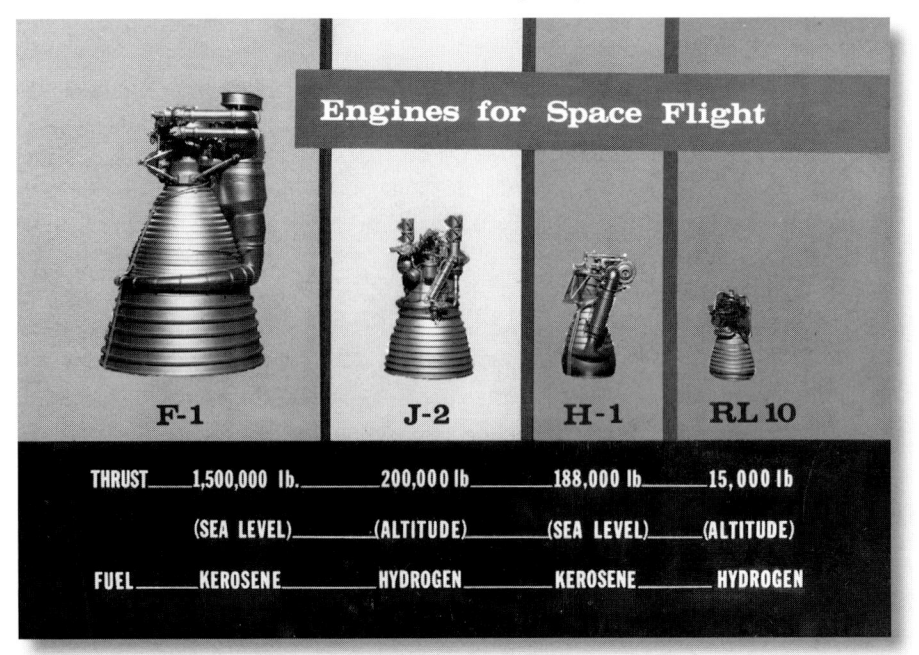

Rocketdyne was the builder of the F-1, J-2 and H-1 engines which powered the booster and upper stages of the man rated Saturn V and Saturn IB launch vehicles. Pratt Whitney provided the RL-10 engine used on the supper stage of the Saturn I.

	F-1	J-2	H-1	RL 10
THRUST	1,500,000 lb. (SEA LEVEL)	200,000 lb (ALTITUDE)	188,000 lb (SEA LEVEL)	15,000 lb (ALTITUDE)
FUEL	KEROSENE	HYDROGEN	KEROSENE	HYDROGEN

46

NASA - Marshall Flight Space Center

Saturn V second stage, S-II, is shown here on a transporter being moved from the North American Aviation Assembly Facility. (North American Aviation photo).

as the "shake stand" allowed an entire Saturn V vehicle with Apollo spacecraft, to be subjected to various stresses and vibration at 800 points on the rocket and spacecraft. The test stand played a major role in studying the natural bending modes of the vehicle and solving the problem of sloshing propellants that threaten to break up the vehicle during powered flight.

Overcoming Manufacturing Problems

Never before in history had so massive and complex a flight structure been built as the Saturn V. New materials, new processes, and the size of the components taxed engineering ingenuity. The 33-foot diameter sections of the S-1C first stage, and the 60-foot long, 25-inch diameter lines that fed liquid oxygen to the combustion chambers presented special challenges in welding techniques. Every S-1C had more than 10 kilometers of welds. Invariably, small deformities accumulated, producing slightly deformed tanks.

Bob Schwinghamer, who later became Director of the Materials and Processing Laboratory, led efforts to find solutions to these and other manufacturing problems. One innovation was a powerful electromagnetic hammer that could be used to correct the deformities without reworking. His group solved welding problems by the use of a special electromagnetic welding technique that was far more accurate than conventional techniques. He recalled that "we all let out a cheer" after a particularly troublesome weld on a fuel tank turned out perfectly.

Program Management

Von Braun's team and the Army shared the belief in the usefulness of the arsenal concept, which called for on-site research, development and initial manufacturing. Saturn I was developed this way, as were its predecessors, Jupiter and Redstone, and Army tactical missiles such as the Pershing. Under this system, power and influence at Marshall resided with the nine laboratories: Aero-Astrodynamics, Astrionics, Computation, Manufacturing, Propulsion and Vehicle Engineering, Quality and Reliability Assurance, Test, Launch Vehicle Operations, and Research Projects. The lab directors were members of the von Braun team who had labored with him at Kummersdorf and Peenemunde, risked their life to escape the oncoming Red Army at the end of the war, and endured the hiatus at El Paso, Texas. Von Braun respected and trusted these men, and relied on their technical expertise. Largely to them went the credit for Marshall's growing reputation as a center of technical excellence. They set the standards for the Americans of many technical and management professions who joined with the German rocketeers to develop rockets for the peaceful exploration of outer space and to operate one of the preeminent research centers in the world.

Political, schedule and other pressures forced Marshall to gradually depart from the early arsenal system. As the Saturn programs progressed into the mid-1960s, an increasing amount of work was performed by the aerospace contractors. News coverage of the growing space program made Boeing, Douglas, North American Aviation, and Rocketdyne familiar names to America. Marshall's task became mostly initial design, program management, systems integration, testing, and the oversight of scores of contractor firms who performed final design and production.

The growing size and complexity of the Saturn I and IB programs led von Braun to establish Program Offices for the various Saturn programs. It signaled a sea change at Marshall, where since the old ABMA days the laboratories held virtually all the power, and reflected the new reality that Marshall simply did not have the resources to simultaneously develop and produce the Saturn V and Saturn IB. "No one had ever managed a program like this. It was a big, monumental project," recalled Alex McCool. With thousands of Marshall employees and contractors and some 200,000 people across the U.S. supporting the Saturn program, a change in the management structure was sorely needed for Marshall to transition to a program management and systems integration role. The Saturn Systems Office, established earlier, became larger and more powerful. But the '62 organizational change did not go far enough.

The following year saw the Center's organizational structure transform into the basic configuration it would maintain through the rest of the '60s. In 1963 von Braun created the Office of Industrial Operations, under which the Saturn programs previously organized under the Saturn Systems Office became separate programs for Saturn I/IB, Saturn V, and engines, but under one overall director. The labs in turn were reorganized under Research and Development Operations, headed by an old von Braun confidant and former Peenemunde, Hermann Weidner.

50 Years of Rockets and Spacecraft

Von Braun recruited a top ranking Air Force general, Edmund O'Conner, to manage Industrial Operations, who brought along with himself several of his subordinate officers. O'Connor and his staff had management experience with large, complex Air Force missile programs; it was felt this experience would be needed as Marshall transitioned to a more program management role. The labs supported the program offices, and when disagreement emerged over who had greater authority, von Braun issued a directive clearly establishing the power of the Program Offices. Many felt it was tough for him to do this, given his abiding affection for the labs, nearly all headed by men he had worked with since the pre-V-2 days of the 1930s.

Under von Braun, Marshall "worked backwards from the launch date, providing a master plan for manufacturing" remembered Jay Foster, at that time a member of von Braun's executive staff and who later became Director for Administration and Program Support. Feeling a need to strengthen planning for the lunar mission, von Braun appointed Hans Maus, who had headed the Manufacturing Engineering Lab, to the new post of Director of the Lunar Program Planning Office.

The new arrangement never quite transitioned to the Air Force style of hands-off management of contractors. Marshall continued to maintain control through the use of "penetration," a practice whereby Marshall program managers and engineers maintained close supervision of their contractors. "Penetration" took several forms, but the most pervasive was the establishment of the Resident Manager Offices at each major contractor location. As Marshall employees, Resident Managers closely monitored contractor activities and reported back to the Center on a daily basis. Years later, Eberhard Rees, who took over as MSFC Director in 1970, referred to them as the "dynamic interface." After some period of time working at contractor sites, some RMO managers knew more about the contractor's role in the project than did the contractor's management.

Due to the size of Saturn V, new methods of transportation had to be developed like the Super Guppy aircraft shown here used to haul the 22-foot diameter Saturn V third stage, S-IVB, and the Instrument Unit (inset) across country.

Besides oversight of contractors, penetration had other advantages. "We always felt," recalled J.A. "Woody" Bethay, another member of the executive staff, "that in order to manage a program, you had to be able to do it. This kept our people sharp, by having in-house projects." Von Braun had another term for it, "dirty hands" engineering. He required that the Center retain in-house expertise in the laboratories that enabled the Center to tackle any problem that occurred at the contractors. He wanted to make sure his engineering managers stayed abreast of the engineering progress of whatever system or project they were managing. Bob Lindstrom, involved in the early development of the Saturn I and IB, recalled that "von Braun felt it was better to keep your people working on something you started…you would get better at it."

There were other forms of interface between NASA and its contractors. An elaborate schedule of weekly, monthly and quarterly meetings kept the airspace between Huntsville and the west coast full as contractor and Marshall officials and engineers shuttled back and forth. At Marshall, many top-level meetings took place in the Program Control Center at Building 4200. In addition, Marshall lower level program managers interfaced with their counterparts at NASA's Office of Manned Space Flight through a setup known as the GEM boxes, named after OMSF Director George E. Mueller (GEM), who originated the concept. GEM boxes provided a means for Marshall program managers to communicate directly with their counterparts at NASA headquarters.

The contractors accepted penetration, but disliked it, much preferring the freedom they enjoyed with the Air Force. While most contractors winced under it, they recognized the benefits of Marshall's close management of their programs. Von Braun and other Marshall directors defended the practice, citing cost control benefits and maintenance of quality. Incidents such as finding a rag left in a rocket motor certainly bolstered their case. By the mid-60s, most of the contractors got used to the close partnership arrangement, and Marshall people quickly forged a solid, working relationship with the contractor firms.

From 1963-65, the pace of activity at Marshall grew. By April 1965 the center could boast more than 6,700 civil service employees, with another 600 employed at the Michoud Facility and the Marshall-managed Mississippi Test Facility. In addition, more than 15,000 contractor personnel worked at the center or nearby. Every fourth worker in Huntsville and Madison County worked at or with Marshall. Huntsville proudly embraced the nickname "Rocket City."

Marshall Space Flight Center was in its golden age. A visitor driving onto Marshall from the north could see the vast expanse of the America's rocket development center stretching out literally for miles. Crowned by the glittering new ten-story Administration Building, dubbed the "Von Braun Hilton", visitors could drive for several miles past lab buildings, fabrication complexes, test areas, and support structures. Many observers noted a decidedly industrial look, especially when compared to the campus-like setting of other NASA centers. But Marshall was built to design, test and manufacture rockets; if the look was utilitarian, people knew important development work was taking place there.

The Power of the Labs

If there was a defining characteristic that distinguished Marshall from other NASA centers during the Saturn years, it was the strong role the laboratories played in the overall arsenal system. The arsenal system supported in-house development of programs, backed by the immense capability of the laboratories to solve technical problems.

"The lab structure was the key to the success of the center," according to Bethay, whose position as Director of Program

Water transportation by barge of the Saturn V booster stage, March 30, 1967.

Planning for the Center Plans and Resources Office gave him an overview of the Saturn program. "Von Braun and the others relied on the lab directors and their people on the technical issue." Laboratory directors had the power to go along with their responsibilities. "If a lab director didn't like what a contractor was doing," Foster recalled, "he would go up the ladder and get the problem solved." "Going up the ladder" might mean going to von Braun himself.

Marshall was a government enterprise, but the nine laboratories made it look and function a lot like an aerospace firm. The labs could design, test and fabricate rocket components or the entire vehicle. One of the largest was the Structures and Mechanics Lab, later renamed Propulsion and Vehicle Engineering, which had responsibility for structural and mechanical design, materials analysis and systems engineering. Closely related to it was the Manufacturing Engineering Laboratory, which fabricated the large prototypes. Its facilities were some of the largest structures at the center. Those prototypes were tested in huge test stands operated by the Test Laboratory. Quality in manufacturing was a major concern of the Quality Laboratory, which verified all significant operations and tested smaller components such as vehicle systems and subsystems.

Wind tunnel testing was the realm of the Aero-Astrodynamics Laboratory, which studied air flows on rockets and spacecraft. The Astrionics Laboratory developed the crucial guidance, communications and electric power systems. The Space Sciences Laboratory provided scientific support to Marshall's vast engineering capabilities, and conducted independent scientific research. The Computation Laboratories supported research activities in all the labs. Launch Operations conducted all launches of space vehicles at immense facilities at Cape Canaveral, later named Cape Kennedy after the president's death in November 1963. Shortly afterward, Launch Operations became a separate NASA center.

The lab directors were tied together by the doctrine of "automatic responsibility," under which von Braun made his top lieutenants responsible for bringing any and all technical or engineering problems to the table. Bethay recalled that, "If you were a lab director and you saw something going wrong, it was your responsibility to raise that issue, whether it was in your area of expertise or not, and cause it to be dealt with." In a sense, automatic responsibility was the natural outgrowth of the types of problems Marshall faced in the development of major programs, given that such complex systems could not be designed, tested or fabricated by one lab alone.

In 1965 the Center's budget was $1.8 billion with a work force of 7,522 civil servants and an estimated 14,000 contractors in Huntsville.

After Saturn and Skylab, the center underwent a major reorganization that reduced the power of the labs. Pressured for years to abandon the arsenal system, by the early 1970s the labs became primarily centers for the oversight of government contractors. Penetration lessened; the labs were no longer "in your britches" as one contractor complained during the Saturn years. And in the budget cutbacks of the early 1970s, the labs were no longer staffed to the extent they once were. The mighty Propulsion and Engineering Lab, for example, was reduced from over 600 employees to less than half that number.

It was the end of an era, one remembered with pride by many who worked on Saturn during the glory days.

The Satellite Centers

The tremendous scope of the Apollo/Saturn program required Marshall to spread tentacles of control to many parts of the United States. Foremost of these sites were the launch operations center at Cape Canaveral, the Michoud Assembly Facility at New Orleans, and the Mississippi Test Facility. At the height of the Saturn program, some several hundred Marshall employees were employed along with thousands of contractors at these Marshall-managed sites.

The Army had initially developed a launch facility at Cape Canaveral, Florida. When the von Braun team became part of NASA in 1960, the Florida launch operations were part of the deal. Television made Cape Canaveral known to millions with its coverage of the first U.S.-manned space flights, the Mercury and Gemini programs, in the early 60s. The growing pace of launch activity prompted NASA to designate the Cape Canaveral as a separate NASA center in 1962, and long time von Braun colleague, Kurt Debus, was chosen to manage it. One of his first actions was to supervise construction of Launch Complex 34B, where in the fall of 1961 the first Saturn I vehicle launches took place. Renamed the Kennedy Space Center shortly after John F. Kennedy's assassination in November 1963, this new NASA center experienced growth that paralleled the growth of the Saturn program itself.

Needing a second site for Saturn I production after the Chrysler contract was signed, Marshall began a search for a location that would combine water access, relative proximity to Marshall and the Cape, and an existing building. Marshall found it at Michoud, just east of New Orleans, where Chrysler had ten years earlier employed about 2000 workers to build engines for Army tanks. Designated the Michoud Assembly Facility, Chrysler and Boeing both occupied the 200,000 square yards of manufacturing space the site provided. Later upgraded with an engineering area, test facilities and a vertical assembly building, the Michoud Facility became the site for manufacture of the Saturn S-IB and Saturn S-IC stages.

The Mississippi Test Facility emerged from the need for a remote test area for Saturn stages that was close to Michoud, provided for water transportation, and sparsely populated. At the time of this decision, the Nova was used as a justification for acquiring a massive 150,000-acre tract, of which 130,000 acres would be used as an acoustical buffer zone. In late 1961 Marshall settled on the Pearl River in southern Mississippi, and after a $300 million investment, the Mississippi Test Facility was born. Only the animals were around to hear the thunderous roar of the Saturn stages from the massive test stands at the center of the complex.

The Lunar Rover

NASA had always realized that mobility on the moon's surface could extend the lunar mission. Despite the weightlessness of the moon, it was difficult for astronauts to move about in spacesuits; pressurization of the suit made it stiff and difficult to move in.

At Marshall, ideas for lunar cars had originated as far back

NASA - Marshall Flight Space Center

Astronauts John Young, Eugene Cernan, Fred Haise, Charlie Duke, Anthony England, Charles Fullerton, and Donald Peterson await development tests of the LRV, November 1971.

schematic, the LRV contract was strictly performance-based. The specifications were clear-cut: the vehicle must be able to operate in temperatures of +/-250 degrees Fahrenheit, it had to carry two astronauts with either astronaut driving from either seat; it must have a navigation system that would not only provide its position on the lunar surface but also display the direction and distance back to the Lunar Lander. In addition, the LRV would have the ability to surmount one-foot high obstacles, be stable on 45-degree slopes, be able to cross craters a foot deep, and climb 25-degree slopes. In doing all this, it must carry double its own weight, and be able to fold into an area four feet square. The last requirement resulted from MSC's rigid refusal to change the lunar lander to accommodate a larger package.

The Marshall project team and Boeing went to work. They had to be responsive to several groups, including Manned Space Center, the Kennedy Space Center, and the astronauts who would actually use the rover. LRV Project Manager Sonny Morea remembered that the astronauts had strong opinions on vehicle configuration. "As pilots, they were used to joysticks and hand controls like an airplane." Issues such as interface with the lunar lander, communications with mission control, and crew training had to be resolved. The low gravity situation on the moon helped in limiting vehicle weight, since the lightweight tubular framework need only support the weight of two people on earth.

Boeing delivered the first rover in March 1971, three weeks ahead of schedule. They met the deadline, though final project costs of $38 million were almost double their original bid but within the parameters of Marshall's original estimate.

But the rover was a success. Apollo astronauts were able to cover more territory and collect more samples than was possible on foot. Astronauts on the three pre-rover missions collected a total of 215 pounds of samples. On Apollo 15, 16 and 17 they collected a total of 635 pounds, and traveled 134 miles over the lunar surface.

as the ABMA days of the 50s. After the Saturn go-ahead in '61, von Braun had proposed an "Apollo Logistics Support System," a device incorporating both transportation and shelter. Several other ideas and designs emerged over the years from the fertile minds of MSFC engineers and contractors. Companies such as Bendix, General Motors, Lockheed, Northrop, Brown Engineering, Grumman and Boeing conducted studies for lunar vehicles based on operating durations of more than a year, with ranges of up to 1,000 kilometers on the lunar surface. Various vehicle types and sizes were proposed, including both wheeled and tracked vehicles, large mobile laboratories the size of a bus and small automated rovers the size of a suitcase.

But as the time for Apollo 11 neared, NASA's concern with the astronauts' safety on the moon overrode size and weight concerns. Also, it was believed that the astronauts should be able to make the most of their limited time on the lunar surface to conduct scientific experiments and collect samples.

NASA headquarters made the decision for the Lunar Roving Vehicle (LRV) development in May 1969, a scant two months before the first manned landing. The project, clearly more engineering than scientific in nature, went to Marshall. Since the directive called for the rover to be available for the Apollo 15, 16 and 17 flights, Marshall had 17 months to get the job done, far shorter than the typical four to five years of most NASA projects. They lost no time getting started, and Requests for Proposals went out on July 16, the same day Neil Armstrong stepped onto the moon.

In October NASA awarded the LRV development contract to Boeing. Given the short time frame for development, the contract was very different from any the Center had ever issued. Rather than the usual practice of designing to a predetermined

A Look Back

Many consider the Saturn/Apollo development years to be the best, the highlight of Marshall's nearly 50-year history. Certainly the level of excitement, the sense of mission, and the level of activity reached a zenith during the years of 1962 to 1967. Following Saturn, the center would continue in its development role for other important missions. Though future funding would never approach, in real terms, what Marshall enjoyed during those years, and future employment levels would never again reach their 1967 height, much was still in store for Marshall. Saturn/Apollo might fade into memory, but Marshall Space Flight Center would remain an American icon.

Chapter 3
Post Apollo and Skylab: 1970-1975

The '60s have long been heralded as the glory days of Marshall Space Flight Center, when a torrent of Federal dollars and rising employment in pursuit of a national goal seemingly would never end. But Marshall's managers knew, as well as did other top NASA officials, that completion of the Saturn vehicles and the end of the Apollo program in the early '70s would put an end to the glory days. The big question was, what would be the Center's post-Saturn mission?

For a time there was no assurance there would even be a mission. As Saturn development began to wind down by late 1966, NASA chipped away constantly at Marshall's huge assemblage of engineering and technical talent. Though von Braun hoped that Marshall strength would be maintained to compensate for contractor layoffs, Reductions in Force (RIFs) began to occur at Marshall. Even the timing of the announcements showed no mercy; a RIF of 700 Marshall employees was announced on November 9, 1967, the day of the first successful Saturn V launch.

Marshall was not caught totally unprepared. Planning for post-Apollo activity at the Center began as early as 1964 with the establishment of the Future Projects Office. Dr. Ernst Stuhlinger, always the advocate for scientific missions for Marshall, had long felt space science to be the logical next step to the lunar landing program. Planning for a probe to Mars, codenamed Voyager, started in the mid 60s, but Congress shut down its funding in 1967. The RIFs came with increasing frequency in 1967 and '68, and as the number of Marshall employees dropped, so did morale. Von Braun described Marshall Space Flight Center as "a tremendous solution looking for a problem."

Apollo 11, landing a man on the moon well within the end-of-decade deadline, produced an exhilarating sense of national pride, totally erasing the sense of doubt that characterized America's space efforts a decade earlier. But that heady feeling was not to last. The early '70s saw a shift in America's mood and attitude toward space exploration. Changing national priorities, such as the war in Vietnam and rising concerns for Federal social programs, drew both money and attention away from the space program. The U.S. President himself had championed the lunar landing program, but after Apollo 11 there was no overriding national space goal.

Aside from a boost in national pride and new knowledge of the lunar surface, many Americans also wondered what real benefit resulted from Apollo/Saturn's massive expenditure of funds and effort. "Von Braun said we were fortunate to have the support of the American taxpayer, but now we had to show that spaceflight would produce useful benefits on earth," Stuhlinger remembered. This realization gave birth at Marshall to space technology programs and other efforts to bring to Earth knowledge gained from space travel.

Development of Skylab, America's first space station, had begun as early as 1966, and carried Marshall through the early 1970s. To support Skylab, NASA delayed the major Reductions in Force until after Skylab development was completed in 1973. But by mid-decade the Center had to fight recurring studies of the feasibility of closing the center altogether. Dr. William R. Lucas, who took over as Center Director in 1974, recalled there was a strong possibility Marshall might cease to exist by 1980. And if it was not outright abolished, it might die a slow death through budgetary restrictions. Rapid inflation in the mid to late '70s mandated steady personnel cuts despite a static budget. "NASA decided money to pay contractors was easier to come by," said Jay Foster, whose department, administrative support, took the brunt of the earlier RIFs.

Woody Bethay, from his vantage point near the top of Marshall's administrative group, noted how the RIFs hit some groups harder than others. "We started with manufacturing and non-technical people. Through it all, the guiding philosophy was the preservation of engineering capability."

Like the false reports of Mark Twain's death, the rumors of Marshall's demise were vastly overstated. NASA headquarters thought long and hard about the real costs of closing down a center with the assemblage of talent Marshall possessed. Shuttle development, well underway by the mid-'70s, would be disrupted. There were doubts that all the development work done at Marshall could be done at the other centers—they simply did not begin to have the vast engineering and development capability so painstakingly developed over 20 years in Huntsville.

Skylab in orbit, taken by the Skylab 4 (the last Skylab mission) crew.

Besides, closing the Center would send a wrong signal about America's commitment to space exploration.

The Program Development Directorate

A major factor in Marshall's salvation was the Program Development Directorate. Established in August 1968 when von Braun and others became fully aware of the threats to Marshall's continued existence, the mission of Program Development (PD) was simple: to find new work for the center and provide an incubator for new programs until they became developed enough to be born. PD's first director was William R. Lucas, who later became Marshall Director after Rocco Petrone.

By almost any objective standard, Program Development was one of the most successful institutions in the center's history. Bill Snoddy, who worked with Stuhlinger at the Space Sciences Lab and later became Deputy Director of PD, remembers that Lucas recruited some of NASA's best people to work in the Program Development Office. Starting with a core group, PD eventually grew to nearly 250 employees.

In accomplishing its mission, PD worked closely with NASA headquarters, other NASA centers, industry and academia. Over its 30-year history, the PD made a major imprint on Marshall and its programs. Many aspects of the Space Shuttle and Spacelab, for example, started in PD, along with Phase A and B definition studies for the Space Station. PD put its stamp solidly on the large observatory program, which eventually included the High Energy Astronomy Observatory (HEAO), the Hubble Space Telescope, and the Chandra X-ray telescope. Even the energy programs such as solar energy systems, fuel cells, and more mundane coal gasification programs got their start in the Program Development Office.

The Program Development office operated in a political and economic environment of steadily declining NASA budgets, in stark contrast to the Saturn years. Saturn had provided a fixed goal and timetable; the budget could vary to meet that timetable. After the successful lunar missions, however, yearly budgets became the only fixed items: NASA and its Centers would have to make do on what an increasingly reluctant Congress would provide.

Skylab: A New Program in Old Hardware

NASA planners, and in particular Marshall's von Braun team, had long envisioned some type of space station program that would allow a team of scientists, not just astronauts, to live in space and perform experiments for an extended period of time. Before the first moon mission had even occurred, NASA's Director of Manned Space Flight George Mueller was encouraging von Braun to begin conceptual studies on a space station. With Saturn work winding down, it was a welcome development for Marshall's assemblage of veteran engineering talent. "The outpost in space," as Skylab manager Leland Belew termed it, would become a reality. NASA headquarters gave the program the official stamp in December 1966 when the Apollo Applications Office released the initial schedule for what would be called the Skylab program. The primary mission would be to make synoptic observations of the Sun over an extended period of time in wavelengths that are not accessible from Earth. Since high-resolution digital cameras were not available at the time, film would have to be used to record the images of the Sun. The crew could periodically change out the film canisters and return them to Earth. As development proceeded other experiments were added including an earth resources package (EREP), a materials processing facility, several exposure experiments to collect space dust and energetic particles, and several secondary observational experiments to be deployed through the scientific airlock. Of course the long exposure to microgravity would also be used to evaluate the crews ability to live and work in space and also to record the physiological changes produced by weightlessness as well as the effectiveness of various countermeasures to mitigate the adverse effects.

In the late '60s, NASA knew there was no budget for an entirely new program, so again the idea of using existing hardware took hold. George Mueller, head of the Office of Manned Space Flight, knew Congress was in no mood to fund a costly new space mission, but he felt he could sell the idea of a "space workshop" built from the S-IVB third stage. Mueller felt that, properly modified, it could accommodate a three-man crew to live in for weeks on end. The ongoing Apollo Applications Program provided a funding mechanism.

So the Center had a new mission. But with Houston, they had a problem. Marshall's newfound mission put it squarely into the business of developing manned spacecraft, a mission Houston's Manned Spacecraft Center felt they owned. To sort out who would do what on the orbital workshop program, NASA headquarters convened a meeting at Lake Logan, North Carolina, in August 1966. When the dust settled, Marshall emerged as the Lead Center for workshop development, while Manned Spacecraft Center trained astronauts and maintained control of mission operations. With Houston in charge of operations, Marshall's engineers found themselves, essentially, working for the Manned Spacecraft Center. But the multi-faceted interface situations that arose in Skylab's design did not help the inter-center rivalry; unlike Saturn's clear-cut distinction between launch vehicle and spacecraft, the orbital workshop was much more a joint MSFC-MSC development project.

So Skylab began, like the Saturn I program of a decade earlier, as a concept for the re-use of existing hardware. The S-IVB stage would be converted to an orbital workshop, with areas for astronauts to live, sleep, and work for up to nine months. The technical challenges were formidable. Providing up to nine months of electrical power from solar cells had never before been attempted; developing systems to guarantee pinpoint accuracy for the telescopes was an entirely new venture. High system reliability was a requirement, as was the mandate that astronauts be able to make repairs in space.

The Wet Workshop Idea

The overriding need to save money on Skylab development led to the concept of having the S-IVB do double duty, the so-called "wet workshop" scheme. This idea, first floated in 1966, would require the astronauts to purge the spent second stage of all remaining gaseous fuels, then convert the stage in earth orbit into the orbiting workstation. Astronauts would for the first time need to perform hours of extra vehicular activity in the weightless condition of space as they performed the conversion.

In Houston, MSC officials believed the wet workshop idea was impractical. They promoted the idea of smaller, self-contained workshops, each outfitted uniquely for each Skylab mission. When NASA ultimately rejected that idea in July 1969, the idea of Skylab as an outfitted S-IVB stage finally took hold. And since the S-IVB would be launched dry, not being used as a propulsive stage, it was strictly a payload that had to be put into earth orbit. But that payload would weigh some 100 tons, and only one American launch vehicle could put that much weight into orbit: the Saturn V. By 1970, the program still called for two launches, with the Orbital Workshop, the converted S-IVB stage, being launched atop a converted Saturn V.

Design and Construction

The S-IVB provided the structural shell and defined Skylab's size and configuration. Within those limitations, Marshall designers had to meet hundreds of design challenges. Scientific equipment and life-support systems had to be designed to withstand the stresses, temperatures and vibration of launch and the shock from docking procedures. Protection from radiation and micrometeoroids was needed, as well as thermal control from temperatures ranging from boiling to far below zero. Marshall engineers developed special insulation and coatings, as well as a system of cooling pumps, heat exchangers and radiators. Electric power, produced by large solar arrays during the sunlit phase of orbit, was stored in batteries for the shaded phase.

Marshall used the Saturn approach in managing Skylab development; most of the actual R & D work was performed by contractors, with the center directing budget and schedule, as well as providing systems integration, technical oversight and contractor management. Leland Belew headed the Skylab Program Office, and, following the Saturn model, established project offices for each major hardware component. Penetration was retained, with Belew setting up Resident Managers at each main contractor site. Marshall maintained a full-scale Skylab mockup, which proved its worth hundreds of times in checking systems and testing possible changes. The center took the lead in design of the docking adaptor, building and testing the prototype, then turning the project over to Martin Marietta for final development.

Much of the S-IVB's structure was reused. The liquid hydrogen tank was converted into living and working quarters for the crew, while the liquid oxygen tank became a sophisticated waste storage and disposal system. Altogether, Skylab astronauts had about the same living space as a small three-bedroom house. Marshall engineers employed Saturn concepts of conservative engineering and redundant systems, especially since weight was not a problem with the more powerful Saturn V launch vehicle. Money and time was saved by the use of proven components from earlier programs, such as hatch latches from the Gemini program of a decade earlier.

As with Saturn, the labs handled technical design, while the project offices administered contracts, and managed schedules and budgets. Each lab received needed support from other labs, which sometimes led to negotiations between lab directors on the best way to approach a problem. Testing could be a problem when components and systems from several labs were involved. But the system worked, as it had with Saturn. Yet Skylab was the swan song of the arsenal system; it was the last major project in which most of the initial design and testing was performed by Marshall people. After Skylab, Marshall became more and more a group of contractor managers, similar to NASA's other centers, although they continued to design and fabricate major sub-systems and performed testing in-house.

Skylab development proceeded steadily, if not always smoothly, through the late '60s and early '70s. There were occasional clashes with Manned Spacecraft Center (MSC) planners and designers over issues such as Skylab livability and other areas in which Houston saw Marshall intruding into their turf. Jim Downey, who worked in Program Development, recalled that Houston was not happy when Marshall got tasked to design the environmental control and life support system. MSC cast a jaundiced eye on the Neutral Buoyancy Simulator, taking the opinion its purpose was more astronaut training than support of scientific experiments. Max Faget, head of the MSC Technical Group and a veteran of the NASA program since the early '60s, had definite opinions that were often at odds with Marshall thinking. "Max always had something to say about everything," recalled Leland Belew. "You might say he was their sparkplug." Yet in the end, said Belew, Marshall managed to satisfy MSC's requirements "and integrate their team with our team quite well."

Mission Training

Many hours of planned extra-vehicular activity, performing tasks from unloading Apollo Telescope Mount (ATM) film canisters to performing repairs on the orbiter, were planned for the Skylab missions. Though astronauts walked on the moon and

Simulation testing Skylab emergency procedures in the MSFC Neutral Buoyancy Simulator, May 1973.

drove on the moon in the last three Apollo missions, the idea of floating in space, tethered to a spacecraft orbiting the Earth, was something American astronauts had not done for long periods of time. Marshall also recognized the need for design engineers at MSFC and the contractors to verify the feasibility of their designs. Taking the initiative, Marshall installed the Neutral Buoyancy Simulator (NBS), a huge tank 75 feet in diameter and 40 feet deep, holding 1.5 million gallons of water. Marshall justified the NBS solely for the use of design engineers. The subsequent use by astronauts was an added bonus. It was a costly project, and when NASA funds could not be secured under the facilities budget, Marshall took a creative approach, using tooling funds to build the tank with in-house fabricators. The tank technically was not a facility since it was not attached to the building and it could be drained and removed from the building. It prompted a GAO audit and reprimand, but contributed to the Marshall lore of creativity and pragmatism in getting the job done.

Marshall trained a number of design engineers to utilize the NBS and additionally trained 15 Skylab astronauts, nine prime and six back-up crew members. The astronauts would be called upon to perform literally hundreds of scientific experiments, both inside and outside the Orbital Workshop. So, in addition to the normal astronaut training, each of the three mission participants underwent roughly 2,000 hours of additional scientific and engineering training, approximating the time and effort needed to earn a college undergraduate degree.

The Apollo Telescope Mount (ATM)

In Marshall's quest for scientific diversification, research on the Sun and solar phenomena, such as sunspots and solar storms, figured prominently. Apollo planners had envisioned a solar telescope from the early years of the Saturn program, but realized Saturn's short missions did not allow enough time for the types of observations scientists had in mind. Skylab, however, functioning as a space station with longer duration manned missions, offered the opportunity to solar scientists and researchers. The ATM was originally assigned to Goddard Space Flight Center for development. But von Braun and Stuhlinger made a strong case for Marshall to do the job, pointing to a track record of earlier Marshall science successes such as Explorer and Pegasus.

The largest and most sophisticated scientific payload Marshall ever developed in-house, it evolved into a cluster of telescopes mounted into an octagonal framework 11 feet in diameter and 12 feet long. The ATM was mounted on a structure above the docking adapter, which was connected to the orbital workshop through the airlock module. The entire apparatus, encased under its protective shroud, formed the top stage of the converted Saturn-Skylab launch vehicle. In orbit, the ATM was rotated 90° to face the Sun, allowing the Apollo command and service module access to the docking port.

Four solar cell "wings" attached to the ATM provided electric power to the telescope's cameras and precise positioning apparatus. Orbiting 270 miles above the Earth, the telescopes could observe the Sun free of distortions and haze caused by Earth atmosphere.

The ATM was designed in-house, a product of the Astrionics,

The Apollo Telescope Mount undergoing vertical vibration testing in a vibration test unit.

Propulsion and Vehicle Engineering, Space Sciences, and Manufacturing Engineering, and the other labs. Gyroscopes and star seekers gave the mount two arc-second pointing accuracy and use of special radiation resistant film was required to operate in the Earth's radiation belts. Marshall engineers felt they were able to keep down the cost through use of in-house fabrication and testing. The ATM was a good example of Marshall keeping a significant in-house project to maintain technical expertise in managing contractors.

The ATM contained eight separate solar telescopes mounted on the main spar. Together, they spanned the electromagnetic spectrum from X-ray, through the extreme ultraviolet, to the long wavelength end of the visible. A chronograph provided continuous observations of the solar corona, which can only be seen from Earth during total solar eclipses. These were the most advanced instruments of their type to have been flown in space. The crew had also received extensive training in solar physics and became highly motivated and extremely interested in the Sun. In addition to their task of changing out and returning the film canisters, they continuously monitored the performance of the instruments and were able to react quickly to alter the observing protocol when solar event occurred. They were also called upon to perform in-flight modifications and repairs to the instruments when needed.

A major concern developed during the development of the observational experiments planned for Skylab, which stemmed from a report from previous astronauts that they were unable to see stars from their capsule. It was speculated that outgassing and other spacecraft effluent creates an induced atmosphere that could interfere with some of the planned observations. There was also concern that some of the outgassing from paints and other polymeric materials could coat on optical surfaces and cause unwanted absorption or scattering. A major research effort was launched in the Space Sciences Laboratory to investigate these possible problems and rigorous contamination control measures were instituted to mitigate possible contamination problems. Apollo astronauts were requested to take star field photographs during their return from lunar orbit. Quartz crystal microbalances were installed near critical optical surfaces. A photometer was deployed through the science airlock to measure scattered light from the gas cloud surrounding the spacecraft and also to obtain measurements of the zodiacal light (sunlight scattered from gas molecules and dust particles in the solar system). It was found that any gas cloud that may have surrounded the Skylab was sufficiently sparse to not interfere with any of the observations. The inability of earlier astronauts to see stars was apparently due to the high background illumination in the capsule. Optical surfaces could become contaminated only if they were in the direct line of sight of an outgassing source. These lessons proved to be useful in the design of the great observatories such as Hubble and Chandra.

The Skylab Rescue

On May 14, 1973, barely five months after the Apollo 17, the last Apollo moon mission, was successfully launched on December 7, 1972, the huge Saturn V carrying Skylab lifted off from Launch Pad 39-A at Kennedy Space Center. Those who followed the mission were treated to one of the most dramatic episodes in NASA history. The countdown and launch proceeded flawlessly, but 63 seconds into the flight, the micrometeoroid shield apparently became detached. Skylab went into orbit as planned, but stripped of the shield, temperatures inside the orbital workshop rose rapidly to over 200 degrees F. There was no electrical power and it was later determined that when the shield detached, it also ripped away one of the solar arrays and jammed the other, preventing its full deployment. Without electricity, the workshop was crippled; without the shield, not only was there danger from micrometeoroids, but the reduction in sunlight protection would cause tremendous heat buildup inside the orbiter. Temperatures inside could reach over 200 degrees; food would spoil, film would fog, and the polyurethane foam insulation could begin emitting harmful gases.

NASA scrambled to find a solution, and with little time to spare. As a first step, the launch of the astronauts Charles Conrad, Paul Weitz and Joe Kerwin on Skylab 2 was delayed ten days, until May 25, providing a 10-day period in which a solution had to be found. Marshall went into a crisis mode with engineers working literally around the clock. Marshall engineers regarded Skylab as "their baby" and refused to let the mission fail. Some relief from the heating problem was achieved by re-positioning the workshop, maneuvering the orbit to a 45-degree angle to the Sun. A compromise, it reduced heat buildup inside the workshop to 130 degrees, but also reduced electrical generation by the ATM solar arrays, now the only power source.

Everyone knew something had to be done to get some type of solar protection for the crippled spacecraft. The fix required a heat shield to be deployed by the first Skylab astronauts when they arrived on orbit, and the solution arrived at was amazingly low-tech. The team formed to deal with the shading problem, as a quick fix, devised a square parasol that could be deployed through the science airlock and erected without having to perform an EVA. This was sent up with the Skylab 2 crew. A more permanent fix, to be added later, deployed a reflective cloth screen, clothesline-fashion, between two poles attached to the ATM. The cloth material was similar to that used for spacesuits, and Marshall flew in two expert seamstresses from the spacesuit contractor to fabricate the shield. After simulating the attachment procedures in the Neutral Buoyancy Tank, Marshall engineers felt the Skylab crew could perform an EVA to attach the shield to the spacecraft.

Looking back over more than 40 years, most observers felt this was one of Marshall's finest hours. Alex McCool remembered it well. "Everyone who could contribute pitched in. It was an exciting time. But we knew we could not let that mission fail."

Skylab 2, with its first 3-man astronaut crew, the thermal screens, and extra medicines and food to replace those spoiled by the heat, was launched on May 25. Upon arrival at the stricken Skylab, the astronaut's first task was to free the stuck solar panel. Conrad maneuvered the Apollo ship within 10 feet of the panel. Weitz opened the hatch and, with Kerwin hanging onto his legs, he leaned out and attempted to jerk the stuck panel loose with a long handed hook. After an hour struggle, Weitz conceded he couldn't break it loose.

The crew docked with Skylab and slept in the Apollo cabin. The next morning, the crew transferred into the sweltering lab. The first order of business was to erect the umbrella-like solar screen. They were largely successful: soon after the 22-by-22 foot cloth screen was deployed, internal temperatures returned to normal.

With temperatures stabilizing, freeing the jammed solar panel became the next priority to increase onboard power. MSFC engineers had quickly developed several long handled tools, similar to those used for electrical line repair, and the crew had tested them in the Neutral Buoyancy Simulator before the launch of Skylab 2. Now able to live inside the Skylab, the astronauts practiced activities designed to free the panel and got advice from their back-up crews who practiced the repair in the Marshall NBS.

On June 7 astronauts Conrad and Kerwin exited the spacecraft into the blackness of space. After several strenuous hours they managed to free the jammed solar array doubling the power to the Lab. It was man's first major repair in space. Skylab was rescued. Its missions would be successful. "We wouldn't have saved Skylab if we hadn't had the Marshall tank. We worked day and night preparing and training that first crew to make the fix. We did everything in the tank that the crew did on orbit. Skylab Commander Conrad told us later it was so much like the simulation in the tank that he had to remind himself that this was the real thing," said James Splawn, who ran the tank during Skylab.

NASA learned much more about astronauts' ability to perform space repairs from the three Skylab missions. Also on the first mission, the Skylab 2 astronauts repaired a faulty solar observatory power conditioner by hitting it with a hammer at the right spot. Later, when gyroscopes used to control the orbiter overheated, Marshall engineers worked out a solution performed by the second Skylab crew.

A Multitude of Experiments

To a great extent, Skylab evolved as a tool for using space travel to provide information useful to mankind. Responding to a growing sense in America that the billions spent on space flight should serve useful purposes, NASA planned the three Skylab missions for round-the-clock experiments and scientific observations, all designed to increase mankind's useful fund of knowledge.

The Orbital Workshop was exactly that: an orbiting physics, biology and manufacturing laboratory. Equipped with an array of telescopes on the outside and TV and radio equipment inside, it amassed tremendous amounts of knowledge, much of which it transmitted back to Earth in real time.

Program objectives included a series of biomedical and physiological experiments designed to study how a person would function under extended periods of zero-gravity. Every physical aspect of human activity—sleep, digestion, respiration, and cardiovascular functions—was studied. The progressively longer missions would give some idea of how long a person could function under these conditions. Other experiments were designed to study how well men could work in space. "It's still probably the best data ever collected, even now in 2002," said Dr. Owen Garriott, Science pilot for the August 1973 Skylab 2 mission. "Those experiments have never been replicated."

Solar observations were a major part of the mission; the eight telescopes packed into the ATM were designed to observe and record photographically every part of the solar spectrum, from visible light to X-rays. Some of these images were relayed back to Earth by television. Operating hundreds of miles above Earth's atmosphere, Skylab was able to detect X-ray and ultraviolet radiation not detectable on the ground, radiation that often effects weather and communications on Earth. Since solar activity follows an 11-year cycle, the timing of the mission as the Sun was getting quieter allowed observations of high as well as low activity.

With the final splash-down of the Skylab 4 crew, the most comprehensive data set on the Sun that had ever been compiled was now available for extensive study. The several solar science Principal Investigators had rich data and remarkable individual results. However, these individual results needed inter-comparisons and joint analyses, and they also needed to be consolidated with simultaneous ground observations taken from the solar vector magnetograph operated at MSFC and other solar observatories and with theory. To accomplish this, the Skylab project office established a series of investigator workshops devoted to the principal solar phenomena — flares, coronal holes and active regions. The scientists in the MSFC Space Sciences Laboratory coordinated the workshops through a contract with the High Altitude Observatory at the University of Colorado. In several sessions over the course of a year, each workshop addressed a single phenomenon and produced a monograph on its topic. This set of monographs became the world baseline of solar knowledge for perhaps a decade, until new data came from the Solar Maximum spacecraft. The monographs are still valued archival references.

The Apollo Telescope Mount was truly another major Marshall success story by any standards. While in space, it expanded our knowledge of solar activities considerably, showing X-ray and ultraviolet emissions never before observed from Earth. These data revealed important relationships between the various layers of the Sun, the photosphere, the chromosphere, and the corona that had previously been hidden. X-ray and ultraviolet observations revealed early stages of solar flares that have forced a reassessment of the older theories of their formation. Bright points of X-ray and ultraviolet radiation seen in the corona and upper chromosphere are now believed to play a more fundamental role in magnetic activity than the sunspots that are visible on Earth. A new realm of solar activity on a cosmic scale was discovered. Huge blobs of corona matter, larger than the Sun itself, were seen blowing out from the Sun toward the planets, a phenomenon that had remained undetected until the ATM observations. Leo Goldberg, Director of Kitt Peak National observatory, stated, "The study of ATM observations has already led to many new discoveries about the nature of the Sun and about the fascinating events that occur on even a very ordinary star. Especially illuminating has been the recognition of the extent to which the Sun's magnetic field is responsible for the structure, dynamics, and heating of the Sun's outer layer. So massive was the harvest of information, however, that it will be many years before the possibilities for productive analysis are exhausted."

Skylab also looked beyond the Sun into the vast reaches of the Universe. Telescopes and other instruments discovered new X-ray, ultraviolet and visible light emissions from far reaches of the Milky Way Galaxy and beyond. Subatomic particle emissions such as cosmic rays were detected and recorded. They even photographed the passing in 1973 of the Comet Kohoutek.

One of the most useful areas of experimentation, however, was the Earth observation program. Skylab's orbital path covered 75% of the Earth's surface, passing over each point every five days. As awareness of the research potential grew, more than 140 scientific teams from across the U.S. and other nations came together in a coordinated scientific investigation of Earth. Skylab's photographic equipment worked flawlessly, producing thousands of photographs of the Earth, accurately recording land features, many of them in remote parts of the planet. Radar accurately measured land and water features.

The Materials Processing facility, developed by the Manufacturing Engineering Laboratory under the direction of Dr. Mathias Siebel, could accommodate a variety of experiments including a combustion chamber, a gradient freeze solidification furnace, and an electron-beam welding apparatus. Their primary motivation was to explore and develop technology that could be used in future, more ambitious space ventures. They were interested in knowing whether welding and brazing would produce different results in reduced gravity. The combustion chamber produced the first information about flame propagation under low gravity conditions, a topic that has been explored extensively since then. The original solidification experiments were prompted by a desire to know how phase-change materials, used for thermal control, solidified in space. However, the opportunity to experiment for the first time in a reasonable controlled microgravity environment attracted experiments some of the top crystal growers in the country. Although the results of these experiments were not spectacular, several intriguing things were discovered. For example, it was found that the solidifying melt in several cases were detached from the ampoule wall and that regions that were not in wall contact have fewer twins and dislocations. The mechanism by which the solid becomes detached is still not completely understood, but the improved quality gave an important clue to the role wall effects play in defect formation. These pioneering experiments, which grew from Hans Wuenscher's early advocacy for manufacturing in space, crude as they may have been, became the foundation for the Microgravity Science and Applications Program and put MSFC in the forefront of this research.

Despite the busy timeline set up for the crew's work schedule, the Skylab 3 became so efficient that they began to run out of things to do and started requesting new activities to make good use of their time. MSFC's Tommy Bannister and Barbara Facemire pulled together a group of simple demonstration experiments that could be performed using items that were onboard the spacecraft. The intent was to produce video tapes for classroom use to illustrate how fluids behave in the virtual absence of gravity. Though unduly criticized by some members of the scientific community as being unscientific and a waste of time, they dramatically demonstrated several interesting phenomena and even discovered some unexpected effects. For example, they showed how a spinning drop bifurcates into a dog bone shape and fissions (a process proposed by S. Chandrasekhar for double star formation). A simple liquid-liquid diffusion experiment using tea and water in a toothbrush tube showed that wall effects can seriously alter diffusion data taken in capillaries (one of the standard methods for measuring diffusion on Earth to minimize convection). One of the most elaborate experiments was suggested by John Carruthers, then at Bell Labs, who had been interested in the stability of rotating floating zones used to grow very high purity silicon. The crew rigged a simple lathe from two socket wrench extensions, driven by pulling a string. A liquid bridge of water was suspended between the two wrench extensions. When rotated, the water bridge deformed into a jump rope-like configuration, known as the "C-mode instability", which had not been predicted by theory. Carruthers was so intrigued by the possibility of doing such experiments in space that he later joined NASA and became the Director of the Materials Processing in Space program in NASA Headquarters. Also the prospect of having a crew on a permanent space station that can conduct experiments, observe the results, and modify the experiments was very appealing to the Materials Science community, which is the way they operate on Earth.

Even high school kids got in on the act. At NASA, someone came up with the idea of having high school science students submit ideas for experiments to be performed on Skylab. Thousands of ideas were submitted, of which NASA chose 17. One of the more fascinating involved the ability of spiders to create webs in the absence of gravity.

A detailed history of the Skylab project with many photographs of the vehicle, mission operations, the crews, and the ATM observations can be found in series of books published by the National Aeronautics and Space Administration, NASA SP-400, 401, 403, 404 and 443.

Managing the Reentry

Following the three Skylab missions in 1973, the spacecraft was deactivated, although NASA had planned to revisit it with the third phase of the space program, the Space Shuttle. By late 1978, however, NASA realized the Shuttle would not be ready in time to interact with Skylab. NASA also misjudged the drag effect of Earth's upper atmosphere on Skylab, largely because unexpected solar activity heated and expanded the atmosphere. Observations of the dormant spacecraft showed it to be falling to a lower orbit earlier than planned. By 1979, NASA realized the need to bring it back to Earth in a way that would minimize the chances of it falling in a populated area.

The first step was to somehow reactivate the spacecraft's power supply, dormant since early 1974. Through tedious, months-long efforts that involved recharging Skylab's batteries through commands from Earth, a team of Marshall and Johnson Space Center (Manned Space Center was renamed in 1973) engineers reactivated Skylab's electrical system, then changed Skylab's flight attitude to minimize drag. Concern mounted as the orbiting workshop made successively lower orbits of Earth, encountering thicker atmosphere that continually slowed its flight

To achieve an atmospheric entry over a sparsely populated region, on July 11, 1979 Marshall engineers activated a flight sequence program that caused Skylab to tumble in the upper atmosphere, where it slowed and lost altitude as predicted. After a fiery descent across the Indian Ocean, the remnant of Skylab crashed into the ocean, with a few pieces falling in western Australia. Though Marshall directed the spacecraft's final moments, media attention focused on JSC.

Skylab B, the second flight article that Marshall built, never got to fly, and is displayed in the National Air and Space Museum, Washington, DC. It was a complete flight article with experiments and everything needed to function in space. After the program continuation was cancelled, it was literally cut into transportable pieces and shipped to the Smithsonian Air and Space Museum and placed on exhibit, where it remains today.

Skylab, America's first space station, came to a close and Marshall people were proud of their contributions. In early June 1973, after the successful rescue of Skylab, Center Director Rocco Petrone read a letter of congratulations from NASA Administer James Fletcher to Marshall's Skylab team. Again, the warm glow of success flooded over the Center, the same warm glow so many felt after Apollo 11. Rein Ise, who headed the program after Leland Belew, remembered Skylab as "the highlight of anyone's career who was associated with it." Skylab made important scientific contributions in space science, engineering, medicine and physics. It demonstrated the extent to which men could live and work in space. And it cemented in the mind of America the excellence of the Marshall team.

Chapter 4
SCIENCE INTO SPACE

Throughout the '60s and Saturn years, Marshall was primarily an engineering center. It could be said that putting a man on the moon involved finding solutions and answers to thousands of engineering challenges. There were, of course, scientists at Marshall Space Flight Center, and in their fertile imaginations were the sparks of hundreds of ideas for scientific exploration and the thirst for knowledge of the vast, incomprehensible universe. But despite their longings, their efforts, by and large, were geared toward the support of propulsion projects. Science, as an end in itself, got the short end of the stick. With fewer than 100 personnel, Ernst Stuhlinger's Research Projects Laboratory was by far the smallest of the eight major labs, while five other labs each could boast more than 500 personnel. He lamented, in a memo to von Braun, that science was "Marshall's poor stepchild."

Of course, science activities had occurred at Marshall, making news for the center as far back as the late '50s. Explorer I in 1958 confirmed the existence of bands of high cosmic radiation in the upper ionosphere. With Marshall's primary mission in the 1960s focused on the Saturn family of launch vehicles, several flights were made of the booster with dummy upper stages. One flight carried sand for ballast and others carried water to simulate sloshing of the liquid fuel. Dr. Charles Lundquist and his colleagues suggested they test the explosive destruct system as a bonus, releasing 86 metric tons of water in the upper atmosphere. Project Highwater, as it was named, added to our store of knowledge about cloud formation and molecular dispersion in the upper atmosphere. Some scoffed at Marshall for launching a payload of water, but Marshall scientists learned much from the experiment.

"I was astounded at the amount of information (Dr. Ernst) Stuhlinger was able to gather from those experiments," recalled Bill Snoddy, looking back on his days as an assistant to Stuhlinger in the Research Projects Division.

A concern in planning for manned space flight was the danger from meteoroids, which could damage a spacecraft as it traveled through space at high speeds. NASA planners felt they simply needed to know more about the odds that a spacecraft might be hit — and what damage would occur. The Pegasus program was proposed by Marshall's Dr. William Johnson. They devised a large satellite — the largest launched thus far — with two large, sensing wings that unfolded in space. When a meteoroid penetrated the panels, it caused a temporary short circuit that registered as a hit. The total area or the detector panels was about 1000 square feet, the size of a small house. "We wanted a really big target, something that would have a good chance of being hit," Snoddy explained. A team lead by Dr. Robert Naumann analyzed the data. The three Pegasus missions were considered major successes, proving that meteoroid activity was less a hazard than originally feared so that a mission to the moon could be completed with a high degree of confidence.

Through the '60s, science moved in a sort of parallel course with the manned spaceflight programs. Not until Skylab in 1973 did these converge in one major program. Their ultimate marriage in the Space Shuttle program would bring to scientists from many nations the opportunity to carry on research in the unique environment of space.

Much of the impetus for Marshall's growing involvement in science resulted from the center's diversification efforts in the late '60s as Saturn development ended. Dr. William R. Lucas, then Director of Propulsion and Engineering, proposed a plan for bringing in and managing new projects for the Center. A group of 24 managers, a mix of German and American members of the von Braun team, went to Jekyll Island and for three days debating the program development proposal and other changes required for the post-Apollo era. This hideaway meeting officially established the Program Development Office (PD) as a permanent Marshall feature, with Lucas as its first Director. Von Braun promptly dubbed him the "vice president for sales" in recognition of PD's role in bringing new work to the center.

Technicians check out the extended Pegasus meteoroid detection surface, 1965.

One of Lucas's prime targets was the scientific community. It was not an easy sale: in the late '60s and early '70s, Marshall had never been known for science, but rather as an engineering and propulsion center. Nonetheless, the focused efforts of Lucas and his staff quickly began to pay off. Despite the difficulties, Lucas' PD group brought to the center important scientific projects in areas such as solar energy, telescopes and materials processing in microgravity conditions. Slowly, Marshall's image began to broaden from its strictly engineering base.

Marshall's top managers realized the necessity of recruiting more highly educated scientists to support these new projects. Throughout the '70s the recruitment of scientists continued, even as a series of Reductions in Force steadily reduced overall personnel levels after Apollo. By 1980, after Charles Lundquist took over as Director of Space Sciences Laboratory, Marshall could count more than 150 scientists — physicists, astronomers, biologists, and others. But Marshall, heavily involved in the Shuttle program by the late '70s, was still an engineering and propulsion center. Within NASA, science was still centered at the Goddard Center in Maryland and the Jet Propulsion Laboratory in Pasadena.

The buildup in scientists at Marshall, unfortunately, was more than offset by the continuing RIFs. In March 1972, Center Director Eberhard Rees informed Marshall employees that just over 200 jobs would be eliminated by June. Rumors flew about other RIFs, and Rees had to quell a rumor shortly afterward that 1,000 Marshall employees would be transferred to Houston. But the Nixon administration continued to cut NASA's budget. Rees retired as Center Director in February 1973. In response to a question upon his retirement, Rees responded that what he liked best about retirement was never having to fire another person. His successor, Rocco Petrone, was an outsider, who many Marshall people saw as a hatchetman sent to preside over the first massive downsizing of the center. One of the first actions he was required to take was yet another RIF, which would cut another 650 positions by the end of 1974, 12% of the Marshall workforce. From an Apollo high of nearly 6,800 civil service positions, Marshall would shrink to just over 4,700 by the end of 1974. Petrone's task required by headquarters was to reduce the size of the center, including retiring most of the remaining German members of the team, stabilized the center's roles and missions. He enabled the center to concentrate and respond better to the agency's new programs and requirements, which required a reduction in Marshall personnel and closing of certain facilities. Rocco Petrone, while tough at home, did succeed in convincing headquarters that Marshall was on the right track in balancing current workload with available manpower and facilities.

Apollo/Soyuz

Until the mid '70s, NASA had conducted all scientific projects without international involvement. But in the early '70s the Nixon administration was beginning to reach out to the major communist states of Russia and China. In May 1972, Nixon set the stage for a joint space project with Russia, our long-time adversary and rival in the space race, when he signed a 5-year agreement on cooperation in space science and technology. A plan for a rendezvous mission in space between a Soviet and American spacecraft was the centerpiece of the agreement; it would be known as Apollo-Soyuz Test Project.

As the only manned orbital mission to bridge the gap between Skylab and the Space Shuttle, the 1975 Apollo-Soyuz mission represented more than anything else tangible evidence of the Soviet-American thaw. "It was important from an international relations standpoint," remembered Bill Snoddy. "It also, you might say, represented the first step towards international cooperation in space. For one thing, we wanted to see if we could work together and communicate together. Could we overcome the language and engineering barriers?"

Apollo/Soyuz was also the last launch using a Marshall-developed Saturn rocket. The Saturn IB, among the last produced, performed flawlessly. Although the mission was described as "a level of sophistication less than Skylab," it did much to set the tone for Spacelab, the International Space Station, and other future international space ventures.

HEAO

What became known as the High Energy Astronomy Observatory (HEAO) began as the brain-child of Ernst Stuhlinger and his staff at the Research Projects Office in the Apollo days. Astronomy had been associated with the Goddard Center in Maryland, but Marshall was getting more and more into astronomy, with the Apollo Telescope Mount, plans for a large space telescope, and HEAO. Stuhlinger credits one of his deputies, Jim Downey, with the idea. "I was trying to raise interest in ion propulsion, but there was little interest in this in Washington.

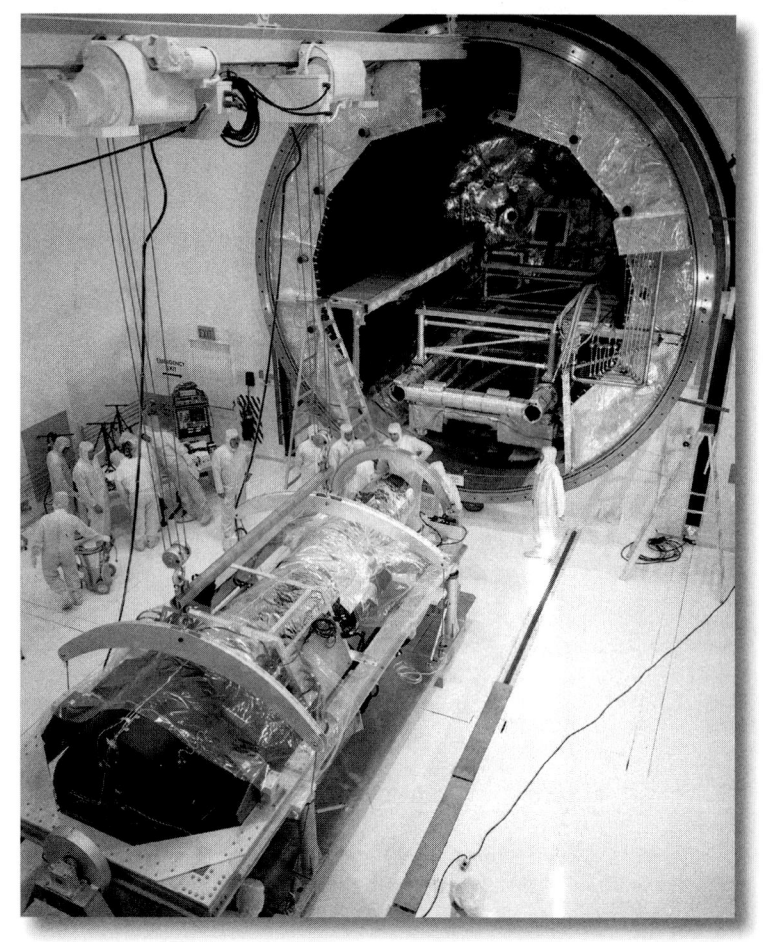

High Energy Astronomy Observatory (HEAO)-2 telescope being evaluated by engineers in the clean room of the X-Ray Calibration Facility at the MSFC.

NASA - Marshall Flight Space Center

The family of High Energy Observatory (HEAO) instruments consisted of three unmanned scientific observatories capable of detecting the x-rays emitted by the celestial bodies with high sensitivity and high resolution. In this photo, an artist's concept of three HEAO spacecraft is shown: HEAO-1, launched on Aug. 12, 1977; HEAO-2, launched on Nov. 13, 1978; and HEAO-3, launched on Sept. 20, 1979. Official NASA Photo.

Jim Downey, my assistant, said we should do a major telescope project in space." Functioning as a long-duration observatory, HEAO's powerful instruments would conduct high altitude and space studies on X and gamma radiation, and cosmic ray particles emitted from the far reaches of space.

Funding for the program was secured, and work on HEAO began in the shadow of the ongoing Saturn program. Stuhlinger put Jim Downey at the head of the HEAO Electromagnetic Radiation Project team, a job that included considerable responsibility for HEAO development. As Downey recalled, "it was tough, interfacing between the telescope and the spacecraft." The team acted as a go-between for the scientists and spacecraft engineers, helping in development of instruments and dealing with flight issues. By March 1970, plans were in place for three HEAO missions, and NASA had selected experiments for the first two satellites. TRW was selected as the contractor for the spacecraft.

HEAO's funding, however, did not last. When the administration cut NASA funding in 1973, HEAO was an early victim. All work stopped on the system, while a major lobbying effort by America's scientific community tried to get the program reinstated. Marshall people, expecting the program would be restored, but at a lower level, huddled with HEAO contractors and scientists to downscale the program. Just as with Skylab, Marshall would have to accomplish more with less. Funding, on a smaller scale, was restored the following year, and the program moved ahead in the mid-'70s.

What finally emerged was a three-satellite program, managed by Fred Speer, HEAOs-A, B, and C, with each version having a specialized mission. To save money, prototype versions were eliminated; the "protoflight" versions that emerged from development were used for both testing and actual flight. The vast majority of HEAO components were built by contractors; by the mid-'70s the arsenal concept at Marshall was retrenching and going underground.

The three HEAO missions were launched between 1977 and 1979. NASA felt their duration would be limited by the amount of thruster gas used on each mission to adjust the attitude of the satellite. They were pleasantly surprised to find out that they had misgauged the density of the upper atmosphere, which allowed greater attitude control with less thrust. Working with the scientists, they developed techniques to minimize the needed attitude changes while still conducting experiments. It was enough to dramatically extend the missions. In February 1978, NASA decided to extend the mission of HEAO-A by six months.

The HEAO series of satellites boasted several major successes. They produced the first images of high-altitude radiation sources, and discovered thousands of new sources of space X-rays. HEAO contributed new knowledge about supernovas, cosmic rays, black holes, pulsars, quasars, and dwarf stars. It was able to precisely fix the locations of X-ray sources.

LAGEOS

Since the days when Marshall old-timers worked for the old Army Ballistic Missile Agency, satellite launches had been part of the Center's mission. For years, scientists had wanted to orbit a satellite in a stationary pattern above the earth. If that satellite were highly reflective of laser beams sent from Earth, it could be useful in fixing positions on Earth with pinpoint accuracy. Thus was born Marshall's Laser Geodynamic Satellite (LAGEOS).

Scientists knew that even at orbital altitudes of more than 100 miles, the very thin atmosphere was enough to disturb a satellite in orbit. The solution: get the mass to frontal or cross sectional area as high as possible. A cannon ball — cross sectional area and high mass — would be a good example, which was the code name for Marshall's initial design effort. But 1971 budget restrictions killed the development of the four-ton Project Cannonball.

Marshall responded by going for a scaled-down version of the satellite, a 24-inch sphere consisting of a brass core overlaid with aluminum, and weighing 900 pounds. The surface of LAGEOS consisted of 426 silica reflectors, giving the satellite the appearance of a giant golf ball. According to NASA calculations, LAGEOS would stay in orbit for 8 million years, although its operational life would be about 50 years.

Design and production of LAGEOS was largely an in-house project. Like HEAO, the original version became the flight model, which, according to James Kingsbury, Director of the Science and Engineering Directorate, was produced at less than half of budget. Marshall directed the 1976 launch, but Goddard Space Flight Center coordinated the research performed with the satellite.

By all accounts, LAGEOS has performed flawlessly. From research centers around the world, scientists have bounced laser beams off its reflective surface. The tremendous accuracy of these measurements, to within two inches on the Earth's surface, has been used to detect shifts in polar ice caps, fault lines and tectonic plates.

The Birth of Microgravity

The need to diversify in the '70s led Marshall to begin long-term studies of a field that had intrigued scientists for decades: how do materials and substances perform in the weightless conditions of space? Both Apollo and Skylab cast light on the subject, but raised more questions than they answered. Scientists did learn, however, that perfect weightlessness was almost non-existent in a spacecraft. Atmospheric drag, crew motion, thruster firings, and other occurrences created small acceleration vectors that simulated minute amounts of gravity. What had been zero-gravity research got a new name: microgravity research.

Much had been learned on Skylab. Dr. Mathias Siebel, Director of the Manufacturing Engineering Lab at Marshall, pushed to have materials studies added to the regimen of Skylab experiments. Marshall scientists directed experiments and investigations into areas such as construction methods in space, and how different materials performed under near-weightless conditions. Skylab astronauts experimented with growing pure crystals for electronics use, and producing alloys from different densities of metals.

Despite all that was learned from Skylab, however, NASA did not see microgravity research as a priority, especially in an era of tight budgets. By the latter part of the decade, microgravity research funding amounted to less than 1/10 of 1% of NASA's budget. Not until the Space Shuttle in the early '80s was there any sustained funding for the program.

But if NASA initially lacked the resources for a space-based program, there were other, if less effective, ways to do research. Marshall engineers fabricated a drop tube inside the dynamic test stand, which provided a scant 3 seconds of weightlessness. A KC-135 aircraft, flying a parabolic or roller coaster pattern, could provide up to 30 seconds of weightlessness. Just as they had been used to train astronauts, they were employed for microgravity research.

In 1975 Marshall secured NASA funding for the Space Processing Applications Rockets (SPAR). The concept was simple: a low cost rocket would be fired into a suborbital flight. Microgravity experiments would be conducted during the five-minute return to Earth. The program was run on a shoestring budget, with each flight costing less than $1 million. This forced Marshall to relax its normally tight technical standards as well as compress schedules. Despite the restrictions, SPAR could fly up to nine experiments on a single flight.

The Center had by the end of the '70s taken the lead in microgravity research. Marshall scientists, working with

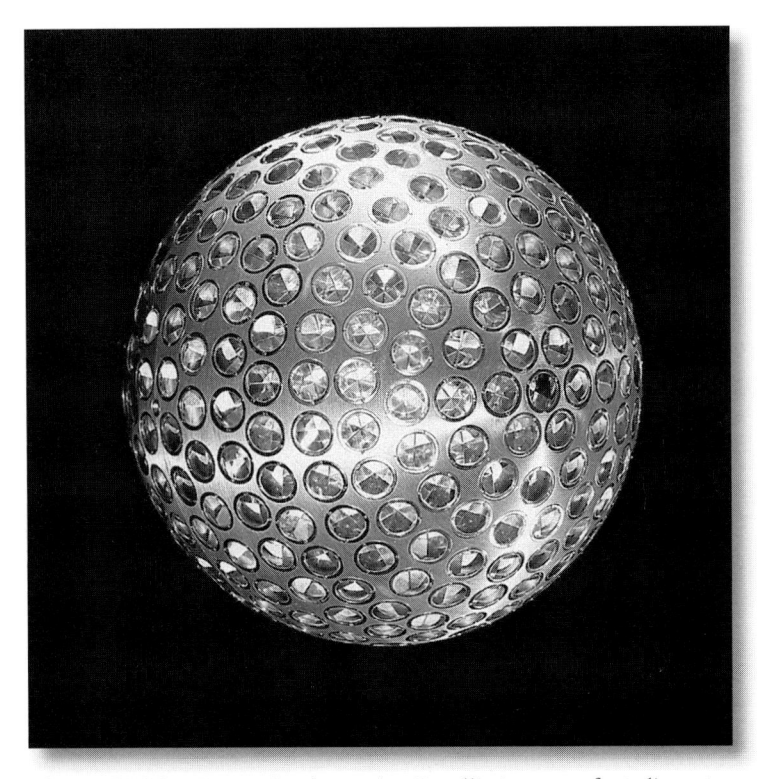

The LAGEOS I (Laser Geodynamics Satellite), a two-foot diameter satellite, orbited the Earth from pole to pole and measured the movements of the Earth's surface.

universities and corporations, were excited over the prospects. It was felt that materials processing in space might lead to new materials, improvements in electronics, medical devices, and other areas. Some envisioned "space factories" where these applications could take place on a large scale.

The Energy Programs

After Apollo and Skylab, Marshall Space Flight Center and, indeed, all of NASA had even more new realities to adjust to. Under the Nixon administration, NASA found itself pressed more and more to justify its existence. Its role changed from the customer being served by other Federal agencies to serving these agencies, as well as corporate America.

Marshall people showed their willingness to accommodate these new realities by getting into the coal business. In what is surely one of the strangest episodes in Marshall history, the center in 1974 found itself teamed with the Interior Department's Bureau of Mines in an effort to develop new technology to stimulate flagging mining productivity. The nation was in an economic slump, facing an energy crisis from the Arab oil embargo. Marshall faced a reduced workload after Skylab, so with the need to show tangible results of Earth from space efforts, Marshall people accepted the challenge.

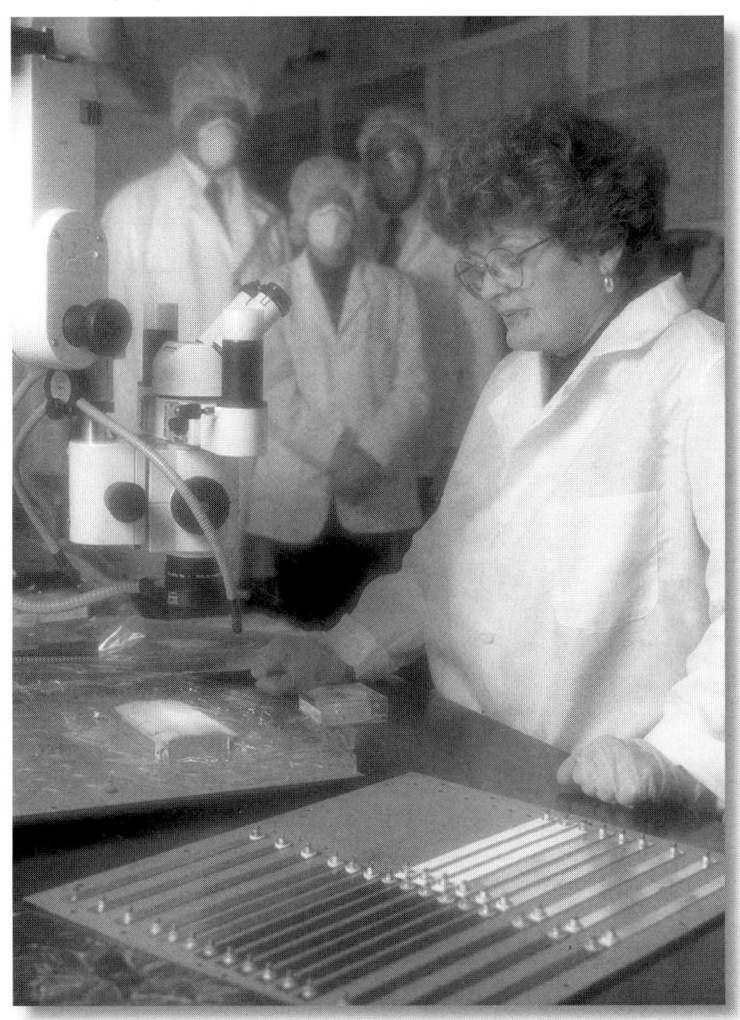

Ann Whitaker, formerly director of Marshall's Science Directorate, is shown here as a researcher examining a sample from the Solar Array Passive Long Duration Exposure Facility (LDEF). LDEF, which flew in space, measured the number, severity and effects of micrometeoroid hits on various materials. The data led to improved spacecraft design.

The Center was not exactly a novice in this field. Since 1970 the center had worked with state governments on satellite land classification, detection of tree pests such as the Southern Pine beetle, and development of satellite-assisted resource management plans.

Assistance to the Bureau of Mines involved using space-derived technologies to improve the efficiency of long-wall shearing machines, an automated mining technique replacing the old room-and-pillar mining methods. Improved sensing devices were needed to make sure the equipment extracted as much coal as possible from a seam without cutting into rock. Marshall scientists and engineers improved shearing equipment accuracy by developing sensors capable of measuring the amount of coal. Coal mining companies using the technology reported significant cost savings. The Center was lauded for its efforts by the newly created Department of Energy.

More appropriate to Marshall's background, however, was solar heating research. Solar heating was another way to respond to the energy crisis, and Marshall had been involved in this activity since late 1972. While it had to share the solar limelight with the Lewis Research Center, its efforts considerably expanded the nation's knowledge of this important technology.

In September 1974, when Congress passed the Solar Energy Heating and Cooling Demonstration Act, NASA named Marshall the lead center to conduct research supporting the agency's responsibilities. Despite considerable jockeying for power among NASA centers, Center Director Lucas fought for and maintained Marshall's lead role. Before the Reagan administration cut solar research funding in 1981, Marshall produced solar demonstration projects, conducted research, and helped reduce the cost to consumers of solar heating and cooling systems. Indeed, Wernher von Braun, now working for the Fairchild Corporation after his long career in government, told a group of southeastern state senators that, "Marshall helped give you the moon, and I don't see why Marshall can't help give you the sun."

Microgravity Glove Box being tested for use on Spacelab missions. Official NASA Photo.

Chapter 5
THE SPACE SHUTTLE

The successful Skylab program was another triumph for Marshall Space Flight Center. It was an opportunity for Marshall people to expand science activity, be successful in handling Skylab experiments, and accelerate the change of the Center from its former propulsion-only role. But propulsion was still the core of Marshall's role in NASA, and the center would continue that role in America's longest-running space program, the Space Shuttle.

The children of the "baby boom" generation, that group born after 1965, have come of age knowing little else of NASA than the Shuttle flights. Since 1981, the Shuttle has been part of the American scene. Millions have watched the liftoffs and followed the flights with fascination and awe; to some, flights of the "space truck" have become routine. No one can deny, however, the major leaps in knowledge of our Earth and the surrounding universe that the Shuttle has given us.

The Shuttle has continued to keep Marshall at the forefront of American space consciousness. Many of the large aerospace firms that today populate Huntsville's Cumming Research Park came here because of the Shuttle program. The orbiter is the key feature of the seal of the City of Huntsville. Many of Marshall's 2700 civil service employees, and over 4,000 contractors, come to work each day in support of the Shuttle program. A look at the shuttle itself explains why. Marshall managed the development of and continues to refine three of its four major components: the Shuttle main engines that power the Orbiter; the Solid Rocket Boosters, and the External Tank. The Orbiter is the responsibility of Houston's Johnson Space Center.

Origins: Gaining Support for a Reusable Spacecraft

The idea of a reusable spaceship goes back to the earliest 20th century dreamers of space travel, from Hermann Oberth's ideas in the 1920s in Germany to *Buck Rogers* comic books of the fifties. In his 1952 *Colliers* magazine series, Wernher von Braun envisioned a logistics vehicle to supply a huge orbital space station. Already attuned to the idea of reducing the cost-per-pound of payload to a minimum, he calculated that a reusable shuttlecraft capable of up to 100 missions could lower that cost to $50 per pound.

From the early '60s NASA planners had conceived the idea of a reusable spacecraft with multiple purposes – one that could carry astronauts into space, deliver materials, and serve as a platform for a wide array of experiments. In 1961, Marshall's Future Projects Office issued a request for proposals for the conceptual development of winged, reusable launch vehicles. These vehicles would include passenger and cargo carriers designed for Earth Orbit, with all stages capable of multiple re-use. Boeing, Lockheed, and North American Aviation – the main study participants – all concluded that such vehicles were possible. A follow-on study by Lockheed and Boeing in 1964-65 provided systems configuration guidelines for spacecraft that could take off and land like airplanes.

Bob Lindstrom, who became head of the Space Shuttle program office at Marshall in April 1973, recalled NASA's thinking as Apollo development reached a climax. "NASA and DoD started looking at fully reusable launch vehicles using flyback boosters and other ideas."

With Apollo development well underway in the mid '60s, NASA planners envisioned orbiting space stations as the next major

The first Space Shuttle External Tank, the Main Propulsion Test Article (MPTA), rolls off the assembly line September 9, 1977 at the Michoud Assembly Facility in New Orleans. The MPTA was then transported to the National Space Technology Laboratories (later named Stennis Space Center) in southern Mississippi where it was used in the first static firing of the three main engines.

step. These stations would need to be built, equipped and sustained in space, which would require some type of logistics vehicle. Even before the end of Apollo, many at NASA were thinking about space station development, which would get the major funds, while the logistics or re-supply vehicle would be relegated to a secondary priority.

In October 1966, officials of NASA's Office of Manned Space Flight, Marshall Space Flight Center, and the Manned Spacecraft Center powwowed in Houston to discuss the post-Apollo future. From these meetings came the basic concepts that would guide Shuttle development. As one possible concept, Houston's Manned Spacecraft Center proposed a fully reusable two-stage spacecraft with fixed wings.

By 1968, Shuttle conceptual planning was well underway. But after the triumph of the moon landing of Apollo 11 in July 1969, America's thinking began to change. Putting a man on the moon by the end of the decade had been accomplished, many argued, so what was the need for another costly project? The war in Vietnam had intensified; by 1968 more than 500,000 American troops were fighting in a far-off land, and after the 1968 Tet offensive, the war was increasingly seen as a stalemate. Racial violence was putting the torch to many American cities; in a dramatic gesture, Ralph Abernathy, a major leader of the civil rights movement, showed up at Cape Kennedy with mules and overalls to protest the American space program. The golden glow of support for the space program was rapidly fading as millions of Americans turned to what they considered more pressing problems.

The diminishing glow was felt throughout NASA. "I remember how the invitations to speak fell off so much after Apollo," recalled James Kingsbury, who headed the Center's Science and Engineering Directorate. "We even began to encounter hostility to the space program."

James C. Fletcher became the new NASA Administrator in 1971. He soon realized Congress and the Nixon Administration would not support both a space station and a space transportation program. Economics, not national pride, would drive American space efforts after Apollo. NASA would be forced to develop the shuttle, in whatever form it would emerge, with a budget a fraction of Apollo/Saturn's, and with constant oversight from the federal Office of Management and Budget.

In 1969 NASA took the next step in Shuttle development when it provided $500,000 to each of four Shuttle feasibility studies. In February, President Richard Nixon charged his new Space Task Group to come up with a plan for American space efforts after Apollo, whose final 1972 launch would occur in his administration. It was that year that the term "space shuttle" became attached to the program, after Manned Space Flight Director George Mueller mentioned that NASA needed a "vehicle similar to a shuttle bus."

Full Scale Development, Battling Program Costs and Management Control

Von Braun was called to NASA Headquarters in 1970. In his new role as Associate Administrator, he was instrumental in convincing the Nixon administration and Congress to authorize and appropriate the initial funding for full scale Shuttle development. Von Braun's testimony to Congress in 1969 and his in-depth cost benefit analysis went a long away toward convincing Washington to proceed with full-scale development, according to Jay Foster, who accompanied von Braun to NASA headquarters. George Mueller, Director of Manned Space Flight, assigned Marshall a leading role in evaluating several Shuttle configurations, a decision supported by the center's success in both propulsion and space science.

A major issue to be settled was that of program responsibility. In the fall of 1969 a series of intra-center meetings produced agreement on the need for separate contractors to develop the orbiter and the booster; and it seemed logical that Houston would manage the orbiter, while Marshall would retain its usual role of booster development. Though similar to the division of duties under Apollo, the complex interface between these two major components would complicate cooperation between the centers during the initial program phases. After the roles and missions issues were settled, however, MSFC and JSC personnel went to work to make the program successful.

To an extent not seen in the Saturn or Skylab programs, the aerospace industry played a major role in the Shuttle program, taking on more responsibility for development as well as manufacture. Along with the depressing stream of post-Apollo reductions in force, this led to a downsizing of Marshall's manufacturing capability. It had always been von Braun's philosophy that prototypes and subsystems be manufactured in-house, which provided program control and hands-on experience to engineers. But with serious cost-cutting pressures, NASA had little choice.

The cost issue eventually forced NASA to drop the fully reusable concept, adopting by 1971 the idea of a partially reusable shuttle. What ultimately developed was a design in which an expendable tank would carry both liquid hydrogen propellant and liquid oxygen for the ascent, which would permit the orbiter to be smaller and lighter. NASA would have to order a new fuel tank for each mission, but the tank was simple to develop, and its size allowed it to serve as the common attachment point for Shuttle main elements.

Budget pressures continued, forcing a reduction of the shuttle development budget from Paine's original 1968 estimate of $12 to $15 billion to $8.3 billion by 1970. After further cuts in early 1972, NASA was left with $5.15 billion for the program.

While NASA debated Shuttle configuration, the issue of program control still remained. Many in NASA felt Houston should be named the lead center for Shuttle development, which would put JSC in charge of financial management, program management and systems engineering, the latter two being management functions held by NASA headquarters under the Saturn program.

After some debate, a compromise was reached that made JSC the lead center, but under which Marshall reported to both Houston and the Shuttle program office at NASA headquarters. Development responsibilities followed a pattern established under Saturn/Apollo. Marshall had responsibility for Shuttle main engine development as well as the Solid Rocket Boosters and the expendable tank, while Houston kept responsibility for orbiter development, vehicle system engineering, and mission operations. Financial control stayed in Washington, a move seen as a way to enforce fiscal responsibility on the centers.

This photograph shows the second full-duration ground static firing test of the Solid Rocket Development Motor-2 (DM-2) at the Morton Thiokol Test Site in Wasatch, Utah. The twin boosters provide the majority of thrust for the first two minutes of flight, about 5.8 million pounds, augmenting the Shuttle's main propulsion system during liftoff. Official NASA photo.

The Decision for Solid Propellants

Since the early Saturn days, Marshall had become synonymous with liquid-fueled rockets. The center had a definite preference for liquid propulsion. Liquid fueled rockets had propelled Mercury, Gemini and Apollo astronauts into space. They could be tested dozens, even hundreds of times, in keeping with Marshall's reputation for conservative engineering. But except for the small escape rockets atop the Mercury and Apollo capsules, Marshall had limited experience with solid rockets.

On the other hand, a strong case could be made for solid boosters. They were much simpler, like huge roman candles, essentially huge metal tubes filled with fuel and oxidizer mixed together. Estimates were that solid fueled boosters would cost $700 million less than liquid boosters to develop; they would also be far less costly to produce. And if they hit the ocean surface at 100 mph after burnout, they would experience far less damage, and thus be far less costly to refurbish for future flights. NASA might not be comfortable with solids, but the Air Force had used solids for years. The Air Force's Minuteman missile used a solid fuel virtually identical to that later used in the Shuttle solid booster.

In the Spring of 1972, after careful review of the advantages and disadvantages, MSFC and NASA made the decision for Solid Rocket Boosters. Along with the decision the previous summer to combine oxygen and hydrogen fuel in the expendable tank, the choice of solids allowed NASA to maintain Shuttle development costs within the narrow budget confines NASA was forced to live with.

Development: The Space Shuttle Main Engine

Since three main engines would be mated to each orbiter, development of these components proceeded in tandem. The Space Shuttle Main Engine (SSME) program was seen as the pacing element, taking precedence over the Solid Rocket Booster and External Tank programs.

In more than a quarter century of propulsion experience, NASA had never tackled anything like the challenges of SSME development. Each engine would produce 470,000 pounds of thrust for eight and a half minutes of the shuttle flight, and would have to do this at least 50 times per engine. It would operate with turbopump pressures over 5,000 PSI and a temperature of over 1,000 degrees, both much higher than any previous engine. "It was the most difficult part of the program," said Science and Engineering Director Jim Kingsbury. "We operated under an extreme set of conditions, and had to feel our way along."

J. R. Thompson was named SSME Project Director, reporting to Lindstrom. Thompson had worked on Skylab in Man/Systems integration, and would later head NASA's Challenger investigation team.

The depressed state of the aerospace industry prompted intense competition between the three contenders for the contract: Rocketdyne, Pratt & Whitney and Aerojet General. In July 1972, Marshall awarded the contract to the Rocketdyne division of North American Rockwell. Though Pratt & Whitney protested the award, claiming a better design, the award was upheld.

NASA funding restrictions complicated main engine testing. Initially, it was planned that testing of engine subsystems would take place on special test stands at Rocketdyne's test center at Santa Susana, California, north of Los Angeles, designed to simulate actual engine and turbopump conditions. However, difficulties with building stands that could truly simulate SSME conditions led Marshall to opt for full engine testing at the Marshall complex at the National Space Technology Laboratory at Bay St. Louis, Mississippi. "It was just too complicated to try to simulate all the temperature and pressure conditions that a component would see," recalled Gerald Smith, then S & E's Assistant Chief Engineer for turbomachinery. There were a host of problems to deal with: turbine blades that cracked, LOX post failures resulting from undetermined powerhead flow conditions, and problems with hydraulic valves, controllers and software. "Marshall laboratories," said Lindstrom, "worked closely with Rocketdyne to correct the problems." He gives the labs a major share of the credit for the SSME's final success.

Since the main engine was the pacing element of the program, its development was underway for more than a year before contracts for the external tank and solid boosters were signed. Despite their considerable experience on earlier NASA engine projects, including the H-1, J-2, and F-1, Rockwell soon found itself grappling with development problems. Delays in construction of facilities, vendor delivery problems, schedule slippages, and cost overruns plagued the program. But the problems were no different from those NASA and its contractors had faced on other major programs, and Marshall and the contractor team dug in and found solutions.

NASA mandated a series of requirements changes to the SSME, essentially requiring Rocketdyne to shoot at a moving target. For example, said Lindstrom, "the engine was originally designed to operate at 100% thrust, but there were times when we felt we needed 109% for emergency situations. Due to the need to improve vehicle performance, 109% became the operating thrust level. Later, we decided to downrate it to 104%."

Cracking of turbine blades used in the engine turbopumps was a serious early challenge, caused by the turbopumps' extreme operating temperatures and rotation speeds of up to 35,000 rpm. A certain explosion would result if a blade were to break off and jam the turbopump. "We could not find a material that could operate in that environment," Kingsbury asserted. With help from Science and Engineering, a new material was found, a super hard material that required new casting and machining methods for its use. But it solved the blade-cracking problem, and today SSME turbines last ten or more flights before replacement.

The turbopump bearings were another source of headaches. The best steel ball bearings available could not perform adequately at prolonged temperatures above 400 degrees. SSME Engineer Bob Schwinghamer's group developed a ceramic ball bearing made of silicon nitride, a superhard material that could withstand 1700 degrees. Center Director Jack Lee at first objected to the use of what he considered "a glass bearing," but was won over when shown how well the bearing held up under the extreme turbopump pressures and temperatures.

Development: The Solid Rocket Boosters

Like the SSME, the Solid Rocket Boosters were to be refurbished and reused after each flight. Since these giant boosters were man-rated for space flight, reliability was the utmost consideration in their development. But they did not present new technological challenges as much as they simply required the use of existing rocket technologies on a larger scale.

Center Director Rocco Petrone appointed George Hardy to head up the SRB development effort. Hardy, a veteran of both the Saturn and Skylab programs, brought solid practical experience and management capabilities to the project.

Two key decisions at the program's outset defined Marshall's role in SRB development. The first was to split the project between development of the solid rocket motor itself and the rest of the SRB components. The second was that Marshall would develop the SRB in-house, incorporating Thiokol's solid rocket motor with a host of SRB components produced by several other contractors. It was a move that avoided further Reductions in Force while allowing Marshall engineers the type of hands-on experience for which the Center had always been noted.

Marshall thus functioned in a systems integration capacity, rather than have one main contractor serve in this role. "You could say that Marshall was the prime contractor for the SRB," Hardy said. When the costs of the program were tallied in later years, it was found that Marshall's experience and expertise in management and integration produced considerable savings in program costs to NASA in the early years of the Shuttle program. In December 1976, after initial development was largely completed, Marshall contracted integration and assembly to United Space Boosters, Inc.

When Marshall released the RFP for solid rocket motor development early in 1973, four firms submitted proposals. Thiokol Chemical Corporation of Utah was awarded the contract, a major factor in their selection being their lower bid. In an era of intense cost scrutiny, Thiokol took advantage of lower facilities and labor costs.

The reusability requirement meant that the SRB, which parachutes into the ocean after burnout at each Shuttle launch, would encounter problems from saltwater corrosion. Schwinghamer's Materials and Processes Laboratory carried on major research into corrosion prevention. They tested various types of metal primers, paints, greases and other materials that could fight the degrading effects of salt water. They studied how the Navy fought corrosion in ships. Hardy described the effort as a "very fruitful activity." Corrosion resistance was also enhanced by use of a special steel to manufacture the SRB casings.

Successful recovery and refurbishment of spent SRBs depended largely on a soft landing; "soft" being a relative term, since the SRB would hit the ocean, tail first, at more than 60 mile per hour. To achieve this, three huge parachutes were used to slow the descent of a spent booster. When the first few Shuttle flights showed severe damage to the aft skirt and solid rocket motor nozzle, Hardy's group beefed up their design.

Development: The External Tank

The only non-reusable shuttle element, the external tank presented technical challenges that would tax Marshall's expertise, largely from the requirement to keep down costs and

The first lightweight external tank (LWT) for the STS-3 mission was moved to the Vehicle Assembly Building (VAB) at the Kennedy Space Center. The LWT, unpainted external (ET) saved 6,000 pounds in the Shuttle liftoff weight. The giant cylinder, higher than a 15-story building, with a length of 154-feet (47-meters) and a diameter of 27.5 feet (8.4 meters), is the largest single piece of the Space Shuttle. Official NASA photo.

weight. But to Project Manager Jim Odom, there was another major issue to contend with. The tank was the expendable element of the shuttle, and early NASA shuttle plans called for up to 40 flights per year. "This was the first major manufacturing program NASA ever did." The shuttle program, therefore, required a factory where tanks could be produced for years.

In an effort to save time as well as costs, Odom also successfully pushed the development of test article tooling that would also serve as manufacturing tooling, a new idea in the aerospace industry. "Initially, we had a rather significant amount of money invested in tooling, and we wanted to be able to use this both up front as well as later in the program." The tooling, said Odom, had to be both highly producible as well as repeatable.

Another major decision was to use NASA's Michoud facility near New Orleans as the site for tank assembly work. The facility's tremendous size, as well as proximity to the Gulf of Mexico which allowed barge transportation of the external tank, offered the same advantages it did twenty years earlier when Saturn I and V stages were manufactured there. But its selection, Odom said, "made criticality of the suppliers very high." Odom and Martin Marietta contractor management visited each of the 40 or so critical suppliers that supplied major tank parts before contracts were signed.

NASA contracted in August 1973 with Martin Marietta Corporation (MMC) to design, develop and test three test tanks and six flight tanks for development purposes. Under the contract, MMC was responsible for the liquid hydrogen and liquid oxygen tanks, an intertank section, propellant feed system, and the thermal protection system. Another critical design responsibility was how to attach the other Shuttle elements to the tank. MMC set up a separate division, managed from a divisional office in Denver, to manage the contract, although the Michoud operations were later elevated to divisional status, which aided program management.

A major consideration in tank design was insulation. The tank would carry more than a million pounds of liquid oxygen that had to be kept at –297 degrees, as well as 250,000 pounds of liquid hydrogen at a temperature of –423 degrees. Improper insulation could result in ice formation on the tank during flight that might break off and damage the Orbiter. With help from Science and Engineering, Marshall and MMC solved the insulation problem through use of a super light ablative material and a special foam insulation that covered the entire tank.

Approximately a mile of welds were required in fabricating the External Tank. The welding technology used produced a tank whose welds were as uniform and perfect as any in modern history, a critical consideration for the tank.

Management Structure

After about a five-year lull, Marshall was back in a whole new chapter of the rocket development business. Project offices were set up for each of the three elements, with the offices reporting through Marshall's Shuttle Project Office to Houston's Space Shuttle Program Manager, as well as the Marshall Center Director. And despite heavier contractor involvement, the arsenal system was not yet dead. Marshall engineers and managers handled their newest mission by hewing as closely as possible to the conservative management style that proved successful during the Saturn and Skylab programs, albeit with much more limited resources.

The management structure changed to reflect the new needs of the Shuttle program. Under the 1964 restructuring, the project offices were consolidated into the Industrial Operations Directorate, and the powerful laboratories jelled into the Research and Development Operations Directorate [R&DO]. But 1972 witnessed a new program, with new challenges, and many new people in senior positions. Under the direction of Rocco Petrone, who became Center Director in 1973, the laboratories were renewed, streamlined, and combined under the Science and Engineering Directorate [S&E]. Richard Smith, who later became Director of the Kennedy Space Center, was the first head of this revised organization. Leland Belew, the former Skylab Manager, and earlier the Saturn Engines Manager was named as Smith's Deputy.

With the labs separate from the three Shuttle program offices, technical evaluations could be made separately from program issues. The result continued the successful matrix form of management, with the Project Managers answering to the Marshall Projects Manager, who reported to the Center Director and to the Shuttle program manager in Houston. Under this arrangement, Science and Engineering could have critical technical points heard outside of strictly program channels.

Hardy, Odom and others who had major shuttle development responsibilities are unanimous in their praise of Science and Engineering. "The quality and quantity of the technical staff was superb," said Odom. "They were the unsung heroes who made the shuttle." Hardy characterizes their help as "extensive and extraordinary." With Marshall playing the integration role in the SRB program, Hardy employed them to do parallel development work with contractors, as well as provide contractor technical oversight.

Historically, it must be said that Marshall and its shuttle contractors operated in the 1970s in a less-than-optimum economic and political environment. Both faced the economic

Pictured is the first Technology Test Bed night firing with Space Shuttle main engine 0208. This test was the first in a series used to demonstrate the merits of the large throat main combustion chamber. This test stand was originally built to test the Saturn V booster stage.

An interior ground level view of the Shuttle Orbiter Enterprise being lowered for mating to External Tank (ET), April 21, 1978.

slump of the mid-'70s, with the aerospace contractors also facing the stark reality of industry cutbacks after Apollo and Vietnam. Another wave of budget reductions hit Marshall after 1973, further reducing its quality oversight ability and its ability to penetrate contractors. As a result, Marshall found itself trying to manage a program in some ways more complex and demanding than Saturn, with far fewer resources at its command. The successful Shuttle development program is a tribute to the Center's strong management and many hundreds of dedicated individuals.

Systems Engineering

By 1979, Marshall's initial development work on the components was largely complete, although major improvements continued well into the 2000s. As development of the components progressed, Marshall grappled with the concerns and problems that their interaction posed: in short, how to integrate these complex systems with the orbiter into an amazingly complex vehicle called the Space Shuttle.

JSC recognized Marshall's vast experience with systems integration, experience gained during the Saturn program of twenty years earlier. The two Centers organized two main integrating groups, the Ascent Flight Systems Integration Group (AFSIG) for flight systems integration, and a Propulsion Systems Integration Group (PSIG) to integrate propulsion systems components. All four groups – main engine, external tank, solid boosters, and the orbiter – were represented. Judson Lovinggood chaired the former, while Charles Wood led PSIG. Both were Marshall veterans who had cut their teeth on similar work in the Saturn program.

AFSIG worked closely with contractor guidance and navigation control people, and received considerable help from Science and Engineering. One of those was Robert Ryan, who later co-chaired the AFSIG group. Their job was complicated. "We worked to integrate aerodynamics, trajectories, guidance navigation and control, and thermal loads to balance out the systems for the best flight," he explained.

Meanwhile, Wood's PSIG grappled with the myriad technical issues associated with a propulsion system consisting of both solid and liquid rockets, the latter drawing propellants from both the external tank as well as tanks in the orbiter. This group tackled issues such as maintenance of sufficient pressure in the super-cold liquid oxygen and hydrogen propellants as they moved to the main engines, how to handle the pressure changes that resulted when the main engines throttled down, and dealing with the ground-based cryogenic fueling systems. Leonard Worlund played a key role in this group from his post as Chief of the Fluid Dynamics Branch of the Propulsion and Vehicle Engineering Lab. PSIG dealt, he said, with several technical issues, one of them being the chill down of all the hardware from propellants hundred of degrees below zero was a major concern. Then there was the problem of propellant blow-off caused by differentials in fuel line pressures. Another concern, he said, was maintenance of positive tank pressures.

For example, fuel line design required a proper balance

between robust design and weight savings. "We had to build them to the lowest possible pressure to avoid the extra weight," Worlund said. "You had to be very careful. You can fracture fuel lines easily." Many years later, Alex McCool, who directed Marshall's Space Shuttle Projects Office, could not over emphasize the importance of PSIG's work. "Wood had a greater responsibility than almost anyone else. There just wasn't any margin for error."

The Marshall Test Program

It was a day of jubilation when, on March 13, 1978, the first orbiter, Enterprise, made its way slowly down Rideout Road towards the Dynamic Test Stand. As it passed the Marshall headquarters complex, hundreds of employees crowded the windows to watch the half-plane, half-rocket as it was towed along at a walking pace. The national press was there; the Center held an Open House that weekend so thousands could catch a glimpse of America's newest spacecraft.

The late 1970s saw final exhaustive testing and verification of the hundreds of Shuttle systems. These took place virtually from coast to coast, with final main engine tests at the former Mississippi Test Facility, now renamed the National Space Technology Laboratories, as well as Rocketdyne's California facilities. Out in Utah, Thiokol conducted tests of the Solid Rocket Motors. Marshall performed the dynamic tests, the massive vibration tests used to test structural dynamics, at the refurbished and modified Dynamic Test Stand on the south side of the Center complex. Later, in October, a completed Shuttle with filled External Tank and Solid Rocket Boosters, weighing over 4 million pounds, was subjected to similar tests. These Mated Vertical Ground Vibration Tests (MVGVT) prompted some minor modifications, but overall the Shuttle responded as expected.

This series of tests also provided assurance that Shuttle bending modes and vibration frequencies would not interfere with electronic signals and controls.

Development of the Shuttle, or Space Transportation System (STS) as it was officially called, took seven years. But finally the STS was nearly ready for its first test flight. This climactic event had originally been planned for mid-1979, but finally took place in April 1981. All systems performed as expected. Former Center Director Eberhard Rees returned to Marshall to witness the moment of triumph with his former employees.

But Marshall's patriarch of propulsion was no longer around to see the first Shuttle launch. Wernher von Braun had died four years earlier, finally succumbing to cancer. As the Shuttle rose into space, Rees reflected on how much this day would have meant to his old boss.

In the years after 1981, Marshall and contractor engineers changed the Shuttle in hundreds of ways. There were changes that modified engine performance, eliminated weight, and extended the useful life of Shuttle components. One example is the External Tank, reduced in weight by more than 7,000 pounds to provide increased payload capacity.

But the ambitious schedules that envisioned more than 40 launches per year by the early 1990s were scaled back dramatically. Inflation, which increased the cost of Shuttle flights, was a factor, as well as the longer than expected refurbishment times, and the extra time needed for post-flight analysis. President Reagan's post-*Challenger* decision to restrict Shuttle flights to scientific missions was another factor.

Marshall learned to rely more and more on its contractors, which were increasingly seen, especially during the Reagan years of the 1980s, as the key to holding down costs. The number of civil service employees at the Center declined steadily, from 5,500 at the onset of the Shuttle program in 1972 to less than 3,300 by 1986.

But by all accounts, the Shuttle program was a huge success. In the 20 years after its first test flight, it would ferry hundreds of tons of payloads to earth orbit. Names that would make headlines, such as Chandra, Hubble and Spacelab, would rise to earth orbit aboard the Space Shuttle, and it would ferry to orbit the components of the International Space Station. Its reliability was superb, marred only by the *Challenger* accident.

The Shuttle Orbiter Enterprise passes the MSFC Building 4200 complex on its way for testing, March 1978.

The Challenger Accident

By January 1986, the Shuttle had flown 24 near-flawless missions over a four-year period. All indications were that STS mission 51-L, set for January 28, would be no different. The orbiter *Challenger* rose from its launch pad that cold January morning, roaring off into a cobalt blue Florida sky. Then, just 73 seconds after launch, about nine miles above Atlantic Ocean, the external tank exploded. Millions watched in stunned disbelief as the solid rockets skewed crazily off on their own trajectories. What the huge vapor cloud obscured was the shattering of *Challenger* into several large chunks. The six astronauts and the first schoolteacher in space were killed.

At Marshall's Operations Support Center, technicians monitoring thousands of data items on the launch were puzzled when those monitors froze shortly after liftoff. Most thought it was a simple telemetry problem. But as the reality of what they were witnessing sank in, they silently began gathering the data they knew would be needed for the investigation.

What they did not realize was that it would be nearly three years before the Shuttle flew again. During that time, through tireless effort, 12-hour days, and endless reams of reports and testimony, Marshall people pinpointed the cause of the accident, led NASA's efforts to fix the problem, and did everything humanly possible to prevent another Shuttle loss.

NASA's investigation began immediately. At Marshall, J.R. Thompson, who had directed Shuttle main engine development, headed the NASA team. Several hundred Marshall people played key roles on the three panels that investigated technical causes, combed flight data, and studied pieces of the Shuttle recovered from the ocean floor.

On February 3, just one week after the accident, President Reagan appointed a Federal panel to investigate the disaster. Headed by former Secretary of State William P. Rogers, the panel was charged with oversight of the entire Federal and NASA investigation.

The media began to descend on Marshall Space Flight Center, with some 25 media organizations setting up shop in and around the Center. In the media frenzy that developed, some even set up camp outside the homes of Marshall officials.

Attention quickly focused on the Solid Rocket Boosters, specifically the area of the aft joint on the right-hand booster. Videotapes of the launch just prior to the explosion clearly showed flame escaping at the joint, and the recovered solid rocket booster had a large hole burned in that area. It was readily apparent that the rocket exhaust from the joint area had ignited the External Tank, causing it to explode and destroy the orbiter. It did not take long to find the culprit. It was clear the o-rings used to seal the SRB joints had failed, allowing flame to escape, which caused the destruction of the External Tank.

The final report of the Rogers Commission included findings of the NASA investigators. There were four main points. First, the Commission found the joint design flawed. Second, NASA's safety and quality systems were cited as inadequate. Third, NASA was blamed for promoting a flight schedule that was too demanding, the planned schedule of 20 or more flights per year was seen as unsustainable. Finally, the commission accused Marshall of having poor communications procedures. But at Marshall, there was considerable disagreement on the last point. Key Marshall engineers and project managers had testified that they always felt they could communicate with management on any issue.

Some key Marshall officials cited other reasons for the o-ring failure. Kingsbury found evidence that the long train ride the SRB sections made from Utah to the Cape was a contributing factor. He believed the slight flattening of the sections during the long trip produced a difficult assembly situation for Thiokol's people at the Cape, one they were not able to totally correct. He was at the Cape before the launch and had the opportunity to observe SRB assembly. "When the segments got to the Cape," he recalled, "they spent little time standing up. The bottom segment did not fit. The assembly crew tried to force it back round, and they worked on it for three days." He added that the only assemblies that failed were those that had inadequate time to return to perfect shape. "The chairman of the investigating committee only pointed to the o-ring failure, which I believe was damaged," he said.

In retrospect, though NASA pinpointed the cause of the accident, it was impossible to fix the blame. Those who were involved point to the near-record cold launch time temperature of 29 degrees, possible improper assembly of the joint, and pressures to conduct the launch despite the low temperature as factors. For everyone who criticized the joint design, there were others who heartily defended it. In Washington, Fletcher told Congress "headquarters was as least as much to blame as other parts of the organization." But Marshall, stung by criticism of its role in the accident, resolved that the Shuttle would fly, and fly safely, again.

Getting Back in the Game

A grim determination pervaded the center, a determination that Marshall people would fix whatever problems, real or imagined, that the Rogers Commission, Congress or the nation perceived the Shuttle to have. In September 1986, J. R. Thompson returned to Marshall after a three-year absence to take over as Center Director. The Rogers Commission report made nine specific recommendations for improvements in management and engineering, and both Thompson and James Fletcher, who had left NASA but returned as administrator that summer, pledged to implement them.

NASA told the nation it would take two years to get the Shuttle ready to launch again. President Reagan saw this as an opportunity to promote private space initiatives. The Reagan Administration ordered most commercial and DoD payloads off the Shuttle, and encouraged private industry to adopt its own space launch capabilities. From now on, said Reagan, the Shuttle's purpose would be to support science and technology in space.

Marshall's main task in getting back into space was to make any needed changes to the solid boosters. But during the more than two years devoted to this effort, the Center took the opportunity to look at improvements to all Shuttle components. Though the main engines had nothing to do with the *Challenger* accident, the Shuttle Projects Office increased not only their reliability but also their performance. For example, where previous launches required the engines to perform briefly at 109% of their rated power, the new standard was 104%. Marshall

Shown at NASA-Kennedy Space Center, Florida, rolling on the crawler to the launch pad is the full Space Shuttle stack, the Orbiter, External Tank and Solid Rocket Boosters. Official NASA photo.

redesigned the old Saturn S-1C test stand for SSME testing, while Rocketdyne conducted still more tests at their facility in California.

Of course, the greatest efforts focused on the SRBs. Marshall had used a test method called Failure Mode Effect Analysis, developed by NASA during the Saturn-Apollo program, which focused on the identification of worst-case problem. But FMEA worked well when budgets would allow the development of different prototype designs, something NASA no longer had. One of the first steps of the Structures and Propulsion Lab was to reevaluate all previous FMEAs. Marshall set up a system of cross checks for the booster redesign process in an effort to assure that no problem would become isolated. A special booster redesign team at Marshall worked separately from a similar team at Morton-Thiokol, and the two teams met frequently to compare ideas. Watching over the entire process was a committee whose members came from all areas of NASA, the aerospace industry, the Army Missile Command, and the National Research Council.

After months of effort, the team came up with a new joint design that improved on the previous design by including a new, metallic capture lip incorporating a third o-ring. The idea was to create a static seal rather than one sealed by pressures generated from launch. Most importantly, the design permitted absolutely no blow-by or erosion of the seals. It also required changes only to the tang portion of the motor casing, thus allowing NASA to use the existing clevis design of the inventory of lower booster cases.

Throughout the entire redesign process, both NASA and Marshall adopted the attitude that they would take whatever time was required to get the job done right. George McDonough felt that "what we did after Challenger was some of the best work the Center ever did." In Washington, Rear Admiral Richard Truly, NASA's new Associate Administrator for Manned Space Flight, voiced Marshall sentiments when he told a Congressional panel, "I don't care if NASA will make February 1988 (the target for the next shuttle launch) or not. If we don't make it, I don't care. I just want to fly when we are ready."

Primarily as a result of increased test requirements, the February launch date was slipped to early fall. But finally, on September 29, 1988, two years and eight months after Challenger, the orbiter Discovery lifted off from Kennedy Space Center. Marshall engineers and scientists watched in silence, almost not breathing, as the shuttle rose. Months of 60 and 70 hour weeks, more than $2 billion, and the hopes and dreams of America's civilian space agency rode on that flight. Cheers and congratulations erupted at solid booster separation. Marshall had resurrected itself.

Columbia *Goes Down*

The U.S. lost its second Space Shuttle Orbiter on February 1, 2003 when *Columbia* and its crew of seven were lost when the spacecraft broke up during descent over Texas. The Columbia Accident Investigation Board (CAIB) conducted a seven-month inquiry and determined the physical cause of the accident was a breach in the orbiter's left wing leading edge, caused by a piece of insulating foam that fell off the External Tank during the climb to orbit. This foam block, weighing approximately 1 1/2 pounds and about the size of a suitcase, struck a portion of the wing covered in a grey substance called reinforced carbon-carbon, part of the Orbiter's heat-shielding thermal protection system. The damage allowed superheated air to penetrate the wing's interior and to weaken the structure, eventually causing *Columbia* to breakup 40 miles above Central Texas 16 minutes before landing.

Return to Flight

Even before the investigative board had laid out its 15 recommendations, the agency had begun the process of returning the Space Shuttle to safe flight by forming a team to plan for Return to Flight. The agency has gone above and beyond the board's recommendations with a list of corrective actions known as "raising the bar" to make the Space Shuttle safer than it has ever been.

Marshall Space Flight Center played a major role in Return to Flight. The Marshall team made changes to the tank design and conducted extensive testing of the changes. First, the large insulating foam ramps that flew on STS-107 have been removed from the forward bipod fittings—part of the tank-to-Orbiter attachment structure—and replaced with rod-like heaters to help prevent ice from forming on the fittings. In the original design, the ramps helped to prevent ice buildup on the fittings—a potential debris source. The fittings themselves—each has two—are the same basic design as before.

Small lipstick-size cameras have been placed in the tank's liquid oxygen fuel line faring, on the External Tank and on each of the two Solid Rocket Boosters to record any possible debris that might be shed during ascent. The thermal protection system on the tank's bellows—joints that allow the tank's fuel lines to flex—also have been reshaped to a squared "drip-lip" that allows moisture to run off instead of building, and heaters have been added.

Changes were made to the bolt catchers that hold the Solid Rocket Boosters to the External Tank. Each shuttle flies with two bolt catchers fixed to the forward, or top area, of the tank where the booster attaches to the tank. The canister-like catchers are designed to "catch" the 40 lb. separation bolts—moving 70 feet per second—when they are severed as the boosters separate from the tank, approximately two minutes after launch. The other half of the bolt remains with the booster. Modifications included changing the bolt catcher from a two-piece welded design to a one-piece machined design. Energy-absorbing material inside the bolt catcher also was improved, the size of the bolts that hold the catcher in place was enlarged, and the outside thermal protection material was changed from a "melt-away" coating to machined cork covered with a protective paint finish.

On July 26, 2005, after two-and-a-half years of standing down in the human spaceflight arena, STS-114 lifted off, putting the U.S. back in space. Return to Flight was more than a single mission. It launched a new era of human exploration that's far reaching, but focused. Return to Flight illustrated the Marshall Center's ability to learn from mistakes and its determination to prevent new ones.

Marshall Center's director, David King, said on July 14, 2005 in part:

"Preparing to return the Space Shuttle to safe flight has been all-encompassing, requiring a commitment to remain steadfast while working through many difficult issues. Your dedication to ensure the Shuttle fleet, its propulsion elements and launch faculties are the safest ever will reap great personal rewards. I want to commend you not only on your hard work, but also on your earnestness toward improving our safety systems and leadership skills. The difference it has made is noticeable and encouraging, and I am proud of your willingness to learn new ways of looking at things—we are better for it. The Shuttle is an iconic vehicle, recognized around the world as a symbol of the pioneering spirit that compels exploration. I am proud that Marshall's propulsion elements lift this amazing reusable launch vehicle into low Earth orbit, and I am very proud of all the Marshall civil servant and contractor hands that have played a role in that responsibility."

Return to Flight was one of Marshall's finest hours.

The seven crew members of the STS-114 mission and two Expedition 11 crew members gather for a group shot in the Destiny Laboratory of the International Space Station (ISS). From the left (front row) are astronauts Andrew S. W. Thomas, mission specialist (MS); Eileen M. Collins, STS-114 commander; Cosmonaut Sergei K. Kriklev, Expedition 11 commander representing Russia's Federal Space Agency; and John L. Phillips, Expedition 11 NASA Space Station officer and flight engineer. From the left (back row) are astronauts Soichi Noguchi, STS-114 MS, representing the Japan Aerospace Exploration Agency (JAXA); James M. Kelly, STS-114 pilot; and Charles J. Camarda, Wendy B. Lawrence, and Stephen K. Robinson, all STS-114 mission specialists.

Chapter 6

SPACELAB

November 28, 1983, dawned clear and cool at Cape Canaveral as NASA engineers and launch control personnel prepared for Space Shuttle flight STS-9. This shuttle mission would be different. Nestled snugly in the cargo bay was a special payload, the fruits of a new effort, a new collaboration. It was both the beginning and the continuation of a new NASA era, one of international collaboration. Spacelab I, a totally self-contained, orbiting scientific laboratory, was a child with two international parents: NASA and the European Space Agency.

It also marked a new role for Marshall's role in space science. Marshall mission controllers would step to the forefront, monitoring and controlling 72 experiments planned for the Shuttle flight.

Spacelab I, a self-contained, modular scientific workshop, carried 40 scientific devices that would be used in five different areas of experimentation: astronomy and solar physics, material sciences and technology, space plasma physics, atmospheric physics and Earth observations, and life sciences. In its maiden flight, a packed, 10-day mission, Spacelab looked outward to the universe beyond as well as downward to view the Earth as it had never been seen before. It could study the tenuous space plasma it was flying through and perform experiments in the microgravity environment unique to orbiting spacecraft, including how man can perform in virtual weightlessness.

Skylab, ten years earlier, had whetted the appetites of scientists who dreamed of the possibilities of conducting experiments in the near-zero gravity conditions of earth orbit. The three Skylab missions had accomplished a lot. But Skylab was limited in the research that could be performed, and the three-man crews were astronauts not scientists. Spacelab would exceed Skylab's capabilities in so many ways. It would allow true scientists to work in space, and do it in a shirt-sleeve environment, with standard scientific instruments. But most importantly, it would be reusable. Spacelab furthered scientific knowledge that could best be gathered from space, and paved the way for the International Space Station of the '90s.

Spacelab carried a six-man crew. Two were pilots, trained astronauts who had underwent years of NASA's extensive training. The other four, two payload and two mission specialists, could be described as surrogate scientists who would conduct the experiments, to act as the "eyes, ears, hands and feet" of hundreds of scientists on Earth.

It might be said that Spacelab was conceived in a Space Task Group report to President Nixon in 1969, just two months after the triumphant Apollo 11 mission. The report suggested the United States, after Apollo, should seek to build a space station and a reusable space transportation system. Significantly, it also called for "international participation on a broad basis," a concept favored by Nixon himself.

Shortly afterward, in October, NASA Administrator Dr. Thomas Paine went to Europe to sound out support for space activities after Apollo. The Europeans were enthusiastic in their support.

Seizing on the space station idea, Paine's Associate Administrator for Manned Space Flight, George Mueller, proposed that the Agency construct a modular space station by the mid-1970s. He later called for Marshall to be the lead center if the project reached approval stage. When Houston was named lead center for the Shuttle in 1971, Marshall's designation as the lead center for the space station idea was seen as a consolation prize.

But by the end of 1971 it was obvious that the space transportation system had funding priority. Though study for space station concepts continued, a new twist to the space station idea began to take hold. Based on an airborne laboratory that NASA's Ames Research Center flew in a converted Convair 990 aircraft, NASA planners saw the possibility of using the Shuttle's cargo bay as the carrier for an orbiting laboratory. Scientists familiar with the Convair lab felt a similar arrangement could work on the Shuttle, and energetically endorsed the added orbital advantages of weightlessness, a much higher altitude, and limitless vacuum. Thus was born the "sortie can" idea.

Spacelab Development

After it was recognized the Sortie Can would not satisfy the minimal experimental requirements recognized by the space science community, a formal Sortie Lab Task Team was established at Marshall. This approach was in keeping with the Phased Project Planning concept utilized in the Program Development Directorate to accomplish the Phase B preliminary design of future programs. The team was lead by Jack Lee and Luther Powell and made up of selected individuals within the Center who had experience and/or expertise unique to satisfying the program objectives.

The design effort first was driven by a good understanding of the user requirements for an on-orbit habitable lab to be flown into low earth orbit in the Shuttle Bay, and then designing that lab to be compatible with Shuttle design that was still in an evolutionary stage. The Sortie Lab was required to meet the user requirements for the astronomy, solar physics, atmosphere and space physics, Earth observations, space processing, and life sciences disciplines. This included identifying the resources (power, environmental control, data handling, etc.) required from the Shuttle necessary for the Lab to meet the user requirements.

From the time the Task Team was established in 1972 until the Memorandum of Understanding (MOU) was signed in August 1973 by NASA and the European Space Research Organization (ESRO), the Task Team proceeded to develop a Sortie Lab preliminary design as if NASA would retain sole responsibility for managing the design, development, test and operation of the Lab.

Once the decision was made by ESRO, later to be ESA, the European Space Agency consisting of ten participating European Countries, to proceed with the Spacelab program, a formal NASA program office was established in Headquarters and Marshall was designated as lead Center. The Sortie Task Team formed the nucleus of the Spacelab Program Office at Marshall with Jack Lee as Manager and Luther Powell as Deputy. Coincidentally, the Sortie Lab name was changed to Spacelab due to the French dissatisfaction with the use of sortie, meaning "exit."

The first order of business was to establish a formal working relationship with our ESA counterparts. Their Project Office, which resided at the European Space & Technology Engineering Center (ESTEC) located in Holland, was headed by Heinz Stower. To show good faith, all the preliminary design data produced by the Sortie Task Team was provided to ESA, which assisted them in their own preliminary design efforts and ultimately their assessment of the competition between ERNO and MBB, the two German prime contractors bidding on the program.

The program was unlike any previous program managed by Marshall. Even though the MOU was approximately 20 pages in length, and covered every aspect from schedules to funding to disputes, considerable flexibility on how to implement the program was prevalent. It was clear that ESA wanted to be a part of the manned program and NASA saw an opportunity to satisfy Nixon's Space Task Group's objective to internationalize space and at the same time get a free Spacelab.

The entire MOU can be paraphrased as: "ESA, you provide a Spacelab to fly on one of the early Shuttle flights that meets all technical and safety requirements, and NASA will not initiate a competitive program, and in addition NASA will agree to purchase one additional lab if the price is reasonable."

One of the major challenges for the Marshall management team was created by the fact that ESA had never developed a manned system, therefore basic specifications and design experience that had been developed in this country since the Mercury program did not exist in ESA or their aerospace contractors. To complicate matters, NASA was only allowed to provide their detailed specs and standards for information, but not require ESA to implement them as NASA specifications. In order to insure that design requirements were met, in some cases it was necessary for NASA engineers to first review the specs that ESA provided to their contractors to determine their adequacy.

Once it was clear that the proper direction was given to the contractors by ESA, then the program management difficulties were associated more with problems inherent with developing the lab to be compatible with the Shuttle, still being developed, and at the same time attempting to meet all user requirements, which were also changing. All aspects of the development program were monitored by Marshall including development, qualification, and acceptance testing up to and including flight readiness. The Marshall management and technical people felt responsible for the program's success and were held accountable by the Center Director.

The agreement for NASA to procure one additional Spacelab if requirements were met gave the Program Office considerable leverage over ESA during the development phase in insuring the hardware would meet design and safety requirements. This was due to the fact that it was the only way each participating country could recoup some of their investment. The reasonable price requirement was less conspicuous however during the development program because of the uncertainty of the final cost. A lot of this uncertainty resulted from ESA's use of the Accounting Unit (AU), which normalized the value of the currency of the

Scientist and Mission Specialist Owen Garriott, a member of the NASA-Marshall Retiree Association, is shown in the center of this photograph aboard the Spacelab I science module with his fellow crew members Robert Parker, Byron Lichtenberg and Ulf Merbold. Garriott, who flew one of the famous long duration Marshall Skylab missions, directed experiments on this first mission of Spacelab.

participating European countries, and the geographical distribution method, which ESA was required to implement. At the time the MOU was signed, one AU corresponded to 1.26 U.S. dollars.

Other factors that contributed to price uncertainty was our ability to barter with ESA, and NASA agreed to procure the various Spacelab elements directly from the development contractors in each country. Each country had variable currency values and labor rates. Fortunately, these worked in NASA's favor, since we were able to procure more hardware for less than we had budgeted. This can be attributed to the innovative planning by the Marshall Spacelab Program Office Business Manager—Jerry Richardson.

The Spacelab development program can be classified as one of the most successful, if not the most successful, international programs of this magnitude. The flight hardware was designed to meet all scientific payload requirements and the lab operated flawlessly on each and every flight. This can be attributed to the dedication and competence of the Marshall and ESA Teams and their cooperative attitude.

Scientific Results from Spacelab Missions

Beginning with OSTA-1 in November 1981, and ending with Neurolab in March 1998, thirty-six Shuttle missions are considered Spacelab missions because they carried various Spacelab components such as the Spacelab module, the pallet, the Instrument Pointing System (IPS), or the Mission Peculiar Experiment Support Structure (MPESS). The experiments carried out during these flights included astrophysics, solar physics, plasma physics, atmospheric science, Earth observations, and a wide range of microgravity experiments in life sciences, biotechnology, materials science, and fluid physics which includes combustion and critical point phenomena.

All told, 764 experiments were conducted by investigators from the United States, Europe and Japan. These experiments resulted in several thousand papers published in journals, and thousands more in conference proceedings, chapters in books, and other publications. Scientists from MSFC and universities from Alabama and the southeastern U.S. contributed a significant number of these experiments. The University of Alabama at Huntsville, which had long historical ties to Marshall, was especially involved. Several of these investigations could be considered as landmark experiments in that they produced results that set the tone for new vistas to be explored, which consequently added greatly to our body of knowledge of the universe, the planet on which we live, how our bodies and other biological systems function, and the science of materials processing.

An example of one of the most significant results obtained from Spacelab 1 was the protein crystal growth experiment carried out by two investigators from Germany's University of Freiburg. It is necessary to obtain crystals of certain proteins and other complex biomolecules involved in diseases in order to obtain their 3-dimensional structure using X-ray diffraction. Knowing the structure greatly facilitates the design of drugs to combat the disease by disabling the target molecule. Modern X-ray facilities and computational technology has made this structure-based drug design feasible, but obtaining suitable crystals of the target proteins remains the rate-limiting step in the process.

Shortly after the Spacelab 1 mission, Gunther Seibert from ESA Headquarters was visiting MSFC and stopped by to see Bob Naumann, Chief Scientist for the MSFC microgravity program. "The German space experiment," he said, "had grown crystals of the proteins lysozyme and beta-galactosidase that were much larger than they had been able to grow on Earth." Naumann was scheduled to brief a group from the University of Alabama at Birmingham (UAB) the next day and casually mentioned the experiment's results. Charlie Bugg, the Associate Director of the Comprehensive Cancer Center at UAB, jumped up and shouted, "I've got proteins I've been trying to crystallize for months. When can we fly?"

Neutral Bouyancy Simulator in operation with the crew training for repair of the Hubble Space Telescope mission.

Naumann and Bugg met afterward and developed a simple set of experiments and a proposal to NASA to investigate the effects of microgravity on the growth of protein crystals. Charlie Walker, the first industrial Payload Specialist, was scheduled to fly with his company's electrophoresis experiment on STS-41D, a few months later. When he heard about the protein crystal growth experiment, he offered to carry a simple hand-held growth experiment as part of a "protein diagnostic package" and thus launched the extensive U.S. effort to grow better protein crystals in space while Bugg's proposal was still being evaluated at NASA Headquarters. Since then, protein crystal growth experiments carried on many Shuttle, Spacelab and SpaceHab flights have demonstrated for a variety of different proteins that the crystals grown in space were superior to the best crystals grown on Earth. These experiments became one of the most successful undertakings of the NASA microgravity program. They demonstrated that improved protein structures could be refined from data obtained from space-grown crystals, and that gravity affects their growth. This has stimulated some of the world's most talented crystal growers and theorists to take a serious look at the role gravity plays in the growth process. Their contributions have led to a better understanding of the crystal growth process and to major advances in growing protein crystals on Earth.

A number of Spacelab flights were designated as "low-gravity emphasis" flights in which the Shuttle/Spacelab was flown in an orientation that would produce a well-defined, minimum acceleration vector of less than one micro-g. Such missions were dedicated to the studies of materials processing, fluid physics, combustion, and life sciences in microgravity. Having trained scientists onboard as Payload Specialists to conduct many of these experiments proved to be a valuable asset. A glove box, which was inspired by the "demonstration experiments" performed on Skylab, provided a level of confinement as well as video and other support functions for hands-on experiments and allowed the Payload Specialists the freedom to do exploratory research when they observed unusual phenomena.

The materials processing experiments included studies of crystal growth, alloy solidification, and the measurement of thermophysical properties, especially in molten materials at temperatures below their normal freezing points. The microgravity environment not only drastically reduces sedimentation and buoyancy-driven convection, but also allows containerless processing—processing of material without physical contact. The results produced a much greater understanding of the role of wall effects in defect formation in electronic and photonic crystals and of the effects of convection in various processes. For example, liquid metal diffusion experiments in space consistently measured diffusion coefficients that were 30–50% lower than had been obtained on Earth, which raises the possibility that most, it not all, of the thermophysical transport coefficients for liquid metals measured on Earth are contaminated by convective effects. Accurate values for diffusion coefficients and thermal conductivities of melts, which are essential for process modeling, will probably have to be measured in a microgravity environment.

By eliminating wall contact, metals can be cooled well below their normal freezing points before the nucleation required to form a solid occurs. Using an electromagnetic levitator to control the position of a containerless metal and some clever non-contacting measurement techniques, the surface tension and viscosity of materials could be obtained in the undercooled state. These properties are important in developing novel metallic glass systems with unique mechanical as well as magnetic properties. One investigator, using data obtained from these experiments, developed a new bulk metallic glass, which is now used in the manufacture of golf clubs as well as in military applications.

Life sciences experiments fell into three major discipline areas: 1) Advanced Human Support Technology, 2) Biomedical Research and Countermeasures, and 3) Gravitational Biology and Ecology. A landmark experiment by Cogoli et al. (Technopark, Zurich) on the Spacelab 1 mission was the first to show a dramatic suppression of the growth of human T lymphocytes, which form a major part of the immune system. Additional experiments on subsequent Spacelab flights confirmed this result and also indicated that the ability of T-cells to recognize antigens appeared to be significantly compromised in microgravity. To make matters worse, it appears from previous studies on Soviet Salyut missions and later Spacelab flights that bacteria tend to grow faster in microgravity. The impact to crew health on long term missions because of this double-edged threat is not known and remains a significant biomedical area to be investigated.

When an object as large as the Shuttle moves through the residual atmosphere at nearly eight kilometers per second, it generates many complex interactions with both the neutral and ionized molecules as evidenced by the glow that was observed to form around the vehicle.

A deployable Plasma Diagnostics Package (PDP) was developed in order to probe the plasma, electric field and wake environment of the Shuttle. A Release/Engage Mechanism (REM) was designed and built in-house at MSFC to allow the Shuttle to transport the probe to orbit and return it to Earth. The PDP proved invaluable in supporting the beam -plasma experiments conducted on several Spacelab missions as well as the measuring ionospheric perturbations caused by the Shuttle. One of the most exciting plasma experiments was a joint endeavor by NASA and the Japanese in which artificial auroras were created by firing electron beams down upon the atmosphere from the Spacelab using a particle accelerator. Some 60 artificial auroras were created over the South Pacific for study from ground as well as from space-based observations. Another exciting prospect for space plasma research is the use of a small satellite tethered to a larger vehicle such as the Shuttle/Spacelab or Space Station by a long conducting wire. The feasibility of generating electricity by the tether wire moving through the Earth's magnetic field at orbital velocity was demonstrated in a joint venture between NASA and the Italian Space Agency.

A number of Spacelab missions were dedicated to using the Shuttle and Spacelab components as an observing platform. A battery of telescopes mounted to the Instrument Pointing System (IPS) conducted surveys in previously unavailable spectral regions. The measurements of singly ionized helium (He II) in the spectrum of the quasar HS 1700 + 64 with the Hopkins Ultraviolet Telescope on Astro-2 provided the first measurement of the concentration of intergalactic helium, which contributed directly to estimates of total matter generated in the big bang, which is essential to the development of cosmological theories. The catalogs generated from these observations will find continuing utility in identifying individual astronomical objects worthy of future detailed study.

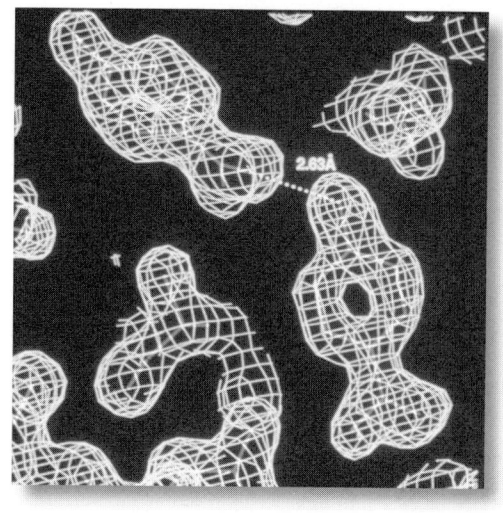

Scientists analyze the molecular structures of insulin crystals growing during Space Shuttle experiments to discover how insulin works.

Several Spacelab missions carried instruments to measure trace gases in the upper atmosphere and focused on the detection of and monitoring of transport of atmospheric pollutants on a global scale. Two Spacelab missions were devoted to mapping of the Earth using C- and X-band synthetic aperture radar. The radar on these missions could penetrate up to 4 m of desert sand and were able to reveal sand covered features in Arabian deserts, suggesting that such system may be able to detect shallow groundwater deposits in arid regions.

Clearly the Spacelab was not an ideal observing platform because of its limited stay time on orbit and the fact that it was intended to serve many diverse disciplines. However, it served as a valuable test-bed for new observation techniques and contributed to the evolutionary development of instrumentation for dedicated free flying satellites. Presumably, the International Space Station may provide a comparable test-bed for instrument concepts yet to be invented.

There were some disappointments in the Spacelab program. There were the few inevitable problems with experiment-specific hardware that prevented some of the science objectives from being met. In several cases the crew was able to devise work-arounds to salvage some of the science. The Shuttle allowed late access to load time-sensitive samples such as protein growth solutions and other biological specimens. However, a number of launch delays due to weather and various technical problems caused degradation of some of these materials as well as less-than-optimal viewing conditions for some of the observational experiments. The MSL-1 mission had to return to Earth prematurely when most of the science experiments could not be performed because of a fuel cell problem on the Shuttle. The problem was fixed and the mission was reflown with the same crew four months later.

However, given the diversity and complexity of the missions, the involvement of the international engineering and science community, the sheer volume of the data produced, and the number of scientific publications resulting from the Spacelab-related research, the Spacelab program has to be considered an unqualified success by any standards. The experimental results and lessons learned from Spacelab should set the stage for the next generation of space scientists who will use the International Space Station as the platform for their research.

Chapter 7
INTO THE '90s...AND BEYOND

With the return to flight of the Space Shuttle in late 1988, Marshall people felt a sense of renewal and pride. The Shuttle would be the defining program for the Center and NASA for the next decade, and if the average of seven yearly missions did not meet with NASA's original expectations, the frequency was enough to quickly restore the confidence of America in the reusable space transportation system. The 1990s saw successful shuttle launches of dozens of scientific and other payloads, including components of the International Space Station.

Marshall had always been seen as a well-managed center, but Marshall's world was changing in the 1990s, reflecting an increased emphasis on strategic planning and organizational development. The Center crafted its first Strategic Plan in 1990. Center Director Jack Lee stressed continuous improvement as a theme, and recognized that Marshall's work continued to support the twin functions of propulsion and space science, as expressed in four major programmatic goals: Space Transportation Systems, the Space Station, Space Explorations Systems, and Large Space Observatories.

The Plan envisioned resumed lunar missions and possibly a trip to Mars. Echoing a speech made by President George Bush on the 20th anniversary of the July 20, 1969 Apollo 11 moon landing, NASA planned for a manned return to the moon sometime between 2000 and 2010, as well as a mission to Mars. Those plans were later shelved, but as the designated Center of Excellence for Propulsion, Marshall saw its role as continuing to provide safe and reliable access to space, while developing advanced space transportation capabilities as well as manned and unmanned spacecraft and participation in space science endeavors.

Workplace safety and quality were stressed during the '90s. When Art Stephenson took over as Center Director in 1998, he moved quickly to put safety near the top of his Implementation Plan. Safety meetings were held monthly, with mandatory attendance by all Marshall employees, civil servant or contractors. Quality of operations and processes became a byword, with increased management attention and refinement of processes.

NASA had always promoted the practical, down-to-earth benefits and discoveries of the space program. The 1990s saw an increased NASA emphasis on technology transfer. Marshall expanded a Department of Technology Transfer, which provided greater focus to Marshall's work with universities and the business community to transfer the discoveries of the space program to everyday human life. The government's Small Business Innovative Research program encouraged entrepreneurs, especially women, disadvantaged and minority businesses, to come up with commercially viable ways to use NASA discoveries for the benefit of the American people.

Spacelab, the joint U.S.–European venture, took a hiatus during the down time after *Challenger*, but blossomed anew when the Shuttle returned to flight in late 1988. Missions of the early '90s focused on Microgravity and Life Science research. Throughout the '90s, until the program's end in 1998, Marshall played a central role in the development of scientific Spacelab payloads. The thousands of discoveries from Spacelab experiments became the substance of hundreds of scientific papers presented at conferences held the world over.

NASA was founded largely on the basis of scientific investigation. In the early '70s, Marshall's Program Development Office began Phase A planning for a trio of space observatories designed to survey the entire electromagnetic spectrum, from gamma rays to visible light. From this effort developed the Hubble Space Telescope, the Advanced X-ray Astrophysics Facility (AXAF), or x-ray telescope, and the Burst and Transient Source Experiment for detection and recording of gamma rays, which flew on the Gamma Ray Observatory. Two major BATSE missions were launched in 1991 and 1992. One of them discovered a possible third radiation belt around the Earth.

These and other projects, in a line going back to the Explorer satellite launches of the late 1950s, solidified Marshall's reputation in the scientific field. From the Saturn days when Ernst Stuhlinger argued for increased attention to science-based programs, Marshall's scientific research capabilities by 1990 would rival those of any major research center. As the decade began, Marshall scientific research had made major contributions in such fields as solar physics, astronomy, Earth science, microgravity applications, magnetospheric physics, and atomic physics. If science textbooks of the year 2000 seemed quite different from those of 40 years earlier, it was due in large part to the efforts of Marshall's scientific people.

NASA thinking during the 1990s recognized that the Space Shuttle would continue to be the main space transportation system for many years to come. Shuttle upgrades continued through the decade, with major improvements in the main engines, Solid Rocket Boosters, and External Tank. Marshall engineers refined the main engine turbopumps, reducing the number of welds, thickening castings, and making other improvements to increase the engine's reliability. A new External Tank made of a lighter aluminum/lithium alloy reduced overall tank weight, allowing payload increases of 7,500 pounds.

Geopolitics influenced NASA's missions and goals, which in turn affected the Marshall center. NASA was born during the Cold War; its missions largely the result of the bitter ideological struggle between the political and economic philosophies of democracy and communism. The Apollo/Saturn race to the moon was motivated in great part by this struggle. That precarious world balance changed after the 1989 fall of the Berlin Wall and the breakaway of the former Soviet satellites of Eastern Europe from the Soviet sphere. No longer was the Cold War the fuel for human space flight. By 1990, NASA planning began to stress international cooperation.

Dan Goldin, who became NASA Administrator in 1993, articulated this new reality in his February 1996 Strategic Plan for the agency, calling for increased NASA interaction with the international space community. Russia, America's old nemesis, was now the newest member of a team that included the U.S., Canada, Europe, and Japan.

The Plan reorganized NASA's efforts around five "enterprises." The Mission to Planet Earth dedicated NASA efforts toward understanding the Earth's global environment. It called for world policies "based on the strongest possible scientific understanding." The Aeronautics Enterprise recognized the importance of NASA aeronautics research in maintaining American competitiveness in the air transportation industry. Human Exploration and Development of Space encompassed NASA's many efforts over the decades to promote manned spaceflight and exploration. The Space Science Enterprise committed NASA to a four-fold mission of science, technology, education and public outreach. The Space Technology Enterprise focused on reducing the cost of access to space, building the private U.S. space industry, and technology transfer.

Marshall Center Director J. Wayne Littles outlined Marshall's role in this ambitious new mission in the November 1996 Implementation Plan. Marshall continued as the lead center for Space Transportation Systems Development, reflecting NASA's designation of Marshall as the Center of Excellence for Space Propulsion. Marshall continued as the lead center for Microgravity Research, a reflection of the Center's lead role in the development, integration, and operation of microgravity payloads and experiments. New programs were added to the continuing Space Shuttle upgrade, space station development and payload integration, and microgravity programs. Reusable Launch Vehicle technology committed the Center to closely working with the U.S. space launch industry to attain radical improvements in launch vehicle performance and cost savings in the first decade of the 21st century. Looking farther into the future, the Advanced Space Transportation Program looked out to the year 2040 and beyond with the goal of putting payloads into low-Earth orbit at 1% of current cost, while dramatically increasing flight safety.

The Reusable Launch Vehicle (RLV) program combined several propulsion efforts, some of which had been underway at the center since the early '90s. The Rocket-based Combined Cycle (RBCC) program involved efforts to develop a launch vehicle that would take off from land using jet engines, attaining Earth orbit under rocket engine power. The RBCC concept was successfully tested in 1999. Included in this effort was the development of miniaturized avionics technology, the so-called "avionics on a chip." There was a program for the development of automated rendezvous and capture (AR&C) hardware for unmanned, low-cost launch vehicles.

A major Marshall program, space station development was seen as supporting the Human Exploration and Development of Space. It's name had been changed in the mid-'90s to the International Space Station (ISS), reflecting the efforts of America's world partners to foot the cost and develop the technology for this ambitious project. 1998, the final year of Spacelab, saw the beginning of ISS construction in earth orbit as several Space Shuttle missions were devoted to assembly of the space station.

Underwater training on the Hubble Space Telescope in Marshall's Neutral Buoyancy Simulator, August 15, 1992.

As the decade of the '90s ended, Marshall underwent another reorganization that reflected the changing priorities and complexities of the Center's missions. Center Director Art Stephenson announced in January 1999 the formation of a new directorate at Marshall. The Space Transportation Directorate recognized Marshall's role as the Lead Center for Space Transportation Development, combining ongoing transportation development programs with future planning. The new Directorate drew together more than 500 employees from Program Development, Science and Engineering, and the Space Transportation Programs Office.

With the 1990s came a major emphasis on the development of technologies and propulsion systems that would permit lower-cost space access. Business uses of space had been a part of NASA planning since the 1970s. Economic studies had shown how the cost and price constraints of space travel would have to be overcome before commercial space transportation could become a reality. The studies showed the potential for a tremendous increase in space travel: a 10% reduction in the cost of space flight should lead to a ten-fold increase in demand. Achievement of this goal became the defining objective of NASA's Space Launch Initiative (SLI). As the decade began, dreams for such things as vacations on the moon, worldwide parcel delivery within a few hours, and mining minerals on Mars were considered distinct possibilities within 20 years.

By the year 2000, Marshall's propulsion efforts had focused on three main missions. First, an ongoing program of Shuttle upgrades worked to ensure safe Shuttle operation through the first decade of the 21st century. The Strategic Launch Initiative focused on development of the "second generation" reusable launch vehicle, with deployment planned for about the year 2012. The Advanced Space Transportation Program planned for 3rd and 4th generation transportation systems, projected for use after 2025. Marshall planners combined management of these efforts under the Integrated Space Transportation Plan. The ISTP would, as Space Transportation Directorate Deputy Director Chris Singer put it, "be an overarching NASA roadmap, creating a roadway for the Highway to space."

Marshall Space Flight Center entered the 21st century leaner, stronger and as focused as ever on its missions. As the Center celebrated its 40th birthday in July 2000, Marshall people could be proud of the Center's accomplishments.

Hubble Space Telescope

On Earth, telescopes have long provided man the ability to look deeply into the heavens. But the limitations of Earth-bound telescopes, peering through hazes of atmospheric dust, smoke, and other matter in Earth's atmosphere, led astronomers to dream of a telescope orbiting in space, far above the confines of Earth's atmosphere. Former Marshall Science Director Ernst Stuhlinger credits 1920s space pioneer Hermann Oberth as the first to suggest such a telescope.

NASA was founded on the basis of scientific research, and soon after the agency took form, a National Academy of Science study group recommended the development of a space telescope. By 1969 astronomers' dreams had jelled into a proposal that NASA construct a space telescope with a 3-meter (10-foot diameter) primary mirror.

That spring, Marshall organized a scientific and engineering conference to take a closer look at the project. By year's end NASA's Office of Space Science and Applications prompted Marshall to push the idea, the way von Braun had pushed for space travel in the 1950s. For the center, which by the late 1960s was winding down the Saturn program and needed new projects, the timing was fortunate. But initially, NASA headquarters was cool to the idea of Marshall leading the project.

The Program Development Office was the incubator for the new project. Jim Downey, a veteran of the High Energy Astronomy Observatory (HEAO) project, was named Task Team Manager for the Phase B definition studies. Over its years of development, it would become the largest and most complex science technology project managed by the Center.

Though Marshall had earned respect for its earlier scientific ventures, it faced competition for space science projects from Goddard Space Flight Center, which had astronomers on its staff and more experience than Marshall with astronomy satellites. In what almost developed into a shootout between the centers for control of the project, Marshall won, in part due to its enthusiasm for the program.

In April 1972, NASA named Marshall the Lead Center for the large space telescope program. Goddard was not left entirely out of the project; it was given the task of developing the crucial Scientific Instrument Package. Despite later differences with Goddard over who would manage contact with the scientific community, and overlapping responsibilities, the two centers reached a working agreement. The program moved into Phase B development later that year, and Marshall issued a Request for Proposals for formal development of the space telescope.

Designing to Budget

News of the space telescope project spread quickly in the scientific world, and soon the project enjoyed widespread support from astronomers, physicists and other scientists. They had great expectations for the project, and placed high performance of the instrument as a top priority. But that hope was soon to encounter the reality of NASA funding limitations.

NASA's budget crunches of the early '70s forced retrenchment on many fronts, and the Large Space Telescope, as it was called, did not escape the budget axe. Congressional funding and support for the LST seesawed through the early and mid-'70s, and what started out as a $700 million project had to make do with $300 million. Just as on the Shuttle program, Marshall was forced to design to cost instead of performance.

A major technique for saving money on the space telescope was the use of a "protoflight" approach to development in which one telescope would serve as both test and actual unit. This approach also saved development time, thus getting the telescope into orbit faster.

Marshall selected two major contractors to build the telescope, contracting with Lockheed Corporation to produce the telescope's structure and most major systems, and a Connecticut firm, Perkin-Elmer, to provide the large mirror and guidance sensors. The Center's role was to manage and support Lockheed's systems integration of this large and complex project.

Costs and Space Shuttle lifting capacity drove a major reduction in telescope size. In 1974, the Marshall telescope team reduced the reflecting mirror diameter from 3 to 2.4 meters, or about eight feet. They reduced from seven to four the number of scientific instruments the LST would carry. These changes enabled a smaller required support system, which reduced the weight of the unit from 24,000 to 17,000 pounds.

New Technical Challenges

Accurate grinding of the 2.4-meter mirror, the largest ever to operate in space, was a painstaking process requiring more than two years. The LST would be no better than its reflecting mirror, and even the smallest flaw in the mirror's curvature would distort images.

But perhaps a greater challenge was the pointing and control system. Making photographs of faint stars would require a pointing accuracy of less than 0.1 arc second, and maintaining that accuracy for hours on end. It would be as if the LST were in Los Angeles and focusing on a traffic light in San Francisco without straying. LST managers felt that development of the pointing and control system "was by a factor of ten" more difficult than development of Spacelab's Instrument Pointing System.

Funding delays and other problems prompted a reorganization of the LST project in 1983. NASA decided the program needed a new name, and a coalition of scientists suggested renaming the LST in honor of Edwin P. Hubble, one of the nation's leading astronomers. But more importantly, the program got a new manager, Jim Odom, a Marshall veteran who had proved himself on the Shuttle external tank project. It was a turbulent time for the program, plagued with money shortages and resulting cost overruns. Odom gets credit for getting a handle on things, putting a program headed for disaster on the road to success. Funding was increased for the project, which had risen to over $1.1 billion in the rapid inflation of the late 70s. A new launch date was targeted for late 1986.

Lockheed completed final assembly of the Hubble components in late 1985, but the *Challenger* accident in January 1986 delayed the launch until April 1990. As the project neared completion, NASA headquarters launched a public relations campaign: Administrator James Beggs even called it "the eighth wonder of the world."

But unknown to the Marshall team, the mirror contained a very slight flaw: it was incorrectly ground near one edge by about 1/50 the width of a human hair. So slight was the flaw that it was not discovered until Hubble had been in space for more than two months. While images of bright objects were superior to any observed by land-based telescopes, faint objects, such as extremely distant stars and galaxies, were blurred. The mirror could not be reground, so Marshall scientists and engineers began planning a fix. Similar to correcting bad vision with eyeglasses, they developed the Corrective Optics Space Telescope Axial Replacement (COSTAR). Installation of COSTAR, which used relay mirrors mounted on moveable arms to refocus scattered light from the main mirror, was the focus of Shuttle mission STS-61 in December 1993.

Images From Space

Subsequent serving missions improved Hubble's capabilities and performed routine repairs. In February 1997, the crew of STS-82 installed the Near Infrared Camera and Multi-Object Spectrometer (NICMOS) and the Space Telescope Imaging Spectrograph (STIS) to detect infrared light from deep-space objects and to take detailed photos of celestial objects. Servicing mission 3A in December 1999 replaced all six of the telescope's aging gyroscopes, which accurately point the telescope at its target. STS-103 astronauts also replaced one of the telescope's three fine guidance sensors and installed a new computer, all in time to redeploy Hubble into orbit on Christmas Day. Servicing mission 3B, came aboard STS-109 in March 2002. Columbia crewmembers installed the new Advanced Camera for Surveys (ACS), which had sharper vision, a wider field of view, and quicker data gathering than the Wide Field/Planetary Camera 2. Astronauts also replaced Hubble's solar panels with a more efficient array and conducted repairs on the NICMOS.

As it continued to return groundbreaking photos of the universe and help astronomers do valuable research, the Hubble Space Telescope's future would once again be thrown into uncertainty by a Space Shuttle tragedy with the lost of Columbia and crew on February 1, 2003. The resulting two-year investigation led to Administrator Sean O'Keefe's decision that future Shuttle missions would go only to the International Space Station, where safe haven was possible, as well as inspection and repair of the Shuttle. Though the telescope was slated for servicing in 2005, O'Keefe concluded that another Hubble repair mission with the Shuttle would be too risky and ultimately rejected the option of sending another servicing mission. In late 2004 the National Academy of Sciences concluded that the risks were acceptable and recommended that a shuttle mission should serve Hubble after all. O'Keefe asked NASA's Goddard Space Flight Center to conduct feasibility studies into a robotic servicing mission.

NASA revisited the Hubble servicing mission decision and rejected the robotic servicing mission option, calling the $1 billion plan too costly. After revisiting the idea of a Shuttle repair flight, NASA sent a repair crew to Hubble in May 2009. The last astronaut to touch the Hubble occurred on May 16, 2009 when the crew of STS 125 launched the space telescope from the cargo bay of the *Atlantis*. Over the course of five spacewalks, astronauts successfully installed two new instruments, repaired two others, and replaced a number of components essential to the telescope's smooth operation such as batteries and insulation. It was NASA's last visit to the 19-year-old observatory that was conceived and developed in the laboratories of the Marshall Center. The repairs should keep the telescope operating for another five to ten years. Hubble is expected to be a window on the universe for many years to come.

Chandra

The July 1999 launch of Shuttle mission STS-93 carried one of the most fascinating cargos ever put into orbit: the world's most powerful X-ray telescope. Measuring 43 feet long, and 64 feet wide with its solar arrays deployed, the Chandra X-ray telescope was the culmination of another long-term Marshall development effort, one whose roots go back to the late '60s.

NASA's Great Observatories program was an ambitious plan to survey the entire electromagnetic spectrum of visible and invisible radiation. Scientists knew that distant stars, star clusters, galaxies, and nebulae emitted x-rays and gamma rays in addition to visible light. The plan envisioned three large, orbiting observatories: a major telescope having light gathering capabilities far surpassing anything on Earth, and two observatories to detect x-rays and gamma rays.

What became known as Chandra began as the Advanced X-ray Astrophysics Facility, a brainchild of the Smithsonian Astrophysical Observatory, which in 1976 had proposed that NASA develop a large x-ray telescope. Headquarters supported the idea, and sought proposals from the field centers. The project got a further lift when the National Academy of Sciences threw their weight behind the project. Marshall responded, as did the Goddard Space Flight Center and the Jet Propulsion Laboratory. Marshall was selected to develop AXAF, says Dr. Martin Weisskopf of Marshall's Space Sciences Directorate, largely due to the Center's success with the earlier HEAO-2 x-ray observatory launched in 1978. In that program, Marshall had forged a close working relationship with the Smithsonian Astrophysical Observatory, which would play a lead role in AXAF development. AXAF would far exceed HEAO-2, with far greater sensitivity and optical precision than its predecessor.

AXAF, like Hubble, was nurtured to birth in the Program Development Office, with Jim Downey as the first Program

Manager. Beset by budget limitations, it had a long gestation period. Work proceeded on development of the science and optics, but ran into snags by the mid-'80s. "Congress," says Fred Wojtalik, Director of Marshall's Observatories Projects Office, "began to scrutinize the program closely. They wanted to know what was the toughest part of the job," says Wojtalik. "We told them it was manufacturing the six concentric mirror pairs."

AXAF development, says Weisskopf, was in many ways a more demanding program than Hubble. Gathering x-rays, which will penetrate rather than be reflected from a mirror, depends on gradually bending their path through the instrument in a series of "glancing reflections" from the mirror cluster. Precise alignment of the six pairs of mirrors, the largest being 1.2 meters in diameter, was absolutely crucial to the proper functioning of the observatory. Marshall developed a one-half size version of one of the mirror pairs to prove that the complex optics could actually be made. The Center also built and operated the 1700 foot long X-ray Calibration Facility, used to aid in mirror test and calibration.

The High Resolution Camera (HRC) for the Chandra X-Ray Observatory (CXO) being integrated with the High Resolution Mirror Assembly (HRMA) in MSFC's 24-foot Vacuum Chamber, March 16, 1997.

The original program proposed a six-mirror observatory with a 15-year life span that would operate in low earth orbit. But funding, says Wojtalik, was insufficient during several crucial years of AXAF development. "By 1992 we realized that, if we didn't do something about cost, we'd lose the entire program." That year NASA descoped the program, cutting the observatory to a four-mirror arrangement and shortening its orbital life to five years. While this did cut yearly expenses for the program, Weisskopf says the constant delays led to higher final program costs. To compensate for the shorter life, AXAF would be boosted to an elliptical orbit that would take it nearly one-third of the way to the Moon, drastically reducing the time the observatory would be in Earth's shadow.

AXAF would go to orbit on the space shuttle, so was subject to the Shuttle's strict size and weight requirements. These limitations required investigation of new materials for construction of the x-ray telescope. To save weight, Marshall engineers chose graphite epoxy, a space-age, super strong, but lightweight material. But, AXAF still tipped the scales at more than 13,000 pounds, one of the heavier loads deployed from the Shuttle.

By 1992 funding for the AXAF program finally stabilized, which, with significant descoping, saved the project. "Development costs," says Weisskopf, "totaled about $1.5 billion, plus another nearly $500 million for launch preparation and costs." And, as for that five-year projected life, Weisskopf hopes Chandra may last ten years or more.

As the July 1999 launch date neared, NASA looked for a better name for its newest space observatory. A naming contest produced more than 6,000 entries, the winner being Chandra, the shortened form of Subrahmanyan Chandranekhar, a world-famous Indian-American astrophysicist.

Chandra was developed by Marshall and launched under its direction. TRW, Inc. was the prime contractor with science and flight operations remaining under the auspices of the Smithsonian Astrophysical Observatory. In conducting research with Chandra, Marshall researchers often team up with their counterparts at major universities, the Smithsonian Astrophysical Observatory, and the Max Planck Institute in Germany.

Since its 1999 launch, Chandra has exceeded astronomer's expectations, its gathered x-rays revealing exciting new information about the universe. Astronomers have gained new understanding about clusters of galaxies millions of light years from Earth, how old they are, and how they might have come into being. Chandra, in allowing the study of so many different types of stars, has given many new insights to researchers interested in the dynamo effect of stars like our own sun.

Black holes, the super gravity entities at the center of many galaxies, have been a continuing subject of study. Chandra can detect their x-ray emissions from thousands of light years away. In early 2000, Chandra detected x-ray flaring indicating the existence of a black hole at the center of our own Milky Way galaxy.

But the most important advances, Weisskopf claims, are in the study of black holes, their formation, and their role in the early universe, as well as the study of plasmas under the extreme conditions of strong gravity and high magnetic fields.

By looking deep into the hottest, most violent parts of the cosmos, providing us with a laboratory that could never be reproduced here on Earth, Chandra is revealing an entire new

NASA - Marshall Flight Space Center

> ### Important Chandra Facts
>
> - The Chandra X-ray Observatory is the world's most powerful X-ray telescope. It has eight-times greater resolution and can detect sources more than 20-times fainter than any previous X-ray telescope.
> - The Chandra X-ray Observatory, with its Inertial Upper Stage and support equipment, is the largest and heaviest payload ever launched by the Space Shuttle.
> - The Chandra X-ray Observatory's operating orbit takes it 200-times higher than the Hubble Space Telescope. During each orbit of the Earth, Chandra travels one-third of the way to the Moon.
> - The Chandra X-ray Observatory's resolving power is 0.5 arc-seconds — equal to the ability to read the letters of a stop sign at a distance of 12 miles. Put another way, Chandra's resolving power is equivalent to the ability to read a 1-centimeter newspaper headline at the distance of one half-mile.
> - If the State of Colorado were as smooth as the surface of the Chandra X-ray Observatory mirrors, Pikes Peak would be less than an inch tall.
> - Another of NASA's incredible time machines, the Chandra X-ray Observatory is able to study some quasars as they were 10 billion years ago.
> - The Chandra X-ray Observatory can observe X-rays from clouds of gas so vast that it takes light more than five million years to go from one side to the other.
> - Although nothing can escape the incredible gravity of a black hole, not even light, the Chandra X-ray Observatory can study particles up to the last millisecond before they are sucked inside.
> - It took almost four centuries to advance from Galileo's first telescope to NASA's Hubble Space Telescope, an increase in observing power of about a half-billion times. NASA's Chandra X-ray Observatory is about one billion times more powerful than the first X-ray telescope, and we have made that leap in slightly more than three decades.

level of detail in the far reaches of space, and will take our minds where our feet may never have a chance to tread.

Gravity Probe B — Testing Einstein's Universe

Marshall Space Flight Center and Stanford University have developed a sophisticated experiment, Gravity Probe B (GP-B), to test Einstein's general theory of relativity. Einstein's theory predicts that space and time are distorted by the presence of massive objects. Launched on April 20, 2004, the GP-B mission is one of NASA's first to address a question of fundamental physics in the new millennium.

Marshall is managing the program. Stanford University conceived the experiments and is Marshalls' prime contractor for the mission. Stanford is responsible for the design and integration of the science instrument, payload and spacecraft, as well as conducing mission operations and data analysis. Lockheed Martin is a major subcontractor on the project and was responsible for designing and constructing the unique spacecraft as well as some of its major payload components.

Once in orbit, the spacecraft underwent a four-month Initialization and Orbit Checkout (IOC), in which all systems and instruments were initialized, tested and optimized for the data collection to follow. The IOC phase culminated with the spin-up and initial alignment of the four science gyros early in August 2004.

On August 28, 2004, the spacecraft began collecting science data. During the ensuing 50 weeks, the spacecraft transmitted over a terabyte of science data to the GP-B Mission Operations Center (MOC) at Stanford, where it was processed and stored in a database for analysis. On August 15, 2005, the GP-B Mission Operations team finished collecting science data and began a planned set of calibration tests of the gyros, telescope and SQUID readouts that lasted six weeks. In October 2005, the GP-B science team began a three-phase data analysis process that was initially expected to yield a final result towards the end of 2006. However, two complications were discovered that slowed and extended the data analysis efforts. During 2006-early 2007, good progress was made understanding the cause of these complications and developing sound methodologies for working through them.

Since that time, the team has continued to improve the results in a number of ways. At least a dozen new technologies had to be invented and perfected to carry out this experiment. For example, the spherical gyros are over a million times more stable than the best navigational gyros. The ping-pong-ball-sized rotors in these gyros had to be so perfectly spherical and homogeneous that it took more than 10 years and a whole new set of manufacturing techniques to produce them. They're now listed in the Guinness Database of Records as the world's roundest man-made objects.

Over its 40+ -year lifespan, spin-offs from GP-B have yielded many technological, commercial and social benefits—e.g. GP-B's porous plug for controlling helium in space was essential to several other vital NASA missions. Most importantly, GP-B has had a profound effect on the lives and careers of numerous faculty and students—graduate, undergraduate and high school—including 79 PhD dissertations at Stanford and 13 elsewhere. GP-B alumni include the first U.S. woman astronaut, an aerospace CEO, and a Nobel laureate.

NASA makes no assumptions about Einstein's theory being right or wrong; rather, they collect the data, and do everything humanly possible to maximize the precision and accuracy of the final results—whether or not they agree with Einstein's predictions. NASA fuels discoveries that make the world smarter, healthier and safer.

International Space Station

Wernher von Braun had captivated the nation with his 1952 series of articles for *Collier's* magazine about his vision for a permanent station in space. His concept, a 250-foot wheel, rotating in space to provide artificial gravity, became the image fixed for decades in American minds of how a space station should look.

Over the 40 years since, NASA planners made halting progress toward a space station that finally came to fruition in the '90s. Marshall people were key to America's progress towards a space station. At the Center, space station planning at some level was an ongoing activity from the time the Center came into being in 1960.

As Apollo wound down in the early '70s, the choice of the Shuttle as NASA's next major program forced the space station off the immediate agenda, but at Marshall, some space station planning continued. Dr. William R. Lucas, Marshall Center Director from 1974 to 1986, had also been the first Director of the Program Development Office, charged with bringing new

This is a von Braun 1952 space station concept. This concept was illustrated by artist Chesley Bonestell.

A concept of the International Space Station (ISS) Phase III with Space Shuttle being docked.

programs to the Center. "The space station had been on our agenda for a long time. After the Shuttle decision, we still had a team of people working on the space station. It was never far from our attention," he recalls.

Formation of the Program Development Office in 1968 gave renewed life to Marshall's space station planning efforts. Lucas and his people proposed a plan to develop a 12-man space station by 1975.

But the era of tight budgets that faced NASA after Apollo made a casualty of early space station efforts. In March 1970, the Nixon Administration, feeling it could not fund both a Shuttle program and the space station, chose the former. It would be more than a decade before work would progress on a true space station. Marshall efforts instead focused on Skylab, the space station built from a Saturn third stage.

During this period, debates raged over the necessity for a space station, and exactly what role it would play in America's space program. NASA wanted a modular space station that could serve as a shuttle depot, space construction base, and orbiting laboratory. Another use, which emerged in the mid-1970s, was that a space station would provide a site for space manufacturing. Detractors countered that many of these activities could be done just as well on the Shuttle, whose development was well underway by the mid-1970s.

Reagan Gives the Go-Ahead

After the lean years of the 1970s in which attention was devoted to the Shuttle, the arrival of the Reagan Administration in 1980 was seen as a possible new lease on life for space station development. The new NASA Administrator, former Navy commander James Beggs, was pro-space station. He and his Deputy, Hans Mark, saw the space station as NASA's next logical step, and campaigned for its approval.

But even in the Reagan Administration, selling the space station idea was an uphill fight. The program had powerful enemies, both in the White House and Congress. Beggs and Mark looked for allies within the administration, and found them in the military and intelligence communities. NASA presented figures that said the space station could be developed for $8 billion, a figure that Lucas later referred to as "very constrained." To help free-up money for the program, Beggs shelved plans to order a fifth Shuttle orbiter.

Finally, Beggs won presidential approval for the space station. In what was now a presidential tradition of announcing major space initiatives, Reagan used his State of the Union speech of January 25, 1984 to "direct NASA to develop a permanently manned space station, and do it within a decade." At Marshall, Center Director Lucas welcomed the announcement as an "exciting new venture."

Competing for Work

The space station design would employ a cluster, in which habitat, operations, and utility modules would form the manned base. For political as well as technical reasons, NASA decided to spread space station development work among the centers in four work packages, with assignments divided along functional lines. This created a situation of complex interfaces between the centers, which Lucas later described as "a management nightmare." Of seven principle areas of technology research, Marshall eventually ended up with three: the Attitude Control and Stabilization System, Auxiliary Propulsion System, and Space Operations Mechanism. JSC was assigned the Data Management System, Environmental Control and Life Support System, and Thermal Management System. The Lewis Center in Cleveland, Ohio would have charge of electrical power requirements. Not to be left out, the Goddard Center retained charge of the unmanned elements.

But as the Phase B stage of development approached, NASA and the centers decided that one of the Centers should have program management responsibility. Houston's Johnson Space Center was awarded this coveted prize, although Marshall won responsibility for about 40% of the space station development budget, with responsibility for the lab modules, as well as the orbital maneuvering system. Marshall's successful Skylab experience had helped the center's case.

Development work proceeded at a brisk pace, with Marshall

again involved in a major national project. The hopes, plans, and dreams of 30 years earlier were finally being realized. Lucas formed a Space Station Program Office, with Luther Powell as Program Manager. Actual work took place in the renovated building 4708, where two decades earlier Saturn rockets had been tested. Boeing, a major space station contractor with nearly 36% of Marshall's space station budget, made a further commitment to the Huntsville area by building a new office complex just south of the new Huntsville airport.

Jim Odom, the Marshall veteran of the Shuttle External Tank program, went to NASA headquarters in 1988 as Associate Administrator for the space station. Though he only spent a year in Washington, he used the time to tackle the interface and budget problems. "I spent a lot of time trying to simplify the contracting and management interfaces," he said, "and tried to implement a solution in which Grumman Corporation would function as an integrator between work package contractors." Numerous changes in the configuration took their toll on budgets and schedules. "There were so many redesigns that you could not get enough critical mass of budget to achieve positive progress over more than a year or two."

NASA wanted international partners on the space station effort. During 1985, the European Space Agency (consisting of the United Kingdom, Germany, France, Italy, Belgium, Denmark, the Netherlands, Norway, Spain, Sweden and Switzerland), the Canadian Space Agency, the Japanese Space Agency and Brazil all agreed to participate in the space station program. When Russia signed on as an international partner in 1993, the program became known as the International Space Station.

When the Clinton Administration took over in early 1993, it ordered NASA to reorganize and simplify the project. NASA had appointed three study teams, one headed by Marshall's Charles Darwin, to develop new, less costly space station concepts. After thorough examination of the teams' efforts, NASA chose the Marshall team's concept, a simplified version of space station *Freedom*.

The station may have been simplified, but Marshall still ended up with a major portion of the development work, although program management remained with Johnson Space Center. Boeing, which had been a Marshall contractor on *Freedom*, won a massive systems integration contract on the newly named *International Space Station*. Under this arrangement, the Europeans developed Nodes 2 and 3 under Marshall management, while Boeing picked up Node 1, the airlock, and the scientific modules, under JSC management. Primarily as a tool in systems engineering and integration, Boeing built a space station mockup in nearby building 4755, which also had the unintended result of aiding Marshall's public relations efforts.

Only a few weeks after the U.S-funded, Russian-built, Zarya module was launched from Kazakhstan, the Space Shuttle Endeavour carried aloft the Boeing-developed Unity connector module in December 1998. Constructed on opposite sides of the Earth, Unity and Zarya met for the first time in space and were joined to begin the orbital station's assembly of the largest spacecraft ever built. The orbital assembly of the space station began with an international construction project of unprecedented complexity and sophistication.

Ten years later, the station's mass has expanded to more than 627,000 pounds, and its interior volume is more than 25,000 cubic feet, comparable to the size of a five-bedroom house. Since Zarya's launch as the early command, control and power module, there have been 29 additional construction flights to the station: 27 aboard the Space Shuttle and two additional Russian launches.

At the end of 2008, 167 individuals representing 15 countries have visited the complex. Crews have eaten some 19,000 meals aboard the station since the first crew took up residence in 2000. Through the course of 114 spacewalks and unmatched robotic construction in space, the station's truss structure has grown to 291 feet long so far. Its solar arrays now span to 28,800 square feet, large enough to cover six basketball courts.

The station is a venture of international cooperation among NASA, the Russian Federal Space Agency, or JAXA, and 11 members of the European Space Agency, or ESA: Belgium, Denmark, France, Germany, Italy, the Netherlands, Norway, Spain, Sweden, Switzerland, and the United Kingdom. More than 100,000 people in space agencies and contractor facilities in 37 U.S. states and throughout the world are involved in this endeavor.

Marshall has developed and managed many systems onboard the space station including:

- Express Racks, which house multiple types of drawer payloads.
- Regenerative Environmental Control and Life Support System (ECLSS) for the space station. This system provides water and oxygen for the crew, recycling wastewater and reducing resupply costs.
- Harmony connector module provides a passageway between three station science experiments facilities, growing the station from the size of a three-bedroom house to the size of a typical five-bedroom house.
- Multi-purpose Logistics Modules carry large racks of supplies to the station and return to Earth with cargo no longer needed on the station.
- Microgravity Science Glovebox developed by the European Space Agency and managed and integrated by Marshall enables scientists to participate in the assembly and operation of experiments in space.
- Window Observational Research Facility will accommodate viewing and Earth imaging using the Destiny's labs optical quality window.
- Microgravity Science Research Rack is used for basic materials research in the microgravity environment of the space station.
- Node 3, a pressurized module headed for the station in late 2009 will house the Environmental Control of Life Support Systems racks and the crew's waste and hygiene compartment and enable the addition of future modules.

Located at Marshall is the science command post for the International Space Station, the Payload Operations Center (POC). The POC manages all U.S. science and research experiments onboard the station and coordinates all payload-related mission-planning work; hardware delivers; and retrieval, training and

50 Years of Rockets and Spacecraft

The Node 1, Unity module, flight article (at right) and the U.S. Laboratory module, Destiny, flight article for the International Space Station being assembled at MSFC, June 1997.

safety programs for station crew and flight controllers. It serves as backup to JSC-Houston during hurricanes or other emergencies.

The International Space Station hosts 19 research facilities, including nine sponsored by NASA, eight by ESA and two by JAXA. Cooperation among international teams of humans and robots is expected to become a mainstay of space exploration throughout our solar system. The 2005 NASA Authorization Act recognized the U.S. orbital segment as the first national laboratory beyond Earth, opening it for additional research by other government agencies, academia and the private sector.

A NASA official said:

"Orbiting above, on board the International Space Station, cosmonauts and astronauts work together in conducting long-term space experiments in a true microgravity environment. The station's capability and sheer size today are truly amazing. The tremendous technological achievement in orbit is matched only by the cooperation and perseverance of its partners on the ground. We have overcome differences in language, geography and engineering philosophies to succeed. We are one on board the International Space Station."

The International Space Station Payload Operation Center at NASA's Marshall Space Flight Center is the world's primary science command post fort he International Space Station. the most ambitious space research facility in human history. The Payload Operations team is responsible for managing all science research experiments aboard the Station. Official NASA Photo.

Chapter 8
Ares Projects

2003: Examining America's Commitment to Human Space Exploration

After seven months of thorough investigations, the Columbia Accident Investigation Board (CAIB) issued Volume 1 of the CAIB Report in August 2003. This investigation not only addressed the causes of the February 2003 accident, but also examined the culture and goals of American's space program. The Board Statement on Page 6 said, "We sought to discover the conditions that produced the tragic outcome and to share those lessons in such a way that the nation's space program will be stronger and more sure-footed. If those lessons are truly learned, then Columbia's crew will have made an indelible contribution to the endeavor each one valued so greatly." The CAIB emphasized that a public debate was needed to decide the future of human space exploration. The report stated, "The loss of Columbia and her crew represents a turning point, calling for a renewed public policy debate and commitment regarding human space exploration. One of our goals has been to set forth the terms of that debate."

2004: Setting a New Path for Space Exploration

In January 2004, President George W. Bush addressed this CAIB finding with a new U.S. Space Exploration Policy that would return astronauts to the Moon by 2020 in preparation for the human exploration of Mars. This policy directed NASA to finish building the International Space Station, retire the Space Shuttle in 2010, and build and fly a new crew exploration vehicle no later than 2014. Although the policy included target dates, it stressed space exploration should be a journey, not a race. The policy called *A Renewed Spirit of Discovery, the President's Vision for U.S. Space Exploration* said, "The fundamental goal of this vision is to advance U.S. scientific, security, and economic interests through a robust space exploration program."

The NASA workforce was energized by this vision and the possibilities it suggested. Marshall Space Flight Center employees could often be heard to say that it was their generation's time to contribute new vehicles and technologies for space exploration. Many had worked on advanced transportation concepts and technologies throughout the 1990s, so much of what they had learned could be applied to the new efforts. They had also learned much about what would not work or would be too costly to develop under NASA's planned budgets.

2005: Developing a Space Exploration Architecture

Based on the successes and failures of past programs, NASA reorganized to better implement these new goals and objectives, and established the Exploration Systems Mission Directorate (ESMD) and the Constellation Program to develop the architecture for space exploration and a strategy for meeting the architecture's requirements. ESMD funded definition studies to determine what sorts of vehicles and technologies could best accomplish the directives outlined in the policy.

In 2005, Congress endorsed the new U.S. Space Exploration Policy and approved funding in the 2005 NASA Authorization Bill, giving the program an even greater foundation of support. That year, Congress also confirmed a new NASA Administrator, Dr. Michael Griffin, who had an extensive background in both commercial and governmental aerospace endeavors. He commissioned the Exploration Systems Architecture Study (ESAS), which built on numerous prior studies. A team of experts from across NASA analyzed thousands of potential vehicles including Shuttle-derived launch vehicles and expendable launch vehicles and, subsequently, recommended an approach for delivering human-rated systems within the target timeline and budget guidelines.

Steve Cook from the Marshall Center was the ESAS Deputy Study Manager. After the study, Cook was selected as the project manager for NASA's Exploration Launch Projects, which later became the Ares Projects. The focus of this project was to develop the Ares I crew launch vehicle and the Ares V cargo launch vehicle based on the concepts recommended by the ESAS study. The Ares Projects are part of the Constellation Program, which is developing technologies and a fleet of vehicles, including the Orion crew exploration vehicle and the Altair lunar lander, to help America extend its exploration to the Moon and beyond. Jeff Hanley, another ESAS team member, leads the Constellation Program. Both Constellation and Orion are managed by the Johnson Space Center in Houston, Texas.

2006 to 2008: Building on History to Develop the Nation's New Launch Vehicles

The Ares Projects team at the Marshall Center leads a multi-Center and nationwide partnership to design, develop, test, and evaluate both the Ares I and Ares V launch vehicles. Other NASA Centers involved with this work include the Ames Research Center, Glenn Research Center, Johnson Space Center, Kennedy Space Center, Langley Research Center, and Stennis Space Center. Marshall manages the Michoud Assembly Facility in New Orleans where the Ares I upper stage and the Orion crew exploration vehicle will be built. By the end of 2007, NASA had awarded all four major Ares prime contracts, as well as the Orion prime contract, and by 2008, hundreds of small businesses joined the effort as subcontractors, vendors, and suppliers.

The Marshall Center's expertise in integrating large, complex space systems ties all the Ares Project elements together. In this second generation of lunar exploration, Marshall is designing and developing the structures and propulsion systems for Ares I and Ares V, as well as integrating the vehicle stacks. This is a change from the Space Shuttle program management style and a return to Wernher von Braun's "arsenal" concept of design, where NASA keeps the design and vehicle integration to maintain the Agency's in-house expertise. NASA engineers are enjoying this "getting-your-hands-dirty" approach of rocket design and development.

One key to making both Ares I and Ares V affordable and available for flight in a reasonable amount of time is developing rockets that use proven technology. Many Ares elements are based on hardware from the Space Shuttle and Saturn vehicles, and the Ares I and Ares V also share common elements. Although heritage hardware increases NASA's confidence in the safety and reliability of hardware, the hardware is still being used in a new configuration, and in most cases, modern technology and materials have been introduced. This requires the launch vehicle to undergo rigorous review, testing, and evaluation.

Ares I Crew Launch Vehicle

Ares I is a two-stage vehicle designed to launch Orion and four to six crew members into Earth orbit for missions to the International Space Station or to rendezvous with the Ares V for missions to the Moon, Mars, and beyond. The Ares I first stage is a single five-segment solid rocket booster derived from the existing four-segment Space Shuttle booster. An in-house NASA team is designing the new Ares I upper stage. The design builds on years of experience with Saturn V stages and Space Shuttle external tanks. This stage is powered by a single J-2X engine based upon the J-2 engine used for the Saturn upper stages. A NASA team is also leading the vehicle integration effort to ensure all Ares I elements work together as an integrated system and function with Orion to meet the Constellation Program's goals.

Ares V Cargo Launch Vehicle

The Ares V cargo launch vehicle will be capable of launching the Altair lunar lander for Earth orbit rendezvous with the Orion crew spacecraft for trips to the Moon. It can also launch automated cargo landers to the Moon or launch other large payloads, such as telescopes or planetary probes to destinations throughout the solar system. The Ares V first stage consists of two Shuttle-derived five-and-a-half-segment solid rocket boosters similar to the Ares I first stage and a core stage powered by six RS-68B engines based on the engine developed for the Air Force Evolved Expendable Launch Vehicle program. The Ares V upper stage is known as the Earth departure stage because its primary job is to send payloads beyond Earth orbit to other destinations. It is powered by a single J-2X engine. It is topped with a composite payload shroud that encloses payloads such as the Altair lunar lander.

Missions To the Moon

On a typical lunar mission, Ares V will launch the Earth departure stage into orbit, where it will await Orion, which will launch aboard Ares I. Orion will then rendezvous and dock with Altair. The Earth departure stage, powered by a J-2X engine similar to the Ares I upper stage engine, will accelerate Altair and Orion on a trans-lunar injection flight toward the Moon. (For lunar missions, either Ares I or Ares V can be launched first.)

Once the astronauts arrive in lunar orbit, they will transfer to the lunar lander and descend to the Moon, while Orion remains in orbit. At first, astronauts will be able to "camp out" on the Moon for up to a week at a time. Later, they will be able to stay for months, once living quarters are established. After completing their mission objectives, the crew leaves the Moon in Altair's ascent module and rendezvous with Orion, which returns the crew to Earth. Missions to the Moon are planned for no later than 2020.

Missions to the International Space Station

For its first missions, the Ares I will send Orion and up to six crew members to the International Space Station. Orion will dock with the Station. In most cases, Orion will bring up new crew members, and return to Earth with crew members who have just completed assignments on the Station. Orion can return small equipment, such as film or experiment samples. Missions to the International Space Station are planned for no later than 2015.

Recent Accomplishments

For the first time in 32 years, NASA has completed critical milestone reviews for a human-rated launch vehicle. As of 2008, the Ares team had conducted a critical design review for the J-2X upper stage engine and preliminary design reviews for the overall Ares I vehicle, as well as for each of the major elements that make up the rocket. Experienced engineers and project managers who worked on the Saturn and Shuttle programs participated and weighed in on the designs and the go-forward plans, allowing NASA to build on its foundation of launch vehicle development know-how. In September 2008, the 31-member Preliminary Design Review board voted unanimously to proceed to critical design on Ares I. Critical design is scheduled for 2011 and will mark the evolution of the spacecraft from the design phase to the development phase focused on manufacturing and preparation for operations.

Vehicle Integration

The Vehicle Integration team brings all launch vehicle elements together and examines concerns that affect the entire vehicle. Working with Marshall's engineering directorate and the Ares chief engineer, Ares vehicle integration organizes, plans, and conducts all the major design reviews (the Systems Requirements Review in December 2006, the Systems Definition Review in October 2007, and the Preliminary Design Review in September 2008). During each review, hundreds of documents and drawings are reviewed and revised. It is a time of assessment and for focusing on issues and top risks. This process includes following up after each review to ensure major actions have been implemented or date collected to resolve issues identified in the review.

The Vehicle Integration team also ensures the necessary tests and analysis to solidify the vehicle design and ensure that the launch vehicle systems work together in a safe and efficient manner. Since 2006, numerous tests have been conducted to help guide designers' decisions about how to build various parts, components and subsystems of the Ares I. By the end of 2008, the Vehicle Integration team completed almost 7,000 hours of wind tunnel tests. These tests were conducted in collaboration with engineers and analysts from several NASA Centers and at numerous facilities across the country. By simulating flight-like conditions in wind tunnels and with analytical fluid dynamic models, designers can determine how the vehicle will fly through the various atmospheric environments encountered on the way to orbit. Tests cover all aspects of Ares I operations from sitting on the launch pad to ascent to separation of the first and upper stage. The test results are helping the team perfect the Ares I aerodynamic design and identify potential risks. A loads analysis conducted in the fall of 2007 revealed thrust oscillations, resonance vibrations in the first stage motors that might be detrimental to the vehicle's structure, and if levels were high enough, dangerous to the crew. A Thrust Oscillation Focus Team completed an in-depth study of the problem, and in 2008, the team recommended several solutions for mitigating the oscillations. The vehicle integration group is leading efforts to further study the options and develop the best solution to mitigate this risk.

Ares I First Stage

The Ares I first stage benefits from more than 20-years of experience flying solid rocket boosters on the Space Shuttle. In the past few years, static firings of solid rocket motors for the Space Shuttle, including one firing of a five-segment motor in 2003, have provided data to improve the Ares I first stage design. New materials for O-rings and insulation have been tested. One static firing test stand, T-97, is being refurbished for Ares I first stage testing. In 2008, a major milestone was reached when ATK Space Systems cast the fuel to be used in the first development motor, which is scheduled for test firing in 2009. This motor is the first full-scale version of an Ares motor. Workers have also fabricated a first stage nozzle and prototypes of the control system. The first stage will be recovered and reused, so a new parachute recovery system is being developed. In 2008, all chutes, including the 150-foot-diameter main parachute—the largest parachute ever developed—were tested successfully. In 2009, the entire recovery system will be tested with all three-parachute systems (the pilot, drogue, and main chutes) working together. For Ares I-X test flight, an identical parachute system will be used to recover the first stage, providing further data for refining the Ares I first stage recovery system.

Ares I Upper Stage

The second, or upper, stage of the Ares rocket provides propellant to send Orion to orbit. It has two tanks: one filled with liquid hydrogen and another filled with liquid oxygen. In 2008, the NASA Design Team for the upper stage began working with The Boeing Company, the new production contractor for both the upper stage and the instrument unit, the "brain" that controls Ares I. Together, they modified processes and refined the design to significantly reduce the time it will take to manufacture and assemble an upper stage tank, which has a complex dome-shaped common bulkhead. To test process and fabrication methods of test articles, the upper stage team at Marshall assembled a world-class friction stir welding facility. Trial welds were performed on several full-scale domes and barrels similar to the parts that will make up the upper stage fuel tanks. Confidence panels were built and evaluated to verify the structural characteristics of the aluminum-lithium materials and to determine the proper thickness for the tanks. The goal is to make tanks strong, but light beause if the spacecraft weighs less, heavier payloads can be carried to orbit. Companies also began delivering and installing welding equipment and tooling at the Marshall-managed Michoud Assembly Facility in New Orleans, where the upper stages and Orion will be manufactured. Tests on other upper stage components were also completed. In 2008, Marshall engineers designed, built, and tested a small solid rocket called an ullage settling motor. The motor settles propellant in the upper stage tanks, which is especially important in allowing fuel to flow to the engine correctly.

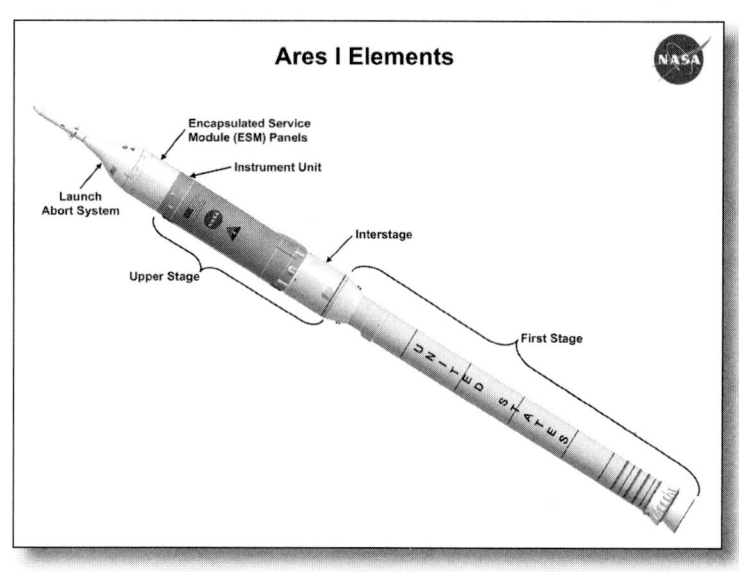

Ares I J-2X Upper Stage Engine

The J-2X will power the upper stages of both the Ares I and Ares V. In 2007, 14 hot-fire tests at Marshall examined how different injector systems channeled propellants into the engine's combustion chamber. Two other test series examined various configurations for the gas generator that powers the engine turbines. In 2008, a series of six hot-fire tests of heritage J-2 hardware from the Saturn and X-33 programs provided more than 1,300 seconds of operating time at power levels equivalent to 274,000 pounds of thrust. The results helped engineers refine the design of the J-2X, which must perform to higher thrust, safety, and reliability standards than its predecessor. More powerpack tests with development J-2X hardware are scheduled for 2010. Pratt & Whitney Rocketdyne, designer and manufacturer of the engine, has ordered basic materials and begun casting and machining parts for the J-2X. In November, the engine became the first element of the Ares Projects to complete a critical design review, which indicates that NASA is ready to begin production of development engines for testing. In 2007 and 2008, workers made substantial progress on clearing the site, pouring the foundation, and erecting the first structural steel for a new engine test stand at NASA's Stennis Space Center in Mississippi. The A-3 Test Stand will allow NASA—for the first time—to simulate starting and operating the J-2X at the high altitudes required for the Ares I and Ares V missions.

Ares V Conceptual Design

Ares V, while still in the conceptual design phase, started wind tunnel testing in 2008. The Ares V design was evaluated during the 2008 Lunar Capabilities Concept Review, an historic review that examined all requirements for the lunar mission, from launch vehicles to surface systems. To increase performance, the Ares V engine cluster was increased from five RS-68 engines to six, while booster length was increased from five segments to five-and-a-half segments. Tests and upgrades to the RS-68 engine were completed in 2007 and 2008 in collaboration with the U.S. Air Force, which uses the engine on the Boeing Delta IV rockets. Because the Ares I and Ares V both use the J-2X, the Ares V benefits from J-2X development work for the Ares I. Work on the Ares I upper stage fuel tanks will be useful when NASA designs the Ares V core stage, which will be larger than the Saturn V first and second stages combined. During 2008, several study groups and workshops explored how Ares V's enormous cargo capacity of more than 400,000 pounds to low-Earth orbit could be used for telescopes, satellites, and other scientific payloads. In December 2008, NASA issued a draft solicitation for industry proposals to refine the Ares V concept and develop requirements. The Ares V team also briefed industry representatives on its procurement plans.

Flight and Integrated Testing

The Ares flight and integrated testing group has been laying out plans for integrated tests of both Ares I and Ares V. This involves determining which facilities are needed to verify spacecraft are qualified and ready for flight. At Marshall, construction workers and engineers are refurbishing the Dynamic Test Stand, originally built for the Apollo Program and later used for the Shuttle. This facility will be used for ground vibration testing of the full-scale integrated Ares stack including Orion and the Launch Abort System designed to take astronauts to safety in the event of an accident during launch or certain phases of the Ares flight. The flight and integrated testing team also plans flight tests and is already at work planning Ares I-Y, the second unmanned flight test for the Ares I. Ares I-Y is scheduled for 2013, and one major objective of this test flight is to test the Launch Abort System at high altitudes during a flight. Ares I-Y will also be the first flight of the upper stage.

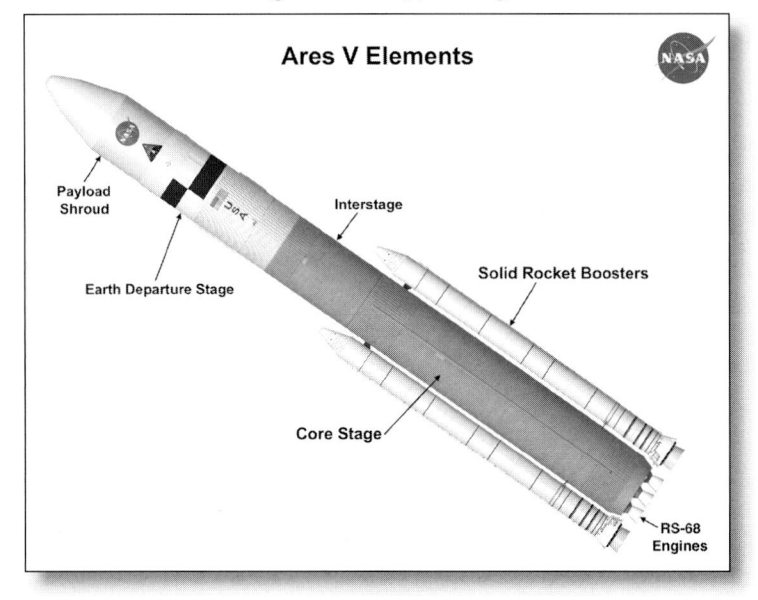

2009: Launching the Ares I-X Test Flight

Marshall employees are excited about the first test flight of Ares I, called Ares I-X, scheduled to launch in 2009. This vehicle's suborbital flight will test the flight characteristics of Ares I from liftoff to first stage separation and recovery. The flight will demonstrate the computer hardware and software (avionics) needed to control the vehicle; deploy the parachutes that allow the first stage booster to land in the ocean safely; measure and control how much the rocket rolls during flight; test and measure the effects of first stage separation; and develop and refine new ground handling and rocket stacking procedures in the Vehicle Assembly Building at Kennedy Space Center in Florida. Developing a test flight this early in a program's development was challenging. However, the flight will provide critical data for the Ares I design that is expected to reduce the development risks associated with a new launch vehicle and increase the efficiency of ground operations and handling.

As of late 2008, all Ares I-X major elements completed their critical design reviews, and were nearing final fabrication. The first stage—a four-segment solid rocket booster from the Space Shuttle inventory—incorporates new simulated forward structures to match the Ares I five-segment booster. The upper stage, Orion crew module, and launch abort system are simulator hardware that incorporate developmental flight instrumentation for collecting essential data during the mission. The upper stage simulator consists of smaller cylindrical segments and ballast plates, which were transported to the Kennedy Center in November 2008. One of the Deputy Mission Managers for Ares I-X, Steve Davis is at Marshall, and he leads teams working on the Ares I-X first stage, avionics system, and roll control system. All Ares I-X hardware arrived at the Kennedy Space Center in Florida in the spring of 2009.

Into the Future: Upcoming Milestones

In the coming year, Ares will continue to stride toward its first operational flight. Upcoming activities include:

- Fabrication and test firing of the first stage development motor, called DM-1
- Testing of hardware to mitigate Ares I thrust oscillation
- Manufacturing of the first J-2X development engine
- A three-parachute "cluster" drop of the first stage recovery system
- Creation of the upper stage manufacturing and demonstration article
- Concept and functional feasibility reviews to refine the Ares V design and integrated design and performance studies of the core stage, boosters, departure stage, and shroud, and
- The Ares I-X launch.

People across the nation are building and testing components, constructing and refurbishing facilities, and installing tooling to build America's new rockets. Many well-built test and launch facilities used for Shuttle and Saturn have gained a new life and are being refurbished and outfitted for Ares. The Ares Projects at the Marshall Space Flight Center is leading the way for new journeys beyond Earth orbit by building rockets to send people to the Moon and beyond.

Chapter 9
A New Age of Exploration

NASA is moving forward with a new focus--to go beyond Earth orbit for purposes of human exploration and scientific discovery. NASA will use the Space Shuttle to complete the International Space Station, before retiring the Shuttle, with an impressive 130 missions. NASA will deliver a new generation of spacecraft to venture out into the solar system.

The International Space Station now has six astronauts aboard orbiting the Earth made possible by the hard work and dedication of the men and women of the Marshall Space Flight Center. It is a laboratory where we are learning how to live and work in space. Building hardware that can survive and function in a harsh environment is a critical step in preparation for long-duration support of life beyond the Earth's protective atmosphere. For more than two decades, the Space Shuttle Program has lifted people, equipment and hardware to the station and into Earth orbit.

NASA is now developing Constellation, an exciting new exploration vehicle system, to expand our capabilities. This new system, powered by Ares 1 crew launch vehicle and Ares V heavy cargo launch vehicle--to be delivered by the Marshall Space Flight Center Team--will increase the number and the amount of cargo that can be transported into space. NASA's plans included using the new system to service the space station by 2015.

As humans go beyond the Earth and learn how to live on other planets, we must develop methods for using the resources found there. The Moon is a crucial stepping-stone along that path--a harsh and distant world, yet one that is only three days by an Ares rocket from Earth. Marshall is already preparing robotic missions to go to the Moon. Marshall's Ares rocket will take humans there by 2020 to begin construction of a permanent lunar outpost.

Completing the space station and constructing a lunar outpost as preparations for further exploration are critical milestones in America's quest to become a truly space faring nation. Marshall Space Flight Center has a vital role in that future.

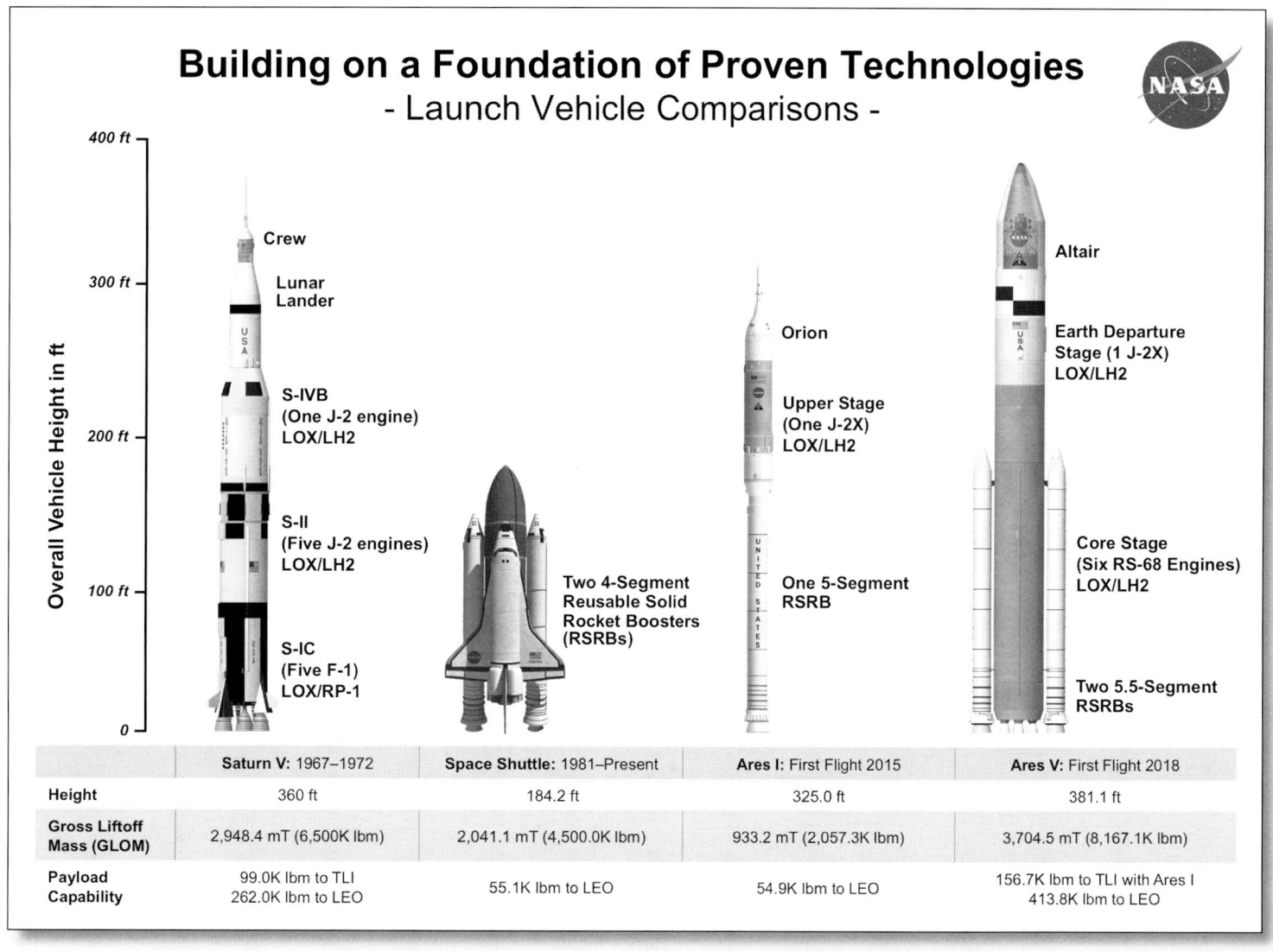

Building on a Foundation of Proven Technologies
- Launch Vehicle Comparisons -

	Saturn V: 1967–1972	Space Shuttle: 1981–Present	Ares I: First Flight 2015	Ares V: First Flight 2018
Height	360 ft	184.2 ft	325.0 ft	381.1 ft
Gross Liftoff Mass (GLOM)	2,948.4 mT (6,500K lbm)	2,041.1 mT (4,500.0K lbm)	933.2 mT (2,057.3K lbm)	3,704.5 mT (8,167.1K lbm)
Payload Capability	99.0K lbm to TLI / 262.0K lbm to LEO	55.1K lbm to LEO	54.9K lbm to LEO	156.7K lbm to TLI with Ares I / 413.8K lbm to LEO

Saturn V: Crew; Lunar Lander; S-IVB (One J-2 engine) LOX/LH2; S-II (Five J-2 engines) LOX/LH2; S-IC (Five F-1) LOX/RP-1

Space Shuttle: Two 4-Segment Reusable Solid Rocket Boosters (RSRBs)

Ares I: Orion; Upper Stage (One J-2X) LOX/LH2; One 5-Segment RSRB

Ares V: Altair; Earth Departure Stage (1 J-2X) LOX/LH2; Core Stage (Six RS-68 Engines) LOX/LH2; Two 5.5-Segment RSRBs

In Memoriam

Archie C. Absher
by the Family of Archie C. Absher

Archie Absher served 17 years continuously with Marshall Space Flight Center from December 1961 until March 1978. While with MSFC he was first assigned with the AST, Experimental Facilities and Equipment Group as an aerospace engineer, then later as supervisor aerospace engineer with the AST Flight Systems/Vehicle Systems Checkout Division. Then in 1975 reassigned to S & E, Test Laboratory, Systems and Components Test Division, Electrical Systems Simulation Branch as supervisor aerospace engineer.

He retired in March 1978 to pursue other interests, which eventually included grandchildren and a 53 acre cattle farm near Athens, AL. Archie Absher died on December 6, 1999 after a long battle with diabetes and was buried in Madison Cemetery on December 7th. He is survived by his wife of 50 years, four children, and seven grandchildren.

Archie Absher was a faithful man of God, his family, and his country. On January 22, 1977, he received a service award from NASA thanking him for 25 years of faithful service which included USAF and NACA service years. We are grateful to a loving God for sharing him with us for so many years, and the promise of seeing him again.

Konrad Dannenberg
by Ed Buckbee

Konrad Dannenberg was a rocket scientist who stayed engaged in the rocket business his entire lifetime. A member of the original Wernher von Braun rocket team, he worked on the V-2 in Germany and was instrumental in the development of the Redstone, Jupiter and Saturn vehicles for the U.S. Army and NASA.

After he retired from NASA in 1973, as a propulsion expert, he became the "resident rocket scientist" at the Space & Rocket Center, conducting tours and keeping management focused on programs that would impact and influence the youth. He was the "father of GAS (Get-Away Special)" a small piggyback payload built by Huntsville students that eventually flew on the Space Shuttle. He conducted training sessions for bus and museum guides.

When Space Camp was conceived, he suggested a program for teachers and contributed to the curriculum that was followed by Space Camp and University of Alabama in Huntsville (UAH). Dr. John Pottenger, UAH Space Teacher Director, said, "Teachers are always impressed with the historical perspective of the times that Konrad remembers so well: from the success and failures of the trial and error approach of the German team to the limited resources placed on them during and even after wartime. Students need to see the process of discovery is not easy and is littered with failures time-after-time before success is achieved. The teachers, of course, know this, but to get a story of this same kind of experience from one of those who ultimately got us to the moon is valuable for them to hear and to use when they return to the classroom. Konrad tells his story so well, and the teachers are always in awe of his experiences and accomplishments. He truly inspires them."

Dannenberg gave immediate credibility to Space Camp. He didn't sign on just to be a storyteller and stand in the classroom and give Space Camp trainees a lesson on propulsion. He took them on guided tours of the rocket engines and rocket stages where they could touch and feel the real thing, causing questions that he could answer right there on the spot. It was the engagement in the subject that enthralled the campers, having Dannenberg describe how the fuel and liquid oxygen was pumped into the engine, mixed and ignited creating the power to escape the gravitation pull of earth. He was the teacher and the Rocket Park and museum became his giant classroom as his students learned the subject of rocketry and human space travel. He challenged them to set bold goals and reminded them, it was time for their generation to go for their walk and this time on Mars. He engaged with thousands of youngsters and teachers during his stay at the center taking time to share with them his experiences and never failing to answer their questions.

As the 50th anniversary of Explorer I, January 31, 2008 was approaching, Dannenberg suggested that the community hold a symposium. He developed the program structure, selected and contacted speakers and was in attendance at the two-day event, celebrating America in Space, 50 Years of United States Space Exploration. It was Konrad Dannenberg who came up with the idea that Huntsville should celebrate its birth into the Space Age, convinced sponsoring organizations and persuading others to participate. He never quit thinking and he never quit dreaming. He was a true space ambassador.

Otto A. Hoberg
by Ted Paludan

Mr. Otto Hoberg was born in Ruethen, Germany on Sept. 5, 1912. After receiving his Bachelor's and Master's degrees in Electrical and Communication Engineering (Dipl., Ing.) from the Technical University of Darmstadt, he remained first at the University's Electronics and Communications Institute as a scientific assistant. In 1939 he transferred to Peenemunde, participating in development of the V-2. In 1945 he came to the U.S., Fort Bliss, Texas. He moved to Redstone Arsenal, AL in 1950, and in 1953 became the leader of the Instrumentation and Communication group. He became a U.S. citizen on April 14, 1955. In April 1967 he became Assistant Director of Astrionics Laboratory, a position he held until he retired on December 31, 1973. He died on January 27, 1991.

Mr. Hoberg was thoroughly familiar with the technical

aspects of sensors, signal conditioners, multiplexors and the several radio transmission methods used for telemetry. As he had in Germany and at Fort Bliss, he was an authority on radar and other tracking systems. But the greatest thing he brought to the rocket programs (Redstone through Shuttle) was the example he set in skill as technical manager of a complex system. He influenced and improved the lives of hundreds, probably several thousands, of people. (His Division was the largest division in the largest laboratory of MSFC).

Fritz K. Mueller
by Ursula Mueller, wife of Fritz K. Mueller

"Old-timers remember Fritz as one of the essential pillars in the evolution of mankind's space-faring ability during this past century," said Dr. Ernst Stuhlinger, chief scientist of the von Braun Rocket Team.

Fritz Mueller was born on Oct 27, 1907 in Schalkau, Thuringia, Germany. He obtained a degree in electrical engineering in 1929 from the Thueringische Technische Staatslehranstalt in Hildburghausen. In 1930 he joined the Aerogeodetic (later Kreiselgeraete) in Berlin-Zehlendorf as design engineer for sea and air navigational instruments. During his first five years with the company he got thoroughly familiar with the gyroscopic development and gyro techniques.

Wernher von Braun, who had a group of engineers in Kummersdorf near Berlin working to develop liquid fueled rockets knew as an accomplished flyer about the control of an airplane by means of gyroscopes. Because of his experience and work on the design and development of gyros for airplanes and ships, in early 1935 Fritz Mueller was assigned to help to solve the guidance problems for the von Braun group. He eventually played a key role in the application of gyros as the key element of a guidance and control system for rockets.

This led to the development of the A-5, which had to be built and flight-tested before the A-4 design was finalized. Mueller designed and constructed a more simplified and rugged guidance and control system with two different types of control means.

In 1940 as chief of a design and construction department at Kreiselgeraete Mueller started the design of a three gimbal stabilized platform with integral provision for sensors and computation for trajectory control.

When Wernher von Braun selected people involved in the development of the A-4 rocket for a transfer to the United States, he also invited Fritz to join the team. Fritz arrived in November 1945 with the first transport via Le Havre, Fort Strong in Fort Bliss, Texas.

In 1950 the rocket team was transferred to Redstone Arsenal in Huntsville, Alabama. The first assignment for Mueller was the design and development of the gyro stabilizer for the Redstone rocket. This was the beginning of a new area of accurate, rugged and reliable gyro stabilizers used in the Explorer and Juno launches. His work led to the design of the ST 124 inertial air bearing stabilizer, which guided the Saturn V rocket to the moon.

Fritz was a member of the von Braun development group. In 1958 he became deputy director of the Guidance and Control Laboratory. In recognition of his outstanding work in helping maintain America's pace in missile development, he received with Wernher von Braun and two other scientists an Honorary Doctor of Science degree from Rollings College in Winter Park, Florida. He was awarded eight patents.

He retired in 1978 at 71 years of age. Dr. Fritz Mueller died on May 15, 2001 at the age of 93.

Hermann Oberth
"The Father of Space Flight"
by Konrad K. Dannenberg

Fifty years of rockets and spacecraft in the Rocket City demonstrated most successfully the benefits of a new technology, which originated during the 1940s in Germany. Next generation rocket improvements at Redstone Arsenal led to the Redstone missile system. A four-stage Redstone named Jupiter C launched the first American satellite into earth orbit. "Man-rated" Redstones launched two astronauts into sub-orbital flights. The Saturn V launched 27 astronauts to the moon (12 walked on the moon). The Space Shuttle conducts routine missions to put satellites and scientific payloads into Earth orbit and is at present assembling, changing out crews and resupplying the International Space Station.

How did this astounding new technology get started?

The Russian Constantine Tsiolkovsky and the American Robert Goddard have often been referred to as the "Father of Space Flight." This credit must actually be shared with the German Hermann Oberth whose writings and activities influenced Wernher von Braun and many of the other initial participants who worked with von Braun in the German Rocket Research and Development Center of Peenemunde. This is where the Space Age began about 60 years ago. In 1942 an A-4 rocket traveled through space to get from one point on Earth to another. This event also opened a new technology that permitted humankind to travel to another heavenly body.

This history making event began soon after Hermann Oberth's publication of his book, *Die Rakete in den Weltenraum*, (Rocket into Interplanetary Space). Based on this well received book a German movie company invited him to come from his native Romania to Berlin to be a consultant for the movie "*Die Frau im Mond*" (the Woman on the Moon). He was also supposed to build and launch an 8 feet tall rocket to publicize the movie. He was the mentor of many amateur enthusiasts including Kurt Heinisch, Rudolf Nebel, Klaus Riedel, Helmut Zoike, and high school student Wernher von Braun. They worked at the Raketenflugplatz (Rocket Airfield) in Reinickendorf near Berlin. When the movie was ready for release, the rocket was not ready to be launched. Hermann Oberth was forced to return for financial reasons to Romania to teach mathematics again. Local members continued the rocket work and eventually joined Wernher later on in Peenemünde. Hermann Oberth was definitely the scientific head of the team during his stay, and provided to the Berlin rocket society greatly important funding as well as some moral support that no other amateur group enjoyed.

Since Hermann Oberth was a Romanian citizen the German Army could not employ him. This restriction was eventually lifted, and he later on worked during the war for a few years in Peenemünde. However, at that time the design and most of the development of the A-4 had been established. Hermann Oberth also joined the von Braun rocket team at the Redstone Arsenal in the late 1950s during the Army years. He had to return to Germany to be eligible for his pension. Wernher invited him and Rudolf Nebel to attend the Apollo 11 launch at the Cape, the first lunar landing.

Dr. Joseph Randall
by Brooks Moore

The untimely passing of Dr. Joseph L. Randall on March 24, 2001 marks the end of an era during which he made many significant contributions to the mission of the National Aeronautics and Space Administration (NASA). Dr. Randall was employed as a Nuclear Engineer by Lockheed and as a Solid State Physicist by the Army Ordnance Command before beginning his brilliant career with NASA's Marshall Space Flight Center (MSFC) in 1962.

Dr. Randall's original research efforts with MSFC focused on the development and adaptation of laser systems for space applications. He pioneered the use of lasers in space communications and was the Principal Investigator on an experiment that demonstrated for the first time the feasibility of two-way ground-to-space laser communications. He was awarded the NASA Medal for Exceptional Scientific Achievement for his work on this and his other laser research.

Dr. Randall was also a pioneer in the design and development of large optical systems for NASA's space observatories. One of his most significant contributions in that regard was the initial design and implementation of the optical system of unprecedented size and resolution employed on the Hubble Space Telescope. In recognition of his significant role on this program, he was awarded NASA's prestigious Medal for Exceptional Service. Dr. Randall also directed and personally participated in early research efforts in the field of X-Ray optics, out of which evolved the systems employed on NASA's other two "Great Observatories," the High Energy Astronomical Observatory and the Chandra Observatory.

During his 35 years of dedicated service to NASA, Dr. Randall was assigned to a number of key supervisory and management positions. He initially headed the Optics Branch and from there progressed through several senior assignments to his final position as Director of the Astrionics Laboratory. During his tenure in this latter position, he directed research and development efforts and provided technical oversight on all MSFC programs in the disciplines of optics, lasers, communication, instrumentation, space power, computers, and other electrical and electronic systems.

He retired from his position as Director of the Astrionics Laboratory in January 1997 after a long and illustrious career. Although Dr. Randall has departed from us, the result of his outstanding ingenuity and dedication lives on through the continuing scientific information being collected daily by the Hubble Space Telescope and NASA's other active space observatories. Each new observation is providing scientists a better understanding of the universe in which we live, thereby providing a continuing tribute to Dr. Joe Randall and his many NASA associates by whom he was universally admired.

Dr. Eberhard Rees
by Konrad K. Dannenberg

The second most important member of the von Braun Rocket Team was without question his long time deputy, Eberhard Rees. He was his deputy and manager of all plant operations at the Heeres Versuchsanstalt Peenemünde (HVP). He played the same role during the employment of the Team by the U.S. Army as well as during the NASA years at the Marshall Space Flight Center (MSFC). He was responsible for all shop operations when the Chrysler-built REDSTONE missiles had to be modified at MSFC for the launch of the first US satellite and when the Redstone had to be "man-rated" to put two astronauts into sub-orbital flights. He played a key role in the Saturn V development and in the redesign of the Apollo command module after the Apollo I fire at Cape Kennedy.

Eberhard Rees was born on April 28, 1908 in Trossingen, Württemberg. He was therefore 4 years older than Wernher and most of the other team members. This fact caused von Braun to look at him always as an "elder statesman" who often cautioned Wernher in many of his daring proposals. He got his BS in mechanical engineering in Stuttgart, and his MS in the same field in Dresden. When von Braun was looking for a person to head the manufacture of A-4s, his Dresden Professor recommended Eberhard to von Braun. By that time Eberhard had also gained some industry experience having worked at a steel foundry in Leipzig.

He joined Peenemünde in 1940, a time when World War II had started and the German Army wanted to speed up the development of the A-4. At that time the growth of this Army Research and Development facility was dramatic. A successful flight of an A-4 on October 3, 1942 had demonstrated that this vehicle can fulfill the military requirements. But many additional units had to be built and launched to improve the performance and especially its reliability and accuracy. Eberhard was at this time in charge of the entire plant operation, to include testing, procurement of parts, quality control, etc. He was de facto the number two man, especially after the bomb raid on Peenemünde, when Dr. Walter Thiel and his entire family were killed, as well as several other plant operations managers.

During the Fort Bliss period Eberhard spent most of his time in the General Electric shops and at the firing range. He came to the United States as a bachelor, but when the opportunity came, he traveled back to Germany to marry Gerlinde Segebrecht, a tennis buddy and a local dentist in Zinnowitz, who served many HVP employees as patients.

Upon the transfer to Huntsville, Eberhard could assist von Braun again when the Redstone production started almost immediately upon our arrival.

Eberhard played possibly his most important role in the Apollo program after the fire of Apollo I at the Cape. Astronauts

Gus Grissom, Ed White and Roger Chaffee had been killed in the fire, which was apparently started by an electrical spark causing a fire in the 100% pure oxygen atmosphere. The astronauts could not be removed quickly from the command module since it featured a slow opening hatch. The capsule had to be redesigned with a quick opening hatch, fireproof wiring insulation, and many other less major changes. To study and accomplish the redesign von Braun lent his deputy Rees to the Manned Spacecraft Center and he and about 80 people from MSFC spent several months assisting North American Aviation in the redesign of the Apollo Command Module, thus assuring a safe continuation of the Lunar Landing program. The astronauts' sacrifice would not be in vain.

When Wernher von Braun left the Marshall Space Flight Center and transferred to NASA Headquarters in Washington, Eberhard Rees became director of MSFC. After his retirement in 1973 he moved from his mountain home on Monte Sano to a downtown Condominium. After his first wife Gerlinde died, he married Maria Haase in his retirement and went with her on several month-long voyages on excursion ships. He liked to play golf for a while until he had trouble to see and hit the golf ball. He and Maria moved eventually to Florida, where he died in a senior citizens home in April 1998 at the age of 90.

Werner Kurt Rosinski
by Konrad K. Dannenberg

Werner Rosinski is one of the original members of what became known as the "Wernher von Braun Rocket Team." He joined the group of young rocket engineers and technicians already at the time when static firings of the first Army rockets got under way at the Army Artillery Arsenal located at Kummersdorf, a small village just about 50 kilometers south of Berlin.

Werner Rosinski was responsible for the electric cables that were installed underground in the "static firing" facility. The extensive set of cables provided the operational control of the rocket and a multitude of additional lines for measuring devices. Wernher von Braun always emphasized the importance of "static firings" before taking the rocket to its launch pad. Only in a "static firing" can one determine any design shortcomings. If taken to the launch site without a static firing, one might never be able to detect any design shortcomings during the rocket's flight.

Werner Rosinski was born in September 1914 in the Berlin area. He studied electrical engineering, a field in which he worked all his life. In 1934 he married his high school sweetheart, Erika. Soon after he started at the Kummersdorf Arsenal, he was hit by a truck riding his bicycle to work. This accident cost him a leg. In spite of this tragic event he executed all his assigned duties like anyone else. He climbed without hesitation the 40 feet high ladders to check-out proper electrical connections in the guidance and control compartment at the upper end of the erected A-4.

During the early part of the Fort Bliss period Werner spent all his time at the White Sands Proving Ground. He worked with Army personnel and General Electric employees. The Army had contracted GE to prepare the V-2's for firings, and to equip them with experiments to conduct research of the upper atmosphere.

Upon the transfer of the rocket team to the Redstone Arsenal near Huntsville, Alabama, Werner was instrumental in the construction and wiring of the first test stand, the Redstone test facility. All Redstone missiles underwent a "static firing" before shipment to the Cape. That included the modified Redstones for the first U.S. Satellite Explorer launch, as well as the ones for the first U.S. launches of manned rockets of Alan Shepard and Gus Grissom.

As a member of the Test Laboratory of the Army Ballistic Missile Agency (ABMA) under General John Bruce Medaris, he participated in the design of the first large test stand for the Jupiter program, the "T-Stand," which had been built to even test the first stage of the Saturn I and Saturn IB vehicles. For the lunar landing vehicle, the Saturn V, a much larger test stand was needed, and Werner again contributed to the cabling and measuring network of all facilities needed for this gigantic project.

Werner retired from NASA in 1970, but continued as a consultant to private industry. He traveled back to the old country regularly to visit his doctor who had made his prosthesis for his missing leg, which had to be re-adjusted. He visited Peenemünde in 1991, right after the unification in Germany. He was surprised to see that the Russians had dug up all the underground cables.

He passed away on May 14, 2000 at the age of 86.

Axel Roth
by James W. Bilbro

Axel Roth passed away on November 12, 2008 at the age of 72. Axel, born in Darmstadt, Germany in 1936, was the son of Ludwig Roth, a member of the Von Braun Rocket Team. Axel was among the first employees at NASA's newly formed George C. Marshall Space Flight Center in 1960. During his long and distinguished career spanning over 45 years, Axel held increasingly responsible positions beginning with the Apollo/Saturn program and continuing through the Skylab and Spacelab programs, Space Station Projects Office, Payload Projects Office, Program Development Office and Flight Projects Directorate. He attained his highest position at Marshall in 2001 when he was named Associate Center Director, a position from which he retired in 2004.

During the course of his career he was the recipient of numerous awards including the Presidential Rank Award. He was a member of the Aerospace Engineering Advisory Council Board at Auburn University and the Auburn Alumni Engineering Council.

As an outstanding engineer and highly respected manager, Axel made significant contributions to NASA and the Marshall Space Flight Center. Perhaps his most significant contributions resulted from his unflappable disposition and calm, thoughtful demeanor which allowed him to defuse many thorny situations throughout his career. He was a friend and mentor to many individuals both young and old, and his calming manner and sage advice continue to be missed by his colleagues and friends.

Victor C. Sorensen
by Ed Buckbee

One of the guys I remember from my Marshall days was Vic Sorensen, the Director of Marshall's Management Services Division during the 1960's. He was one of the key ABMA people who helped organize and transfer Army personnel to the new Marshall Space Flight Center. He set up the first Management Services organization, which consisted of over 1,000 civil service and contractor personnel responsible for institutional support consisting of everything from security, food service, and technical documentation to telecommunications.

Vic's office was down the hall from mine in Public Affairs. I remember Vic as one of those dedicated managers who was committed to supporting this research and development oriented organization and its highly technical work force which often had demands difficult to fulfill. He took his job seriously and I think Vic enjoyed going to work every day. He was a kind man and a true gentleman. He always had a smile and a greeting for you, plus he had a great sense of humor. He enjoyed a good joke and often times played "gotchas" on my boss, Bart Slattery, Jr. whom he referred to as the "the Center's propaganda czar."

Vic left his mark on this unique organization as Center Director Eberhard Rees said, "…during the 1960's Sorensen skillfully and effectively directed an office, which was so vital to the success of the Center's mission of landing man on the Moon."

I was proud to have known Vic Sorensen. Marshall lost one of its charter members in 2002. We will miss him.

Friedrich Graf von Saurma
by Ruth G. von Saurma

Born February 1908 in Dahsau in the former German province of Silesia, Fred was an enthusiastic and active private pilot and radio amateur even before his enrollment at the Technical University of Breslau. He graduated with a master's degree in engineering, and in 1940 subsequently became a development engineer for rockets in the German Air Force. After being involved in the development and deployment of jet-assigned take-off units for heavy fighter aircraft, he was assigned as Technical Director to the V-2 Testing and Training Center in Zempin on the island of Usedom. There he developed and realized the concept of a mobile launch platform and also pursued aerodynamic improvements of the wing structure of the V-2.

In April 1945, as a Lieutenant Colonel in the German Air Force Engineering Corps, he was taken prisoner of war. Released from a prisoner-of-war camp in West Germany in early 1947, he started a small enterprise manufacturing radios and loudspeakers and later became a development engineer for electronic precipitators in a German industrial company.

In 1953, he accepted a contract with the U.S. Department of Defense under the so-called paper-clip project and emigrated to the United States with his family in October of that year, staying initially for several months in New York City. After personal negotiations with Dr. Wernher von Braun, Fred joined the Huntsville rocket team in January 1954.

Some of his assignments with the Army Ballistic Missile Agency and the NASA-Marshall Space Flight Center included his being technical assistant to von Braun and overseeing development projects in the Saturn System Office. Closest to his heart and interest was his position as project engineer for the ambitious Dyna-Soar project, a piloted space plane, the pursuit of which was later cancelled in favor of the more expeditious Saturn-Apollo lunar program.

During his tenure with ABMA and MSFC, Fred received various distinguished service awards and was issued a number of U.S. patents, one of which was for the development of a pilotless rocket-powered projectile for accurate cargo delivery.

In his leisure time in Alabama, Fred renewed his private pilot licenses and enjoyed flying, sailing on the Tennessee River lakes and keeping contacts with radio amateurs in far away places.

His untimely death in December 1961 was a great shock not only to his family, but also to his friends, many of whom may still recall his engaging and distinguished personality, great sense of humor and bright intelligence. Most of all, his passing was a serious and lasting loss for my daughter, Lilly, and myself. We both cherish his memory with greatest fondness.

Ernst Stuhlinger
Courtesy of Frederick I. Ordway III

Years ago, Ernst Stuhlinger and I would debate where and when we met. It turned out that the "where" was New York City's Hotel Edison and the "when" December 1955. Our encounter took place during a meeting of the American Astronautical Society at which he lectured on "Electrical Propulsion Systems for Space Ships" As it turned out, his first technical article on this subject was published in the winter 1955 issue of the AAS' *The Journal of Astronautics* whose editor I just happened to be!

Our next get-together occurred on the morning of 6 April 1956 when I visited him in his Army Ballistic Missile Agency office on Redstone Arsenal in Huntsville, Alabama. He patiently explained in great detail what was going on there in those uncertain pre-Sputnik/pre-Explorer satellite days and outlined some of his dreams of the future. I was fascinated by what I heard; then and there began what became a lifelong feeling of admiration, kinship, and warmth.

Stuhlinger soon laid out a series of tasks to be undertaken by a small group of colleagues drawn from Republic Aviation, Republic's Guided Missile Division (where I had worked), Grumman and Bell Aircraft by then under ABMA consulting contract. Soon afterwards we received our preliminary task assignments along with Stuhlinger's verbal caution that we were never to refer to artificial satellites as part of our scientific instrumentation, power supplies, telemetry and other survey assignments, only to high-altitude research vehicles—the code word! In those days, spaceflight was not officially part of the Army agenda and mention of the subject was frowned upon by the Eisenhower Administration and its Bureau of the Budget. How that would change!

So who was this man who was to help mold the careers of our small group of space-flight enthusiasts? The son of a school teacher, Ernst Stuhlinger was born in 1913 in Niederrimbach in southern Germany. In Tuebingen, he received his high-school training before moving on to the town's well-known university. There, he pursued studies in physics under Professor Hans Geiger, receiving a doctorate in 1936 upon completion of a dissertation in cosmic-ray physics. He then moved on to the Technical University of Berlin; and, by 1939, found himself involved in the German atomic-energy program under the guidance of Professor Werner Heisenberg.

Despite his outstanding scientific training, a few years later Stuhlinger was called up for military service and sent to the Russian theater (he later wrote up his experiences in a 17-page manuscript "Russia—Winter 1942-1943"). Fortunately, after a year and a half of incredible hardship, he was transferred, still in uniform, to the German ballistic-missile research and development center located at Peenemuende on the Baltic Sea. Under the technical direction of Dr. Wernher von Braun, he was assigned to work on guidance and control systems.

Following the collapse of Germany in 1945, Stuhlinger joined some 120 colleagues hired by the U.S. Army to pursue missile research and development in the United States. For the first five years, he worked at Fort Bliss, Texas and was involved with V-2s fitted with scientific instruments that were launched from the Army's White Sands Proving Ground to the north in New Mexico.

Then, in 1950, the team was transferred to the Army's Redstone Arsenal near Huntsville in northern Alabama. For the next ten years, its assignments expanded to include the Redstone, Jupiter and Pershing ballistic missile weapon systems and other programs. On 31 January 1958, Stuhlinger was intimately involved in the successful orbiting, by a modified Redstone referred to as Juno I, of America's first satellite, Explorer 1. In 1961, another modified Redstone placed astronaut Alan B. Shepard into space along a ballistic trajectory that ended with the recovery of his Mercury capsule in the Atlantic Ocean.

In July 1960, the von Braun team became the nucleus of the National Aeronautics and Space Administration's brand new George C. Marshall Space Flight Center located on Redstone Arsenal. For the next seven years, Stuhlinger directed the Space Science Laboratory and its myriad scientific activities. In his new civilian role, he became involved with three High Energy Astronomical Observatories, the Pioneer 4 lunar by-pass probe, the Skylab space station and many other space projects. In 1964, his book on the potential of electrical propulsion for space travel [1] appeared to worldwide acclaim--including the award of the Electric Rocket Propulsion Society's Medal for Outstanding Achievement in Electric Propulsion. Four years later, he was appointed the Marshall Center's Associate Director for Science from which post he continued to supervise a broad array of space research programs. Stuhlinger retired from NASA in 1976, but his career was by no means over. He was soon named adjunct professor and senior research scientist at The University of Alabama in Huntsville where he taught astrophysics and space sciences and worked on electric-car projects. Later, he became a senior research associate at Teledyne Brown Engineering, all the while continuing his writing and lecturing.

He spent several months during this period at universities in Munich and Heidelberg, Germany under the Alexander von Humboldt Research Award Program. Meanwhile, another publishing milestone occurred in 1994 with the appearance of the two-volume biography of his boss and mentor for so many years, Wernher von Braun [2]. Then, in 2001, his definitive chapter "Enabling Technology for Space Transportation" was published in the two-volume *The Century of Space Science* [3].

As the New Year broke in January 2008, Stuhlinger, now fully retired but ailing, was looking forward to participating in the celebration of the 50th anniversary of the Explorer 1 satellite. On 31 January, celebration symposia coordinator Ralph Petroff turned the Explorer 1 symposium over to co-chairmen Frederick I. Ordway III and National Space Society executive secretary George Whitesides. From the start, we had counted heavily on the participation of "old-timers" Ernst Stuhlinger and his Peenemuende-era colleague Konrad Dannenberg. Representing a younger generation that day were Dwayne Day from the Space Studies Board, National Academy of Sciences; Steven J. Dick, NASA chief historian; and Roger D. Launius, Division of Space History, National Air and Space Museum, Smithsonian Institution. Our special guest was Natalya Koroleva, daughter of Sergei Korolev, Soviet-era "chief designer" and father of Sputnik.

The success of the Explorer 1 symposium was marred by the absence of Ernst Stuhlinger who, on 31 January 2008, found himself bed-ridden in the Huntsville Hospital. A number of us, including Natalya Koroleva, visited him while other participants sent him best wishes for a speedy recovery.

Alas, the great man's time was running out; and, less than four months later, on the 25th of May, he left his family, friends and colleagues to embark on the ultimate voyage across space and time.

References

1. Stuhlinger, Ernst, *Ion Propulsion for Space Flight*. New York, 1964: McGraw-Hill Book Company.

2. Stuhlinger, Ernst and Frederick I. Ordway III, *Wernher von Braun Crusader for Space: a Biographical Memoir*, and *Wernher von Braun Crusader for Space: An Illustrated Memoir*. Malabar, Florida, 1994: Krieger Publishing Company.

3. Stuhlinger, Ernst, "Enabling Technology for Space Transportation," Chapter III in Volume I of *The Century of Space Science*, edited by Johan A.M. Bleeker, *et al*. Dordrecht, Netherlands, 2001: Kluwer Academic Publishers. The Stuhlinger Chapter III runs from pages 59 to 116 in the 960-page first volume; Volume II continues on to page 1,846. The set lists at $495!

James D. Willard
by Chester B. May

James Willard retired from NASA on January 21, 1978 after 30 years of service with the United States Government. While with NASA he worked as a drafting engineer helping to put the ideas and concepts of the Redstone and Saturn rockets as well as the Mercury, Gemini and Apollo spacecraft into design form.

Among some of his awards during his service with NASA, he received a Certificate of Participation for his contributions to the successful launch of the first Saturn rocket on November 9, 1967. He received an Award of Achievement for his contributions to the first moon landing on July 20, 1969. He also received a "First Shuttle Flight Achievement Award."

J.D., as his friends called him, passed away on January 11, 1997. He is deeply missed by his family and friends.

Marshall Space Flight Center Directors

Dr. Wernher von Braun
1960-1970

Wernher von Braun, the first to direct the newly created NASA Center, brought Marshall into existence and directed the tremendous Saturn endeavor with energy and foresight. Dr. von Braun's challenge was to form an organization of the best talent to invent and prove the technology for super launch vehicles for peaceful exploration of space and capable of sending American astronauts to the moon. The technological effort of the Saturn years is without parallel in our history. Never before nor since in peace time have so many people in government, industry, and universities been orchestrated to work together on a project of such complexity and originality as the manned lunar landing much of which was managed out of the Marshall Center.

Dr. Eberhard Rees
1970-1973

Dr. Eberhard Rees, who was von Braun's long time deputy, served as the Center Director from 1970 to 1973 during a period of transition. Under his leadership, the Saturn-Apollo program was completed, the Skylab program was implemented, and formulation of the Shuttle concept began which Marshall played a major role. During the tenure of Dr. Rees, the agency and the Marshall Center in particular, experienced fiscal and manpower reductions that presented new managerial challenges. Although many valued employees were lost, the Center retained its capabilities in all disciplines and emerged from this period with renewed energy.

Dr. Rocco A. Petrone
1973-1974

Dr. Rocco A. Petrone served as Marshall's Director from 1973-1974 and presided over the extraordinary Skylab program. In three successful missions, including the dramatic rescue operations, Marshall demonstrated convincing expertise in scientific payloads and manned systems. The Center reapplied experience gained in the development of space launch vehicles to this ambitious systems engineering project and succeeded admirably. During this period, the Center's organization was restructured to accommodate Marshall's changing roles and responsibilities.

Dr. William R. Lucas
1974-1986

Dr. William R. Lucas began his term as Center Director in 1974 and held that position longer than any other Marshall Center Director retiring in 1986. Marshall bears the stamp of his influence. As Lucas advanced through key positions at all levels from the laboratory onward, he insisted on the virtues of competence, discipline, and the commitment to excellence. During the Saturn era, von Braun named him the first director of Program Development and from that position Lucas introduced diversification to the Center and it has remained one of the foremost technical and managerial elements of NASA. It has been said that the systems engineering capabilities of the Marshall Center could be applied to any large-scale technological challenge in the nation. The multidisciplinary expertise is rare and truly a national asset. Exciting new programs occurred under Lucas' leadership—the first Shuttle launch, Spacelab, HEAO, and the beginning of Hubble.

J.R. Thompson
1986-1989

James R. Thompson became the fifth Center Director in 1986 and returned the Space Shuttle to flight status after the Challenger accident. Thompson not only introduced a cultural change; he reorganized Marshall to improve rocketry engineering and management. He reemphasized the philosophy: "I want young engineers to get their hands dirty working on fire and smoke projects." On the laboratory side of the Center, he brought propulsion specialists together and pooled rocket engineers from several laboratories into a new Propulsion Laboratory. On the project side he restructured the Shuttle Project Office for the recovery and successful return to flight. For the first time Thompson introduced Shuttle upgrades like the alternate turbo pump to the Shuttle main engine not only to Marshall but also to the entire agency.

Thomas J. Lee
1989-1994

Jack Lee took over the Center Director's position in 1989 when Thompson moved to NASA Headquarters to become the Deputy NASA Administrator. Lee had served as Deputy Director since 1980. Under his management, Marshall became heavily involved in the impressive Space Station program, which became an international program. Lee was one of the more experienced people in the Agency in dealing with international partners having worked with the Europeans for many years on the SpaceLab program. Culminating more than 30 years of Marshall work in manned space flight systems, Lee applied the legacy of Apollo, Skylab, Shuttle and SpaceLab to the new Space Station program. Marshall engineering talents were tested with the station's challenging technology not to mention the unprecedented political, budgetary and congressional oversight.

G.P. (Porter) Bridwell
1994-1996

Porter Bridwell became the Center Director in 1994 having a long history at Marshall in the management of key human flight systems beginning with Saturn and continuing with major elements of the Space Transportation System. He served as manager of the Space Shuttle Projects office at Marshall. He managed the Heavy Lift Launch Vehicle, the National Launch Systems programs and was a senior official on the Space Station Redesign and the U.S.–Russian Space Station Integration teams. During his tenure Marshall became the Center of Excellence in Space Propulsion, Transportation Systems Development and Microgravity. Marshall's ASTRO 2 mission, the longest mission flown on Shuttle thus far, occurred under his management.

Dr. J. Wayne Littles
1996-1998

Wayne Littles assumed the position of Center Director in 1996. Littles was another experienced manager of human space flight systems having spent 24 years in key positions at Marshall. He became NASA's Associate Administrator for Human Exploration directing both the Space Shuttle and Space Station programs and restructuring those programs. He worked with the Russian Space Agency to assist in their participation in the International Space Station program and served as NASA's Chief Engineer. While Center Director, Marshall's contractors produced the first lightweight External Tank and delivered the first U.S. made Space Station component, Node 1. Marshall's FASTRAC engine passed a critical test series and Littles was instrumental in putting in place the management teams for the X-33 and X-34 programs.

Arthur G. Stephenson
1998-2003

Art Stephenson became the ninth Center Director in 1998. With 35 years experience in space industry, his experience includes engineering and systems management responsibilities with Apollo spacecraft, Pioneer-Jupiter spacecraft and the Shuttle Orbiter. He managed Oceaneering Technologies, which has become a major space systems contributor to the Nation's space program. As Center Director of NASA's largest field installation he was responsible for all propulsion elements of the Space Shuttle, advanced launch vehicles, microgravity, earth and space science projects, and payload operations of the International Space Station. Stephenson accepted the challenge to become the world pre-eminent propulsion research center breaking ground on a new Propulsion Research Lab and continuing to commit the resources and personnel to study advanced space transportation systems.

David A. King
2003-2009

David King served as the Center Director beginning in 2003, having assumed that position after serving as deputy director of the Center. While deputy, he played a key role in the recovery operations of NASA's Space Shuttle Columbia. Prior to his appointment as deputy director, he was at Kennedy Space Center, Florida, where he held several critical positions as Space Shuttle launch director, and director of Space Shuttle processing, managing and coordinating all Shuttle launch operations. He managed a center that was generating a new stable of rockets that will carry crews of explores to the Moon and beyond. King has been directly involved in human space flight since 1983 when he joined NASA as a main propulsion systems engineer.

Robert M. Lightfoot, Jr.
2009-present

Robert M. Lightfoot was named to the post of director of the Marshall Center in March 2009. He had served as the deputy since 2007. He began his career in 1989 as a rocket testing engineer at Marshall working on the Space Shuttle main engine and the Russian RD-180. He has held critical positions involving propulsion testing at Marshall's Test Laboratory and at Stennis Space Center, where he was named director of the organization. He held leadership positions in the Space Shuttle Program Office at NASA Headquarters and became the manager of the Space Shuttle Propulsion Office at Marshall, having responsibility for all propulsion components of the Space Shuttle. As the deputy director of Marshall, he shared in the responsibilities for managing the center and its propulsion development programs.

A Tribute to Wernher von Braun

by Ernst Stuhlinger

Editor's Note: Dr. Ernst Stuhlinger, the chief scientist of the von Braun team, wrote this tribute. The following illustrations represent the significant achievements and accomplishments that occurred during the years that von Braun led the rocket team in Huntsville from 1950 to 1970. This section of the publication was researched and produced through the courtesy of The University of Alabama in Huntsville on the occasion of the dedication of the von Braun Research Hall. The tribute to Wernher von Braun can be viewed by visitors during business hours at the von Braun Research Hall on the campus of the University of Alabama in Huntsville.

Wernher von Braun died 25 years ago. He would have been 90 this year (2002).

"When a man dies," Ann Morrow Lindbergh said, "it is like when a tree falls, and when its roots, its powerful trunk, all its branches and limbs, its crown with its leaves and blossoms, are lying before us, every part closely and clearly visible...."

How does von Braun live in the memories of us old-timers who have been close to him for more than three decades? He was a visionary who, at the age of 14, saw the exploration of space as part of man's destiny, and he was an achiever and accomplisher who made this dream of his early youth come true in a life filled with hard work, with an unshakable belief in his mission, and with a fabulous string of successes. During his lifetime, he enjoyed much recognition, admiration and thankfulness from those whose lives he brightened and enriched. After his death, attempts were made by young writers to ascribe atrocities committed by members of the Nazi government to him—to the dismay of those who knew better.

How do those old-timers who had the privilege of being close to von Braun for many years remember the features that made him so extraordinary? There are many: his superior intelligence, supported by a phenomenal memory; his almost unlimited interest in all kinds of human activities; his capability to define the technical projects he wanted to bring to life very exactly, and then to work tirelessly until they were accomplished; his almost magic ability to form a team, to make it grow and to keep it together, even for years after his death; his way of conducting joint technical discussions so that all participants could understand what the problem was, and of continuing such discussions until a reasonable solution had been found; his ability to make every one of his team feel proud and privileged to work for him; his intent to make sure that a counterpart in a technical dispute (which he usually won) the other party did not feel to be the loser, but just the second winner; his gift of feeling equally at ease whether he conversed with his driver, or with the President of the United States; his proverbial carefulness as a practicing engineer (all through his lifelong career of testing and launching rockets, there was not a single life lost in an accident while he was in charge of operations!); his way of giving presentations to Congress that always had a standing-room-only attendance; his fabulous success as an ardent communicator and public relations man for the idea of space exploration.

Perhaps the best words describing how von Braun will continue to live in our memories were spoken by Riccardo Giacconi, at that time the scientific director of the Hubble Space Telescope Project: "He had a nobility around him that set him apart from all other people in the space program.... He believed in the decency of man."

Wernher von Braun

NASA - Marshall Flight Space Center

1950

Members of the German rocket team

Bumper rockets are launched

The rocket team moves to Huntsville, with the first contingent arriving on April 1. To overcome any unease or lingering bad feelings after World War II, members of the German rocket team are encouraged to become active in community organizations.

"Within 90 days of our arrival in July 1950, my wife and I were licking stamps and envelopes for the community concert association membership drive."
— Walt Wiesman

Upon Dr. Wernher von Braun's arrival, The Huntsville Times publishes a front page article about the scientist, along with a photo of von Braun, his wife and their youngest daughter, Margrit. Von Braun takes advantage of the opportunity to promote not Army missile development, but something else dear to his heart:

Dr. von Braun Says Rocket Flights Possible to Moon
Foremost World Authority Explains That Dollars Are Main Hurdle to Overcome, As Technicians Know Exact Trajectory, Flight Path Of Ship To Be Used

By Bob Axelson

Rocket flights to the interplanetary world are not figments of the imagination, but will be made during the present generation, Dr. Wernher von Braun, project director for the guided missile center at Redstone Arsenal, asserted yesterday ...

"The rocket to the moon is in a similar stage of development at the present time as the trans-Atlantic flight by airplane was in 1912," he explained.

"We scientists know exactly what such a rocket would look like, its size, trajectory, flight path, power plant needed and other requirements. The only thing it takes is development," the 38-year-old scientist declared.
— The Huntsville Times, May 14, 1950

The Bumper Round 8, the combination of V2 and Corporal missile systems, becomes the first Army rocket launched at Cape Canaveral.

The University of Alabama's Huntsville center opens its first classes on January 6, in rooms at West Huntsville High School. There are 137 students. Ten classes of seven basic freshman subjects are taught by five part-time faculty. By the Spring quarter, enrollment has risen to 234. The average age of the students is about 30 and many students use their G.I. Bill benefits to help pay for tuition.

In September, George Campbell is named the center's first permanent director, and the center's first full-time faculty member, John McCormick, is hired to teach English.

Hermes rocket prepared for launch

Engineering area

First launch from Cape Canaveral, Florida

101

50 Years of Rockets and Spacecraft

1951

Honest John rocket

Corporal rocket

Josiah Gorgas Laboratory

First Redstone launch

Honest John rocket

Army personnel secure Nike missile

The rocket team performs design studies and engineering activities for a new ballistic missile system, which would become the Redstone.

Test facilities at Redstone Arsenal are designed, including the static test stand, later designated a national historic site. Using the same shovel he used to break ground for the Redstone Ordnance Plant in 1941, Col. Carroll Hudson, Redstone Arsenal's commanding officer, breaks ground for the $1.5 million Josiah Gorgas Laboratory.

> The British Interplanetary Society hosts its second International Congress on Astronautics in London. The official subject: An Earth-satellite vehicle.
>
> The delegates attacked nearly every angle of designing, launching, supplying and utilizing satellites, and none had given the matter closer study than Dr. Wernher von Braun ... now hard at work for the U.S. Army in Huntsville, Ala. ...
>
> In considerable detail, von Braun sketched out a full-dress flight to Mars. It could be done, he wrote, by using two satellite stations as intermediate refueling and supply bases.
>
> The fuel required — 5,356,600 tons — is a lot of fuel, but he pointed out that one-tenth as much was burned up during the Berlin airlift "just because of a little misunderstanding among diplomats." He hoped that when mankind enters the cosmic age, "wars will be a thing of the past ... and people will be ready to foot the fuel bill for a voyage to our neighbors in space."
> — TIME magazine, September 17, 1951
> "Astronaut von Braun, to Mars and back in 969 days"

In October, the Redstone Arsenal Institute of Graduate Studies is established in Huntsville on a contractual basis with The University of Alabama. The goal is to promote the effectiveness of rocket and guided missile programs by improving technical training through refresher courses, seminars and graduate studies.

NASA - Marshall Flight Space Center

1952

V-1 on display at Redstone Arsenal

von Braun's space staton concept

Workers inspect Redstone

von Braun poses with Mars lander model

Nike-Ajax missile

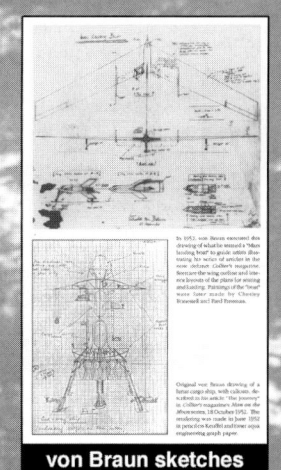

von Braun sketches

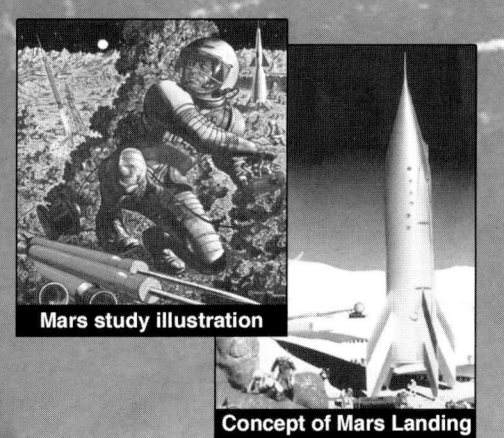

Mars study illustration

Concept of Mars Landing

Von Braun's concepts for space flight, including a space station, receive their first national coverage in a series of articles in Collier's magazine.

Here is how we shall go to the moon. The pioneer expedition, fifty scientists and technicians, will take off from the space station's orbit in three clumsy-looking but highly efficient rocket ships ...

Thirty-three minutes from take-off we have it. Now we cut off our motors; momentum and the moon's gravity will do the rest.

The moon itself is visible to us as we coast through space, but it's so far off to one side that it's hard to believe we won't miss it ...

The Earth is visible, too — an enormous ball, most of it bulking pale black against the deeper black of space, but with a wide crescent of daylight where the sun strikes it.

—Dr. Wernher von Braun,
"Man on the Moon: The Journey"
Collier's magazine, October 18, 1952

Some scientists, including some members of the Huntsville team, question the propriety of promoting space exploration through the popular press.

"We can publish scientific treatises until Hell freezes over. But unless we make people understand what space travel is, and unless the man who pays the bills is behind us, nothing is going to happen."
— Dr. Wernher von Braun comment to Dr. A.K. Thiel
quoted by Mike Wright in Alabama Heritage magazine, Summer 1998

... Dr. Wernher von Braun, is still only 40 and is the major prophet and hero (or wild propagandist, some scientists suspect) of space travel. As a boy, Dr. von Braun wanted to go to the moon. He still does.

The practical rocket men fear that their gradual march toward space may disappoint the public ... But Dr. von Braun ... who would hurry the cautious missile men along, says that manned space flight "is as sure as the rising of the sun."

— TIME magazine, December 8, 1952

The Provisional Redstone Arsenal Ordnance School is created to teach missile technology to soldiers. This school is the forerunner of what is now the U.S. Army Ordnance, Munitions and Missile Center and School at Redstone Arsenal.

At The University of Alabama's Huntsville center, Dr. Wernher von Braun presents a current affairs seminar, "A Proposed Trip to the Moon."

1953

von Braun shows off the V-2 model

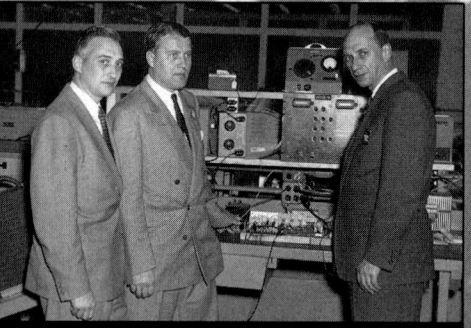

von Braun visits with international contractors

A rocket is prepared for transport

First Redstone launch

In August, the first successful launch from Cape Canaveral, Florida, of a Redstone missile incorporates the first inertial guidance system developed in the U.S.

The U.S. Patent Office issues the first two patents to Redstone Arsenal employees: Fritz Kraemer for "Body for Rockets & Rocketlike Vehicles" and Dr. Johann Tschinkel for "Apparatus for Heating Hydrogen Peroxide."

The first Hermes guided missile is put on display in Huntsville as part of Armed Forces Day.

Philip Mason is named director of The University of Alabama's Huntsville Extension Center.

Hermes test missile

Redstone rocket sits in hangar

NASA - Marshall Flight Space Center

1954

von Braun explains details of Mars craft

Plans made for Disney program

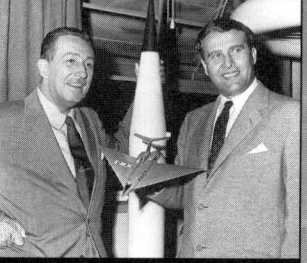

von Braun demonstrates "bottle suit"

Walt Disney, Wernher von Braun

In September, through the Department of the Army, von Braun submits to the U.S. Committee for the International Geophysical Year the report, "A Minimum Satellite Vehicle Based on Components Available from Missile Developments of the Army Ordnance Corps." It proposes using the Redstone missile as the main booster of a four-stage rocket. This concept later becomes the joint Army-Navy Project Orbiter, and still later the Explorer 1 satellite. The Redstone rocket project survives despite President Dwight Eisenhower giving responsibility for launching the first U.S. satellite solely to the Navy.

> Will man ever go to Mars? I am sure he will, but it will be a century or more before he's ready. In that time, scientists and engineers will learn more about the physical and mental rigors of interplanetary flight and about the unknown dangers of life on another planet. Some of that information may become available within the next 25 years or so, through the erection of a space station above the Earth ... and through the subsequent exploration of the moon . ...
>
> — Dr. Wernher von Braun
> "Can We Get to Mars?"
> Collier's magazine, April 30, 1954

First contacts were made with Dr. James van Allen, with the suggestion that he should prepare a scientific experiment for the Explorer 1 satellite.

On November 11, the first 41 members of the team were awarded American citizenship.

Dr. Wernher von Braun becomes a national television spokesman for space exploration, working with Walt Disney to develop three programs which air in 1955 and 1957. Disney's objective, he said, was to combine "the tools of our trade with the knowledge of scientists to give a factual picture of the latest plans for man's newest adventure." It also gives Disney much needed material for the Tomorrowland section of Disneyland.

"Mars Expedition" at Bohran's Observatory

Rocket team members gain U.S. citizenship

Corporal rocket

Gen. Toftoy and Wernher von Braun

105

1955

Rocket team members become American citizens

von Braun consults with technician on a Redstone rocket

Redstone rocket launch

Missile test stand at Redstone Arsenal

On July 29, President Dwight Eisenhower announces that the U.S. will launch a man-made satellite as a contribution to the International Geophysical Year. All three military services submit alternative plans for accomplishing the satellite launches.

U.S. TO LAUNCH EARTH SATELLITE

This country plans to launch history's first man-made, earth-circling satellite into space during 1957 or 1958.

Tentative plans envision an unmanned globular object about the size of a basketball. The satellite will flash around the earth about once every ninety minutes at a speed of 18,000 miles an hour in a fixed path 200 to 300 miles above the ground.

— The New York Times, July 30, 1955

The first satellite may have a mouse aboard, but scientists said they could not foresee the time when human beings would be able to go into outer space as passengers ...

Defense Department research scientists pleaded with reporters to repress any tendency to exploit speculations that have been popularized in recent years by fiction writers. They said the possibility of human passengers in a man-made satellite and its use for military purposes were so remotely in the future that speculations about it were practically useless.

— The New York Times, July 30, 1955

The first two of three television programs from the collaboration of Walt Disney and Wernher von Braun are broadcast in March and December. An estimated 42 million people see the first program, "Man in Space."

On April 14, a group of 103 former members of the German rocket team and their families receive U.S. citizenship. Among those taking the oath is Dr. Wernher von Braun.

Designed by von Braun rocket team member Hannes Luehrsen, Huntsville's Memorial Parkway opens to traffic. At this time, Memorial Parkway is a bypass around the congested courthouse square in downtown Huntsville.

Huntsville celebrates the sesquicentennial (150th anniversary) of its founding by John Hunt.

The faculty at The University of Alabama's Huntsville center grows to 21 part-time instructors.

NASA - Marshall Flight Space Center

1956

Gen. John Medaris

Gens. Medaris, Toftoy

On Sept. 20, a Jupiter C vehicle — a Redstone missile first stage with two solid rocket cluster upper stages — achieves a deep penetration of space, reaching an altitude of more than 682 miles and a range of 3,355 miles. With one more solid propellant rocket on top, this rocket could have launched an American satellite in 1956.

The Huntsville team is so close to launching a satellite (an honor reserved for the Navy's Vanguard program) that the Pentagon orders the Army to launch only dummy payloads. At each launch, a Department of Defense official is on hand to make sure the Army missile's fourth stage rockets are not fired — on purpose or "accidentally."

The Ordnance Missile Command becomes the U.S. Army Ballistic Missile Agency (ABMA), with Major Gen. John Medaris as commander and Dr. Wernher von Braun as director of the Development Operations Division. ABMA's mission is to develop, build and field an intermediate range (up to 1,500 miles) ballistic missile, the Jupiter. The Chrysler Corporation was contracted to produce the Jupiter missiles.

The first Redstone missile battalion is activated at Redstone Arsenal and attached to the U.S. Army Ballistic Missile Command.

Rocket City, U.S.A.
Going Behind the Riddle of Redstone Arsenal

The yellow, one-story building sprawls gracelessly in the Alabama sun. Yet an inconspicuous sign identifies this one-time warehouse as one of the most important buildings in the world. "Guided Missile Development Division," the sign says. This is where, next week, the new Army Ballistic Missile Agency will begin a "crash" program to produce a guided missile with a 1,500-mile range.

Chief of the Guided Missile Development Division at Redstone is a 43-year-old genius named Dr. Wernher von Braun, who directed development of the German V-2 and led the ablest of his staff to the U.S. after the war ... U.S. authorities consider him so valuable that when he returned to Germany to marry his childhood sweetheart, they assigned an Army company to protect him against possible kidnapping by the Reds ... When the Germans arrived, Huntsville was a friendly, busy little city ... At first, Huntsville was wary of the German newcomers.

But, as the newcomers bought land and built houses, proved friendly neighbors and good credit risks, the tensions relaxed. Huntsville citizens began to take pride in the fact that if a power motor wouldn't start, you could call over a world-famous rocket expert from next door to see what was wrong ... This year the new citizens will vote for the first time in an American national election.

"They are respectful of politicians," says Mayor (R.B. "Speck") Searcy, with a certain amount of awe in his voice. "They must have been brought up to respect the "burgermeister."
— Newsweek magazine, January 30, 1956

At The University of Alabama's Huntsville center, Dr. Dennis Nead becomes full-time director of graduate programs in physics, engineering, mathematics andmanagement. Upper level and graduate courses in mathematics, physics and engineering are offered in response to the continuing need for technical personnel at Redstone Arsenal. Funding in the early years is provided by the U.S. Army at Redstone Arsenal and later by NASA.

Canaveral test stand

Redstone rocket launch

Hermes rocket display

Test stand on Redstone Arsenal

ABMA test stand

The observatory dome being lifted into place

Drs. Ernst Stuhlinger and von Braun at Monte Sano observatory

107

50 Years of Rockets and Spacecraft

1957

Life features the von Braun family

Russians launch Sputnik

President Eisenhower poses with Jupiter nose cone

von Braun Astronomical Society

Jupiter launch

Vanguard missile

von Braun astride a Redstone rocket

Rocket test at Redstone Arsenal

The first successful firing of a 1,500-mile-range Jupiter missile from the Atlantic Missile Range is made at Cape Canaveral. The first man-made object recovered from space is a Jupiter C reduced-scale nose cone, proving that Army Ballistic Missile Agency (ABMA) has met the challenge of developing a vehicle that can survive re-entry. President Eisenhower shows off the nose cone during a nationally televised speech from the White House. It later is placed on permanent exhibition in the Smithsonian Institution.

> On the night of Oct. 4, 1957, von Braun was called away from a Redstone dinner honoring Defense Secretary-designate Neil McElroy.
> Voice on the wire: "New York Times calling, doctor."
> Von Braun: "Yes?"
> Timesman: "Well, what do you think of it?"
> Von Braun: "Think of what?"
> Timesman: "The Russian satellite, the one they just orbited."
>
> Von Braun hurried back to the dinner table, broke the news of Sputnik I, turned earnestly to Neil McElroy. "Sir," he said, "when you get back to Washington you'll find that all hell has broken loose. I wish you would keep one thought in mind through all the noise and confusion: we can fire a satellite into orbit 60 days from the moment you give us the green light." Army Secretary Wilber Brucker, who had accompanied McElroy, raised a hand of objection: "Not 60 days." Von Braun was insistent: "Sixty days." General Medaris settled it: "Ninety days." ... two weeks after taking office (McElroy) made his decision.
> — TIME magazine cover story, "Missileman Von Braun," Feb. 17, 1958

With the president's approval the Department of Defense directs the ABMA to use a Jupiter C rocket to boost into orbit a satellite designed by the Jet Propulsion Laboratory and Dr. James van Allen from the University of Iowa.

> The very next week, (von Braun) reserved Cape Canaveral range time for the night of Jan. 29, 1958, between 10:30 p.m. and 2:30 a.m. Jupiter C had been ready for months. Says von Braun: "All she needed was a good dusting ..." But the satellite itself, with its delicate instrumentation, might well have held the whole project up for months or years — had not Wernher von Braun, during most of the period that he was barred from engaging in satellite work, been in what he calls "silent coordination" with Caltech's William Pickering and the University of Iowa's James Van Allen in planning Explorer and its instruments.
> — TIME magazine, Feb. 17, 1958

Huntsville, Redstone Arsenal and the ABMA rocket team begin a remarkable sprint into history.

Three years toward the completion of an undergraduate degree can be completed without leaving The University of Alabama Huntsville Center, which publishes its first student newspaper, "The Twilight Times."

NASA - Marshall Flight Space Center

1958

Charles Lundquist briefs von Braun, Hermann Oberth

Jupiter ballistic missile

Drs. Pickering, van Allen, von Braun celebrate first U.S. satellite launch

After many hard years of work, on January 31, 1958, using the Jupiter C — a modified Redstone rocket— von Braun and his team put the first American satellite, Explorer 1, into orbit. Von Braun says of this accomplishment: "It was one of the great moments of my life. I only regret we didn't do it earlier."

VOYAGE OF THE EXPLORER

A bright, waxing moon rode through the racing cumulus clouds above Florida's Cape Canaveral.

Only the week before, the Navy had babied its slender satellite-laden Vanguard. Day by day the tension tightened as the Vanguard countdown crept steadily toward zero... Each time the launching was scrubbed...

In the Pentagon at that moment, Army Secretary Wilber Brucker and the Jupiter's top scientist, Wernher von Braun, joined a score of other military and civilian officials in the Army's telecommunications room... Elaborately, von Braun lectured the attending brass...

Then: 10:48, T minus zero — Jupiter C was fired... After 14 seconds, flamebelched from the rocket base...

Fire Dance. The rocket moved. The Pentagon teletype sang: IT'S LIFTING . . . IT'S SOARING BEAUTIFULLY . . . Like a flame-footed monster it kicked upward ...

— TIME magazine, February 10, 1958

The first American satellite, carrying Geiger counters for cosmic ray measurements prepared by Dr. van Allen, discovers belts of radiation encircling the Earth.

At Redstone Arsenal, the Army team's success in space exploration builds upon itself. The Army's first lunar trajectory probe, Pioneer III, reaches an altitude of 66,654 miles on its way to the moon. An Army Jupiter C launches Explorer 4 to measure artificial radiation belts generated by the explosions of three nuclear bombs launched aboard Redstone rockets high above the Pacific Ocean.

With funds provided by the Advanced Research Projects Agency, von Braun initiated the rocket development program that led to the Saturn-Apollo moon landing project. The first steps in this program were the development of the huge F-1 rocket engines, five of which were to propel the Saturn V rocket, and the development of engine swiveling and engine clustering technologies. These were first studied with Saturn I and Saturn IB rockets before being used on the Saturn V.

The year also sees the first firing of a Redstone missile by combat troops at Cape Canaveral. A full-scale Jupiter intermediate-range ballistic missile nose cone is recovered. Jupiter missiles are delivered to the U.S. Air Force for overseas deployment.

Congress creates the National Aeronautics and Space Administration (NASA).

The city of Huntsville gives 83 acres of farmland to The University of Alabama's Huntsville center. The land is to be used for developing a university campus.

Gen. Medaris anticipates Explorer 1 launch

Jupiter C heads to Cape Canaveral

Crane lowers Explorer 1 into Jupiter C rocket

Moon Day celebration

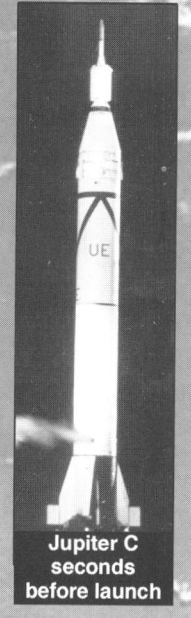

Jupiter C seconds before launch

Huntsville leaders attend first satellite launch

1959

Jupiter rocket greets Redstone Arsenal visitors

Von Braun talks to Mercury astronauts

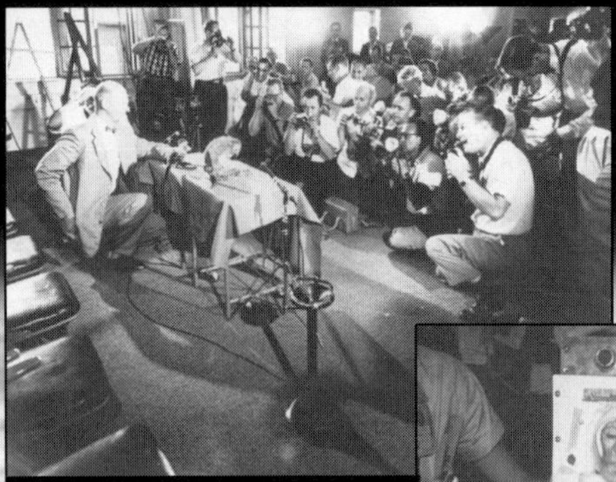

'Miss Baker' and 'Able' made their debut for the press and in space

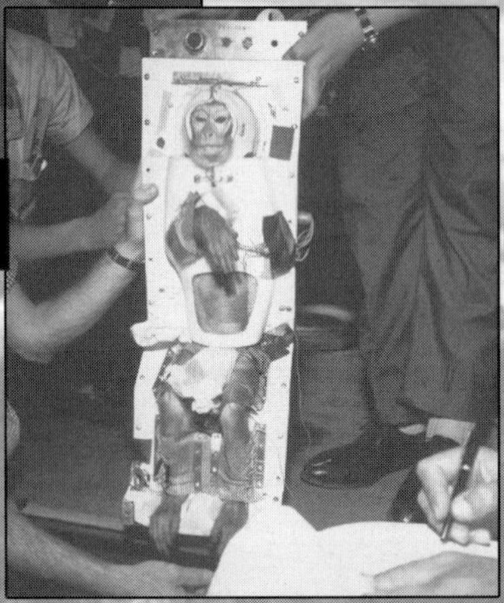

President Dwight Eisenhower presents the Distinguished Federal Civilian Service Award to Dr. Wernher von Braun.

"Practicality, what wonders have you denied man?" von Braun asks at a ceremony honoring Dr. Robert Goddard. "We must open our vision to the unknown. We must expect the unpredictable. We must value knowledge for its own worth, and we must cease to measure the new in terms of its usefulness along."

The Army's success in space exploration continues with the first successful, suborbital space launch and recovery of two monkeys, Able and Miss Baker, on May 28. They are the first space travelers to return safely to the Earth.

A Juno II, the second Army vehicle to launch a NASA probe for lunar exploration, lofts Pioneer IV onto a trajectory past the moon and into orbit around the sun. Explorer 6 returns the first television views of Earth. Explorer 7, also launched by a Juno II rocket, measures the Earth's magnetic field and observed solar flares.

REDSTONE'S GENIUSES PREPART TO GO CIVILIAN

One of the richest prizes in all the feuding over the U.S. space and missile program has long been the Development Operations Division of the Army Ballistic Missile Agency at Huntsville, Ala. Stationed since 1950 at the Redstone Arsenal, the division has never been noted for possessing a gaudy array of missile development equipment. For the brilliance of its staff, however, it has gained worldwide recognition …

Now a change is to come into Redstone's life. The Army's role in the space program has been increasingly controversial, and last October President Eisenhower announced plans to transfer the Development Operations Division to civilian hands — the National Aeronautics & Space Administration.

"Up to now," says von Braun, "the group's bread-and-butter jobs have been military."

His men have sneaked in work on a few non-military projects, such as theJupiter C and the Explorers, by various bootlegging techniques. "Now," von Braun adds, "we have the chance to go honestly after the main objective. The team has been waiting for this moment for many years."

— Business Week magazine, Nov. 28, 1959

Von Braun and astronaut John Glenn discuss space vehicle model

Redstone rocket displayed at Smithsonian

Technicians man Pioneer 1V launch console

Von Braun receives the Distinguished Federal Civilian Service Award

Redstone rocket launch

NASA - Marshall Flight Space Center

1960

Saturn 1B undergoes static fire testing

ABMA employees are transferred to NASA

von Braun explains rocketry to Eisenhower

Test of Mercury/Redstone escape system

During transfer services on the first of July, 4,670 civil service employees rom the Army Ballistic Missile Agency become the nucleus of NASA's new Mashall Space Flight Center. On September 8, President Dwight Eisenhower dedicates the center, which is named in honor of Gen. George C. Marshall, author of the Marshall Plan for the reconstruction of Europe after World War II.

"I find that the leaders of the new space science feel as if Venus and Mars are more accessible to them than a regimental headquarters was to me as a platoon commander 40 years ago.
To move conceptually, in one generation, from the hundreds of yards that once bounded my tactical world to the unending millions of miles that beckon these men onward, is a startling transformation."
— President Dwight Eisenhower

The first full, eight-engine static firing of a Saturn I booster is conducted in Huntsville.

Pioneer 5, launched atop a Juno II, measures radiation and magnetic fields between Earth and Venus.

The space race between the U.S. and the Soviet Union accelerates.

**TWO SOVIET DOGS IN ORBIT
RETURNED TO EARTH ALIVE AS SATELLITE IS RETRIEVED**
MOSCOW - Living creatures have returned safely to earth from an orbit in space for the first time in history, the Soviet Union announced today.

It said its "second cosmic space ship" landed on target today after circling the earth for twenty four hours with its cargo of two dogs, some rats and mice, flies, plants, seeds and fungi.
— The New York Times, August 21, 1960

Huntsville's Lyric Theatre hosts the premier of the movie, "I Aim at the Stars," featuring the life story of Dr. Wernher von Braun, starring Curt Jurgens in the title role.

Huntsville, Madison County and The University of Alabama's Huntsville Center each contribute $250,000 toward the construction of a central building on the new campus. Morton Hall is named in honor of Dr. John Morton, the center's first acting director.

President Eisenhower visits Huntsville

Launch control at Cape Canaveral

Army exhibits missile systems

Air Force training on Jupiter rocket

Saturn 1B, Juno 11 sit in test stand

111

1961

Saturn I is hoisted onto test stand

Huntsvillians watch static test

Saturn I sits in test facility

Armed Forces Day - Huntsville

Saturn booster rockets are fired

First launch of Saturn I

von Braun prepares to address the Alabama legislature

Alan Shepard's flight sets off a celebration in downtown Huntsville

The first flight test of a Saturn I booster is successful.

The Soviet Union launches Major Yuri Gagarian into Earth orbit. The flight of the first man in space is a reminder that the U.S. space program is still behind.

> MOSCOW - The Soviet Union announced today it had won the race to put a man into space ...
> — The New York Times, April 12, 1961

On May 5, Alan B. Shepard, Jr., becomes the first U.S. astronaut in space, launched atop a Mercury Redstone vehicle on a suborbital flight. In the months leading up to the launch, Marshall Space Flight Center scientists, engineers and technicians obsess about making sure everything works as planned. "We could not get it out of our heads that there was a man in that spacecraft," says Dr. Ernst Stuhlinger.

> CAPE CANAVERAL, Fla. - A slim, cool Navy test pilot was rocketed 115 miles into space today. Thirty-seven-year-old Cmdr. Alan B. Shepard Jr. thus became the first American space explorer. Commander Shepard landed safely 302 miles out at sea fifteen minutes after the launching. He was quickly lifted aboard a Marine Corps helicopter. "Boy, what a ride!" he said as he was flown to the aircraft carrier Lake Champlain four miles away.
> — The New York Times, May 6, 1961

> "Our opponents across the ocean, behind the Iron Curtain, thought about a month ago they had slammed the door to the universe in our face, but Shepard has let us out of our dilemma and embarrassment.
> "We will go farther and farther, eventually landing on the moon."
> — Wernher von Braun
> public celebration of Freedom 7 flight
> Courthouse Square, Huntsville

On May 25, with a single manned spaceflight under NASA's belt, President John F. Kennedy raised the stakes: "I believe that this nation should commit itself to achieving this goal, before this decade is out, of landing a man on the moon and returning him safely to the Earth."

Alabama deals with desegregation. A leader on so many fronts, von Braun joins others urging a progressive, anti-racist stance, a position taken to heart by many.

> "We aren't Yankees yet, mind you, but if the governor tried to shut our schools to keep the colored out, he'd be badly mistaken. Huntsville just won 't stand for it."
> — Fifth generation Huntsvillian,
> quoted in U.S. News & World Report, 1961

Von Braun addresses the Alabama Legislature, requesting funds to build and equip a research institute on the university campus, saying:

> "Opportunity goes where the best people go, and the best people go where good education goes. To make Huntsville more attractive to technical and scientific people across the country — and to further develop the people we have now — the academic and research environment of Huntsville and Alabama must be improved and improved immediately.

> "It's the university climate that brings the business. Let's be honest with ourselves. It's not water, or real estate, or labor or cheap taxes that brings industry to a state or city. It's brain power. Nowadays, brainpower dumped in a desert will make it rich."

The legislature approves $3 million in revenue bonds. Huntsville and Madison County buy an additional 200 acres for the campus and the Research Institute is built.

NASA - Marshall Flight Space Center

1962

Saturn booster prepares for trip to Cape

President Kennedy, VP Lyndon Johnson visit Marshall

**GLENN ORBITS EARTH 3 TIMES SAFELY;
PICKED UP IN CAPSULE BY DESTROYER;
PRESIDENT WILL GREET HIM IN FLORIDA**

CAPE CANAVERAL, Fla. - John H. Glenn Jr. orbited three times around the earth today and landed safely to become the first American to make such a flight.

The 40-year-old Marine Corps lieutenant colonel traveled about 81,000 miles in 4 hours 56 minutes before splashing into the Atlantic at 2:43 P.M.
— The New York Times, Feb. 21, 1962

On March 23, von Braun's 50th birthday is commemorated by a book, "From Peenemunde to Outer Space," containing technical and scientific papers by his associates. Among the authors is Dr. Rudolf Hermann, first director of the Research Institute at The University of Alabama's Huntsville center. One of the editors is Dr. Ernst Stuhlinger, who later becomes the Marshall Space Flight Center's associate director for science, a UAH researcher and one of von Braun's biographers.

Following is from the transcript of a briefing for President Kennedy given by Dr. von Braun at Huntsville, Ala., on September 11, 1962:

The President: What is the time schedule for each of these?
Dr. von Braun: This [C-1] vehicle has performed two successful flight tests with the first stage alone. The third one, as I said, is under way. The fourth one will be static-fired today ...

The advanced Saturn will be ready to fly the lunar mission in approximately 1967, if everything works very well. But we like to keep a little time-padding for ourselves, too, so I think this statement should be taken with a grain of salt.

But we will definitely do it in this decade.
— U.S. News & World Report magazine, October 1, 1962

At Cape Canaveral, NASA's Launch Operations Center is created from a former element of Marshall Space Flight Center. The director is Dr. Kurt H. Debus, a long-time von Braun associate and former Huntsville resident.

Across from the The University of Alabama's Huntsville center, Cummings Research Park is being developed. Von Braun insists the group developing the park form a non-profit foundation, which later becomes The University of Alabama Huntsville Foundation. National companies come to Huntsville to support the space program.

Federal grants totaling $114,687 are awarded to the Research Institute to support graduate studies.

von Braun watches Saturn launch from Cape

John Glenn steps into Mercury capsule

von Braun helped plan Cape Canaveral facilities

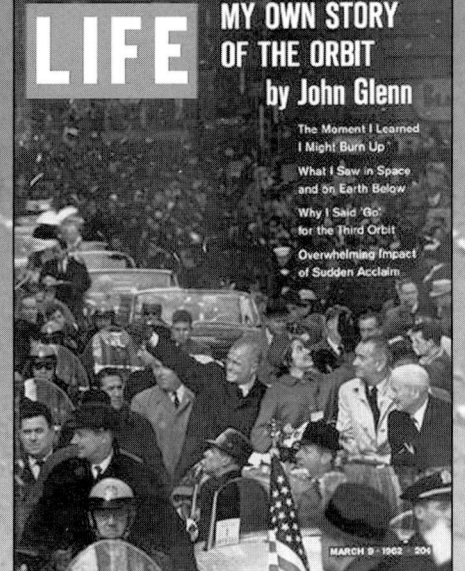

LIFE
MY OWN STORY OF THE ORBIT
by John Glenn

The Moment I Learned I Might Burn Up
What I Saw in Space and on Earth Below
Why I Said 'Go' for the Third Orbit
Overwhelming Impact of Sudden Acclaim

Saturn shown in scale to Statue of Liberty

Saturn components were regular cargo on "Super Guppy" aircraft

50 year old Wernher von Braun displays missile development legacy

113

50 Years of Rockets and Spacecraft

1963

von Braun makes supersonic flight at Edwards AFB Marshall Space

President John Kennedy visits Marshall Space Flight Center

The Mercury program ends with the successful flight of Gordon Cooper. Preparations for the Gemini program continue.

On May 19, President John F. Kennedy visits Redstone Arsenal and the Marshall Space Flight Center for a tour and briefings on Project Apollo. He repeats his pledge that the U.S. will explore space:

"I know there are lots of people now who say, 'Why go any further in space?' When Columbus was halfway through his voyage, the same people said, Why go on any further? What can he possibly find? What good will it be?'

"I believe the United States of America is committed in this decade to be first in space, and the only way we are going to be first in space is to work as hard as we can here and all across the country ..."

Building 4200, Marshall Space Flight Center's headquarters, is completed.

In January, von Braun becomes a monthly contributor to Popular Science magazine. On the splashy front cover, beneath a headline that asks, "10,000 miles Between Grease Jobs? How Safe Is Extended Lubrication?" the magazine trumpets: "Starting This Month: Dr. Wernher von Braun Answers Your Questions about Space"

"Space science isn't like geography, or astronomy, or physics, or chemistry, or medicine. It is a little bit of all of them and more. That is what makes it so fascinating.

"But it is this kaleidoscopic aspect of space science that makes it almost impossible to 'organize' a monthly column such as this. Mr. Crossley and I have therefore agreed not even to try to arrange the questions and answers in any systemic way. If the result is a bit disjointed, it should at least be colorful."

– Wernher von Braun
"Why I am writing for Popular Science"
Popular Science magazine, January 1963

H. Clyde Reeves, vice president for Huntsville affairs, and Dr. Charley Scott, the first director of instruction, oversee rapid growth of the University of Alabama's Huntsville campus.

von Braun serves as spokesman for space travel

von Braun prior to flight in a T-38 aircraft at Marshall Space Flight

Saturn hardware awaits next step of assembly

Kurt Debus, von Braun anticipates launch at Cape Canaveral

Construction of West Test Area at Marshall Space Flight Center

NASA - Marshall Flight Space Center

1964

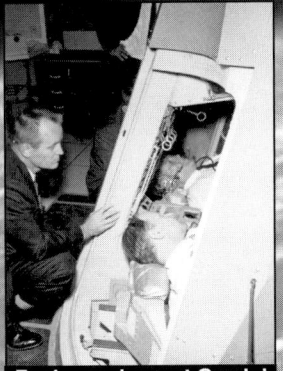

Engineers inspect Gemini Capsule

von Braun studies data at Cape Canaveral

Dynamic test facility at Marshall Space Flight Center

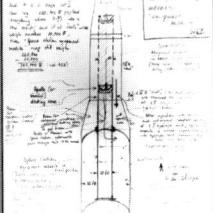

von Braun's early concept for Skylab Space Station

Saturn V's first stage is lowered into test facility from Cape Canaveral

The Dynamic Test Facility is built for testing Saturn rockets. It was the tallest structure in Alabama for years.

Q. Dr. von Braun, is it still important to go to the moon?
A. It is just as important as it ever was. The real purpose of going to the moon is, of course, not to go there to put up a sign, "Kilroy was here," or maybe bring back a handful of lunar sand. The purpose is to develop a true, national space-flying capability.
Q. Why is that so important?
A. Because it will vastly extend our knowledge about the universe, and because it will enable us to develop new technologies from which we are bound to reap many benefits in our everyday lives.
Q. But the moon is ... hardly important to everyday life.
A. Look at the moon as more of a rallying point than an objective in itself. The moon plays the same role in our manned space-flight program that the city of Paris played in Lindbergh's memorable flight. Paris served as a goal, but surely Lindbergh had a more important objective in mind than to go to Paris.

Our objective is to develop a broad, manned space-flying capability. You simply cannot develop such a capability ... unless you have a clear goal — a focusing point like Lindbergh's Paris — something that is universally understood ...

Now, when the president of the United States says, "Let's land a man on the moon in this decade and bring him back alive," then you have such a clear goal. Everybody knows what the moon is, everybody knows what this decade is, and everybody can tell a live astronaut who returned from the moon from one who didn't.
— U.S. News & World Report, June 1, 1964

With an enrollment of 208 students, the first full-time freshman class begins a regular full-day teaching program at The University of Alabama's Huntsville center. The Huntsville center awards its first master's degree to Julian Palmore. Total enrollment at the center tops 2,000 part-time students.
With an enrollment of about 900 students, the center's graduate program is reported to be the largest in the South.

Saturn 1 is launched from Cape Canaveral

Saturn is loaded onto barge on the Tennessee River

Marshall officials watch launch from Cape

Missile and Space Museum at Redstone Arsenal

115

50 Years of Rockets and Spacecraft

1965

Strategic Planning Group: Dr. Ernst Stuhlinger, William Mrazek, Dr. Wernher von Braun, Dr. William Lucas, Ray Kline and Jim Daniels

Space walks added drama to manned space flight

F1 engine-Workhorse of the Saturn V launch vehicle

A Saturn I rocket is launched in February carrying Pegasus I, a five-ton micrometeroid sensing satellite. Pegasus 2 is launched in May and Pegasus 3 in July. All three flights were successful, providing information on the size and frequency distribution of micrometeroids in space.

The Apollo Applications Program is initiated to extend the use of Apollo and Saturn hardware.

Plans for Skylab, the first large U.S. space station, are laid out by the Army Ballistic Missile Agency and NASA.

The Strange World of Zero Gravity
by Dr. Wernher von Braun

When a human astronaut becomes a human satellite, floating freely in the vastness of space, he offers a dramatic example of the weird things that can happen in the realm of zero gravity — a strange world just beginning to be explored ...

Living and housekeeping under zero gravity ... pose curious problems: Before going to sleep, astronauts will need to strap themselves down, or they would float about the cabin, propelled by the thrust of their own breath.

Even the most luxurious space cuisine will avoid needless food wastes: All meat will be boneless and free of undesirable fat; potatoes will be peeled in advance; cherries will be stoneless. If a Martini is ever served aboard a spacecraft, it may come without an olive.

— Popular Science magazine, June 1965

First stage of Saturn 1B

Saturn engines are test fired

On August 5, von Braun interrupts a meeting so he can watch from the roof of Building 4200 as the Saturn V booster's five engines undergo their first full-duration firing. Windows rattle across North Alabama and some plaster cracks. NASA employees a mile or more from the test stand are buffeted by the blast wave.

During earlier, shorter tests in April and May, some frightened area residents wonder if a nuclear bomb has been dropped. It is the middle of the Cold War. Redstone Arsenal is routinely on the Soviet Union's top ten "hit list."

"... a continuous plume of flame blasts from the base of a mammoth concrete structure," Marshall Space Flight Center Historian Mike Wright says in an article for Alabama Heritage magazine. "Thunder rolls. Smoke billows. For two and a half minutes hell unfolds. Alabama has become, as writer Bob Lionel later wrote, 'the land of the Earth-shakers.' "

In November, the state legislature approves $1.9 million in bonds to finance construction of the Alabama Space & Rocket Center.

"If our education system can instill in our youth an early desire for perfection, it has taken a major step in training to meet the space age challenge."

— Wernher von Braun, April 1965
comments to Alpha Phi Alpha fraternity

The first named scholarship exclusively for the University of Alabama's Huntsville campus is designated by the will of Octavia May Palmer to honor the memory of her father, Samuel Palmer. The gift includes land in Madison County which has been in the Palmer family for more than 100 years.

Saturn 1B looms tall on test stand

Launch personnel observe Saturn 1 launch

NASA - Marshall Flight Space Center

1966

Saturn 1B rocket sits at launch mode

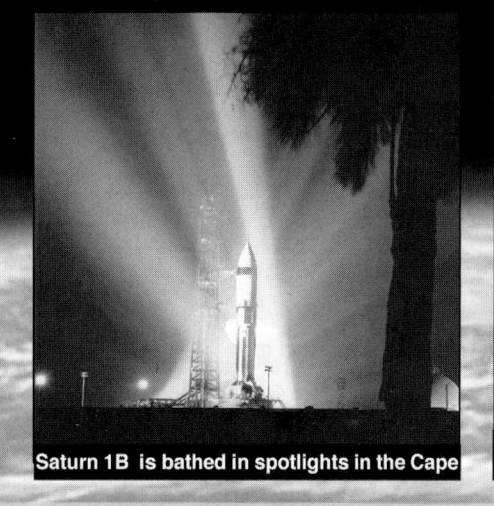

Saturn 1B is bathed in spotlights in the Cape

Saturn V rests in Vertical Assembly Building

Rocket motors undergo tests at Redstone

Saturn ascends into space following launch

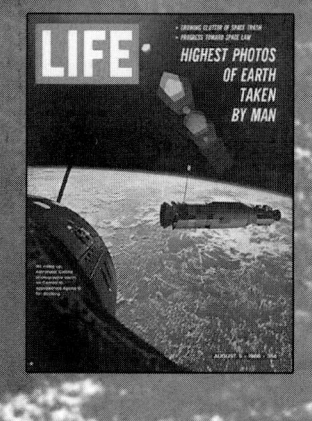

The first flight test of the Saturn I-B with a Saturn IV-B upper stage (liquid oxygen and hydrogen propellants) and an Apollo capsule recovery.

To promote space exploration, Dr. Wernher von Braun continues to work with the press. This results in a variety of coverage, including this question-and-answer session with U.S. News & World Report magazine:

Q. Beyond (the objective of reaching the moon), what is our purpose?
A. I think that even when President Kennedy made that announcement, he made it perfectly clear that, while this is an objective that everybody understood and one that we could use to construct a program for fulfillment, what he really had in mind was to develop a broad, national space-flying capability. He put it very well when he said, "We have to learn to sail on the new ocean of space."

The problem is, when you have an objective as vague as learning to sail on a new ocean and you try to present this, say, to the Congress, then it isn't surprising that everybody interprets this on his own terms.

What does that mean — to develop a capability to sail on an ocean? Theres no objective; there's no goal in space, and no goal in time attached to it.
Q. Getting back to the moon program: Do you think it is so important forus to go there, at such great cost?
A. My firm conviction is that our space-flight program ... (has) helped tremendously to invigorate our sciences, our educational system and our industry. And I think the question is not so much: "Can we afford the space program?" It is: "Can industrial countries which do not have a space program continue not to have one?" — because they are falling hopelessly behind in technologies that spin out of the space program.
Q. Weren't there some disappointments in the Gemini flights?
A. If, in a pioneering program of this magnitude, you run into a surprise, you should not call it a disappointment. You see, the purpose of Gemini was to learn the kinds of things that you can't learn on the ground. And the process of learning inevitably produces surprise discoveries. And we did make some surprise discoveries, for which we are more than grateful, because now we know more than we did when we went into it.
— U.S. News & World Report, December 12, 1966

Madison Hall is built with $900,000 raised by the Huntsville community and a $420,900 federal grant. It is called the Graduate Studies Building, but is later renamed to honor the citizens of Madison County.

Concepts for lunar rover

Officials raise money to build Madison Hall on UAH Campus

1967

Apollo astronauts Gus Grissom, Ed White, and Roger Chaffee

Saturn V in Vertical Assembly Building at Cape Canaveral

Astronauts Gus Grissom, Ed White and Roger Chaffee die when fire guts their Apollo spacecraft at Kennedy Space Center in Florida in January. Marshall Space Flight Center Deputy Director Eberhard Rees is appointed to chair the group that initiates corrective actions.

The first launch of a Saturn V rocket is on November 9, when a Saturn V rocket carries the unmanned Apollo 4 spacecraft into Earth orbit. "No single event since the formation of the Marshall Center in 1960 equals today's launch in significance," says von Braun. "For MSFC employees — more than 7,000 strong — this is their finest hour."

NASA faces budget cuts. Years later, von Braun says that while it knew budget cuts would slow work in the Apollo program "... Congress was even more aware that there were simply not enough funds to satisfy all the requirements of Vietnam and the many urgent demands for domestic programs."

The University of Alabama's Huntsville center becomes a branch of The University of Alabama System and is officially named The University of Alabama in Huntsville (UAH). UAH's Research Institute receives funding from NASA's Marshall Space Flight Center to prepare the official history of the Saturn Program, "Stages to Saturn."

Huntsville's population in 1967 tops 125,000, a more than sevenfold increase since 1950.

"I used to be able to say the name of everyone I met on the street — and if they had a dog with them, I knew the dog's name, too. Now it's different."
— Huntsville Major R.B. "Speck" Searcy, quoted in Fortune magazine

von Braun conducts research

von Braun monitored details of every element of the space program

Saturn V seconds after lift off

NASA - Marshall Flight Space Center

1968

Satellite communication was critical for space

Earthrise - Apolo 8 astronauts show a different perspective of Earth

Dr. Kurt Debus (center, foreground), and staff at Launch Complex 37

Astronauts run initial test of lunar module

Command module undergoes "dunk" test

von Braun experiences weightlessness in a KC-135 and in a neutral buoyancy tank

After successful unmanned flights of Apollo 5 and 6, Apollo 7 is launched by a Saturn I-B rocket, with astronauts Wally Schirra, Walt Cunningham and Donn Eisele. Their Apollo capsule lands and is recovered safely after 163 Earth orbits.

A three-stage Saturn V booster launches the Apollo 8 mission, which circles the moon. On Christmas Eve, the three astronauts send a television broadcast from lunar orbit back to the Earth.

> As the telecast neared its end, Colonel (Frank) Borman said, "Apollo 8 has a message for you." With that, Major (William) Anders began reading the opening verses from the Book of Genesis about the creation of the earth.
> "In the beginning," Major Anders read, "God created the heaven and the earth. And the earth was without form and voice; and darkness was upon the face of the deep ..."
> Captain (James) Lovell then took up with the verse beginning, "And God called the light day, and the darkness He called night."
> Colonel Borman closed the reading with the verse that read:
> "And God called the dry land Earth; and the gathering together of the water He called the seas; and God saw that it was good."
> After that Colonel Borman signed off, saying: "Good-by, good night, Merry Christmas. God bless all of you, all of you on the good Earth."
> — The New York Times, December 25, 1968

> If all goes well, our goal of landing men on the moon and returning them safely by the end of this decade as planned in 1961 will be fulfilled — that is, if we are not held up by having to pass through Russian customs ...
>
> Speculation on the potential benefits from space made only a few years ago were vague and indefinite. Today they are more visible, credible and attractive to potential users. And yet, while we are largely within the exploration phase of space exploration and still looking forward to the exploitation phase, it is impossible to make predictions with certainty.
>
> History teaches us that the greatest payoffs — the ones which will be the most revolutionary, the most significant and have the widest application — may well come from some unforeseen aspect of the program and from discoveries yet to come.
>
> — Dr. Wernher von Braun in a speech to the National Space Club, quoted in U.S. News & World Report magazine, March 18, 1968

> Q. Dr. von Braun, is the U.S. going to beat Russia to the moon?
> A. I am beginning to doubt that we will. It is very important that we are there first, but in view of the spectacular performance of the Soviet spacecraft Zond 5, in late September, I am beginning to wonder. It will undoubtedly be a photo finish.--
> Q. Does the U.S. have no program beyond a moon landing?
> A. It may surprise you to hear this, but for the last two years my main effort at the Marshall center has been following orders to scrub the industrial structure that we had built up a great expense to the taxpayer, to tear it down again. The sole purpose seems to be to make sure that after 1972 nothing of our capability is left. That's my main job at the moment. And we haven't even put a man on the moon yet.
> — U.S. News & World Report magazine, October 14, 1968

1969

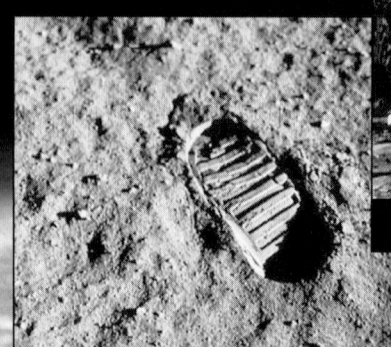

"That's one small step for man. One giant leap for mankind"

MEN WALK ON MOON

HOUSTON - Men have landed and walked on the moon. Two Americans, astronauts of Apollo 11, steered their fragile four-legged lunar module safely and smoothly to the historic landing yesterday at 4:17:40 PM.

Neal A. Armstrong, the 38-year-old civilian commander, radioed to earth and the mission control room here:

"Houston, Tranquility Base here. The Eagle has landed."
— The New York Times, July 21, 1969

"To be borne by a rocket to the moon is one thing, but to be borne up these steps by admirers is almost as impressive."
— Dr. Wernher von Braun
Apollo 11 celebration
Courthouse Square, Huntsville

The Marshall Space Flight Center begins developing the Lunar Roving Vehicle, which is used by astronauts on the last two Apollo missions, in 1971 and 1972, to explore the moon's surface.

The Harvest of Operation Paperclip

Recently, Walter Dornberger, who headed German rocket development in World War II, visited the launching complex at Cape Kennedy. "There was an old friend standing as a supervisor at the rocket take-off site," recalls Dornberger, "and I asked him to show me what was really new about the missile. 'Doc,' he replied, 'it is larger, heavier, more reliable, but in the main part it is the same old cucumber."
— Newsweek magazine, July 7, 1969

If teamwork and a sense of shared responsibility were crucial factors in the U.S. effort to land men on the moon, so were the contributions made by a number of individuals ...

Dr. Wernher von Braun, 57, director of the Marshall Space Flight Center in Huntsville, Ala. ... helped develop the ablative heat shield, which dissipates the searing heat of re-entry by flaking off in harmless fiery pieces. His Huntsville group can also claim credit for what has become known in the space agency as "cluster's last stand" — the grouping of several smaller rockets in a cluster to provide as much thrust as would a single, far larger rocket engine.

... Von Braun, perhaps more than any other man, has been the driving force behind the moon program.
— TIME magazine, July 18, 1969

UAH's Science and Engineering Building is built. It will later be named in memory of Dr. Harold Wilson, dean of the College of Science.

Construction begins on the first of four planned phases of a permanent UAH library, which is later named in honor of M. Louis Salmon, one of the founders of the University of Alabama HUntsville FOundation and its chairman from 1986 to 1993.

UAH's Student Union Building is built. It includes a bookstore, cafeteria, lecture rooms, auditorium and student organization offices.

The journey begins....

Officials relax after a successful launch

von Braun, George Mueller

von Braun, Huntsvillians celebrate historic flight

NASA - Marshall Flight Space Center

1970

Sen. John Sparkman addresses farewell crowd

U.S. Sen. John Sparkman, Alabama Gov. Albert Brewer honor Wernher von Braun

von Braun family: Wernher, Maria, Peter

von Braun bids a tearful farewell

von Braun photographed in his office at Marshall Space Flight Center

von Braun passes leadership of Marshall Space Flight Center to Eberhard Rees

History, forever changed-1950 to 1970

Critics of the National Aeronautics and Space Administration have long maintained that the agency's greatest fault has been its failure to plan for the long-range future after landing men on the moon. ... there can be little doubt that NASA has lost the once-captive imagination of the public. Last week, in an obvious attempt to recoup, NASA announced that it has a new chief planner: Dr. Wernher von Braun, the best known rocket expert in the U.S.

"If we tried to do all of the proposals that are floating around in NASA today ... it would cost the agency anywhere from ten to 100 times its present budget." As the Deputy Associate Administrator for Planning (a newly created post) and the No. 4 man in NASA, von Braun's task is formidable.
— Newsweek magazine, February 9, 1970

Dr. Wernher von Braun leaves Huntsville for NASA headquarters in Washington D.C. As deputy associate administrator, he works on space plans for NASA.

"My friends, there was dancing in the streets of Huntsville when our first satellite orbited the Earth. There was dancing again when the first Americans landed on the moon.

"I'd like to ask you, don't hang up your dancing slippers."
– Wernher von Braun
Farewell speech
Courthouse Square, Huntsville

The name of Wernher von Braun is almost synonymous with space exploration ... As director of the George C. Marshall Space Flight Center at Huntsville, Ala., he became the mastermind for the giant Saturn V ...

Today, von Braun has a new job. His task is to define U.S. space goals ...
Q. What types of projects do you think we will be doing by the late 1970s to 1980s?
A. I consider the shuttle the most exciting program that NASA is involved in at the moment. ... the shuttle will reduce transportation costs to orbit. At the moment, it still costs anywhere between $500 to $1,000 to put a pound of payload into a low earth orbit. I think there's reason to believe this cost can be reduced to like $50 to $100 per pound in orbit.
Q. Do you expect ... big, multi-manned space stations in orbit toward the end of the 1970s?
A. Yes, I think so.

Von Braun's Weekly Notes

by Ed Buckbee

History has embraced few uncommonly gifted scientists––think in the caliber of Galileo and Thomas Edison, as well as few uncommonly gifted managers. So reasonably, history has rarely welcomed an individual who is equally gifted in both disciplines. Wernher von Braun, considered by many to be one of the most successful and effective mangers of a U.S. peacetime technology program, was certainly such an individual.

During the height of the Saturn-Apollo program, von Braun implemented a particularly effective management tool: the Weekly Notes. The notes were his direct channel with his laboratory directors and project managers. The subjects covered in the notes can be categorized as programmatic, strategic, institutional, political or sometimes humorous.

Here is how it worked: Each person would turn in a one-page report—they were restricted to one page—on Friday or early Monday. The report, not a bureaucratic form, was to document what occurred in that laboratory or project office during the prior week. Input was collected from each of the subordinate organizations. The document dated October 25, 1961, authored by Jerry C. McCall, von Braun's special assistant, outlines the procedures to be followed and the persons responsible for preparing the notes. As the document was passed through Marshall's chain of command, additions or deletions were made based upon the importance of the issue. Bonnie Holmes, von Braun's secretary, collected the reports and prepared them for her boss to review. Once von Braun had read and made notations on each of the reports (in most cases within 24 hours) they were returned by courier to the directors and managers.

Von Braun's notations would always include his initial, a 'B', the date that he read the report and a check mark at the end of each paragraph to indicate he had read every word. His notations were addressed to the sender—mostly abbreviated—within the paragraph of the subject being discussed. Von Braun corrected misspellings and any other information submitted incorrectly. His handwritten notes were clear, concise and grammatically correct. Often, his notes were directed to someone else to take action or returned to the sender with specific instructions. The sender would respond with his personal notations and return the same report to von Braun.

Von Braun's comments included more than corrections. He also provided encouragement and suggestions regarding progress in his rocket factory. Sometimes, he requested briefings in order to be better informed on subjects new to him. He constantly advised his people of changing strategies and politics of the day and encouraged them to go directly to a higher level of management to expedite problem solving. He kept abreast of new and innovative technologies in the fields of rocketry and space flight. Especially, he loved to examine new equipment and operate man-in-the-loop simulators. It was common knowledge that if someone were peddling a new space-related prototype and it could be driven, flown or floated; they could bring it to Marshall. If von Braun gave it a test run, there was likely a sale.

Von Braun understood and valued the talents and skills of his people and often gave the same problem to more than one to solve, thus establishing more than one solution. He sought opinions and advice, always prefacing his requests with, "Please," and offered apologies for being short, harsh or too abrupt with a colleague. His ire could be raised when people failed to report a problem in a timely manner or offered sloppy workmanship or "goofs" as he might have call them, but he took care not to "shoot the messenger." He readily offered guidance to improve procedures and prevent a re-occurrence and was quick to congratulate those organizations and individuals who succeeded and surpassed milestone events. He never failed to write letters of condolences to families who lost loved ones.

He often asked to be kept informed, "because I'm greatly interested," or "lay-on a one-hour briefing for me on that subject through Bonnie." Or, he might use the expression "presto-prompto" regarding the urgency of a matter. "How can I help you with this problem," was one of his favorites. He often offered to write letters to contractors who were falling behind schedule, reminding them they were on the "critical path" of the lunar landing program.

Some of the most interesting notes are between von Braun and Heinz H. Koelle, the head of future projects. During the early 60s, Koelle and von Braun discussed exciting new programs like Nova, space station, manned missions to Mars and electric and nuclear powered rocket stages. These were real projects being studied by the von Braun team at the same time the Saturn family of space vehicles was being developed.

One of von Braun's challenges during the early 60s was the transformation of the Marshall Center from an arsenal system to a research-development and industrial operations center. The von Braun team brought to the U.S. an in-house rocket design and fabrication capability that was accepted and expanded by the U.S. Army at Redstone Arsenal. Upon transfer of the team to NASA in 1960, von Braun began an effort to change his one-stop rocket factory to a more diversified organization.

With the growing interest by U.S. industry in building space hardware for the lunar landing program, von Braun found it necessary to add an industry management team. He opened assembly and test facilities in Mississippi and Louisiana, and began awarding contracts across the nation to the big aircraft and missile corporations. This required a major change in the way many of his laboratories functioned. As can be seen in the Weekly Notes, it was painful for many of his people to relinquish to contractors the fabrication of rocket components previously done in-house. Further, it required the labs to have "penetration people," Marshall engineers residing at the contractor plants, serving as the eyes and ears of Marshall. Von Braun was constantly reminding his lab directors to devote more time and effort to advancing the state of the art of rocket technology and developing new vehicles. Von Braun was successful in transforming his rocket factory into a premier space vehicle research and development organization with a strong industry management team.

What follows are reprints of the Weekly Notes collected from the National Archives by Mike Wright, Marshall's historian. We are indebted to Wright for his efforts in locating and collecting these valuable pieces of Marshall Center and von Braun history. My colleague Nancy Guire, who proudly served 39 years with Marshall, joined von Braun's office in 1965 as secretary to assistant directors J.T. Shepherd, Jay Foster and Ed Mohlere. She has read and interpreted several years of the von Braun Weekly Notes and I wish to acknowledge her valuable contribution in making these documents available.

I was fortunate to have worked in the Marshall Center's public affairs office for nearly 10 years where I interacted with von Braun on many public affairs-related events. When he asked me to become the director of the new Space & Rocket Center, I was honored to accept. We spent a lot of time together working on the early science center concepts. He wanted a hands-on place, not a dusty museum. Upon the center's opening in 1970, I proposed to the von Braun family that his papers, honorary degrees and memorabilia be given to the center for scholars, students and historians to review and study. The family agreed and for the past four decades I have had the opportunity to investigate and study von Braun's papers and interview numerous members of his team. For me, it was like re-living the exciting 60s, the decade of the moon landings.

Von Braun continues to receive many accolades long after his death. Here's one I believe would have gotten his attention: In 2003, worldwide aerospace industry professionals sponsored by *Aviation Week & Space Technology*, named the *Top 100 Stars of Aerospace*. Wilbur and Orville Wright were at the top of the list. Wernher von Braun was named the second most important aerospace pioneer in history. Ten others on the listed were astronauts who flew on von Braun's rockets.

One shortcoming of this collection is the lack of responses from the managers to whom von Braun directed his personal remarks. Those documents have not been recovered as of this writing. Due to the quality of the copies, the editor has added interpretative information to clarify notations, titles and organizations. Von Braun is noted as "B."

After reading nearly 10,000 documents—eight years of von Braun's career--there is no question he was the leader of the Marshall Team, forever the mentor, mediator and decision-maker. His leadership was never questioned. Von Braun's superb management skills enabled him to keep abreast of the technical issues while directing and managing the institutional business of the Marshall Center. He was clearly attuned to the political climate within the NASA family, the U.S. Congress and the White House. He demonstrated confidence in his people and set bold and challenging goals. He was highly respected and revered. Today, I still marvel at his vision and sense of purpose. He didn't just dream and wonder if we could launch man into space and land on the moon, he knew we could do it.

Von Braun truly had uncommon abilities to visualize and organize a project to its completion—conceiving, designing, developing, fabricating, testing and launching--an entirely new rocket system within one organization. His superior judgment in all engineering questions, his brilliant leadership, his ability to instill enthusiasm in others to include government officials and the general public, his wisdom in keeping the rocket team together and focused, and simply his exuberant joy of life, made us, an incredible team of over 200,000 employees and contractors, realize that Wernher von Braun was a true crusader for space travel.

NASA Historian Eugene Emme said it well:

"Cruel fate denied Wernher von Braun the chance to buy his ticket as a passenger bound for an excursion in space—his boyhood dream and lifetime goal. Because of Wernher von Braun, however, almost everyone has been brought to the realization that we have been passengers on a spaceship all along—Spaceship Earth. Posterity will not forget Wernher von Braun."

October 26, 1961

MEMORANDUM TO: See Distribution
From: J. C. McCall
Subject: Weekly "NOTES"

In a meeting today of the persons responsible for preparing the weekly NOTES, it was decided:

a. The deadline for delivery of NOTES (original and two copies) to Office of Director will be 11:30 a.m. each Monday. It was emphasized that late NOTES will not be accepted.
b. Attached is a list of names and phone numbers of all persons preparing NOTES.
c. After Dr. von Braun has written his comments on the NOTES, a copy will be made for our files, a copy will be made for Mr. Rees and for Mr. Gorman, and the original will be returned to the author. In order to preserve the personal flavor, no other copies will be made. If some person other than the author is addressed in comments, the question will be phoned by us to the addressee. Co-ordination can then take place between addressee and author for answering the question.
d. The NOTES will not be open for inspection. If information from one author is desired regarding another author's NOTES, the contact must be made direct, not through this office.
e. If answering questions from Dr. von Braun, copies of the old NOTES must be attached. If this is done, state at the end of page on current NOTES: "Attachment 1, NOTES 6 - 2 - 61 MRAZEK".
f. The NOTES are bound at the top. Please allow room for binding.
g. It was emphasized that a "NEGATIVE" NOTE is required if no information is to be presented.

PERSON RESPONSIBLE FOR WEEKLY "NOTES"

M-AERO	Mr. Larsen	876-1301
M-COMP	Mr. Prince	876-3147
M-DEP-ADM	Mrs. King	876-1764
M-F&AE	Mr. Heim	876-1735
M-FPO	Mr. Huber	876-4714
M-G&C	Mr. Chase	876-4705
M-LOD	Mr. Heiser	876-4520
M-L&M	Mr. Stone	876-3829
MICHOUD	Mr. Wible	876-4125
M-QUAL	Mr. Buhmann	876-4731
M-RP	Mr. Bucher	876-4935
M-SAT	Mr. Lindstrom	876-3448
M-S&M	Mr. Rieger	876-4340
M-TPC	Mr. Smith	876-4119
M-TEST	Mr. Rivers	876-2696

1961

5. A meeting has been set for November 15 between Maintenance, Inc. (our janitorial contractor) and the steel workers union. This meeting is to discuss wage demands by the union. There is a possibility of a strike sometime in the future. Mr. Styles is keeping abreast of the situation. There is nothing we can do as far as we know.

11-13-61 (Gorman)

→ *Get me a broom! I'll sweep my own office.* B

11-20-61 (Grau)

CENTAUR: Heat transfer through the intermediate bulkhead separating the LO₂ and LH₂ tanks is intolerable. Mr. K. J. Bossart has been assigned to this problem for a solution.

Hans Hueter → *Who goofed here? I've seen tons of reports studying this problem.* B

4. **P&W Aircraft coverage by Quality Division:** Per your request for comments concerning NOTES 9-29-61 (copy attached), the reduction in the number of personnel stationed at P&W Aircraft was primarily a result of your decision to allow P&W development people as much freedom as they needed to deliver the first 6 or 8 engines. This decision re-

10-16-61 (Grau)

These Notes will be held in suspense. When reply is ready, please send to Dr. McCall w/cy of these NOTES as reference.

Dieter Grau — Do you think the time has come to tighten up again? P&W told us at that time they could keep the new schedule only if we'd "call off our dogs for awhile". Please prepare joint reply & recommendation with Lee Belew. B 10-17

3. **NATIONAL TEST SITE:**

a. T. E. Edwards and Dr. W. H. Sieber made final presentation in Washington, October 19, 1961, on National Test Site. Dr. Sieber made sound presentation. He was questioned quite thoroughly. One question was, "What Effect would the Firing have on Funeral Procession?"

→ *I guess a feller can't be deader than dead!*

10-23-61 (Helmburg)

A draft for the proposed NOVA study effort has been completed and I would like to discuss it as well as the initiation of the "NOVA definition effort" at your earliest convenience.

Let's first get the C-5 issue out of the way. "One emergency at a time"! But why don't you draft something, meanwhile. B

12-18-61 (Koelle)

10-13-61 (Kuers)

Mr. John Leshko, Space Task Group, visited this Division to become familiar with the facilities, mockups, and support equipment planned for use in the preliminary space maintenance and repair exercise scheduled for the latter part of this month. Construction of the airbearing chair/platform for this exercise has been completed.

Mr Kuers — Good. Please keep me posted (with Jim Carter) how this program is shaping up. I am very interested in this whole thing. What can I do to expedite it? B

8. The meeting of November 1 with the people to be displaced at Pearl River went very smoothly; thanks to Senator Stennis, who strongly endorsed our project. About 1,000 people attended. *Let's prepare a nice, personal letter to Sen. Stennis for my signature. He sent me a nice note recently.* B Harry/Bart

9. A GAO team will make an initial survey of our activities the week of November 20. The detailed audit will follow.

11-06-61 (Gorman)

been held in reserve for staffing for management of Pratt & Whitney Convair contracts, Pearl River and Michoud operations, additional positions for Centaur, etc.

2. Nothing new has developed in the Gurtler-Habert issue except an unfortunate article in the Huntsville Times which begins "MSFC Refutes Drew Pearson". As you know, we are trying to avoid any further involvement in this case. *Bart Slattery — Why couldn't we prevent this?* B

3. Our personnel office has received word of my confirmation to the deputy's job. *Harry G — Good. Please cut necessary papers for my signature. Place your picture...*

4. A meeting has been scheduled for November 14 between Styles, myself, and a

12-18-61 (Stuhlinger)

Was discussed, but issue is still open. Meanwhile, suggest you discuss with Haeussermann the new "Lady Richard Concept" of orb ops. B

4. ORBITAL OPERATIONS: RPD is in the process of formulating an 8 million dollar Orbital Operations Supporting Research Program for MSFC. Have we received any indication from Mr. Holmes whether we will obtain funds for such a program?

3. Saturn Launch Photos: Color films of the Saturn Launch are available showing many details of ignition, lift-off and tracking. Technically outstanding. Zeiler can arrange for showing at MSFC if you are interested. *I am!! B*

4. NASA Hqs. Organization: Copy of the new Headquarters organization was furnished me while in Washington. Snyder tells me that there will be 5 individual Launch Operations segments...

11-07-61 (Debus)

Kurt — We discussed this with Brainerd Holmes and told him that this was impossible. We agreed that there should be a special meeting on this entire organizational problem in the near future. Please lay it on with Holmes' office. (Coordinate suitable date with B.)

5. INTRODUCTION OF APOLLO SHAPE: A discussion was held with representatives of STG and Langley Research Center concerning the payload of SA-5. There is evidence of a substantial amount of aerodynamic vibration produced by the escape tower and by the nose shape, which may at transonic speeds lead to structural damage. Further tests with a 2% scale model are expected in the near future simulating accelerated flight through Mach 1. If the results are encouraging, an effort will be made to introduce Apollo shape into SA-5 and schedules are presently being checked to see whether a test flight of this shape can be made on SA-4. It will also be necessary to increase the stiffness of the escape tower, especially in view of the possible use of an α-meter on top of it.

Can't we put a cover over the tower? How much more would it really weigh? B

11-13-61 (Geissler)

3. M-1 (Y-1) ENGINE: Further action has been delayed by approximately 30 days. This information from Del Tishler on 10-19-61, appears to have something to do with reorganization, 11-1-61. *Good!! We've got enough more urgent problems. B*

4. J-2 AND F-1 FACILITIES: Funds have still not been released as of this date. *Please keep me posted. I made a lot of steam on F-1 facilities with Dixon. B*

5. J-2 ENGINE: Initiation of Engine Systems testing has been rescheduled for 2-5-62, a delay of six weeks. This is another admittance by Rocketdyne that the program is marginal and optimistic. ✓

10-23-61 (Mrazek)

5. SPACE ENVIRONMENTAL CHAMBERS: Mr. Jim Carter, FPO, contacted AERO-E during this week for background information about space environmental chambers. He feels strongly that large chambers will be required for "Orbital Operations" R&D and full scale vehicle checkout. Mr. Carter was given copies of the Aeroballistics Division's proposal for formation of a Space Simulation Group plus the NASA-DOD survey on existing U. S. facilities and proposed facilities. Warning: Space simulators are too costly to be dealt with at the Divisional level, centerwide treatment required.

Hans Maus: Suggest you get in touch with Jim Carter to coordinate this program. He is making preparations in the area of "manned outside repairs in orbit" etc. B

11-06-61 (Geissler)

1. NAA required funding was 36.000 million versus 25.000 million available. ✓

2. To gain a more definitive scope of work by allowing MSFC and NAA engineers to work together for 90 days. Open areas include:
   a. Diameter
   b. Electrical networks
   c. Control system

Also: they projected layout for Seal Beach looks very plush to me. (See model in F&AE mockup shop and discuss matter with Kuers!) B

3. NAA proposed schedule was 4 months ahead of our present C-3 or C-4 schedules. ✓

II. C-133 AIR TRANSPORT FOR S-IV STAGE - On the basis of Langley investigation, Seamans has said no. *How's the blimp plan coming along? B ✓*

10-23-61 (Lange)

III. PERT - The S-I personnel of this office have been meeting with TPC personnel in preparation of using the PERT programs developed by TPC for the S-I SA-5 as a firm management tool. This present network indicates the stage is 18 months off schedule; our first work will be to try to work this time out of the program.

10-23-61 (Lange)

WHAT?? B

Dr. von Braun: I remind you, we just lost our Div. of the PERT man from company. If the information is really useful, this loss is quite significant. Sub-249 (Request more info + soon! This delay would be a catastrophe!!)

1962

mainly of 2 huge bulge forming dies, and fabrication of 12 each segments of the apex and knuckle parts was $209,205. After changing the tool design layout on our request twice, Boeing changed the price on October 17, 1962 to a total of $609,805. This week we were informed that the last quotation was in error and that the price of this job would be $2,700,000. We had a meeting with Mr. Coenen and his people where a break-down of the figures was discussed.

11-05-62 (Kuers)

Jim Bramlet / Matt Urlaub — What's going on here?? If Boeing keeps operating like this, we'll be broke in no time! B

The old creeping NASA disease again! B (The "headquarters octopus" strangling all activity if not checked continuously.)

1. HEADQUARTERS ADP MEETING: The Ad Hoc Group Meeting on administrative ADP procedures called by Mr. Seipert was held last week in Washington and the four members from MSFC indicated in last week's NOTES (copy attached) attended. ment instruction dealing with ADP was prepared by Headquarters and discussed. The critical point is, as always, how deeply Headquarters should become involved in center acquisitions and programs. MSFC representatives (as did other centers) made it clear that they felt NASA Headquarters should be kept informed as to the center ADP plans for NASA-wide dissemination. They strongly objected to Headquarters control over ADP equipment procurement and center applications placed on this equipment. Headquarters should approve ADP equipment procurement procedures in the center and the over-

10-15-62 (Hoelzer)

O.L. → (please take up w/ Shepard) There why in h... don't we get moving??? B

09-04-62 (Lange)

MSFC requested from OMSF authority and C of F funds for re-roofing the Manufacturing Building ($955,000) and the storm drainage construction ($700,000). M-SAT understands that appropriate authorization and funds are available from FY-62 supplemental budget.

09-04-62 (Rudolph)

Heck, I don't know what the moon is like! Never been there! B

Office of Systems is very interested to receive MSFC comments on Issue No. 1 of a statement titled: "Office of Manned Space Flight Requirements for Data in Support of Project Apollo".

3. QUALITY REQUIREMENTS FOR STANDARD SPACE LAUNCH VEHICLE: A meeting was held with Space Systems Division (Air Force) personnel and Mr. Howard Weiss, Office of Reliability and Quality Assurance, NASA Headquarters, at Aerospace Corporation, Los Angeles, California, to discuss Quality Requirements for the Standard Space Launch Vehicle. Air Force Specification DCAS 62-10 is to be included as a part of the contract. Since this document is not as comprehensive as NPC 200-2, there were certain provisions which were requested by NASA to be included. A list of these provisions is being prepared for distribution.

D.G. Is that Pentagon-Chinese for Titan III? B

08-06-62 (Grau)

6. Centaur: During the course of Centaur checkout this past week, it was disclosed that considerable harness wiring is not flight worthy. An overall inspection and replacement of wiring harness is in process. The schedule is being revised to reflect an overall slippage of one month from the present official launch date.

02-12-62 (Debus)

Hans Huefer — I'm about ready to suggest to blow up the whole darn project. How long is this kind of thing to continue?? B 2-12

No report. H.H.
Hope you are still alive. How is Slidell? Michoud? B

10-22-62 (Hoelzer)

W.K. Request a briefing on this subject by the most knowledgeable people we have. Please arrange B 10/29

within 10 seconds--tolerances of the die, uniformness of material properties, etc., were: "We have experience with this process and we are confident that we will solve these problems". The confidence is there, but the answer is generalized! Because of uncertainty of success with this method S&ID has been directed through Saturn Systems Office not to proceed with construction of 33 feet diameter dies of this type for the complete bulkhead, but to go ahead with application of this method for sizing of gore segments. The problem of sizing a complete bulkhead of 33 feet diameter to tolerances of ± .020" all over the contour is still unsolved--at least no method with a reasonable confidence factor for success is known at this time. Therefore, we strongly recommend consideration and evaluation of the membrane-type bulkhead design or to bond upper gore segments in place and seal of gaps by the use of doublers.

10-29-62 (Kuers)

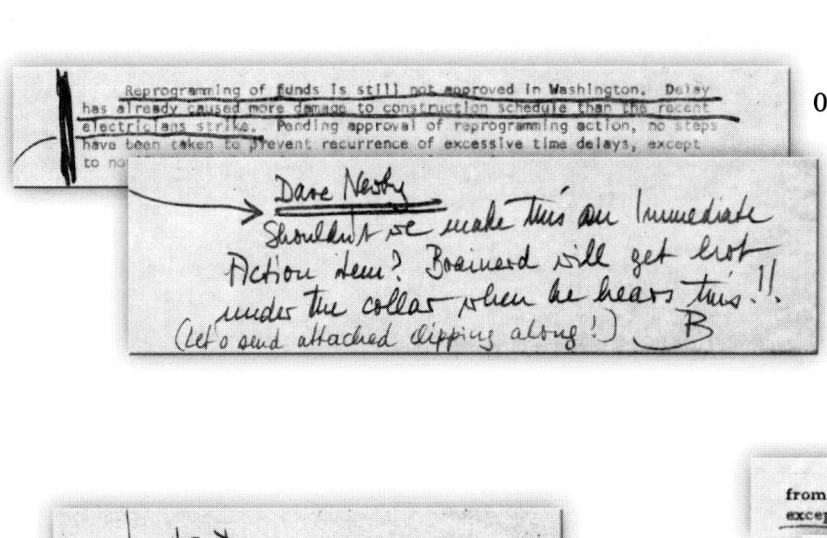

08-27-62 (Helmburg)

> Reprogramming of funds is still not approved in Washington. Delay has already caused more damage to construction schedule than the recent electricians strike. Pending approval of reprogramming action, no steps have been taken to prevent recurrence of excessive time delays, except to no...

Dave Newby — Shouldn't we make this an Immediate Action Item? Brainerd will get hot under the collar when he hears this!! (Let's send attached clipping along!) B

08-27-62 (Constan)

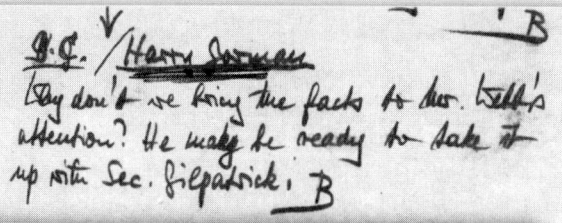

D.C. / Harry ____ — Why don't we bring the facts to Mr. Webb's attention? He may be ready to take it up with Sec. Fitzpatrick. B

> Difficulty is being experienced in obtaining desirable equipment from industrial reserves. Permission to screen has been denied except for those items which are obsolete or worn-out.
>
> ing building. When project approval and money is received from NASA Headquarters, immediate action will be taken to commence work. This item becomes more critical ea...

What in h... is holding it up?? B

03-05-62 (Debus)

K.D. — I only want to hear about troubles. In other words, I don't want to hear anything! B

> 2. SA-2 Arrival: During the week, the SA-2 configuration was installed on Pad 34. Daily Status Report will be published by LOD. Available in your office if you desire it. ✓

04-16-62 (Debus)

> go along with this but that we should continue our attempts to retain title to the land. Incidentally, General Davis had forwarded a letter to the C of E, Jacksonville, instructing Col. Sollahub to turn over the title of the new property to the Air Force. Col. Sollahub, however, forwarded the request to me. We advised them to continue turning the title over to us.

Sounds like war drums! B

> they were pleased with the plans and execution. The Acting Commander AMR was deeply disturbed by the protocol arrangements furnished to LOD by the State Department. The Acting Commander insisted on changing of plans shortly before arrival of party and threatened to "bump" the personal representative of the President of the U.S. Rough going and continued future problem. ✓

K.D. — Here we go! (It surprises me that no such problems exist between AOMC and MSFC. We swap out VIP's freely. How come?) B

02-12-62 (Rudolph)

> No NOTES received by 12:00 Noon.

Arthur — How's your staffing coming along? If I don't hear about your problems I can't help you! B Any news from Shea?

06-04-62 (Hueter)

1	MsnIR	Dr. von Braun	A C T I O N	I N F O R M A T I O N
2				
3				
4				

REMARKS
We have worked this plan of action out with M-P&VE and TEST Division.

Robert E. Lindstrom

Pending in today's NOTES. Jan-29

Bob — Is Heimburg satisfied with this solution or did you have to cram it down his throat? B 1-29

Venus won't wait! B

> b. Mariner R: Shipment of the Atlas booster (145-D) to AMR for the first Mariner has been delayed. Shipment is normally accomplished by C-133 cargo plane; however, the type plane has been temporarily grounded.

Bob Lindstroem / Dr. Lange — Let's stop this nonsense, we're short of funds!! What corrective action has been taken. (See also Kuers' Notes of 2-5-62) B 2-5

Cys 2/5 for Constan furnished Bob Lange Dr. Rees, Jeff 2/8/62. No wonder that. Please let me know. RUSH.

01-29-62 (Lindstrom)

02-05-62 (Constan)

> Chrysler indicated that further detailed layout of the production area and that the space allocated to them was too small. Review of the layout by MSFC personnel showed a number of departures from what we understand to be the basic Chrysler assignment, as follows:
>
> a. Fabrication of nearly all items in Michoud instead of using MSFC developed vendors and sources.
> b. Extensive R&D activities as evidenced by layout of specialized laboratories (such as spectrographic labs.) and environmental test facilities.

1963

LUNAR LANDING SIMULATION PROJECT: A breadboard flying spot scanner system is operating which permits simulation of a portion of the descent to the lunar surface. While the resolution [...] to be desired, the system is satisfactory for [...]

03-02-63 (Hoelzer)

> H.H.
> Please call me when you'd like to give me a little demonstration
> B

1. FUTURE OF THE SATURN FAMILY
You expressed some concern about the lack of activities on the "marketing" of the SATURN launch vehicles. Here is a summary of what we are doing and planning to do:

the "old reliable SATURN's." Thus a 10-year operational lifetime, and production numbers of at least 50 for SATURN IB and 100 for SATURN V, seem virtually to be assured. ✓

08-26-63 (Koelle)

> HHK
> You're an optimist, but then, so am I.
> B

01-07-63 (Mrazek)

> L.M.
> This report is a bit thin considering P&V's tremendous tasks!! B
> (Hope next time it'll be more, and that this is just New Year's eve hangover)

and a task submitted by Aeroballistics Division, for Aerodynamics Instrumentation Research, both to be performed by AEDC-Tullahoma. Two additional tasks, Self Sealants, submitted by Special Assignments Office, and Thermal Control in Space, submitted by Research Projects Division, are to be performed by Wright-Patterson Air Force Base. ✓

These four proposals have been referred to the laboratories/divisions concerned for technical review and approval. ✓

> H.M.
> Please see to it that this doesn't bog down in bureaucratic difficulties. I think we should use these first 4 tasks as "ice-breakers" to get this cooperation with the AF established. It may be helpful and useful in 1000 ways in the future!!
> B

09-23-63 (Maus)

1. RIFT: The projected FY-64 funding level for RIFT was the reason for a Lockheed Nuclear Space Programs Division reorganization. We understand that cognizance over the manufacturing group will be returned to the central manufacturing group, and the directorates for Product Assurance and Test Operations will be combined. Official notification of this rearrangement has not been received from Lockheed.

1. Saturn V, S-IC Stage:

[...] ing a sufficient number of people in purchasing and follow-up activities, etc. Immediately affected is the schedule for hardware for qualification testing and for components for single engine testing for Test Division needed by the end of this year. We had several meetings with the Vice President of Arrowhead and their key people and hope at least to avoid further slippage and possibly catch up some of the lost time.

> I understand this may cost us 2 to 3 months!!! (per Lange Notes 4-1-63)
> W.K.
> This is the kind of thing we simply cannot tolerate!

04-01-63 (Kuers)

1. ST-124 SLED TESTS: A total of 7 runs was made with the first ST-124 Stabilized Platform delivered from Eclipse-Pioneer on the high speed track at Holloman AFB between 11/27/62 and 2/21/63. The first three runs were [...] The reliability was conclusively demonstrated by the series of seven runs.

> W.H.
> Very good, congratulations to everyone involved
> B

03-04-63 (Haeussermann)

Boeing Y-Ring [...] have initiated a [...] vacuum chamber [...] of Y-Ring welding [...] been quite [...] however, have hit [...] was found that [...] level for the [...] ists. As a [...] (1 Amp, [...] thode and anode [...] have been made [...] a 80,000 volt [...] the gun at a [...] vaporized metal

> I hope this process has really been chewed out good! Shall I send him a fiery follow-up letter? If so, please draft one
> 2, we will present a condensed form to 1 presentation
> B

02-04-63 (Rudolph)

RUSH
A.R.

Flight Mission Assignments

1. Fl*[...]* I will have a final review of those with Shea and Gautraud, etc. on 12 Feb 63 (before Shea's submission to Holmes for official approval and distribution). I will provide a set of slides for you.

We reached agreement on Sat I and Sat IB missions but Sat V is still unresolved. Latest Shea plans still meets w/ violent MSFC opposition. We agreed in Mgmt Council 29 Jan to hold a separate meeting on Sat V mission schedule!
B

02-25-63 (Mrazek)

6. S-IV STAGE PROBLEMS: Problems which this actuator-acumulator-reservoir design has encountered are (a) Servo valve body forgings rejected; (b) Markite potentiometers rejected; (c) Dirty servo valves; (d) Excessive leakage across bypass valve -- new bypass valves made; (e) High-pressure drop through filter cavity, caused by restriction--spacer and snap-ring employed as fix; (f) Piston rods scoring during velocity-load tests.

W.M. Anything good at all??
B

04-22-63 (Haeussermann)

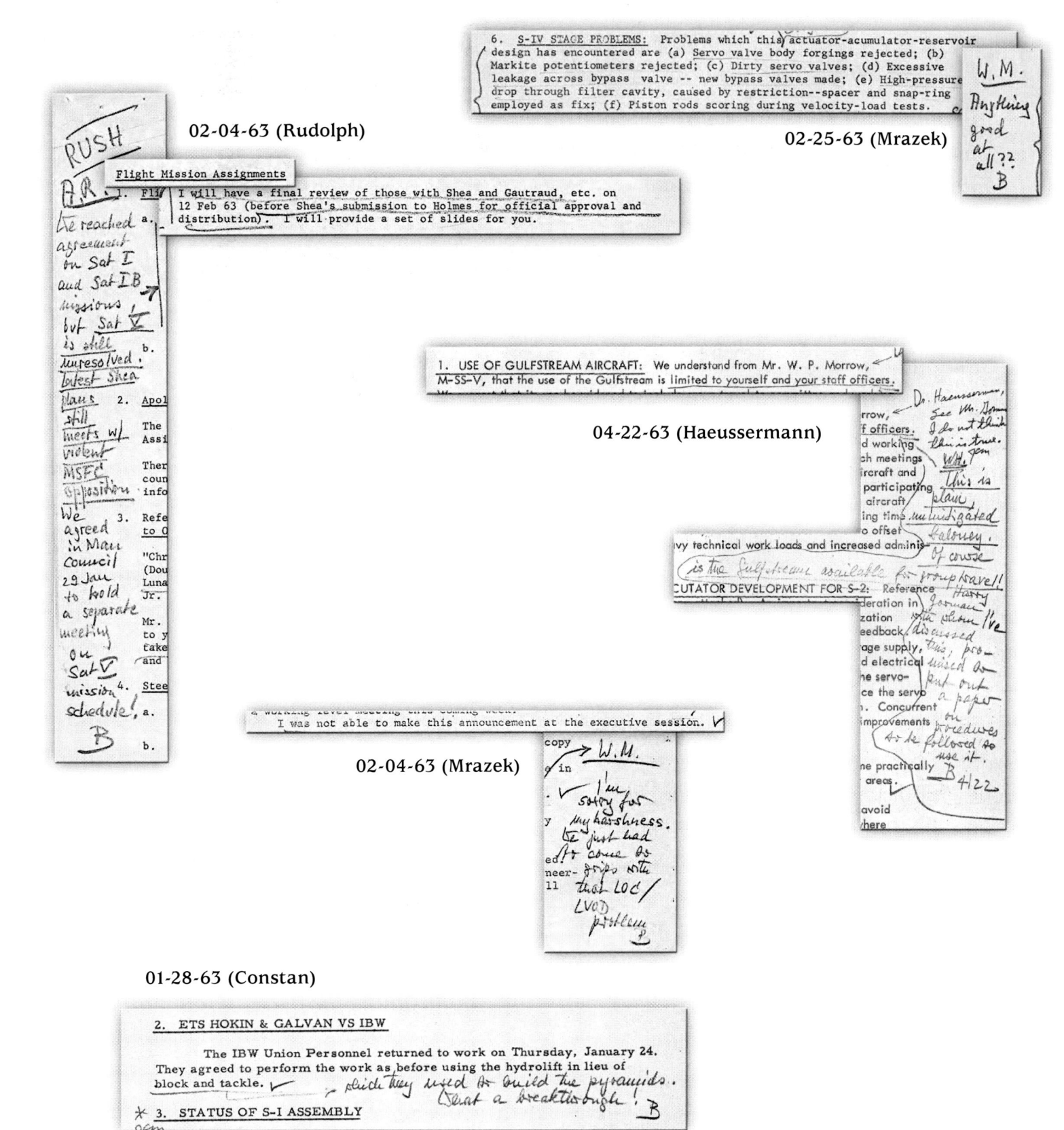

1. USE OF GULFSTREAM AIRCRAFT: We understand from Mr. W. P. Morrow, M-SS-V, that the use of the Gulfstream is limited to yourself and your staff officers.

[...] heavy technical work loads and increased adminis[...]

CUTATOR DEVELOPMENT FOR S-2: Reference [...]

Dr. Haeussermann, See Mr. Morrow. I do not think this is true. WvB

this is plain unmitigated baloney. Of course is the Gulfstream available for group travel! Harry in Germany via phone I've discussed this, promised to put out a paper on procedures to be followed to use it. B 4/22

02-04-63 (Mrazek)

I was not able to make this announcement at the executive session.

copy → W.M.

I'm sorry for my harshness. We just had to come to grips with that LOC/LVOD problem.
B

01-28-63 (Constan)

2. ETS HOKIN & GALVAN VS IBW

The IBW Union Personnel returned to work on Thursday, January 24. They agreed to perform the work as before using the hydrolift in lieu of block and tackle. — slide they used to build the pyramids. What a breakthrough! B

* 3. STATUS OF S-I ASSEMBLY

12-09-63 (Gruene)

4. TV Coverage of SA-5: As you know, Headquarters tries to make a big TV splash out of the SA-5 launch. Due to the numerous difficulties encountered during our three loading test tries, my personal confidence in getting SA-5 off on the first try, is very low. Dr. Debus is trying, through Dr. Mueller, to discourage any commercial TV coverage. If a question should come to you, I would appreciate your backing us up.

H.F. This was Pres. Kennedy's personal suggestion. Will be hard to turn down, particular now, after the tragic event.
B

1964

> 4. PARTICIPATION IN LOCAL POLITICAL CONTESTS: A legal opinion has been requested from the Chief Counsel's Office concerning candidacy of MSFC employees in local political contests. By the end of last week, four R&D Operations representatives had requested permission to enter local political races.

I thought we had enough politics without the Carter! B

07-06-64 (McCartney)

> E.S.
> That's a very good idea. I've always felt that the lack of visual display (photos, models, samples) of past SRT achievements has hampered our sales efforts in the new programs. Let's discuss a hard-hitting plan. B 2/7

> 1. RESEARCH PROGRAM INFORMATION FOR MSF: Ed Gray from MSF requested that MSFC prepare a collection of narrative and exhibit material to demonstrate objectives and results of the MSFC supporting research and technology programs. Fiscal years 63, 64, and 65 should be covered. This material will be needed in forthcoming congressional committee hearings. Unfortunately...

01-27-64 (Stuhlinger)

> WASHINGTON VISIT BY UNIVERSITY OF ALABAMA/TUSCALOOSA BUSINESS PEOPLE: A group of Tuscaloosa business people and University of Alabama officials, including Dr. A. Pow (Dr. Rose joined later), visited with Mr. J. Webb and the Headquarters staff on January 10, 1964. Their intention: to invite NASA to place business in a Tuscaloosa Research Park area of some 700 acres. The proposal was apparently a strictly "Tuscaloosa local" backed by the University of Alabama. Mr. Webb appeared puzzled, first that the University of Alabama and the Tuscaloosa business people came to see him directly for such local interest, and second they had not seen the NASA "Rep." in their area, the MSFC (more details available to you). Dr. A. Pow called today to make arrangement for a MSFC-Tuscaloosa business-University of Alabama mutual briefing here at Huntsville.

O.L. I'll be the coldest fish they've ever seen! B

01-20-64 (Lange)

> for implementation on various contractors.
> 3. CONFIGURATION CONTROL BOARD ACTION: The material used to prevent galvanic corrosion between the GOX Diffuser and the bulkhead of the 105-inch LOX Tank has been determined to be incompatible with LOX. Correction of this problem on SA-7 is impossible without a schedule delay. A change of material on SA-9 may also cause a schedule delay. Investigation of the problem will continue.
> 4. F-1 FLIGHT RATING TEST (FRT) INJECTOR HA...

Fred C. Who goofed? Please see to it that procedures are tightened B

P.S. I'm not interested in name of culprit. I am interested in steps to prevent recurrence B

08-24-64 (Cline)

> No submission this week.

*W.H.
I guess I haven't had any notes from Astrionics for 3 or 4 weeks. Have you stopped working, has your place burned down, or is it that you simply have no problems? B*

11-30-64 (Haeussermann)

02-03-64 (Gruene)

2. **SA-6 Launch**

I would like to make you aware of the fact that in SA-6 we carry, in addition to the SA-5 instrumentation systems, three more telemetry links on the spacecraft, two C-Band Beacons on the spacecraft, and ODOP (Offset Doppler) System in addition to the UDOP in the launch vehicle. The interference history of SA-5 scares us with this number of active RF systems. It might have to be decided, in coordination with Dr. Speer and Mr. Hoberg, that some of the systems have to be eliminated in case we run into serious problems.

H.F. Suggest the three of you get together on this at once. B

12-28-64 (Belew)

Karl Heimburg — Just to remind you that you are on the Critical Path. B

F-1 ENGINE

F-1 production is now at a rate of two deliverable engines a month. The two "December" engines (F-2007 & F-2008) have been acceptance tested and have completed their final checkouts. Both engines will be air shipped via the "Pregnant Guppy" aircraft next week.

02-03-64 (Kuers)

Saturn V, S-IC Stage:

The structural assembly of the T-Vehicle is not making good progress and is already falling behind schedule VI. There are three major reasons for this delay:

Karl Heimburg — When can I get that promised status and problem briefing on SIC-T? B

02-10-64 (Heimburg)

4. **MARINE TRANSPORTATION:**

The last barge trip (SA-6) under Test Laboratory's jurisdiction, from MSFC to the Cape, started at 4 p.m. on 2/7. From now on, operation of the barge will be under the jurisdiction of the Project Logistics Office.

KH! Test deserves a big pat on the back for setting up and running a very efficient transportation system all these years! B (Please pass this on)

12-28-64 (Constan)

1. **MICHOUD FATALITY**   *B 12/28*

On December 21, 1964, Mr. Edward Williams, sheet metal worker employed by Erectors, Inc., subcontractor to Martin K. Eby, the construction contractor for the stage test facility for the Boeing Company, was fatally injured at Michoud Operations. Mr. Williams was hoisting steel siding when the hoisting device overbalanced, throwing a board down on top of Mr. Williams. Boeing is investigating the accident by means of a board of inquiry.

Frank — Please prepare a condolence letter to widow or other relatives. Inform Constan & Boeing. B

01-20-64 (Fortune)

→ Again??? Oh no! B

5. **Visit Of NAS New Orleans Personnel:** Captain Tracy and other personnel from the Naval Air Station visited MTO looking for space for a Navy bombing range. We pointed out

1965

04-26-65 (Koelle)

2. CENTER SELECTION CRITERIA: Our second iteration on the development of suitable project selection criteria resulted in the following weighted priority list. This was obtained by polling the 45 members of the Executive Board and the Center Planning Working Group, of which 39 elected to participate. We are very happy to have this high degree

11. Does this project make us a "rich" Center	7.82
	100.00%

I never knew we had so many idealists around here!

Or are we just snobs who take our having money for granted? B

11-15-65 (Grau)

2. IU-500-FS: IU-500-FS was transferred to R-QUAL on October 25, 1965. After analysis of the assembly status revealed that no systems were complete, representatives of QUAL, ME, and the Project Office, decided to return the unit to ME for incorporation of outstanding EO's and addition of available items that were missing. The IU is scheduled to return to QUAL on November 29, 1965.

Kerner, Kuers and D.G. — Somebody apparently tried to impress someone, but it backfired! Let's stop phony deliveries just to make a schedule a milestone! B

01-25-65 (Stuhlinger)

1. WEEKLY NOTE SYSTEM: In response to your plea to reduce the number of problems brought to the attention of DIR by having more of these problems handled on the R-DIR level, may I suggest that Labs and Offices write two Weekly Note sheets, a short one for DIR, and a more elaborate one for R-DIR? I believe that the "Weekly Note" system is

E.S. Misunderstanding! I don't mind having those problems brought to my Offices attention. I only feel that R-DIR should always be called to help resolve them before I am brought into the act. No. Let's leave the Weekly Note System as it is. I like it. B

SATURN IB/CENTAUR TERMINATION: On November 4 we recommended that $1.1 million be authorized for continued effort on fabrication and test of honeycomb panels, feasibility demonstration of a non-contaminating retro-rocket, and Saturn V-Voyager aeroballistic studies. We received a teletype from John Disher on November 18 which stated that Dr. Mueller had disapproved our request to spend the additional money. Action has been taken to terminate all Saturn IB/Centaur contracts. It is anti-

That's correct B

11-29-65 (Reinartz)

Stan R. ✓ All Sat IB/Centaur funds reverted to Saturn IB, and some of them may find their way into AAP, i.e., to you! B

Ground Computer Program Problem at KSC: On Thursday, Dec. 2, 1965, a problem with the ground computer operating system programs occurred during checkout operations. This program error caused erroneous outputs to be issued to the vehicle electrical system. No known damage occurred as a result to these erroneous signals. result of an error made in a previous change to correct a different situation. This shows again that very close coordination is required in making changes and/or corrections to these programs. MSFC (R-ASTR) and KSC are working with IBM to continually improve this coordination and to improve the procedures.

12-06-65 (Richard)

LR Please keep KSC constantly aware of this!! B

01-25-65 (Kuers)

1. S-IC Manufacturing Problems: The intention of my reporting on manufacturing problems is to show with some examples that we are not engaged in a routine manufacturing task in building these stages and to give you a feeling of the degree of extension of the art in manufacturing that we and the Prime Contractors are involved in. At the same time these examples, I hope, might serve to further the understanding why some delays have occurred in our program.

W.K. You need not be so apologetic! I think your guys at ME have done an excellent job, not only in house but also in your cooperation with our stage primes. B

12-06-65 (Heimburg)

S-IC-T: Test S-IC-14, approximately 145 seconds duration, is scheduled for December 9, 1965.

Sieb — If possible, I'd like Mr. George Bunker of Martin to witness this test. He'll be in HSV for a visit. Bonnie has details. B

12-27-65 (Kuers)

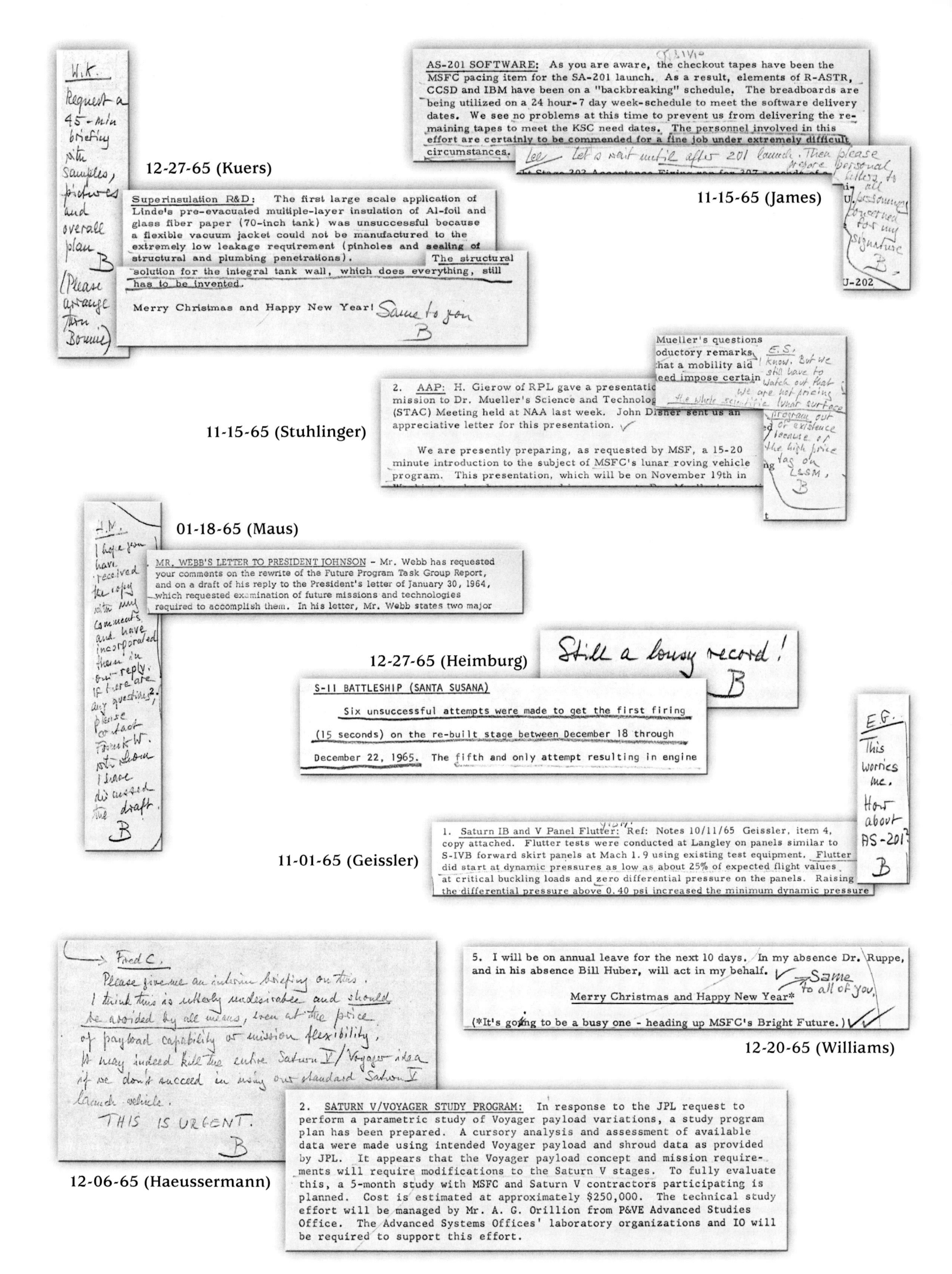

W.K.
Request a 45-min briefing with samples, pictures and overall plan. (Please arrange thru Bonnie)
B

11-15-65 (James)

AS-201 SOFTWARE: As you are aware, the checkout tapes have been the MSFC pacing item for the SA-201 launch. As a result, elements of R-ASTR, CCSD and IBM have been on a "backbreaking" schedule. The breadboards are being utilized on a 24 hour-7 day week-schedule to meet the software delivery dates. We see no problems at this time to prevent us from delivering the remaining tapes to meet the KSC need dates. The personnel involved in this effort are certainly to be commended for a fine job under extremely difficult circumstances.

Lee, Let's wait until after 201 launch. Then please prepare personal letters to all personnel concerned for my signature.
B
J-202

Superinsulation R&D: The first large scale application of Linde's pre-evacuated multiple-layer insulation of Al-foil and glass fiber paper (70-inch tank) was unsuccessful because a flexible vacuum jacket could not be manufactured to the extremely low leakage requirement (pinholes and sealing of structural and plumbing penetrations). The structural solution for the integral tank wall, which does everything, still has to be invented.

Merry Christmas and Happy New Year! Same to you
B

11-15-65 (Stuhlinger)

Mueller's questions
oductory remarks,
hat a mobility aid
eed impose certain

E.S. I know. But we still have to watch out that we are not pricing the whole scientific lunar surface program out of existence because of the high price tag on LSSM.
B

2. AAP: H. Gierow of RPL gave a presentation mission to Dr. Mueller's Science and Technolog (STAC) Meeting held at NAA last week. John Disher sent us an appreciative letter for this presentation. ✓

We are presently preparing, as requested by MSF, a 15-20 minute introduction to the subject of MSFC's lunar roving vehicle program. This presentation, which will be on November 19th in

01-18-65 (Maus)

H.M.
I hope you have received the copy with my comments and have incorporated them in our reply. If there are any questions, please contact Frank W. I have not shown I have discussed the draft.
B

MR. WEBB'S LETTER TO PRESIDENT JOHNSON - Mr. Webb has requested your comments on the rewrite of the Future Program Task Group Report, and on a draft of his reply to the President's letter of January 30, 1964, which requested examination of future missions and technologies required to accomplish them. In his letter, Mr. Webb states two major

12-27-65 (Heimburg)

Still a lousy record!
B

S-II BATTLESHIP (SANTA SUSANA)
Six unsuccessful attempts were made to get the first firing (15 seconds) on the re-built stage between December 18 through December 22, 1965. The fifth and only attempt resulting in engine

11-01-65 (Geissler)

E.G.
This worries me. How about AS-201?
B

1. Saturn IB and V Panel Flutter: Ref: Notes 10/11/65 Geissler, item 4, copy attached. Flutter tests were conducted at Langley on panels similar to S-IVB forward skirt panels at Mach 1.9 using existing test equipment. Flutter did start at dynamic pressures as low as about 25% of expected flight values at critical buckling loads and zero differential pressure on the panels. Raising the differential pressure above 0.40 psi increased the minimum dynamic pressure

12-20-65 (Williams)

5. I will be on annual leave for the next 10 days. In my absence Dr. Ruppe, and in his absence Bill Huber, will act in my behalf. ✓ Same to all of you.

Merry Christmas and Happy New Year*

(*It's going to be a busy one - heading up MSFC's Bright Future.) ✓✓

12-06-65 (Haeussermann)

Fred C.
Please give me an interim briefing on this. I think this is utterly undesirable and should be avoided by all means, even at the price of payload capability or mission flexibility. It may indeed kill the entire Saturn V/Voyager idea if we don't succeed in using our standard Saturn V launch vehicle.
THIS IS URGENT.
B

2. SATURN V/VOYAGER STUDY PROGRAM: In response to the JPL request to perform a parametric study of Voyager payload variations, a study program plan has been prepared. A cursory analysis and assessment of available data were made using intended Voyager payload and shroud data as provided by JPL. It appears that the Voyager payload concept and mission requirements will require modifications to the Saturn V stages. To fully evaluate this, a 5-month study with MSFC and Saturn V contractors participating is planned. Cost is estimated at approximately $250,000. The technical study effort will be managed by Mr. A. G. Orillion from P&VE Advanced Studies Office. The Advanced Systems Offices' laboratory organizations and IO will be required to support this effort.

1966

09-19-66 (James)

KSC RESIDENT OFFICE: I think you are aware that General Shinkle, in a letter concerning Field Engineering Changes (FEC's), has questioned the role of our resident office in the configuration control that we exercise at KSC. announced his resident office at KSC under the leadership of a Mr. Kapryhan. It is interesting that each of the design centers strongly see the need for resident offices at KSC and are able to make them work but KSC has continued to resist them. ✓ *LBJ — Let's hold the line. B*

03-07-66 (Geissler)

4. Key Personnel Losses: Mr. John Winch, Chief of our Applied Guidance & Flight Mechanics Branch, has resigned effective the end of March. He accepted a position with T.R.W. (Houston), performing the same function for T.R.W. (MSC) as he performed here. His stated reason for leaving is "no future at MSFC." His Branch is responsible for Guidance work, Mission planning and Flight Mechanics for mainstream Apollo and AAP. His talents will be very difficult to replace. ✓ Another extremely valuable key employee in our Unsteady Aerodynamics Branch, has indicated he will most likely leave MSFC to accept a teaching job; he has made extensive contributions in the difficult field of unsteady aerodynamics on IB and V. He is a PhD, GS-13, without hope of getting a promotion under present personnel policy. He will make more money teaching than in his present position.

E.G. After our discussion today (Mr. Dahm's presentation) I'd like to have another private talk with you on this personnel morale problem. Please arrange thru Bonnie. (I'm going on a trip, so our talk should be after 3-27) B

Hans Maus — Do we get reimbursed for this work? Houston misses no opportunity to put its hand into our pocket. I think we should reciprocate. B

4. Gondola Tests for Human Centrifuge at Houston: Agreements have been reached for performance of pressure tests for this Gondola, which is 12 feet in diameter, in our big Autoclave. Tooling for this task has been designed by us and will be fabricated by Hayes. Tests will be started at the end of July. ✓

*Copy attached for Dr. von Braun.

05-09-66 (Kuers)

09-26-66 (Constan)

VISIT OF INTERIM COMMUNICATIONS SATELLITE COMMITTEE

A group of fifteen international members of the Interim Communications Satellite Committee, accompanied by Mr. Richard Colino, Director of International Arrangements for the Communications Satellite Corporation, were given a briefing and tour of the Michoud Assembly Facility on Friday, September 23, 1966. ✓ *Good. Sold 'em a Sat. V? B*

04-04-66 (Williams)

Herm, Weidner and F.W. — I couldn't agree more!! This is precisely the situation. Where can I help more? B

2. Impressions from Gray and Mueller Visits to MSFC: In several instances over the past few months, and again during the Gray and Mueller visits last week, a "message" seems to weave its way through the discussions and meetings. That message is:

(1) We don't have much money to go out for new activities.
(2) With no (or very little) money, we (Headquarters) can't make decisions to start or approve new activities.
(3) New activities must be started - but don't ask us for approval because we can't say yes in this climate. Just go ahead and do things. ✓

AS-201 POSTFLIGHT EVALUATION: Analysis of the AS-201 flight data has revealed only two minor problems: (1) At about 79 and 81 seconds of The high noise levels on these signals are of concern because the EDS vehicle angular velocity abort limits are ± 5 1/2 deg/s in pitch and yaw and ± 20 deg/s in roll. Both of these conditions are currently under investigation and are expected to be corrected prior to the next flight test. The overall performance of all other Astrionics systems was satisfactory. ✓

W.H. Congratulations! A splendid record. B

03-07-66 (Haeussermann)

SUBJECT: S-II Battleship

In this same attempt the gas generator for engine no. 3 indicated an over temperature condition as well as a very low turbine speed. The pump was entered on Saturday, July 23, 1966, and found the first stage turbine wheel was installed backward by the Rocketdyne field crew. This pump is currently being replaced by Rocketdyne.

The next attempt is scheduled for Friday, July 29, 1966.

— Samuel Yarchin

URGENT

I.O. Suggest you prepare a personal letter to Sam Hoffman for my signature. See Page 2 — George D. Wallace, Jr.

"How can we ever hope to go to the moon if something like this is still possible??" B

07-25-66 (Rudolph)

S-IC-4 – very little testing accomplished during the week with "change outs" in progress and scheduled to continue thru September 29. Two deep scratches were found in the lower bulkhead of the fuel tank. They are approximately one half inch apart, and it appears as though a tool had been dropped in the tank. Further investigation is in progress and an engineering evaluation is forthcoming.

09-19-66 (Constan)

*George C.,
Let's find out why the man who dropped it did not report it! If people start hiding goofs like this, we'll find ourselves in endless trouble! B*

Lee James — So we still discover surprises during static testing of SIB stages!!! B

H-1 ENGINE During the first static test of S-IB-208 on 11-16-66, Engine H-4078 (position #6) ran significantly lower in thrust than expected (19,000 lbs. less than the other seven). A quick look at the data pointed to the turbine as the suspect component. Disassembly of the turbine showed that approximately 20% of the first stage blades were missing, second stage blades were dented and scored, and there were dents in the heat exchanger coils. The cause of the failure cannot be determined until the engine is disassembled for further investigation. An apparently

11-21-66 (Brown)

Arthur Rudolph $$! B

utilizes about 5 Service Propulsion Burns and phasing via elliptical orbits to achieve rendezvous and the MSFC proposal consists of putting the CSM into an elliptical orbit and the LEM into a circular orbit utilizing ground phasing and the iterative guidance capability in the yaw plane. (2) MSC apparently is having rather severe weight growth problems, particularly in the CSM at this time. It is anticipated that MSC may have to request a payload commitment increase for Saturn V. MSFC position was that we would support analysis of the S/C requirements and L/V capabilities; however, any commitment change is a Program Office (Dr. Shea & Dr. Rudolph) function to initiate. We are also initiating studies to determine a

04-25-66 (Geissler)

---

1. **S-II Welding Meeting:** You might get tired hearing so often about welding problems. But all of the welding techniques are compromises for the many overlapping aspects in the areas of metallurgy, weld equipment, tooling, quality control, and last but not least, management. In ___ year, a meeting

04-04-66 (Kuers)

W.K. No, I'm not. A lifetime in rocketry has convinced me that welding is one of the most critical aspects of our whole job!! B

10-03-66 (Maus)

H.M. When quoted out of context this text can be damaging. Please put it under wraps, somehow. B

ECONOMIC IMPACT STUDY – The first phase of our on-going economic impact study has been completed and reviewed by Mr. Koenig of NASA Headquarters. As was generally expected, the phase-down in MSFC's operation as Apollo is completed will have an extensive effect upon the Huntsville work force and area economy. A similar effect will be felt by MTF and its surrounding area while New Orleans, on the other hand, with its extensive metropolitan area and diversified economy, will feel little effect as the Michoud Operation phases down. The New Orleans area is singularly different from Huntsville and MTF in that it was experiencing a decided boom in its economy before the activation of Michoud.

1967

11-13-67 (Rudolph)

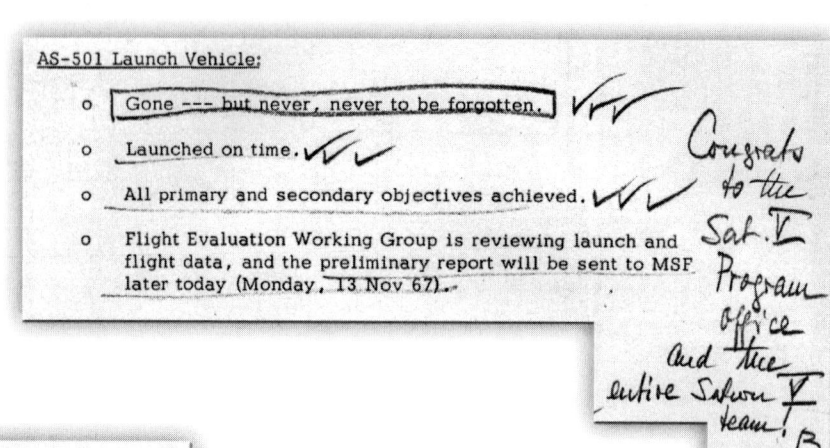

AS-501 Launch Vehicle:

o Gone --- but never, never to be forgotten. ✓✓
o Launched on time. ✓✓
o All primary and secondary objectives achieved. ✓✓
o Flight Evaluation Working Group is reviewing launch and flight data, and the preliminary report will be sent to MSF later today (Monday, 13 Nov 67).

Congrats to the Sat. V Program Office and the entire Saturn V team! B

03-06-67 (Williams)

*F.W.
I think we should go all-out on these studies. Voyager should have a multiple capability: while near-optimum for Mars mission, modular changes should permit Venus, Jupiter, Jupiter-Sun (Thomae profile) and Grand Tour missions. B*

4. Unmanned Planetary Spacecraft: We have been requested by OSSA, (Mr. Pitt Thome of Mr. Oran Nicks' office), to manage system studies of unmanned planetary spacecraft. Specifically, OSSA is desirous of investigating requirements for Voyager-type space-

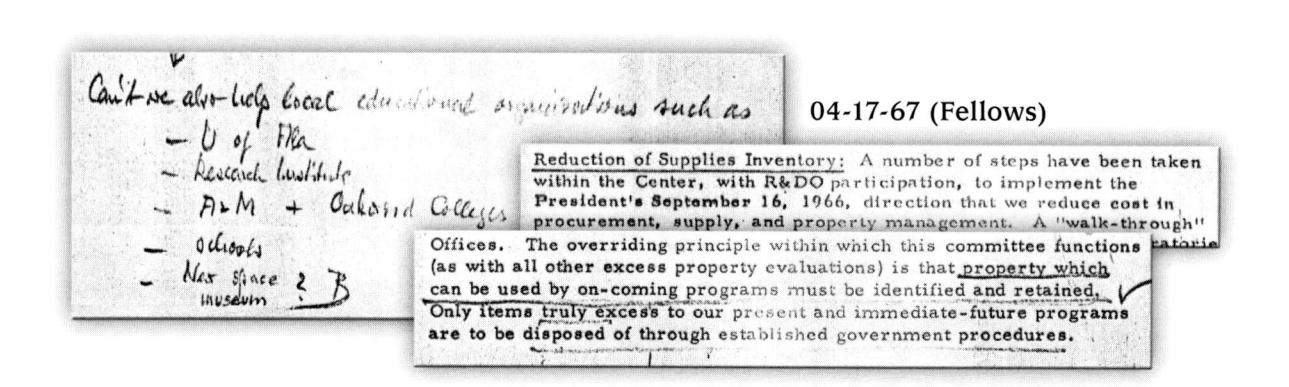

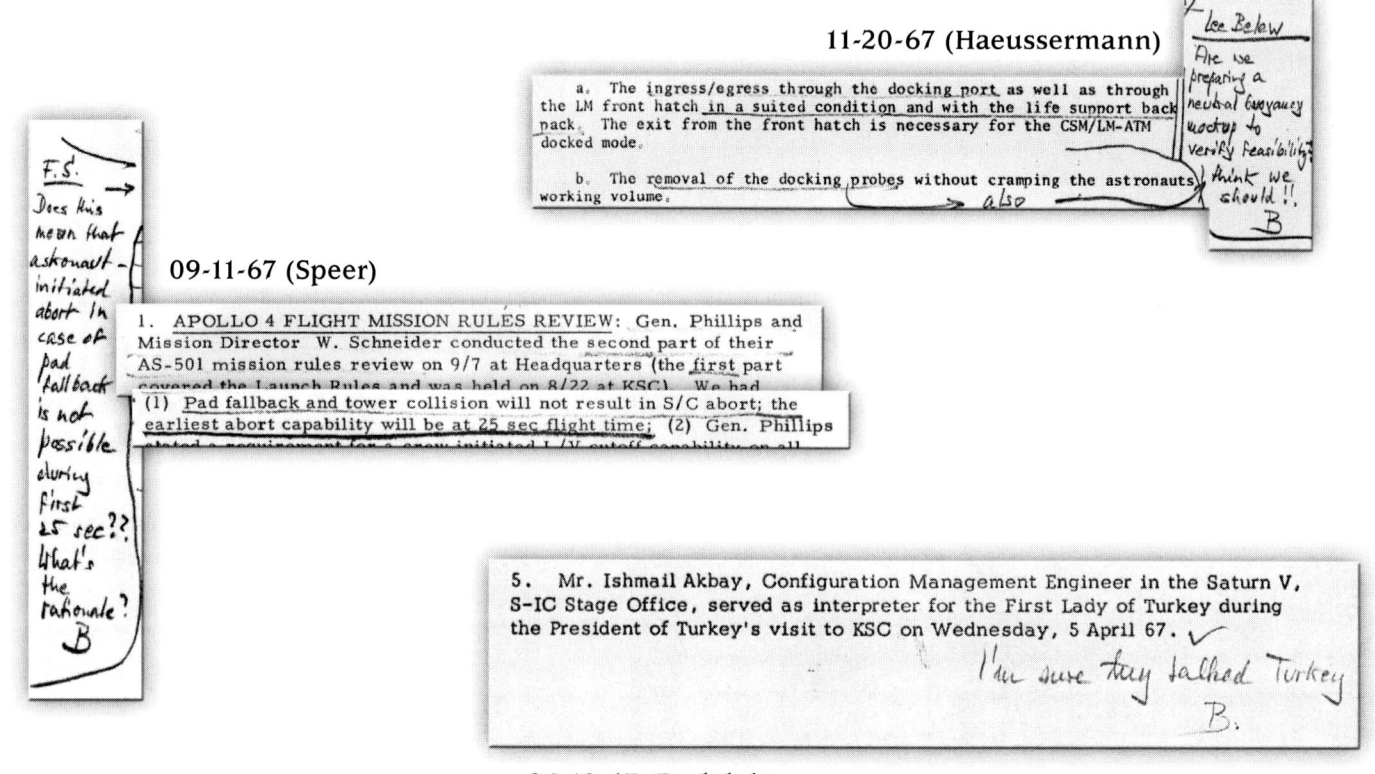

04-17-67 (Fellows)

*Can't we also help local educational organizations such as
- U of Ala
- Research Institute
- A&M + Oakwood Colleges
- schools
- New space museum? B*

Reduction of Supplies Inventory: A number of steps have been taken within the Center, with R&DO participation, to implement the President's September 16, 1966, direction that we reduce cost in procurement, supply, and property management. A "walk-through" Offices. The overriding principle within which this committee functions (as with all other excess property evaluations) is that property which can be used by on-coming programs must be identified and retained. ✓ Only items truly excess to our present and immediate-future programs are to be disposed of through established government procedures.

11-20-67 (Haeussermann)

*See Below
Are we preparing a neutral buoyancy mockup to verify feasibility? I think we should!! B*

a. The ingress/egress through the docking port as well as through the LM front hatch in a suited condition and with the life support back pack. The exit from the front hatch is necessary for the CSM/LM-ATM docked mode.

b. The removal of the docking probes without cramping the astronauts working volume. → *also*

09-11-67 (Speer)

*F.S.
Does this mean that astronaut-initiated abort in case of pad fallback is not possible during first 25 sec?? What's the rationale? B*

1. APOLLO 4 FLIGHT MISSION RULES REVIEW: Gen. Phillips and Mission Director W. Schneider conducted the second part of their AS-501 mission rules review on 9/7 at Headquarters (the first part covered the Launch Rules and was held on 8/22 at KSC). We had (1) Pad fallback and tower collision will not result in S/C abort; the earliest abort capability will be at 25 sec flight time; (2) Gen. Phillips

5. Mr. Ishmail Akbay, Configuration Management Engineer in the Saturn V, S-IC Stage Office, served as interpreter for the First Lady of Turkey during the President of Turkey's visit to KSC on Wednesday, 5 April 67. ✓

I'm sure they talked Turkey. B

04-10-67 (Rudolph)

10-23-67 (Lucas)

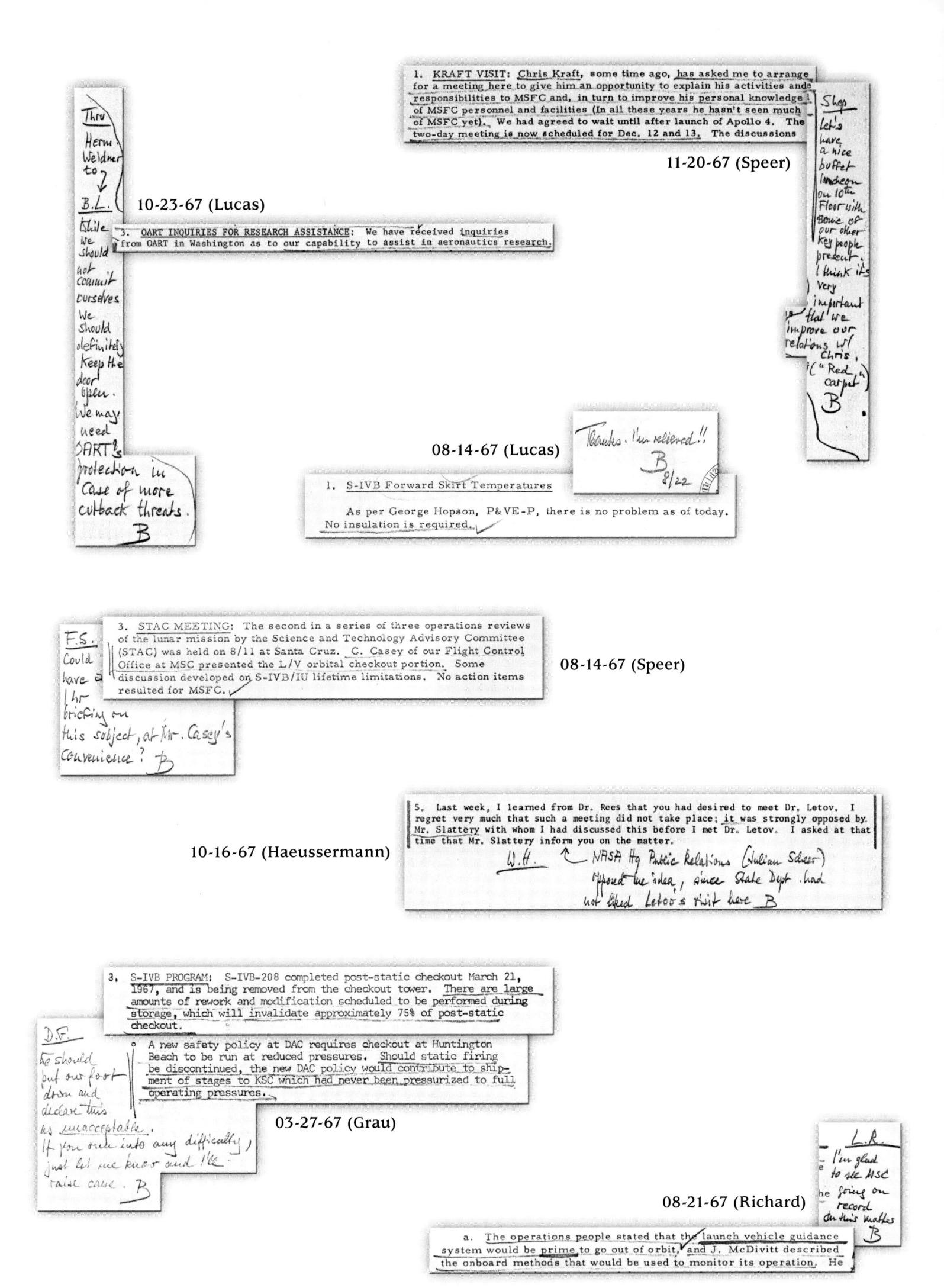

Thru Herm Weldner to B.L.

While we should not commit ourselves we should definitely keep the door open. We may need OART's protection in case of more cutback threats. B

1. KRAFT VISIT: Chris Kraft, some time ago, has asked me to arrange for a meeting here to give him an opportunity to explain his activities and responsibilities to MSFC and, in turn to improve his personal knowledge of MSFC personnel and facilities (In all these years he hasn't seen much of MSFC yet). We had agreed to wait until after launch of Apollo 4. The two-day meeting is now scheduled for Dec. 12 and 13. The discussions

11-20-67 (Speer)

Shep — Let's have a nice buffet luncheon on 10th Floor with some of our other key people present. I think it's very important that we improve our relations w/ Chris. ("Red carpet") B

3. OART INQUIRIES FOR RESEARCH ASSISTANCE: We have received inquiries from OART in Washington as to our capability to assist in aeronautics research.

08-14-67 (Lucas)

1. S-IVB Forward Skirt Temperatures

As per George Hopson, P&VE-P, there is no problem as of today. No insulation is required.

Thanks. I'm relieved!! B 8/22

08-14-67 (Speer)

F.S. Could have a 1 hr briefing on this subject, at Mr. Casey's convenience? B

3. STAC MEETING: The second in a series of three operations reviews of the lunar mission by the Science and Technology Advisory Committee (STAC) was held on 8/11 at Santa Cruz. C. Casey of our Flight Control Office at MSC presented the L/V orbital checkout portion. Some discussion developed on S-IVB/IU lifetime limitations. No action items resulted for MSFC.

10-16-67 (Haeussermann)

5. Last week, I learned from Dr. Rees that you had desired to meet Dr. Letov. I regret very much that such a meeting did not take place; it was strongly opposed by Mr. Slattery with whom I had discussed this before I met Dr. Letov. I asked at that time that Mr. Slattery inform you on the matter.

W.H. — NASA Hq Public Relations (Julian Scheer) opposed the idea, since State Dept. had not liked Letov's visit here B

03-27-67 (Grau)

3. S-IVB PROGRAM: S-IVB-208 completed post-static checkout March 21, 1967, and is being removed from the checkout tower. There are large amounts of rework and modification scheduled to be performed during storage, which will invalidate approximately 75% of post-static checkout.

o A new safety policy at DAC requires checkout at Huntington Beach to be run at reduced pressures. Should static firing be discontinued, the new DAC policy would contribute to shipment of stages to KSC which had never been pressurized to full operating pressures.

D.G. We should put our foot down and declare this as unacceptable. If you run into any difficulty, just let me know and I'll raise cane. B

08-21-67 (Richard)

L.R. — I'm glad to see MSC is going on record on this matter B

a. The operations people stated that the launch vehicle guidance system would be prime to go out of orbit, and J. McDivitt described the onboard methods that would be used to monitor its operation. He

137

1968

05-13-68 (Williams)

2. Chrysler "National Space Booster Study": On 4/30/68, Dan Schnyer (MSF Manager of the Chrysler study contract) met the "Study Team" at MSFC, including representatives from MSFC (IO and R&DO), KSC, MSC, OSSA, OART, and MSF. Study Team participation was discussed, and an orientation meeting with Chrysler was planned. We were later informed by Mr. Schnyer that plans have been changed, based on a 5/3/68 meeting of Messrs. Webb, Mathews, Schnyer, and Lowery (CCSD).

Frank W.
I'd like to have a personal appraisal from you on merits and objectives of this Chrysler study. No formal briefing. Just a chat between the two of us. B

Harry S.
I hope we are taking adequate steps to protect privacy of medical information. B

1. AUTOMATED MEDICAL ANALYSIS SUPPORT: As a first step in the automation of medical records at MSFC, the scheduling and notification of NASA employees for periodic physical examinations is now being accomplished mechanically. Additional support areas have ...
They have a keen desire to mechanically massage the medical data of MSFC personnel in an effort to determine the incidence of various diseases and, hopefully, to attain the ability to detect at an earlier time the onset of disease or its symptoms. They ...

04-08-68 (Hoelzer)

05-13-68 (Balch)

A very nice & constructive item! B

Mr. Richard Lewis, a well-known aerospace writer of the Chicago Sun Times, will visit MTF on 5/27/68.

ATM PRELIMINARY REQUIREMENTS REVIEW (PRR): The ATM PRR was successfully conducted last week. A large number of Review Item Discrepancies were written of which the majority are meaningful comments which deserve serious consideration. However, the incorporation of some could have a significant impact on the ATM design, such as additional redundancy, automation of astronaut functions, redesign of control and displays, etc. The most significant impact could be in the control and display area wherein use of the digital address system has been seriously questioned by MSC. Since the comments in the control ...

L.B.
looks like this display business (ATM) is growing into a major potential problem area between MSFC and MSC. [illegible] brought this up in a recent meeting with [illegible] and myself. B

10-07-68 (Lucas)

01-29-68 (Belew)

RUSH
B.L. Here it is again! Waiting for some "failure history"! I'm ready to buck the decision for 504/505. If you concur, please draft a suitable polite but strong letter to Phillips for my signature (Refer to my previous gripes in this matter). B 10-11

probability of losing most of the J-2 engine data is high again. General Phillips has disapproved these changes for AS-503 and is holding up the incorporation of the changes into AS-504 and 505 until after the AS-503 flight. Unless every effort is made to immediately implement these changes for AS-504 and 505, time limitations will probably again prevent their implementation.

ME Laboratory Internal Information Program: In November, 1967, we initiated this program in ME Laboratory as a result of the studies of Dr. Tompkins and Mr. Richetto from Purdue University which revealed a need for improvement of communications mainly from Division level down to the first line supervisors. The objective of the program is to improve the team spirit in the organization, to enhance the feeling of participation of all supervisory personnel in our challenging programs and to improve the awareness of responsibilities for building flight hardware.

Shep
Suggest we follow this example in all labs and offices. B

05-06-68 (Kuers)

AS-502 Flight Performance: Once the shock of having significant failures on AS-502 is over and those failures have been identified and fixed, we intend to make a compilation of all the things we have learned from this flight. We demonstrated significant characteristics and capabilities that might never have been demonstrated on a totally successful shot. Such areas as EDS, flight control, separation dynamics, guidance, performance reserves, etc., really were exercised.

04-08-68 (Richard)

Ludie
But don't feel encouraged to have a "repeat performance" of 502!! B

National Launch Vehicle Stable Study: We have recently received some unofficial information on the way NASA/DOD plans to implement some closer planning in the review of the National Launch Vehicle Stable. It is expected that the Aeronautic and Astronautics Coordinating Board (AACB) will again be requested to conduct such a study.

H.M.
Maybe this will offer us a new opportunity to promote the SIB/SIVB/SM configuration. Please discuss with Frank Williams. B

02-12-68 (Maus)

OFFICE OF DIRECTOR – MSFC

CODE	NAME	INIT.	A C T I O N	INFORMATION
	Dr. von Braun			
	Cp to Lucas & Belew 11-13			
	Weidner & Murphy 11-14			

REMARKS

Lee Belew and Bill Lucas would like to get your opinion on how they should pursue this matter of different specifications. We can do one of the following:

a. Accept the Headquarters-issued Houston specification (this is considered unsatisfactory). *Shep Please tell them.*

b. Hold out strongly for our spec. *Yes, but*

It would appear that we should take course b. above and try to get Headquarters to negotiate a position between our spec and Houston's spec which would permit the utilization of our specification on the Workshop and not disturb the Houston spec on the Apollo program.

I suggest the following approach: In our disagreement with the MSC spec, let us avoid any language which questions the safety of the Apollo hardware. But take the position that OWL is second generation hardware, and a new start that permits superior solutions all the way through. ("You don't use DC-3 specs in the jet age.") This position doesn't question the safety of

CODE	NAME	DATE
DIR	J. T. Shepherd	11-8-68

MSFC - Form 495 (Rev August 1963)   *B 11-9*

11-08-68 (Shepherd)

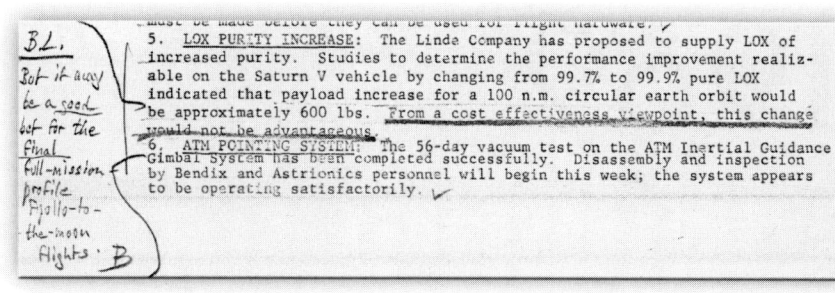

---

5. **LOX PURITY INCREASE:** The Linde Company has proposed to supply LOX of increased purity. Studies to determine the performance improvement realizable on the Saturn V vehicle by changing from 99.7% to 99.9% pure LOX indicated that payload increase for a 100 n.m. circular earth orbit would be approximately 600 lbs. From a cost effectiveness viewpoint, this change would not be advantageous.

6. **ATM POINTING SYSTEM.** The 56-day vacuum test on the ATM Inertial Guidance Gimbal System has been completed successfully. Disassembly and inspection by Bendix and Astrionics personnel will begin this week; the system appears to be operating satisfactorily.

B.L. But it may be a good but for the final full-mission profile Apollo-to-the-moon flights. B

06-10-68 (Lucas)

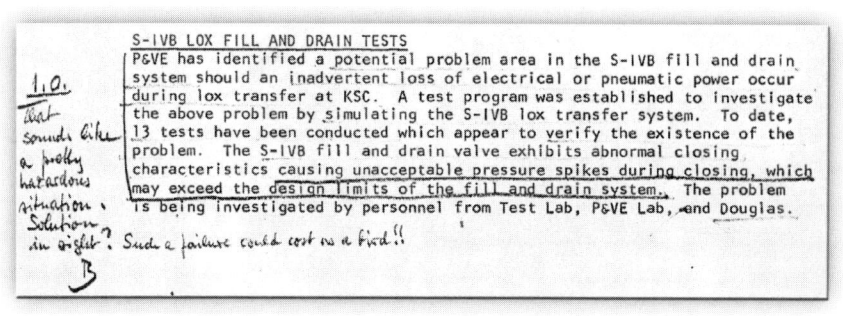

S-IVB LOX FILL AND DRAIN TESTS

P&VE has identified a potential problem area in the S-IVB fill and drain system should an inadvertent loss of electrical or pneumatic power occur during lox transfer at KSC. A test program was established to investigate the above problem by simulating the S-IVB lox transfer system. To date, 13 tests have been conducted which appear to verify the existence of the problem. The S-IVB fill and drain valve exhibits abnormal closing characteristics causing unacceptable pressure spikes during closing, which may exceed the design limits of the fill and drain system. The problem is being investigated by personnel from Test Lab, P&VE Lab, and Douglas.

L.O. That sounds like a pretty hazardous situation. Solution in sight? Such a failure could cost us a bird!! B

02-12-68 (Heimburg)

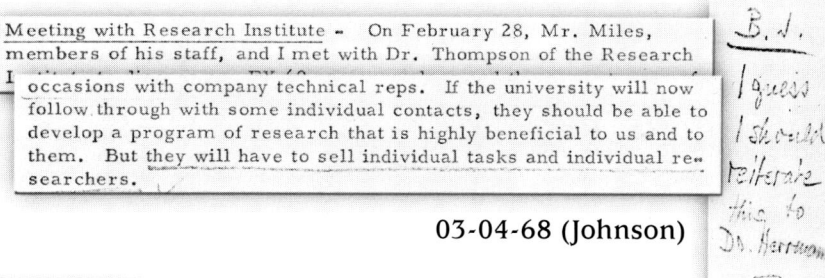

Meeting with Research Institute – On February 28, Mr. Miles, members of his staff, and I met with Dr. Thompson of the Research occasions with company technical reps. If the university will now follow through with some individual contacts, they should be able to develop a program of research that is highly beneficial to us and to them. But they will have to sell individual tasks and individual researchers.

B.L. I guess I should reiterate this to Dr. Herrmann B

03-04-68 (Johnson)

---

2. **AS-502 Data Problems:**
...adequately supported by raw and processed data deliveries. A few shortcomings of the types mentioned above probably cannot be completely avoided. However, action is being taken to avoid repetition of these specific failures, and I would like to endorse the continued need for a strong MSFC in-house data reduction capability.

F.S. I'm all for it. This is one of our most important functions in the forthcoming Apollo flight series! B

06-24-68 (Speer)

---

1. **Stratoscope II.** Stratoscope II was launched at 8:55 P.M. on May 18, 1968 for its 6th flight. The balloon system and the mechanical and electronic equipment of the telescope functioned perfectly. The photographic image of first. Our review committee has essentially contributed in establishing the flight worthiness of Stratoscope II. Mr. Boehm, who chaired the review team, several other team members and I were in Palestine, Texas for the launch event.

My most cordial personal congratulations! I know you deserve much credit for this fine success. Could I get a briefing on the scientific results when available? B

06-03-68 (Haeussermann)

04-08-68 (Heimburg)

S-II-5 (MTF)
S-II-5 is presently installed in the A-1 Test Stand at MTF undergoing systems hook-up and checkout. A cryogenic proof pressure test is scheduled for April 16, 1968, to be followed by the acceptance static firing test.

L.O. Can we conduct the static firing during Dr. Paine's visit to MTF, 22 April, afternoon? B

Effects of Sub-Orbital Trajectories of S-IVB-IU on Passenger Experiments – Decisions appear to have been reached to fly the AS-206 (AAP-I) in a mode such that the S-IVB-IU does not attain orbit. The plan was to fly a P&VE orbital trajectory of the IU is being planned. Prof. Kraushaar and the University of Wisconsin Group now have four years invested in the experiment -- which is OSSA sponsored. We will try to work out with the AAP Office alternatives which will permit flying the experiment but there are

Lee Belew Was this commitment to Bill Kraushaar (indeed a faithful supporter) overlooked when the mission profile was changed? What do you propose to do? I think we shouldn't simply pull the rug from under him! B

10-07-68 (Johnson)

1969

2. FUTURE MSF FILM: A pre-production meeting was held on May 28, 1969, with Capt. Freitag on a new MSF movie. This film will cover space station, space shuttle, lunar exploration, and space science and applications. It will be produced by the AV Corporation in Houston, a contractor to MSC. Presently planned MSFC contributions are: simulated space station operations in a full size mockup, animated space station and space base model operations, astronomy module animations, and lunar surface mobility operations. The film is planned to be available after the lunar landing mission in July.

06-02-69 (Huber)

Stu — I think we should appoint someone to see that MSFC's inputs are properly used and that this doesn't come out as a completely MSC-slanted pitch. B

11-24-69 (Stuhlinger)

EXECUTIVE LEADERSHIP CONFERENCE ON PUBLIC AFFAIRS FOR SCIENCE EXECUTIVES: For six days last week, I attended this conference at Williamsburg, arranged by the Brookings Institution. Ten ... It became painfully clear to us that our government is more complex, decision making is more involved, pressure groups are more influential, opinions are more divided, passing a law is more difficult, expediency is more absent, parochialism is more present, committees are more sterile, the dangers of pollution and poisoning are more imminent, and the fact that "it still works" is more surprising than all of us had thought before. The Kennedy way ("get American astronauts to the moon and back in this decade") seems to be the only way in which significant progress in important fields can be achieved.

E.S. Anything else wrong? B

01-20-69 (Brown)

Bill B. Could this still become a show-stopper for Apollo 9? B

J-2 ENGINE – As you heard at the Center Staff and Board Meeting last Friday, the J-2 engine oscillations on S-II 503 have not been fully explained and detailed studies by Science and Engineering Directorate, NAR/Space Division and Rocketdyne are continuing.

01-06-69 (Brown)

F-1 ENGINE – Based on our experience to date, it is highly probable that the temperature under the thermal insulation would drop below the launch redline value were the weather to be cold and windy (as was experienced at KSC a few days before the launch of AS-503). In an effort to preclude a delay in CDDT or launch under...

B.B. Florida weather extremes ... under ... engine have been ... known for centuries. How come? B

03-10-69 (Newby)

4. EVALUATION OF MSC SPACE SHUTTLE CONCEPT: Attached is an artist's conception, picked up at the Flight Research Center illustrating the probable evolution of the MSC Space Shuttle Concept.

NASA 1969 — This shuttlecraft will do everything according to our paper study.

NASA 1970 — Preliminary wind tunnel studies indicate this shuttlecraft may do everything.

NASA 1971 — More detailed studies indicate this shuttlecraft can do most of the mission.

1972 — After exhaustive studies we have found the shuttlecraft to perform the mission.

MSFC'S PROGRAM HISTORY OF PROJECT SATURN:

Dr. Barton C. Hacker of the University of Houston, Senior Contract Historian for NASA's Manned Spacecraft Center, has orally accepted an offer from the University of Alabama at Huntsville to serve as Chief Historian for MSFC's Program History of Project Saturn, under Contract NAS8-21321. Formal confirmation ... immediate future.

Dave N. I hope Hacker is not a completely brainwashed MSC historian who views the spacecraft as a Lady Godiva and the launch vehicle as her horse! B

01-06-69 (James)

That's good to know. Surely reassuring. B

Apollo 8: From time to time, we have naturally worried as to whether critical technical personnel have drifted away from their specialties on Saturn V. During 503 preparation, we had occasion to examine many random, critical subsystems. We always found the right people hard at work on any concerns we had. I would like to take this occasion to thank all who made 503 such a fine success.

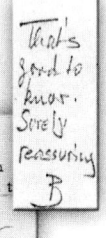

09-29-69 (Geissler)

140

08-19-69 (Haeussermann)

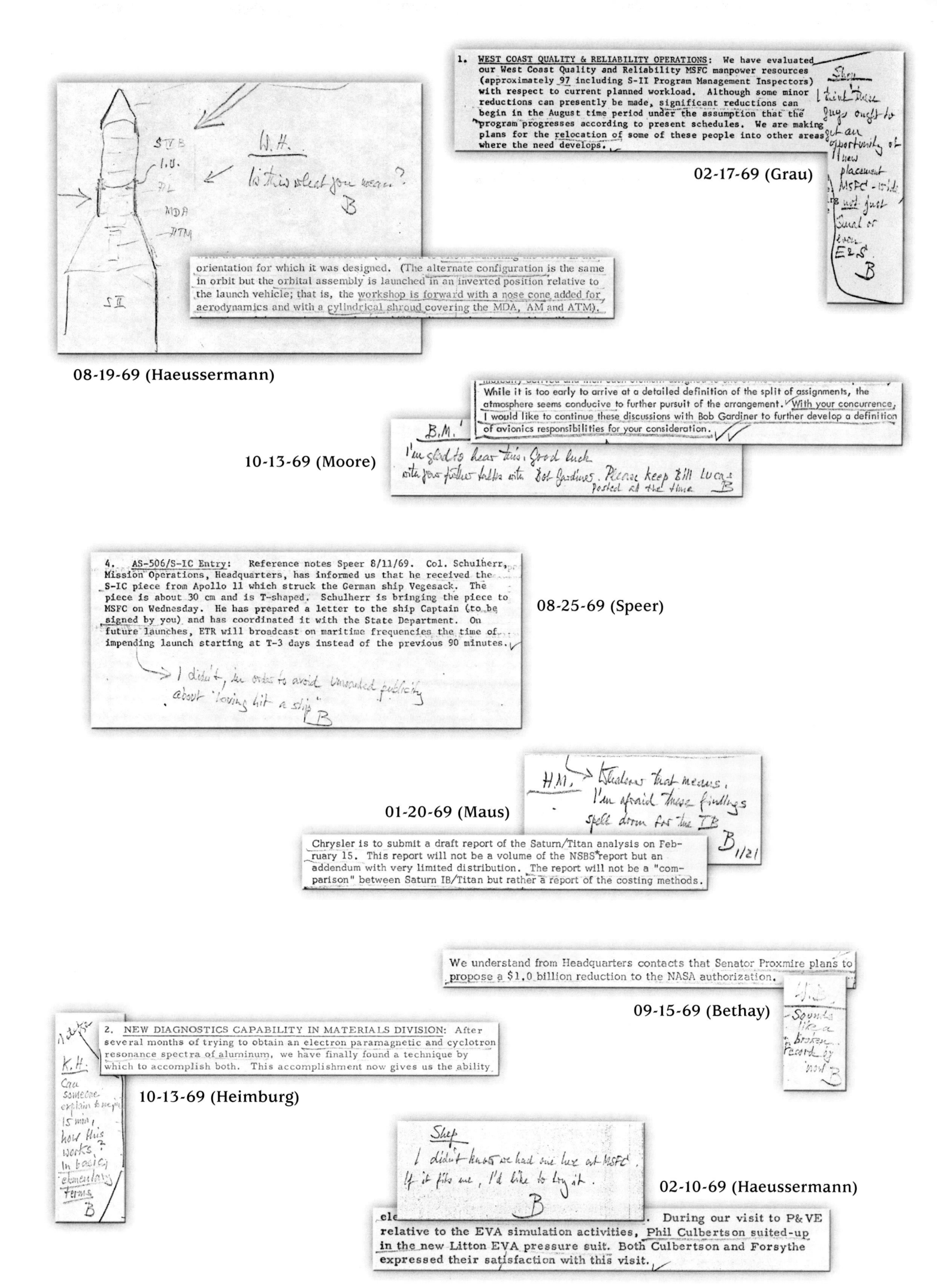

[sketch of Saturn vehicle with labels: STU B, I.U., ?L, MDA, ATM, S II]

W.H.
Is this what you mean?
B

orientation for which it was designed. (The alternate configuration is the same in orbit but the orbital assembly is launched in an inverted position relative to the launch vehicle; that is, the workshop is forward with a nose cone added for aerodynamics and with a cylindrical shroud covering the MDA, AM and ATM).

---

1. WEST COAST QUALITY & RELIABILITY OPERATIONS: We have evaluated our West Coast Quality and Reliability MSFC manpower resources (approximately 97 including S-II Program Management Inspectors) with respect to current planned workload. Although some minor reductions can presently be made, significant reductions can begin in the August time period under the assumption that the program progresses according to present schedules. We are making plans for the relocation of some of these people into other areas where the need develops.

02-17-69 (Grau)

Shy —
I think these guys ought to get an opportunity of new placement MSFC-wide and not just Qual or even E&D.
B

10-13-69 (Moore)

While it is too early to arrive at a detailed definition of the split of assignments, the atmosphere seems conducive to further pursuit of the arrangement. With your concurrence, I would like to continue these discussions with Bob Gardiner to further develop a definition of avionics responsibilities for your consideration.

B.M.
I'm glad to hear this. Good luck with your further talks with Bob Gardiner. Please keep Bill Lucas posted at the time. B

08-25-69 (Speer)

4. AS-506/S-IC Entry: Reference notes Speer 8/11/69. Col. Schulherr, Mission Operations, Headquarters, has informed us that he received the S-IC piece from Apollo 11 which struck the German ship Vegesack. The piece is about 30 cm and is T-shaped. Schulherr is bringing the piece to MSFC on Wednesday. He has prepared a letter to the ship Captain (to be signed by you) and has coordinated it with the State Department. On future launches, ETR will broadcast on maritime frequencies the time of impending launch starting at T-3 days instead of the previous 90 minutes.

→ I didn't, in order to avoid unwanted publicity about "having hit a ship". B

01-20-69 (Maus)

H.M. → Whatever that means. I'm afraid these findings spell doom for the IB B 1/21

Chrysler is to submit a draft report of the Saturn/Titan analysis on February 15. This report will not be a volume of the NSBS report but an addendum with very limited distribution. The report will not be a "comparison" between Saturn IB/Titan but rather a report of the costing methods.

09-15-69 (Bethay)

We understand from Headquarters contacts that Senator Proxmire plans to propose a $1.0 billion reduction to the NASA authorization.

H.B. — Sounds like a broken record by now. B

10-13-69 (Heimburg)

K.H. Can someone explain to me in 15 min. how this works in basic, elementary terms. B

2. NEW DIAGNOSTICS CAPABILITY IN MATERIALS DIVISION: After several months of trying to obtain an electron paramagnetic and cyclotron resonance spectra of aluminum, we have finally found a technique by which to accomplish both. This accomplishment now gives us the ability

---

02-10-69 (Haeussermann)

Shep
I didn't know we had one here at MSFC. If it fits me, I'd like to try it. B

ele... relative to the EVA simulation activities, Phil Culbertson suited-up in the new Litton EVA pressure suit. Both Culbertson and Forsythe expressed their satisfaction with this visit.

Backdropped by the blackness of space and Earth's horizon, the International Space Station is seen from Space Shuttle Discovery as the two spacecraft begin their relative separation. Earlier the STS-119 and Expedition 18 crews concluded 9 days, 20 hours and 10 minutes of cooperative work onboard the shuttle and station. Undocking of the two spacecraft occurred at 2:53 p.m. (CDT) on March 25, 2009.

Stories from the Insiders

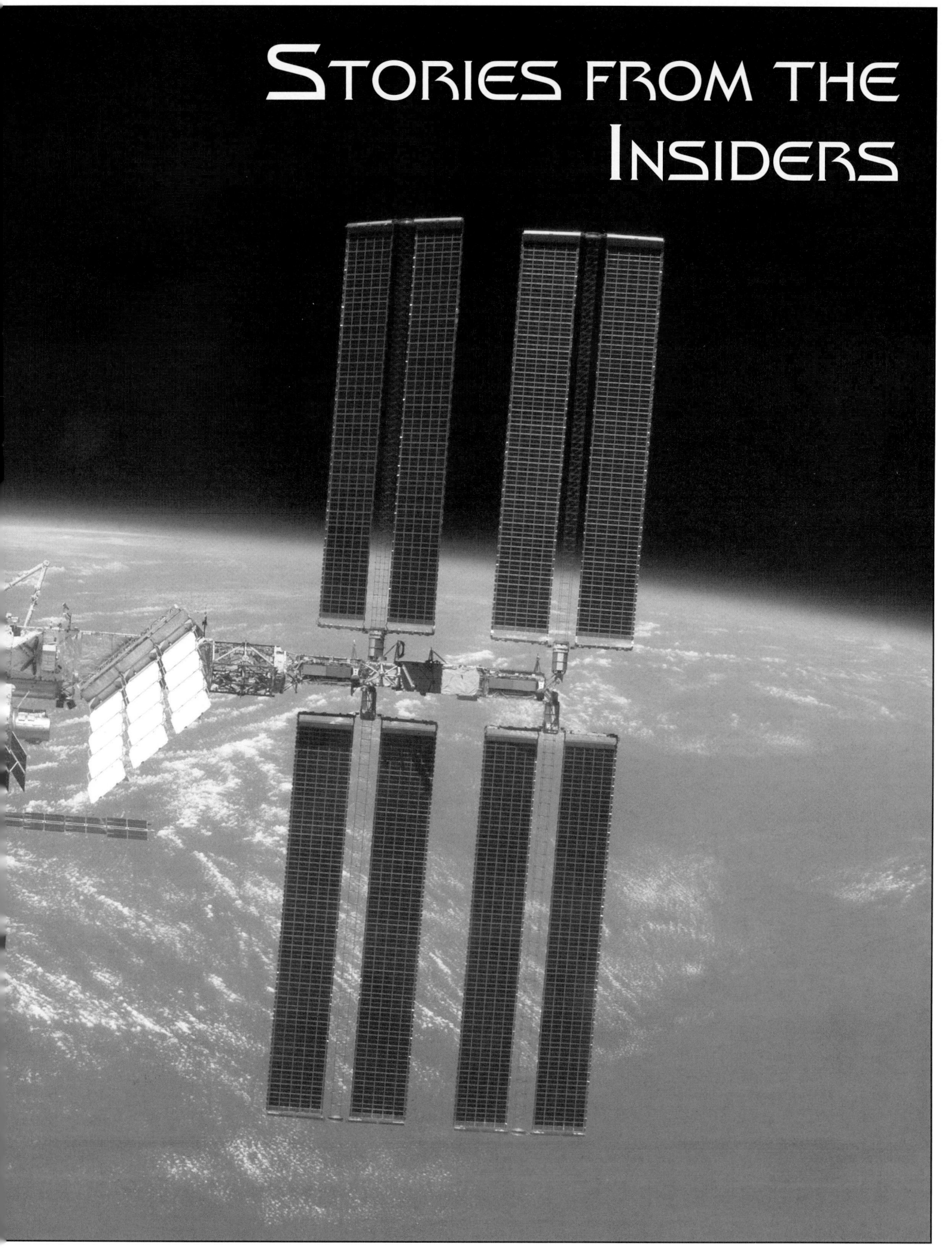

Peenemünde, the Cradle of Space Flight
by Konrad K. Dannenberg

The 50 years of rockets and spacecraft in the Rocket City were the most successful ones in the history of space flight. A modified Redstone missile launched the first American satellite into earth orbit and put two astronauts into sub-orbital flights. The Saturn V launched 24 astronauts to the moon. The Space Shuttle conducts routine missions to put satellites and scientific payloads into Earth orbit and is presently assembling, changing out crews and resupplying the International Space Station.

How did it all start?

On October 3, 1942, the German Rocket Research and Development Center of Peenemünde opened space for humanity. This date records the first time that a machine of human construction had invaded space. The rocket had used space as a bridge between two points on earth. A new means of transportation had been created and had proven rocket propulsion practicable for space travel. I was one of the fortunate participants in this event to experience first hand the "Beginning of the Space Age." Its location "Peenemünde" has often been referred to as "The Cradle of Space Flight".

Here are some pertinent facts of this event:

The A-4 was the first large liquid-propelled rocket. It had succeeded after a number of launch failures to project itself and a dummy payload over a distance of almost 200 Kilometers in a five-minute flight. During its ballistic trajectory it had obtained an altitude of more than 50 miles (84 Kilometers) and had therefore traveled through space.

This feat could be accomplished even though rocket system requirements are an order of magnitude more complicated and difficult to achieve than airplane requirements. There are two primary reasons that create almost insurmountable problems. These can be summarized as:

(1) Rockets need to take-off straight up in order to get out of the atmosphere as quickly as possible. That means the rocket thrust has to be larger than the total weight of the system. Airplanes use their wings to build up lift gradually on a long runway to obtain take-off speed. This requires engine thrust levels which can be much lower than the airplane weight.

(2) Airplanes carry only fuel and burn it with the oxygen in the air, while rockets have to take their oxidizer as well. This at least doubles the rocket take-off weight when standard fuels are used. For liquid hydrogen, the most efficient fuel, the take-off weight will be five times heavier than it would for a hydrogen-powered aircraft.

These are the primary reasons that most early rocket amateurs did not succeed to enter space. Wernher von Braun was an original member of the amateur group in Berlin. He realized that any space vehicle development requires major support in many fields of activity. He therefore left the amateur rocket group and joined Walter Dornberger, a captain in the German Army. His department had indicated interest in ballistic missiles as a replacement for heavy artillery, which had been prohibited to Germany by the Treaty of Versailles.

In the early 1930s I was an amateur rocketeer in Hannover/Germany. In the Spring of 1940 I was called to Peenemünde based on this previous amateur experience. At that time, I had been a soldier in the German army. My unit went after my transfer to Peenemünde to Russia, and only one seriously wounded person came back. Eventually the A-4 propulsion system became my "baby." The first successful launch of the apparently "gigantic" A-4 was the highlight of my career. Our team had shown that good engineering and proper application of extensive resources could overcome the problems summarized above. Although I had been assigned important responsibilities for later outstanding launches, none of those missions can surpass the pride and exhilaration of this first successful launch in Peenemünde in 1942.

Later in the United States, I was in charge of the Chrysler contract to manufacture Redstone and Jupiter missiles in the 1950s. Chrysler built the launchers for the first U.S. satellite and Alan Shepard's and Gus Grissom's suborbital Mercury flights. The boosters had to be modified at Redstone Arsenal to meet the new space launch requirements. When I transferred in 1960 to MSFC, I became deputy-manager of the existing Saturn program reporting to Manager Oswald Lange. When Arthur Rudolph left his Army employment and joined the Marshall Space Flight Center, Wernher von Braun assigned the Saturn program management to him due to his experience in Germany with V-2 production and with the development and manufacture of the Pershing missile. I was not "in charge" of the programs for the most impressive circumnavigation of the Moon and later on, for the astounding landing on the moon. Therefore, none of these major events can surpass for me the elation of this first successful launch of the "huge" A-4 rocket. At the time of the Lunar Landing, I was working on Post-Apollo programs and early versions of a Space Station.

First Launch
by James Bramlet

Wow! What a spectacular sight!

That was my impression of the first V-2 launch that I witnessed on the night of Dec. 17, 1946 at White Sands Proving Grounds (WSPG), New Mexico. The desert was perfectly clear with many stars in the sky and the weather quite cold as winter nights are in the desert. The launch took place a little after 10:00 p.m.

The alcohol/oxygen flame of the V-2 was like a huge inverted Bunsen burner, very bluish with perfect diamonds formed in the exhaust. The boost phase lasted about one minute. The flame disappeared at cutoff, and the carbon jet veins glowed red as the only remaining visible indication of the flight. This flight set a new altitude record of 116 miles. A fleeting memory recalled a trip I made through England in the fall of '44 as a member of a regiment headed for the European theater during WW II. News reports described a new weapon, the V-2, delivering explosive warheads on England. What a welcome change to be viewing the same delivery vehicle carrying scientific payloads. I had just arrived on the scene a couple of weeks earlier, but this was enough to tweak my interest and curiosity and to sense I had come to the right place at the right time.

Many spectacular launches occurred at WSPG, but none more spectacular than the famous missile "Zero." This launch took place on a warm sunny afternoon in May 1947. Several delays pushed the actual launch time to shortly after 7:00 p.m. The sun had already gone behind the Organ Mountains and the desert floor was in shadow. The lift-off was routine. The flight climbed enough to come into the sunlight and the missile seemed to glow.

The Peenemunde Rocket Team reunited on the steps of Marshall Space Flight Center's Headquarter Building 4200 for a reunion, March 1, 1987.

NASA - Marshall Flight Space Center

The Von Braun Rocket Team, poses for a group photograph at Fort Bliss, Texas in 1946.

I was observing the flight from "C" station three miles due south of the launch site. As the missile gained altitude, my neck became more and more strained until finally I had to change directions 180 degrees to follow the bright "sun-bathed" vehicle. The flight was easily observed for the entire sun exposed trajectory. As the vehicle descended to near impact, it went into the mountain shadows again and was not visible. Strangely though, it went south—directly over the south "C" station, when it should have gone north from the launch site.

As our carpool left for El Paso sometime after impact, we heard on the news that the missile had landed in a cemetery in Juarez, Mexico, some 60 miles south of the launch site. Did that ever create an international outburst! But the outburst was short-lived, as some fast work by several officials brought peace and order to U.S.-Mexico relations.

I began my civil service tenure with the von Braun team as design engineer in the combustion chamber development group of the ramjet project. Dr. Gerhard Heller and Krafft Ehricke headed the project with Dannenberg, Palaors and Poppel heading the chamber development. Magnus von Braun, Wernher's brother, supervised the test facility. We were evaluating, among other things, propellant combinations, fuel injectors, and flame holders, throat and nozzle configuration, and heat transfer. The von Braun team consisted of 117 German engineers and scientists that made up part of the Peenemunde development group who came to the U.S. at the close of WW II. We had a few communication/translation problems early, but most were of the humorous nature.

Work was interrupted in early 1950 because of the decision by the Army to move the development team to Huntsville, Alabama. Many of our families were upset because we were just getting adjusted to the dry, arid southwest. However, arriving in Huntsville in mid-June 1950, the change from the brown desert to green vegetation with trees was welcome. The only problem—housing was in very short supply.

Traveling with Dr. von Braun, as I occasionally had the privilege, brought some unusual surprises and cherished "perks". On another trip to the west coast von Braun was invited to dinner at Donald Douglas' home in Santa Monica. Four of us from Huntsville, von Braun, Ludwig Roth, Konrad Dannenberg, and I spent a cordial and interesting evening in the beautiful home of Mr. Douglas. The dinner was excellent. Each place setting had a matchbox with the silver pattern inlayed. I took the box as a souvenir. I plan to pass the souvenir to my grandson some day. He will no doubt ask, "Who is Donald Douglas?"

Professor Hermann Oberth was invited to visit Huntsville in the fall of '57 to view some of the technology development in our laboratories and in our contractor's facilities. Professor Oberth was well known for his pioneering work in the theoretical treatment of earth orbital flight and space flight in the early 1920s. The visit to our west coast contractors occurred in November/December of '57. The Russian Sputnik had just been placed in orbit. Local viewing times were being published and carried on news reports. Professor Oberth, Mr. Dannenberg and I were staying overnight at the Knickerbocker Hotel in Hollywood this particular night and arranged to gain access to the roof area to search for the orbiting vehicle. Sure enough, there it was, visible to the naked eye, streaking across the sky. As we watched this strange sight for about a minute and a half, I couldn't help but think how small our earth had suddenly become and how privileged I was to have shared this moment with the person who defined the possibility of such an event some 25 to 30 years earlier.

How I Stumbled Into the Early Days of the Space Program
by Lucian Bell

Upon graduation from the University of Alabama in 1950 with a degree in electrical engineering and having received a job offer from Seagrams Distillery in Lawrenceburg, Indiana as a plant trainee, the wife and I decided to spend two weeks with my family in Tampa, Florida before reporting for work at Seagrams. While in Tampa, I decided that I would look for a job more in line as an engineer and not as a plant trainee. This decision might seem strange for a person who had been raised in Kentucky on Bourbon toddy's as a young boy. This would usually cure any childhood colds or flu you might come down with. I thought I would try the Florida Employment Office to see if they had any job listings for EE's in the state of Florida. The office told me they had no listings for EE's but had heard of a company on the East Coast that might be hiring new graduates in engineering. I was told to see the Ralph M. Parson's Company at Patrick Air Force Base. I drove over the next day and had an interview with Mr. Charles Feast and was hired to go to work with them as an EE. As a result, I became involved in the electrical design and layouts for the first Launch Pads, Blockhouse and Central Control building to

A V-2 rocket takes flight at White Sands, New Mexico, in 1946.

145

be built at Cape Canaveral, then called the Joint Long Range Proving Ground, which was a joint effort of the Army, Navy and Air Force. Later on I was assigned to the Matador Expediting Group, which was responsible for making sure the range was ready for the testing of the MATADOR missile system. Meanwhile, North American Aviation had been testing the SNARK missile system. It was a larger version of the German "Buzz Bomb" of World War II. It was not achieving much success, because quite a few of them had fallen into the ocean a short distance from the launch ramp. One morning we noticed that during the night someone had posted a sign on the beach that read: BEWARE OF THESE SNARK INFESTED WATERS. Shortly after the testing of the MATADOR System had gotten underway, I transferred to the Air Force Armament Center at Eglin Air Force Base, Florida, where I worked on bombing and navigation systems for the Air Force. In January 1956 I went to work with the Wernher von Braun Team at Redstone Arsenal, AL.

One of the more memorable characteristics of Dr. Wernher von Braun was his concern for his fellow workers. During the Apollo Program there were meetings at NASA Headquarters that Dr. von Braun and some of his engineers would have to attend. Sometimes these meetings would last until 9 or 10 p.m. and by the time the NASA plane landed at the Huntsville Airport it would be 2 to 3 o'clock in the morning. Some of the engineers would have their own cars there and others would have family members there to take them home, but there would be a few who did not have transportation. Of course, Dr. von Braun had a NASA car and driver there to deliver him to his home, so he and his driver would take us home before he went home himself. He made sure that all of his people had a ride home before he left the airport himself. It is no wonder that his people had a deep sense of loyalty to this man.

During the Apollo Program, Marshall Space Flight Center maintained a Launch Control Center in Huntsville to monitor the vehicle systems such as Guidance and Control, Propulsion, Structures, Instrumentation and Computer Programs. If a question or problem occurred, then the senior engineer for the system in question would make a determination whether to hold, cancel or go ahead with the launch. On each Apollo mission the Saturn V vehicle would place the 3rd stage, instrument unit and spacecraft into orbit; after two orbits if all systems were performing correctly then the third stage would propel the spacecraft on its lunar trajectory. In the event the launch vehicle was slightly off in its timing or guidance, then the spacecraft would have to make a mid-course correction to achieve the proper lunar orbit. On one of the launches the third stage burnout of orbit was so accurate that no mid-course correction was required. This would be like making a thousand mile golf shot for a hole in one.

Then the senior engineer for guidance and control, Jerry Mack, sat up in his chair, threw out his chest and said, "You know, that was pretty damn good for government work."

Part of the Team
by Brooks Moore

I was greatly surprised on my first interview visit to Huntsville and Redstone Arsenal 50 years ago (December 1951) to see very austere surroundings and sparsely equipped laboratories. While I was being interviewed and shown the workplace, a young engineer in the Guidance and Control Laboratory on "Squirrel Hill" showed me the lab's sole oscilloscope. He proudly explained its function and capabilities to me. I patiently listened, although I was employed at the time in a very well-equipped US Navy Lab. Even though I felt like I was stepping back in time technologically, I accepted the job at Redstone. I somehow sensed that I was at the right place at the right time. Joining the "von Braun Team" is a decision that I have never regretted.

I spent the first few weeks of my employment in Huntsville in a dilapidated building in a remote area of the arsenal. My first encounter with Dr. Wernher von Braun came at that remote location only a few days after I reported to work. A senior technician and I were assigned to evaluate an antiquated helium leak detector that had been borrowed from another government agency. Our goal was to see if it could be applied for use in checking the integrity of propellant tanks. With no prior notice, we were greatly surprised when in walked one of our supervisors accompanied by Dr. von Braun.

After brief introductions, the supervisor went into his sales routine espousing the potential of the equipment that we were testing. I sensed that Dr. von Braun was very skeptical about the practicality of the intended application. Suddenly Dr. von Braun turned to me and said, "Brooks, what do you think?" I stammered while I tried to sort out the hazards of siding with my third-level supervisor or with the "top-level boss." I decided that one can never go wrong by telling the truth. I told Dr. von Braun that I was too new in the job to have a useful opinion on the subject.

I tell the above story to illustrate the genesis of the "team atmosphere" that existed within the group. Dr. von Braun cultivated that atmosphere by his uncanny ability to remember names and technical details, and by his practice through the years of asking everyone involved for their opinion. His approach proved to be a successful management technique, as everyone was made to feel that they were involved and that their opinion was valued.

One of the early lessons I learned at Redstone was that ingenuity and dedication can more than compensate for a lack of fancy, expensive laboratories. The German team obviously possessed a firm foundation in the major disciplines required to develop guided missiles. The fact that their budgets were severely constrained may have been an impediment, but certainly was no roadblock. They were dedicated and determined gentlemen who were focused on getting the job done. They were open-minded and patient in their mentoring of young native-born American engineers like myself. I can recall on more than one occasion when I came up with what I thought to be a bright idea on a problem resolution, I was politely told that "we tried that already in Peenemunde, and it didn't work!" Out of this austere funded and understaffed operation of the early 1950s emerged the Redstone guided missile, which was our first stepping-stone to outer space. The Redstone was an extremely simple vehicle compared to the space systems of the Year 2000 - a humble beginning to the first 50 years of the space age.

The Move
by Ken L. Rossman

When World War II ended there was a move by Col. H.N. Toftoy to bring Wernher von Braun and as many of his associates as possible to the United States. This was known as "Operation Paperclip." The von Braun team ended up at Fort Bliss, Texas, adjacent to El Paso in late 1945. The "Team" occupied a hospital complex known as Beaumont Hospital Annex on the outskirts of El Paso. The main Beaumont Hospital was a U.S. Army Hospital in El Paso. It had been there for many years.

I was transferred as a private in the US Army to the 9330 TSU (Technical Service Unit) located at Fort Bliss, Texas, in March 1949. This was three months after my 19th birthday and low and behold here were the Germans who had developed the Rockets fired against the Allies only four short years earlier. The 9330th was the support group for the "Team."

Even though I transferred to the Civil Service in September 1951 and stayed until my retirement in September 1979, I am only going to address the military part of my service since there were very few who followed their military service from Fort Bliss to Redstone Arsenal then to NASA.

Having completed the Army's Electrician's School in Ft. Belvoir, VA, I was assigned to work for Mr. Rosinski in the Steering Section. We became good friends (Rossman is German). Mr. Rosinski knew a German General named Rossman.

We were working on, among other things, some V-2's that had been brought from Germany by The Army Ordnance Corp. In October 1949, we were sent to White Sands Proving Ground, New Mexico to fire a V-2. I had the responsibility of maintaining the batteries that supplied the electricity to the missile and on Oct. 6, 1949, I participated in the launch.

The move to Redstone Arsenal started in April 1950. I volunteered to stay at Fort Bliss until my discharge date in September 1950. The Korean War started in June 1950 and President Truman froze discharges and extended all enlistment's one year. This turned out to be a blessing for me. I transferred to Redstone Arsenal in September 1950, served one more year, was discharged and immediately became a civil servant.

New York to Huntsville in the 1950s
by Frederick I. Ordway III

My journey to Huntsville and what was then the Army Ballistic Missile Agency began in New York City back in October 1952. It was at the Hayden Planetarium's Second Symposium on Space Travel that I first met Wernher von Braun, introduced by symposium coordinator Willy Ley. Von Braun, then technical director, US Army Ordnance Guided Missile Development Group at Redstone, described "The Early Steps in the Realization of the Space Station." Such a station, he explained, would not only be "The most fantastic laboratory ever devised," but "the springboard to man's further ventures into outer space, to the moon and the nearer planets." His audience hung on his every word.

Before 1952, I really did not know much about the man. Having joined the American Rocket Society in early 1941 when barely a teenager, from time to time beginning around 1947, I would read about V-2 test from the White Sands Proving Ground near Las Cruces, New Mexico. Typical of the times was Richard W. Porter's five-page "Analysis of the First Two American V-2 Flights" that appeared in the March 1947 issue of the *Journal of the American Rocket Society*. Quite a bit about the rockets; nothing, not even a mention, about von Braun. I had, of course, read Willy Ley's 1944 Rockets and subsequent, expanded editions, and other publications, but just who Wernher von Braun was remained hazy at best.

Then came the famous *Collier's* space series, which began with the 22 March 1952 issue published a little more than six months before I first met von Braun at the Hayden Planetarium. He now became alive in my mind as I devoured his *Crossing the Last Frontier* and learned about launch rockets, recoverable winged upper stages, space stations, space taxies, and lunar reconnaissance vehicles. And those spectacular paintings by Chesley Bonestell, Fred Freeman and Rolf Klep. Heady stuff, indeed.

Living and working in the metropolitan area, I had become active with the New York Section of the American Rocket Society and with the newly formed American Astronautical Society. We at the latter were fortunate in having convinced such well-known personalities as Wernher von Braun, Krafft Ehricke, Fred Singer, and Frederick C. Durant III to join the board, on which I had the honor of serving also. In December 1955, I chaired the second annual meeting of the society at the Hotel Edison in New York that featured von Braun's luncheon talk on "Problems of Guided Missile Development," followed by an all-afternoon series of technical presentations. Among them was Ernst Stuhlinger's lecture on "Electrical Propulsion System for Space Ships." I believe that was the first time we met—and we've been friends ever since. That evening, we all toured the Viking Rocket Hall at the Hayden Planetarium and listened to Singer lecture on the "Economic Aspects of the Unmanned Satellite Vehicle." On the same occasion, I got to know Hermann Oberth, who was the recipient of the society's Space Flight Award.

Three months later, on 28 March 1956 over breakfast at the Harvard Club of New York City, von Braun suggested my joining him at the newly formed Army Ballistic Missile Agency on Redstone Arsenal. My notes remind me that we talked for more than two and a half hours. That led to my first trip to Huntsville, by plane via Atlanta, early the next month. After meeting with von Braun the morning of the 5th of April, I was given a tour of ABMA facilities by Hermann Koelle, then over lunch with von Braun I learned more about his missions and plans for the future. The next day, I spent much of the morning with Ernst Stuhlinger who shed further light on ABMA programs and activities. Little did I realize that four decades later we would collaborate on a two-volume biography of von Braun himself.

I made my third Huntsville trip, in August 1956, not by plane but by rail on the "Birmingham Special" out of New York's Pennsylvania Station. On that occasion I met with Gerhard Heller, Hermann Koelle, Hermann Oberth and Ernst Stuhlinger, dined at Wernher and Maria von Braun's home until my notes tell me, 11:30 p.m. On my next trip, in mid-October, von Braun mentioned that he would like to take Maria up to New York in early December and that he had made some professional appointments on Long Island. "Where, on the Island, would I suggest they stay?"

I blurted out, "Why, you can stay with us"—meaning wife Maruja and my home in Syosset. Before I could collect myself, he said, "Fine, Fred, we'd love to!" So the von Brauns were going to be our house guests! I couldn't believe it. Our home was rather large, so space was not a problem. But how should we arrange for their visit? What should we do to entertain them? After all, von Braun was coming with his wife so we wouldn't be talking about rockets and spaceflight all night!

Well, on the 1st of December we took them to the highly acclaimed Terrence Rattigan play "Separate Tables" at the Music Box Theatre in New York City and attended the Carnegie Hall New York Philharmonic concert. We toured the von Brauns around Long Island's North Shore and its many impressive estates, winding up the day with a dinner party at home for 20 or so friends. I'll never forget how especially the ladies would encircle him during the pre-hour as they listened to his every word and admired his handsome physique and engaging personality.

And thus it happened. On the 15th of February 1957, Maruja, sons Fred and Albert, and collie Rocket—yes, Rocket, left our home on Berry Hill Road, Syosset, Long Island, and drove the nearly thousand miles to Huntsville. There, we settled into our new home on Monte Sano's Lookout Drive that had recently been vacated by Lee and Kathleen James.

Like many others before and after us, we had been drawn there by the vision, magnetism and charisma of Wernher von Braun. And learned to love what at one time had been called watercress capital of the nation. Or was it of the world?

Working in Electrical
by Charles Souther

I stood on the loading ramp of building 4492 and watched Sputnik I pass over the Huntsville area in the fall of 1957. I could see two pinpoints of light, one behind the other, as they passed over. It was a mystery to me why there were two objects passing over, because I did not know at the time that the upper stage of the launch vehicle also went into orbit. I had been with the Electrical Equipment Section, Guidance and Control Laboratory, Army Ballistic Missile Agency about a year during this time.

When I joined the Electrical Equipment Section in late 1956, they had recently moved from Squirrel Hill to Building 4492. I was informed by Mr. Rosinski that the required work schedule was 58 hours per week and additional overtime would be required as needed. This proved to be the case for about three and one-half years. Quitting time being after dark gave me the advantage of observing Sputnik I, which probably could not have been seen during daylight hours.

One of the factors that influenced me to apply for a transfer from Charleston Naval Shipyard to Redstone Arsenal were articles in *Collier's* magazine by Wernher von Braun. I happened to see the articles during different trips to a barber shop on Remount Road near Charleston, SC. The articles talked about a space station, travel to the moon etc., which was very interesting to me.

The Electrical Equipment Section was engaged in the fabrication of ground support equipment and some flight equipment. Much of the ground support equipment for

Redstone, Jupiter, Saturn and other vehicles was built in-house during the first few years I was there.

The Electrical Equipment Section had its share of practical jokesters. On one occasion that I recall, a rack full of ground support equipment was ready to be checked out. Unknown to those conducting the test, one employee had hid behind the rack, and as the test switch was thrown he blew a huge puff of cigarette smoke through a piece of plastic tubing into the rack, which came out the front between the panels. Needless to say, the switch was quickly turned off.

Coming from the humble background that I came from, I consider it a great privilege to have played a small role in projects such as Redstone, Jupiter, Explorer I, Saturn I, IB, and V, Apollo, Skylab, Space Shuttle and others.

One of the most challenging projects on which I worked was the development of watertight reusable cable assemblies and underwater connectors for use on the Space Shuttle Solid Rocket Boosters. Several possible designs were considered and development testing was done. One test involved aluminum shell flight type connectors, which were exposed to the harsh corrosion saltwater environment equivalent to the required twenty-mission use. This resulted in the aluminum shells being ate up with corrosion and literally crumbling apart. The final design of the underwater connectors utilizing a stainless steel shell was a modified version of connectors used by the Navy on underwater mines and diving gear. There were many other design requirements that had to be met, such as the water-tightness of the cable assemblies for 20 flights, unusual high vibration and shock levels, temperature variations, wind buffeting during descent, and proper connector mating to prevent bent pins. It has been said the Solid Rocket Boosters are a cross between a space vehicle and a submarine. This has some element of truth, since they go to approximately 40 miles altitude and drop into the ocean with some parts submerged to perhaps as much as 150 feet. I do know that my submarine experience at Charleston was helpful on this task. This difficult project was carried out under the leadership and inspiration of my friend and Section Chief, Wayne J. Shockley.

I shall always remember the years 1956-81, the stress, the pressure, the fatigue, the disappointments, the failures, the successes, the triumphs, but most of all the wonderful, talented and creative people I came to know at ABMA and MSFC.

From Co-op to Rocketeer
by Willie E. Weaver

In 1956 I was in physics class at the University of Alabama, trying to keep my mind on the lecture, but I kept worrying that this might be my last semester in college. My dad was very sick and all of our meager family income was needed for household and medical expenses. The professor ended the lecture early as the head of the Physics Department entered the room. He told us about a co-op program that the university had with the U.S. Army. Under this program, physics students could alternate semesters of study with semesters of work with the von Braun rocket team at Redstone Arsenal. Well, I was not sure what a co-op was and I didn't know where Redstone Arsenal was, but I did know who Wernher von Braun was and I needed a job. I signed up as a co-op.

On October 4, 1957, as a student member of the missile checkout team, I was inside the tail section of a Redstone Missile heating temperature sensors with a small hand-held hair dryer. This process was to verify the proper installation and operation of the many temperature sensors that were flown on these military test vehicles. Various other methods were used to check the many other types of sensors that were required to track the performance of these vehicles before they would be used to deliver warheads or hopefully to place satellites into earth orbit.

On the night of January 31, 1958, another co-op and I drove through rural Alabama straining to listen to the static filled radio newscasts for some hint of news about the launch of a Redstone Rocket carrying a satellite. After the launch of Sputnik I and the failure of the Navy's Vanguard rocket, Dr. von Braun assured President Eisenhower that our team had the hardware available and could put up a satellite in short order. This day a launch attempt was under way. It was with reluctance and regret that we turned in our badges at 4:00 p.m., checked out of Redstone Arsenal and headed back to Tuscaloosa for the new semester, because the university's registration process waits for no one. I had spent my last work day in the telemetry checkout station where I had worked many long shifts helping to prepare the Redstone for this attempt to pull the U.S. space program even with the Russians. What newscast we could hear made no mention of a rocket launch, but when we arrived in Tuscaloosa, everyone was very excited and quickly herded us to the dorm living room to see what was on the TV. The launch had been successful and Explorer I was in orbit and sending back data. Our Team had entered the Space Age.

In May 1959 the postman delivered a letter for which I had anxiously waited. The letter was from the U.S. Army and was postmarked "Redstone Arsenal." With graduation just a month away, I had not yet accepted a job offer. I had before me two offers, one from New Jersey and the other from California; neither involved with rocketry. With my wife expecting a baby, I would have had to take one of the other jobs unless this letter offered me a job at Redstone Arsenal. I ripped the letter open and — Hurrah! I was going to be a permanent part of the von Braun group.

In 1961 President Kennedy had just set NASA the goal of placing a man on the moon before the end of the decade. I was sitting on the floor of the Recorder Room with my feet in a cable trench. Members of my crew were in the basement feeding long black electrical cables up from below into the trench. The cables were stiff and heavy. I really needed someone to help with this end of the task. As I started to pull up another cable, I was aware of someone entering the room and taking hold of the end of the cable and pulling it across the room. The last cable seemed to be stuck. Without looking, I assumed that my helper was one of my crew, so I told him to get down in the trench and help. He dutifully complied. We broke the cable loose and he pulled it across the floor. Still sitting in the trench trying to catch my breath, I heard my helper ask with a German accent, "Well, Mr. Weaver, what is the purpose of these cables?" I jumped out of the trench and faced the Center Director. As I searched for words, Dr. von Braun extended a hand, now very soiled by our task. As we shook hands, he quickly put me at ease and again inquired about the use of these cables and other equipment in the room. We walked about the room as I described the equipment and told how it was to be used in the testing of the Saturn vehicles. As we passed the doorway, I saw an assistant to the Laboratory Chief looking into the room. He quickly disappeared. Within minutes he returned with the Laboratory Chief and several other minions of lab management. Much to my dismay, they took away my new helper, dirty hands and all.

Adequate Housing
by Patrick F. Keyes

As many no doubt remember, the concern for available space in Huntsville was an issue to accepting a job at Redstone Arsenal in early 1951.

In order to provide some insight into the local housing market, Madison County's probate judge, Ashford Todd, was afforded an office on the arsenal for the purpose of providing the job seeker the names and locations of sale or rental properties in the area.

As I recall, this courtesy was made a part of the overall interview process. Judge Todd's portion of the interview lasted longer than the interview for the job and proved to be of vital importance in my decision of a job offer at Redstone. His knowledge and feeling for this community was boundless and infectious.

The U.S. Army 9330th Technical Service Unit
by Charles W. Shoun and Charles A. Lundquist, Huntsville, AL

The history of the U. S. Army 9330th Technical Service Unit parallels the history of research and development of guided missiles as weapons for the U. S. Army. Throughout its existence, the 9330th TSU was a guided missile organization.

This history begins in 1945 during the aftermath of World War II, when a group of German scientists, engineers and technicians arrived in Fort Bliss, Texas under the provisions

NASA - Marshall Flight Space Center

Standing, from left: Max Brown, Walter Landau, Bernie Wiesenmaier, Al Presto, Carl Winkler, Chuck Shoun, Joe Scollard, George Mandy, Jesse Harris, Roger Hunt, John Griessen III, Lee Hale, Dick Kesti, Bob Lindstrom, Bob Troxell, Tom Horeff, Leland Lee, Morton Mitchell, Leonard Morgan, Bob Wyne, Tom Macomb, Al Shuskus, Bob Shymkus, Bill Barnes, Cecil Ardrey, Bob Doheny, Jerry Kassell, Jim Hess, Vernon Meister, Gil Rehm, Ray Sabatino. Sitting: Don Jessick, Frank Machnikowski, Ernie Pierce, Harry Dawsey, Benny Gardner, Bill Hollingsworth, Andy Karabinos, Don Kittrell, Dick Lusk, Tom Ellis, Frank Ferringno. Front row: Andy Moll, Chuck Lundquist, Buck Cannon, Dick Bigott, Dean Albrecht, Alvin Owens, Bill Riehl, Clark Lemley, George Noyes, Bill Barron, Kats Horiuchi, Jim Kingsbury, Sam Zeman. Courtesy of Charles Shoun.

of Project Paperclip. Rocket hardware and documents captured in Germany by the Army also arrived in Fort Bliss. The German rocket team was the nucleus of a U. S. Army ballistic missile development effort, initially using captured V-2 rocket hardware.

To support this development effort, on October 19, 1946, War Department Circular No. 320 established a special Ordnance Project at Fort Bliss. Major James P. Hamill was assigned as officer-in-charge of the project, and on November 1, 1946 he assumed command of the newly-created 9330th Technical Service Unit, which then included one other officer and 19 enlisted men.

On April 16, 1946, personnel of the 9330th TSU participated in the launching of the first V-2 rocket fired in the United States. This unit also assisted in the first night launching of a V-2. Major Hamill received the Legion of Merit and the unit was awarded the Meritorious Service Plaque for their work on the pioneer phases of guided missile development.

In 1949, the Army chose to concentrate its rocket and missile activity at Redstone Arsenal, adjacent to Huntsville, Alabama. Subsequently, the rocket team led by Dr. Wernher von Braun, moved from Ft. Bliss to Huntsville in 1950. The 9330th TSU also moved to Huntsville to support the new organization.

The Redstone missile projects, when initiated in Huntsville, were short of technical personnel, in spite of vigorous recruiting. The scientists from Germany suggested a solution that had worked for them in a similar situation. The German Army had searched its ranks for soldiers with scientific and engineering training, and then assigned many of them to the German missile program. Following this suggestion, the U. S. Army, under Operation Pipeline, made a search of its ranks and of incoming draftees and recruits. Soldiers with appropriate credentials for rocketry were designated Technical Specialists and assigned to the 9330th TSU. This brought to Redstone badly needed enlisted personnel with bachelors, masters and doctorate degrees. They were integrated into the development organization with responsibilities commensurate with their education, where they served with distinction. Many of them stayed on in civilian status when their Army tour was over, and reached leadership roles in the Army Ballistic Missile Agency and in the NASA Marshall Space Flight Center, spun off from ABMA in 1960.

On October 6-8, 2000, the alumni of the 9330th TSU held a reunion in Huntsville, commemorating the 50th anniversary of the unit's arrival in Huntsville. More than 50 alumni attended, out of a total personnel count of more than 300.

The legacy that the people in the 9330th Technical Service Unit gave to the nation, though not fully understood at the time, probably is that the 9330th TSU was one major factor that made the United States of America the space and missile leader in the world.

Payroll Number 13
by Mildred B. Bobo

My earliest introduction to the National Aeronautics and Space Administration was in late 1959. At the time, I was employed by the U.S. Army Ordnance Missile Command, Redstone Arsenal, Alabama as a Supervisory Freight Traffic Officer.

A Joint Committee, sometimes known as the Committee of One Hundred, was formed and charged with the task of planning for establishment of NASA's George C. Marshall Space Flight Center, to be located at Huntsville, Alabama. I was asked to serve on that committee. The project was given a high security classification, especially in the beginning. Matters discussed in the Committee meetings and Sub-Committee meetings were not to be discussed outside the meetings.

On February 4, 1960 at 1400 hours, my supervisor, Mr. Stanley C. Wilkins, gave me a handwritten note indicating that Luther Adams, from the Personnel Office of Redstone Arsenal, had advised him that Mildred B. Bobo would receive a call from NASA.

When I received the call, it was from a Mr. Whitbeck and a position as traffic manager at the George C. Marshall Space Flight Center was offered to me. I think the caller was Mr. Phil Whitbeck. At first, I declined the offer.

That evening I discussed the matter with my husband, Archie C. Bobo, who had worked at Redstone Arsenal for about eight years in the 1950s. In his various assignments he had the opportunity to observe the ability of Dr. Wernher von Braun and other members of his team. He felt that the proposed space center would be a success and that I would enjoy the assignment.

The following morning, I called Mr. Whitbeck, using my own telephone at home, and asked that my name remain on his list. Due to the difference in time zones, I was able to talk with Mr. Whitbeck before leaving the house for work.

On March 17, 1960, I received a phone

call from Mr. Victor C. Sorensen who stated that I would have the "top job" in traffic management. He stated that Mr. Morris and Mr. Whitbeck had approved. A current Form 57, Application for Employment would be required from me and the official release would be received soon.

The Personnel Action was dated March 27, 1960 (a Sunday) and my first day of work on the new position was Monday, March 28, 1960. (It was Archie's birthday.) My supervisor was Mr. Victor C. Sorensen. His secretary was Betty Foxworthy Houston. Betty was the step-daughter of Grace Foxworthy, one of the first co-workers I met the day I entered the Civil Service work force at Redstone Arsenal early in 1945. Mr. Sorensen was a very able supervisor and his secretary was always courteous and helpful. I was assigned a desk in the building across a courtyard from my supervisor, and set to work writing job descriptions and interviewing applicants.

A number of my co-workers from the Transportation Office of Redstone Arsenal applied for and accepted transfers to the new organization. NASA was a civilian agency and Redstone Arsenal was a military agency, but experience in the military organization was adaptable to the needs in the civilian agency. Good working relations and loyalty were the trademarks of that early organization. I will always remember the dedication and accomplishments of my co-workers. We had many compliments concerning their service and know-how in arranging necessary transportation for people and goods. In retrospect, it should be noted that the commercial carriers (airlines, truckers, etc.) were cooperative and anxious to provide the best service possible for the Government.

Early on, Marshall Space Flight Center established a strong and active Public Relations Office. All employees received the *Marshall Star*, which was published regularly, on schedule. That publication made it easy for every employee to understand the work that was being performed, and the importance of the work that he or she did to aid in the goals of the organization even though we were not scientists or engineers.

In the establishment of Marshall Space Flight Center, the Redstone Arsenal relinquished a share of government owned vehicles. In due time, Marshall Space Flight Center contracted for operation of the Motor Pool. As soon as feasible, Marshall Space Flight Center acquired use of a Gulfstream aircraft, and later a very large cargo plane nicknamed "the Pregnant Guppy," and a special barge for water transportation to Cape Canaveral.

Soon after my transfer to the Marshall Space Flight Center, a gentleman from the Finance Office, or the Payroll Office, called me and asked if I would accept the payroll number "13." I stated that I would be glad to do so, inasmuch as I considered "13" a lucky number. Sometime later, I was promoted to the Civil Service Grade of GS-13.

In August 1962, Mr. William P. Morrow, who had been with the Transportation Office of the U.S. Army Missile Command at Redstone Arsenal, Alabama, was assigned as Chief of the Transportation Branch, Marshall Space Flight Center. I served as Deputy Chief of the Branch until my retirement May 12, 1978.

I shall always have fond and special memories of my association with the dedicated members of the National Aeronautics and Space Administration.

How Early Space Launch Vehicles Got Shiny Tails
by Thomas M. Wade

We were launching early Redstone missiles and the equipment in the tail assembly was overheating. Radiation from the bright flame would build up heat in the flat bulkhead at the tail-end. We built that bulkhead with layers of stainless steel sheets and fiberglass insulation that added considerable weight.

Ted Paludan was collecting data on the heat problem. His research produced a precise pattern and method of installation for thermal pickups and signal amplifiers. These were miniature items and he insisted that an engineer install them, using a homemade spot welder that would not damage them. I was assigned to the Fabrication Lab at that time. After shop personnel would depart, I would get Paluden's suitcase from under my desk and go install the instruments.

Soon we began modifying a Redstone missile to launch our first satellite. The tail was too heavy for stability. The design labs removed a layer of stainless steel and insulation. That remedy alarmed Mr. Paluden and he alarmed me. Since that particular vehicle was no longer a weapon, I believed there was no point in keeping the tail a dull gray. Before it could be moved out of the assembly shop, I wrote a quick work order to polish the end bulkhead of stainless steel to a mirror finish. The missile mechanic signaled completion of the job by laying down his polishing machine and combing his hair.

Someone resented this undocumented change and called Dr. von Braun. A friend warned me. His automobile skidded to a stop in the big shop door. I was apprehensive, but ready, with my Weston Master light meter in hand, and was studiously measuring reflected light from the bright tail and comparing it to the dull gray one on the nearby missile. He rushed in and asked, "What's the problem?"

I said, "See, the tail of the space research vehicle reflects four times as much infrared energy as the weapons model." The professor did a quick look and said, "Looks like a damned good idea to me." He turned and raced off to his automobile. We documented that.

Stories and Anecdotes
by Glen A. Deuel
Enough Fuel For A Cigarette Lighter?

I was a member of the ABMA Experimental Firing Laboratory in 1953 under the Directorship of Dr. Kurt Debus. One facet of my job as supervisor of launching and handling equipment and propellants was to oversee the mixing of the two alcohols, methanol (which is poisonous) and ethanol (which is drinkable), as a fuel for the Redstone missile. When my crew, mechanics and technicians, were loading the Redstone for the second or third firing (launching), I ran out of fuel. The tanker was not full as it should have been. While the launch was delayed and my crew was sent back to the storage area for more alcohol, Dr. Debus came to see me. Instead of saying something harsh to me, Dr. Debus just held out a cigarette lighter and asked if I would fill it for him. I was doubly embarrassed, but learned a vital lesson from that experience.

First Encounter with an MP

My crew and I had to go to Titusville, Florida, to fill our liquid oxygen (LOX) tanker and bring it back. I was driving an Army truck as an escort. When I got back to the entrance at Cape Canaveral, a military police (MP) held up his hand for me to stop. I motioned to him that I wanted to park just inside the gate, so I eased on for about 75 feet. When I got out of the truck to return to the guard to explain that I was leading the way for the LOX tanker, he motioned for me to back the truck to the point where he was still standing. Of course I obliged. The MP had some dire warnings for me, but he let me and my crew continue on to the storage area.

Second Encounter with an MP

I had spent more than 4-1/2 years in the Army Air Corps during World War II, much of it overseas as a master sergeant, so I didn't have the proper respect for some military ways. During the first or second launch of the Redstone, I climbed on top of one of my missile handling vehicles to get a good view of the firing. An MP on the ground told me to come down. I didn't think he had that much authority so I declined. The MP pulled out his ".45" pistol and again told me to come down. The ".45" had enough authority, so I reluctantly returned. Fortunately, he did not see me later when I was in a ditch much closer to the launch.

A Sleepy Team

Before the Saturn days a team, including me, took a government airplane to visit airplane companies around the U.S. to determine their potential capabilities as future contractors. One day after meeting and plant tours in Los Angeles and San Diego, we boarded our airplane to go to Denver to meet with the Martin Company the following morning. We were pretty tired, but the team persuaded the pilot to stop at Las Vegas before proceeding to Denver later in the night. Some of us took in a few shows and the others went their different ways. When we got to the Martin Company the next morning, we went to the conference room immediately so as not to be too late.

Some of our crew had to give presentations to Martin Company as an introduction. We who watched did not know whether to laugh or to hide. Our speakers were struggling to stay awake and to look and sound intelligent. The Martin group enjoyed it as much as we did.

A Space Program Memoir
by Bill Riehl

I was a very fortunate chemist to be sent (via the U.S. Army) to the "von Braun team" in 1951, and consider myself to probably be one of the few still living American pioneers in the space program. I was substantially technically involved with almost all of the U.S. space vehicles until my retirement in 1996. This included the Redstone, Jupiter, Explorer, Centaur, Saturn I, IB and V systems, Skylab and the Space Shuttle. In order to further clarify, and within editorial space limitations, I have selected the first U.S. satellite—the Explorer program—as an example. This launch system used the Redstone missile as the first of four stages. Due to the fact that the Redstone was the U.S. development of the German V-2 missile into a longer range American system, many technical issues were involved in changing from the German materials, processes and propellants to the U.S. capabilities and technology.

Both the V-2 and Redstone used the decomposition of concentrated hydrogen peroxide to power the turbopumps for the propellants, which was stored in steel tanks. Review of the available German files disclosed that they had used a "bituminous" coating in the steel tanks to prevent premature decomposition on the launch pad. A wide variety and number of U.S. "bituminous" coatings showed none to be suitable for contact with peroxide. Several of the most nearly acceptable coatings were purified in our laboratory to remove trace contaminants, resulting in a satisfactory U.S. substitute coating.

The V-2 and original Redstone vehicles used an alcohol/water mixture as fuel. It was found that a mixture of amines (known as "U-DETA") could be used in the same hardware and provide significantly greater thrust. However, this fuel was highly detrimental to most of the "soft" goods in the Redstone (O-rings, gaskets, sealants, plastics, etc.) These had to be replaced with commercially available materials, which were tested and found compatible, or in many cases, new materials formulated in our labs. A typical example was a sealant, which had been used for many years in oxygen systems, and was taken off the market. We formulated a successful replacement after considerable R&D (research and development).

Dr. William R. Lucas (later center director) and I developed a method for evaluating the compatibility of materials with liquid oxygen (LOX), which we published in 1957. This became known as the ABMA Impact Sensitivity Tester, and has been, and still is, used as the primary method for acceptance testing throughout the aerospace industry of all materials for contact with LOX in all space vehicles, and was a major factor in my receiving the NASA Exceptional Scientific Achievement Award later.

The gasket material used in our LOX plumbing systems suddenly began failing acceptance tests in the latter part of the 50s. It was suspected, but not proven, that this was due to a "minor" change by the vendor. After thorough R&D, we developed a process by which the commercial material was treated in our lab to make it acceptable for LOX and mechanical applications. Numerous other materials and/or processes were developed for the Redstone system, such as dye markers for ocean impact locations, chemical plugs to control the floating times of jettisoned instrument containers, etc.

The Explorer program was begun by the von Braun team in 1955-56 and was planned and expected to be the world's first satellite placed in orbit. Two launch systems were prepared, using as a first stages, Redstone missiles nos. 27 and 29. The upper three stages were combined in 14 solid rocket motors, using ten, three and one as the 2nd, 3rd and 4th stages. Naturally, weight reduction was significantly more important on this vehicle. Our group developed and applied a new lightweight paint for the Redstone, which saved roughly 75 lbs. in takeoff weight. In order to provide guidance and stability, all three upper stages were rapidly rotated during flight. This necessitated centrifugal balancing beforehand, with an inert propellant and the same center of gravity as the flight unit. We formulated and molded such a material. Thermomechanical problems during the molding operation arose, and were solved, and utilized.

We were also responsible for section and application of external coatings on the satellite to withstand aerodynamic heating during ascent, as well as thermal control, and protection from suspected micrometeorite erosion once in orbit. Two surfaces were selected after testing: a ceramic coating for temperature resistance and thermal control in orbit, and pre-simulation of micrometeorite erosion by very fine sandblasting. Since the satellite was to be rotating in orbit, these could be laid out in stripes on the payload case. Thermal engineers provided the proper width of the stripes, and the coatings were applied in our laboratory to both the flight and backup units.

It should be mentioned that the Explorer satellite displays in the National Air and Space Museum in Washington, as well as in the Alabama Space and Rocket Center are NOT exact duplicates of the flight units. Immediately after the successful launch and orbit, the Public Relations Office wanted 12 duplicates for display within less than a week. We informed them that was not possible, since application of the ceramic coating required almost a week, and we had only one application system. If we could substitute a white paint for the ceramic coating we could have them ready as requested. This was approved and done. The only truly authentic Explorer display was the backup unit, which I understood was kept by JPL in Pasadena, California.

The Explorer launch system was ready for launch in late 1956 or early 1957. However, we were directed by Washington, DC NOT to put it in orbit*. Consequently, the trajectory was changed. The system worked perfectly, setting three world records at the time—for velocity, altitude and distance—approximately five to six months before the Russians put Sputnik in orbit. Naturally, this was very frustrating to the von Braun team. Not until after the failure of the Navy Vanguard system on the pad were we given permission to launch, which we did successfully 84 days later, January 31, 1958.

Two years later, the von Braun team was combined with the former National Advisory Committee on Aeronautics (and its centers) to become the National Aeronautics and Space Administration (NASA). This management change subsequently seriously affected my morale and pride in involvement in the space program from then on. The new NASA began contracting our development of space vehicles to the aircraft industry (with whom they had previously worked) which became the "aerospace" industry, and in which we were assigned the role of "monitoring" contractor programs, rather than doing it ourselves. By the time of the Space Shuttle program, less than ten years later, even the design of the Shuttle was done by the aerospace industry—with NASA review. I was heavily involved with some of the contractor programs, especially the External Tank. However, in my opinion many contractors did not want, or accept, our suggestions or input. Some were quite cavalier in their relations—especially with Marshall Space Flight Center, since we were the only organization in the government with previous "hands on" and successful experience in space programs.

It is understood that this was a political decision directly by President Eisenhower, on the basis of advice by then Secretary of State John Foster Dulles that because our system used the Redstone missile as the first stage, the Russians could interpret this as a U.S. intention to go into space for military, rather than scientific, purposes.

Not All Space Problems Had Simple Straightforward Solutions
by William Tom Escue

In the late 1950s, I was assigned the task of correlating the activities of the monkey, Able, with the time in flight. I first chose a Veeder Root counter and a synchronous motor to drive the counter. Both had a 1/8-inch shaft. There was a problem finding a coupler for joining the two shafts. With only a few days before flight hardware delivery, I still could not find a coupler and was about to panic. This was about the same time plastic refills for ballpoint pens became available. I purchased a refill, cleaned it thoroughly, cut it to the correct length, and coupled the two shafts. A hair dryer was used to heat the refill and make it pliable, then it shrunk when cooled providing a tight grip for the motor to drive the counter. It worked.

This combination was mounted where a camera could view both it and the monkey. Each count on the digital counter represented a second in time; therefore, correlating the monkey's activity to time in flight. The unit is on display at the Space and Rocket Center in Huntsville, Alabama.

An Engineer Remembers
by Ralph A. Burns

I was a mechanical engineering graduate from Auburn, December 1955, hired into the Army Ballistic Missile Agency, ABMA, by Hans Paul, one of the original members of German von Braun rocket team, in September of 1956. This was my second professional job, the first being seven months with the General Dynamics Corp in Fort Worth Texas working on the environmental control system for the B-58 Hustler aircraft. Initially I worked at ABMA performing aerodynamic heating studies for the Redstone missile nose cone. This was short lived as I was soon initiated into design and testing work for the Redstone instrument unit cooling system and later the Jupiter instrument unit cooling system.

I became a charter member of the Marshall Space Flight Center (MSFC), transferring to MSFC in July of 1960 at its inception. I was responsible for the mechanical design of the Saturn V Instrument Unit, IU, Environmental Control System. This system consisted of a pre-launch and a flight cooling and I was involved with the in-house design/test of the flight system.

During development of the IU, Wernher von Braun chaired regularly scheduled IU status meetings in the 10th floor conference room of building 4200, commonly known as the von Braun Hilton. I was fortunate to be able to attend most of these meetings. Dr von Braun was always a little late even though his office was near the conference room—this allowed every one to get settled in and also meant the meeting could get under way as soon as he arrived—he was a very busy man. You could hear the sharp click of his heels as he entered the room in a brisk walk, looking straight ahead with a stern look on his face. He would walk around to the head of the table, where there was a gavel and sounding board. He would, without recognizing anyone and still with that stern look, take the gavel and just about drive the gavel and sounding board through the table with a loud "thud". Then he would immediately produce this ear-to-ear "boyish" grin and say a very pleasant greeting such as "Good Morning" or whatever greeting was appropriate for the time of day. He was a very charismatic person who liked people.

Some people who headed up meetings seemed to take some joy out of embarrassing people in front of large groups to get control of a situation, but von Braun never resorted to that tactic, he was always in control anyway. On another occasion, I was in a meeting with von Braun in a small conference room, a point of discussion arose and von Braun said, "Am I the only one in this room who does not understand this?" and with a quizzed look, looked around for comment. He then went to the chalkboard and began making a diagram relative to the discussion.

Von Braun was known to walk around, without announcement, and make informal visits to offices just to see what was going on. On one such occasion when I was still a young engineer, I was operating the cooling system for a launch vehicle checkout in Building 4708 and von Braun walked up and started a pleasant conversation wanting to know what I was doing. Needless to say, I was somewhat startled that Wernher von Braun, the center director, was interested in what a little junior engineer was doing. This is a typical example of how Dr. von Braun interacted with people.

50 Years of Rockets and Spacecraft in Huntsville, Alabama – The Rocket City
by John H. Kastanakis, Sr

I became an Instrumentation Engineer in Test Branch, later known as Test Laboratory, of the Guided Missile Development Division (GMDD) on May 7, 1951. Heinz Scharnowski was my immediate boss under Dr. Werner Sieber. Dr. Martin Schilling was Director of Test Branch.

Hermann Weidner, Rudie Beichel, Heinz Scharnowski, Leland Belew, and John Kastanakis were in two connecting rooms of Bldg. 432 performing the pre preliminary design for the East Test Area, which was under architectural-engineering design with Ralph M. Parsons-Aerojet General—a joint venture. Leland Belew was doing the mechanical drafting for Weidner-Beichel, and John Kastanakis was doing the electrical-instrumentation drafting for Heinz Scharnowski. Gordon Artley worked for Aerojet-General and was the interface/representative of Aerojet-General with Weidner, who was in overall charge of us. I performed in this activity the first four months that I was here.

My next six months assignment was to go back and forth to White Sands Proving Ground (WSPG) as the lone instrumentation engineer on the Graphite Vane Tests. The V-2 Graphite Vanes were tested to see if they would withstand 110-120 seconds "burntime" required to steer the Redstone missile because the V-2 "burntime" was only 60 seconds. Heinz Scharnowski took me to WSPG for the first series of tests, but left me "high and dry" for the next and remaining series of tests. The "Whiz Kids", G-I's who were former engineering students at the University of Illinois, and who were "written up" as "Whiz Kids" in an article in *Collier* magazine, operated the strip chart recorders and oscillograph for me. The G.I.'s were cleaning the glass strip chart recorder pens between tests and "lo and behold" broke every one of them. Good old Test Branch /Laboratory and Tom Greenwood airmailed me some new pens and we did not miss or delay a test. I assume William Mrazek, with the able assistance of Bob Lindstrom (who was a G. I. then), was in charge of those tests. Bob was a ceramic engineer from the University of Illinois prior to his G. I. service. Karl Sendler, Werner Rosinski, Joe Pass, Antonio Beltran (all Guidance & Control Branch of GMDD), Mrazek, Rudie Beichel and Lindstrom, and myself were the GMDD personnel at all these tests. Sendler was only at the first test series to set up and see that the firing console worked properly, since he had V-2 experience. I enjoyed watching the corporal missile launches while I was at WSPG.

The only other person who attended the Graphite Vane Tests at WSPG was the most important one, the civilian director of GMDD, Dr. Wernher von Braun. We stayed at the Hotel Del Norte in El Paso each trip and drove and rode 50 miles thru the desert to WSPG. On von Braun's trip to the tests, von Braun, Mrazek, Joe Pass and John Kastanakis were in the same suite at the Hotel Del Norte, and all four of us were in the same car from El Paso to WSPG. As a young instrumentation engineer, I "relished" being with Dr. Wernher von Braun.

My next assignment from September 1953 to 1956, was to be the lead instrumentation engineer at the IITS (Interim Ignition Test Stand). The East Test Area was activated in 1956. As stated elsewhere by test engineers, we had wonderful accomplishments during the Apollo program. So, I will attempt to dwell only on instrumentation. The data acquisition systems in each Blockhouse East and West Test Areas were second to none in the entire United States. The strip chart recorders, recording oscillographs, digital data acquisition systems, and multi-channel magnetic tape recorders were "top of the line". The layout, routing of cables, etc. at the Static Test Tower East and West, Power Plant Test Stand, S-IC Test Stand and West Area F-1 Engine Test stand were equal to any other missile/space vehicle test facilities in the United States. I spent many days and hours witnessing and monitoring the data acquisition equipment in plant checkout in Philadelphia and at Aerojet-General in Azusa, California. Ray Proffitt witnessed the in plant checkout, attended a two-week class, and finished number one (top of his class) in his class on the MilliSadic Data Acquisition Systems.

Static Test Tower East/East Test Area Test Laboratory, site for static testing S-1 stage of Saturn I, 1958-59. MSFC personnel at board in hardhats include (from left): John Kastanakis, Gordon Artley and Harry Johnstone. Courtesy of John Kastanakis.

I was the Contracting Officer's Instrumentation Representative on the $10 million (1960 dollars) East Area Blockhouse addition contract with Lear Siegler, Inc, of Anaheim, California.

One of my most proud accomplishments was when Mr. Heimburg called me in and said, "John, I want you to go to the Corps of Engineers/Los Angeles, McDonnell-Douglas at Santa Monica and Sacramento, and Corps of Engineers/San Francisco to be my representative in negotiating, evaluating and selecting an instrumentation contractor McDonnell-Douglas instrumentation addition for the S-IVB stage test area at Sacramento". It really felt good that Mr. Heimburg had that much confidence in me.

I was very fortunate to have an instrumentation branch in Test Laboratory with the best instrumentation engineers and electronic technicians in the country, plus some very good dedicated data reduction personnel (ladies) and very good secretaries. At peak Apollo program "days" I had an Instrumentation Section, later Branch, later Divisions were Branches then of 30 engineers, 50 technicians, 10 data reduction personnel, and 3 secretaries. Plus, I had five contractor support engineers.

Contrasts
by T. Gary Wicks

I joined the ABMA's Structures and Mechanics Laboratory in the '50s as a student trainee out of Georgia Tech. One of the first tasks I was given was to calculate the number of lead discs needed in the nose cone to balance out the center of gravity of a Redstone vehicle being assembled in the Fabrication Laboratory. Today, lead discs are still being used to ballast the Ares I-X demonstrator launch vehicle being assembled at the Kennedy Space Center.

While much has changed in 50 years, including the people, tools and processes, one thing has not — the physics. The same equations that guided Saturn will guide the space vehicles of the future back to the moon and beyond. The tools then were mechanical desktop calculators and vacuum tube computers used for example, to characterize Saturn vehicle stage separation dynamics — analyses that we performed in the Advanced Studies Office of MSFC's Propulsion and Vehicle Engineering Directorate. Today the tools are the desktop PC and super computers.

Then, one could write their initials on the ring frame of a Redstone being assembled in Fab Lab as a personal "bon voyage". Today's quality, security and safety processes prevent activity near a vehicle not governed by a procedure.

The processes we used for developing new programs in the early '70s, also have changed — as well as the success ration of new ideas approved and implemented. Then, new program ideas came from bottom-up, existent organizations like the Program Development Directorate. Today, they come from ad hoc think tanks and task forces. One highly successful program that came out of the former business processes was Spacelab, which I was fortunate to participate in from in conceptualization as the Sortie Can, through its cooperative European development as Spacelab, and its maiden mission as Spacelab One. Its string of successful missions conducted by the Payloads Project Office were the scientific, engineering, operational, and management bridge from Skylab, America's first space station, to the International Space Station, the world's permanent space station.

I'm confident one other thing that will not change is the continuation of new program opportunities for the next space generation to pursue, from program cradle to grave — opportunities like I have had.

Project Orbiter
by Eugene J. Buhmann

My wife, Renee, and I arrived in Huntsville on October 14, 1954, towing a mobile home from Wisconsin. I had been drafted by the Army after two and a half years as an electrical engineer with Allis Chalmers Manufacturing Company. I was assigned to the Systems Analysis and Reliability Laboratory (SA&R Lab), which performed the inspection, test, and checkout of the Redstone missile that the Army was developing. In February 1956, four months before my discharge, I was asked by Mr. Erich Neubert, the Director of SA&R Lab, to be the Design/Quality interface with the Jet Propulsion Laboratory (JPL) for Project Orbiter.

Project Orbiter was the Army's program to put a satellite into orbit using a Redstone with an elongated fuel tank as the first stage and a three-tier solid propellant upper stage provided by JPL. The payload was a joint Army/JPL design provided by the Guided Missile Development Division headed by Dr. Wernher von Braun.

Trips were made to Pasadena, California for meetings with JPL and to White Sands, New Mexico to observe an Aerobe flight that used the same solid rockets intended for Project Orbiter.

When my discharge from the Army on June 20th occurred, I was so involved in Project Orbiter that I went home, changed into civilian clothes and went back to my desk for five hours. Next week, I was back in Pasadena.

In July 1956, the SA&R Laboratory conducted the simulated flight test of the instrument compartment and the bucket containing the inert upper stages and the payload on the Redstone Arsenal airport. Security was very tight, with Military Police at all the entrances to the airport. The test was run after dark so we could see the firing of the flares from the payload. The bucket containing the three upper stages was spun up to over 400 rpm to check the speed controller provided by Messrs. Bill Greever and Bob Lindstrom.

One month before the scheduled launch, I joined the JPL checkout team at Cape Canaveral, Florida, where the upper stage bucket was being balanced. Since the Redstone is not held down at launch, the balancing was critical. The second and third stages had been replaced with live solid propellant rockets, but the single rocket fourth stage was inert. Apparently, the Army was not going to put the payload into orbit. Rumor had it that Major General John B. Medaris told Dr. von Braun that he must personally check to make sure that the last stage could not be ignited. Since next year (1957) was the Geophysical Year, the honor of putting the first satellite into orbit was given to the Naval Research Laboratory, a semi-civilian agency.

On September 20, 1956, the JPL Checkout Team observed the night launch from the top of the lighthouse overlooking the launch pad. Through binoculars, we watched Redstone Missile RS-27 wind up the bucket and then rise majestically off its launch pad and head out over the Atlantic Ocean. The payload reached an altitude of 682 miles and its range was 3,355 miles. With a live fourth stage, the United States could have been the first to launch an artificial earth satellite. A little more than a year later, the Soviets launched Sputnik.

But we were given another chance. On January 31, 1958, less than four months after Sputnik, the Army launched another Redstone Missile, RS-29 (with a live fourth stage and a JPL payload), that put Explorer-1 into orbit; the first United States artificial earth satellite.

Old-timers
by Dieter Teuber

My first knowledge of Huntsville came in early 1952. At that time I had finished my exams at the Technical University of Darmstadt in Germany and was employed as assistant professor.

One day our chief engineer at the electrical institute asked me if I would like to go to Heidelberg in his place and contact a Dr. Eberhard Rees from Huntsville, Alabama. He also told me that he did not want to leave Germany. Maybe I would like to hear about work in Huntsville. "Why not?" I thought.

During WWII, the institute for electrical engineering of the Technical University Darmstadt had cooperated with rocket developments in Peenemunde. Right after the end of the war my dream was "space research" and I was familiar with certain projects at the Darmstadt University that had been carried out during the war.

In a hotel in Heidelberg, Dr. Rees talked to me about rockets, Dr. von Braun, Huntsville, the South, and if I would consider assisting Professor Theodor Buchhold in his work there. Of course, it would require coordinating immigration from the U.S. consulate in Frankfurt, the U.S. Army Command in Heidelberg and certain security requirements. "Yes, why not?" I thought.

I had studied English for many years before. However, a new concept to me was "red tape." It still took many months before everything was in order.

Then a short stop in Landshut, Bavaria followed and finally departure from Bremerhaven.

On my 25th birthday, 1953, I found myself on a ship passing the Cliffs of Dover on the way to New York. The USS *General Hodges* was a 10,000-ton military ship carrying mostly G.I.s on their way home.

At times it was a miserable crossing because of a storm system peaking at force 11. There were lots of seasick people among the more than 2,000 troops, dependents, and those few German civilians on their way to various destinations. Finally, New York. After a few days I got a train ticket from an Army Office to proceed from Grand Central to Huntsville. The ticket was a few feet long, the train trip in a Pullman roomette took about 30 hours, by electric, steam (near Roanoke), and diesel power the rest of the trip. I changed trains in Chattanooga. Dr. Walter Haeussermann expected me at two in the morning at the Huntsville railway station. I checked in and started work the next day on what was then known as "Squirrel Hill."

A few months later I would repeat a train roundtrip from Huntsville to New York on a business trip in order to escort several families who would also settle in Huntsville. The men had all been working in Peenemunde and were added to the "real" Peenemunde group that had arrived before in Huntsville from El Paso.

Work at the Guided Missile Development Division (GMDD) proved to be fascinating: Delta minimum guidance, mixing computer with magnetic amplifiers, gyro stabilized platform, rotating inverter to be replaced with transistors that just came up. Professor Theodor Buchhold and his coworkers guided all these developments. Work evolved to the first U.S. satellite, the lunar landings, Skylab, Solar Maximum Mission. That is all history now.

Once there was the "old-timer" group that goes back to rocket development at Peenemunde, Fort Bliss, Texas and those who came to Huntsville, 1950. Then others joined the group during the following years. It became known as the "Wernher von Braun Team." Now, almost 50 years, five children, eight grandchildren later, I feel like a "real old-timer and Huntsville has become home sweet home.

From Diesel Locomotives to Rockets
by George B. Doane III, Ph.D.

In September of 1955 my friends and neighbors at the draft board decided that I should stop my efforts on behalf of developing advanced diesel electric locomotives and join the Army of the United States. The Army assigned me to the 9330 Technical Detachment (Ordnance) at the Redstone Arsenal, which at that time was under the overall command of MG H.N. Toftoy. My principal work assignment was with the Guidance and Control organization (G&C), which was a part of the Guided Missile Development Division (GMDD). That, in turn, was under the direction of Dr. Wernher von Braun. Dr. Walter Haeussermann directed G&C. My other assignments included kitchen police, retreat parades and other such important military duties. With the arrival of MG John B. Medaris to command, the whole was wrapped into a new organization, i.e. the Army Ballistic Missile Agency (ABMA).

My early military time was spent looking into transistors and how to keep them from overheating or going into self-destructive oscillation. Because I had done graduate work in controls, I was next assigned to the control group directed by Mr. Brooks Moore. There I collaborated in various stability studies of control systems mainly for the Jupiter family of missiles and boosters, e.g. Junos. The Redstone missile had been completed by that time and had launched Explorer I. There was a big difference in the thrust vector control systems as implemented on the V-2 and Redstone and subsequent vehicles. An electrical actuation system, developed by Dr. Haeussermann, was used on the V-2 and Redstone. An hydraulic system was used on the Jupiters and subsequent vehicles. I had done some graduate work in this technical arena. This helped considerably in my understanding this type of actuation system.

Another G.I. and I did some rigid body stability analysis of the Jupiter flight control system. Until a Jupiter landed on the beach at Cocoa, the effects of propellant slosh were not modeled (as they later were, very extensively, by Bauer). This work pretty well wrapped up my military period activities. As an extra curricula activity during my military period (circa 1956-1957), I started teaching Electrical Engineering, in the evening, at the then University of Alabama Center in Huntsville.

In 1960 I received a Civil Service position on the staff of the Gyro and Stabilizer Division. By this time the Division was under the direction of Mr. Karl Mandel. The technical challenge in which I was mostly engaged at first was the servo stabilization of the air bearing gyros. This bearing was a pressurized gas one which supported the output or precession axis of the gyroscope. This should not be confused with the way the term is usually used, i.e. to describe the self-pumping spin axis bearing of a gyroscope. This output axis gas bearing was used on the gyroscopes, which sensed the orientation of the stabilized platform (relative to inertial space) upon which the accelerometers were mounted. The accelerometers were of the forced precession type of gyroscope and also incorporated a pressurized gas supported output axis. The servo problem hinged around the fact that a good model of the dynamics of the gyro loops assumed zero damping of the relative motion of the output axis relative to the gyro housing. This being the case, all the stabilization damping had to be supplied electronically and applied through the action of the associated torque motors. At the time management forbid us to publish anything about this work and, hence, it is not widely known. The competitor technology, the liquid floated gyro, was developed by the Massachusetts Institute of Technology's Instrumentation Laboratory under the direction of Dr. Charles Stark Draper and was widely disseminated in various published sources. As each new platform system was developed, it required new servo designs to accommodate the new dynamic parameters and new component designs and changes e.g. the change to direct versus geared torque motors. This kept me busy with new loop designs. Platforms of the inside-out gimbal type (which required rotating the missile to accomplish a pre-launch aim) were developed for the Redstone (before I arrived), Jupiter and Pershing I missiles (each platform smaller and lighter than its predecessor). The requirement to rotate the launch vehicle prior to launch was obviously not feasible when it came to the mammoth Saturn vehicles. Accordingly, a more conventional platform was developed using either three or four ring gimbals, depending upon the particular mission to be flown. The latter suite of components made considerable use of Beryllium because of its lightweight and other structural properties. After Saturn the Division took on overseeing the development of the Control Moment Gyros (CMGs), the star tracker and the various rate gyro packages for the first space station, Skylab. With the coming of the Shuttle program all navigation hardware work came to an end, and the Division was transformed into one with five Branches. Its interests then ranged from instrumentation development, hydraulic actuation system development, laboratory instrument calibration, the design of various electronic circuits for widespread uses and monitoring laser gyro developments and other sensing device development e.g. solid-state accelerometers. I eventually became the Deputy Division Chief and that is the position from which I retired in the early July 1981.

The Pegasus Project
courtesy of Max Sharpe

In the early days of the "space age" there was no knowledge on which to build. Optical astronomers were limited to the visible spectrum of their telescopes, normally located on the highest mountains to minimize optical distortions and aberrations caused by the atmospheric water and pollution. Very little was known about comets and other celestial objects. Meteors and meteorites were largely unknown phenomena that frequently came in swarms across the sky.

The Pegasus was the first U.S. spacecraft specifically designed as a micro-meteoroid detector and solar energy collector. It is now known that meteors travel at a velocity of 12 to 15 meters per second...40,000 miles per hour. The detectors were made of 0.020" thick aluminum, still a top-secret war material, i.e. airplanes and high explosive material. The physicists knew that in order to generate a flow of electricity on the solar collectors, each panel had to possess varied epsilon on a ratio of the solar energy absorbed versus the energy emitted by the panel, i.e. a phenomena that could be measured. The coating variants were developed in the ABMA surface treatment plant. Aluminum was an emerging Aerospace material nearly as strong as iron and much lighter; little was known about this material. It later became the primary aerospace material, the use of which was refined by DOD and NASA.

These products/processes developed in 1957 became the industry standard and is still being used by NASA, DOD and the airline industry. Military and space standards were developed and are still in use today in the MSFC metal finishing plant. Critical processes for all space material are still in use in 2002.

Recollections of the Director's Secretary
by Bonnie G. Holmes

What an overwhelming experience! What an education! Having been the secretary and executive assistant (stenography) to Dr. Wernher von Braun and three of his successors as Director of the NASA Marshall Space Flight Center was an enjoyable career spanning almost 30 years. There were many long days, always a staggering workload, sacrifices made by me and my family, but never any regrets.

When I became Dr. von Braun's secretary in the spring of 1952, it was a challenge to teach myself shorthand for all the technical words associated with missiles, rockets and, later, the space program.

These were not taught at the business university from which I graduated a few years previously. Because he didn't like "talking into a machine," Dr. von Braun dictated all his correspondence and reports to me.

In reflecting on my personal association with all members of "the team" I can say that I had the utmost admiration and respect for them.

Because of my association with Dr. von Braun, I met some very notable people, including President Dwight D. Eisenhower, President John F. Kennedy, his long-time secretary Evelyn Lincoln, (as vice-president) Lyndon B. Johnson, (as member of Congress) Gerald Ford, Vice President Hubert H. Humphrey, Lady Bird Johnson (as First Lady), Walt Disney, Walter Cronkite, heart transplant surgeon Dr. Christian Barnard, oceanography pioneer Jacques Piccard, Aviation Hall of Fame's woman pilot Jackie Cochran, Boston Pops' Arthur Fiedler, Metropolitan Opera basso Jerome Hines, famous "B.C." cartoonist Johnny Hart, movie and television personalities Hugh Downs, Hugh O'Brian, Gloria Swanson, John Denver and belly dancer "Little Egypt." It was somewhat surprising to me that people from all walks of life had such a tremendous interest in the space program.

To have been considered a part of "the team" by Dr. von Braun and others is an honor I treasure. This honor was confirmed to me by many of "the team" over the years and upon my retirement. Here I quote from some of my memorabilia these words: "...in appreciation of the outstanding job you do (as unemotional as possible) for all of us..." "...the true 'Mrs. Space,' with sincerest thanks for everything you have done for the space program, and for us at MSFC during the past 25 years..." "...great admiration of your excellent work and with thanks for your unfailing support..." "...with my compliments and highest esteem for your outstandingly capable and dedicated contributions to mankind's first step on the Moon..." "...the 'Hub' that makes the 'wheels' go around..."

One of my most cherished possessions is a silver Paul Revere bowl sent to me by Dr. and Mrs. von Braun their first Christmas in the Washington area. The bowl is inscribed "To Bonnie Holmes who helped us all the way to the moon and back. Wernher von Braun."

Dr. Wernher von Braun and secretary Bonnie Holmes try to find a place to work in one more conference in his already over-crowded schedule, 1960. Courtesy of Bonnie Holmes.

Stories From the Early Years
by the family of Charles P. Trapalis, Jr.

Dr. Wernher von Braun and my father were working on future projects for the program. For a change of pace, my father asked Dr. von Braun to our home to work on the projects. After several hours of little progress, the tension of no success was growing. My mother (Catherine Trapalis) made lunch and suggested that the men take a break and eat. My mother asked what was troubling Dr. von Braun and my father. "We can't seem to get anywhere today with our ideas," was the reply. With a queried look on her face my mother stated, "Well, all you have to do is look at the comic strip of *Buck Rogers* in the paper and you'll get plenty of ideas!" Dr. von Braun broke into laughter.

The Mercury astronauts were in Huntsville during one of their many trips to the city. The von Braun staff asked what the astronauts wished to have for dinner. The astronauts had decided long before coming on this trip that they wanted to try the southern dish of chitterlings (chitlin's - intestines). The staff tried to change the astronauts' minds, but to no avail. So the group headed outside of Huntsville to eat at the Greenbriar. After placing their order, the unusual smell of the dish began to enter the room. The astronauts barely took two bites of the intestines and asked to re-order.

My father left Charleston, SC and worked for awhile in Huntsville prior to my mother and I moving there. The early housing areas were evidence of a future city. For many months shopping had to be done in Decatur. There was no grass or trees in the new subdivisions. Everyone felt as if they were pioneers in a new land.

One day my mother drove to downtown Huntsville and was shocked to find cows, ducks and pigs all in the center road of town. When my father came home, she cried that both of their parents were immigrants and they never had to experience livestock in the middle of the road. "What kind of place have you brought me and your child to?" My father, knowing how much she missed Charleston, said he would make it up to her and bought her a washer and dryer. My mother became the envy of the entire neighborhood because she had the first washer and dryer for miles. A few days later, my father was reading the newspaper and discovered that the 4-H club had a parade of the farm animals in downtown Huntsville.

Lockheed Agena B
by Spike Field

Early in December 1959, while the von Braun team was still a part of ABMA, a small group from the newly formed ARPA/NASA Project Office flew to California to "survey" the hardware capabilities of west coast rocket engine and missile contractors at San Diego, Los Angeles, Canoga Park, and Sacramento. Lee Belew, Bill Griever, Shirley Hale and I made the trip. That was the week before the Silverstein Committee recommended a long-range development program calling for Saturn upper stages to burn liquid hydrogen and liquid oxygen propellants. Dick Canright from the Advanced Research Projects Agency (ARPA) instigated the trip, supported by eager contractors and the few NASA Headquarters officials who knew what was about to be recommended. Of special interest was the hardware produced and tested under Department of Defense contracts by Convair, Douglas, Rocketdyne, and Aerojet. Not only were hardware and test facilities seen, but the Huntsville visitors also got to meet top company executives. It was a whirlwind trip.

On the last night of that west coast foray, I was preparing to return to Huntsville from Sacramento when I got a call. Fred Ordway was on the line. He was a rocket and space author/editor working with von Braun. He said that the Air Force intended to transfer management of a second-generation *Agena* upper stage to NASA, and that current information about the program was needed immediately. Ordway also said that a top-level meeting was scheduled the following Tuesday in Washington to decide which NASA Center would be assigned to manage the transferred *Agena* Program. Fred asked one of us to get down to Lockheed-Sunnyvale and pick up any information that might be available.

After changing flight reservations, I arrived at Sunnyvale on a Friday afternoon. The rest of the "survey" team returned to Huntsville. Since my aunt and uncle lived in Saratoga, they put me up for the night and, luckily, they knew Lockheed's marketing director, who arranged a meeting at the facility the next day. I was briefed on *Agena* and given a short tour of the plant, as well as a classified report that had just been issued the day before. It was so new that Washington had not yet seen it. The report was given to Dr. von Braun on Monday, and he flew to Washington for the meeting with NASA leaders. It was reported that von Braun slapped the report on the conference table to the great surprise of some very high level NASA people, and subsequently convinced NASA, including Hugh Dryden, that Huntsville should manage the *Agena* program.

Von Braun Story
by Cary H. Rutland

Many stories have been written about Dr. Wernher von Braun, but one of the most amazing things about him to me was his ability to delve into technical details on a complex subject but not "bog down" in them. He was a man of vision and never lost sight of the "forest" while digging around the "trees." He just loved the technical detail. He had the uncanny ability to dig into the detail on a subject, give an action that would take two weeks to work, and when you came back he would pick up on the details where you had left off as though the first meeting had never ended.

My favorite example of working the details occurred in the 1969 to 1971 time period, which was in the midst of the Apollo program. As a member of the Future Projects Office, whose charter was to look ahead and define new programs for NASA and the Marshall Center, we were making a presentation, to Dr. von Braun and the other members of the MSFC board on the "Large Space Telescope." This program grew into what became the Hubble Telescope launched in 1991. During this presentation, several references were made to this telescope as a three-meter, "defraction-limited" instrument.

Dr. von Braun said, "Hold it!" Then as he pulled out one of the large blackboards in the conference room, said, "Ernst (Dr. Stuhlinger), come up here, I need a refresher course in my optical physics on this "defraction-limited" business. The rest of you guys will just have to bear with me for a few minutes." After about 10 minutes of sketching on the blackboard and discussion with Dr. Stuhlinger, he said, "Now I've got it, you can go on with the presentation." We never had to worry about this again; he was with us and, of course, was very instrumental in selling this program to the science community and NASA management.

It's Not That Simple
by George C. Bucher

All of the professionals at MSFC hoped that they would be able to describe their work to Dr. Wernher von Braun in person. When that opportunity arose, some of them made presentations that were more technical or longer than necessary, perhaps to impress Dr. von Braun with their expertise.

One such occasion happened early in the lunar exploration program, when knowledge of the moon was minimal. An unmanned space probe carrying a lunar atmosphere experiment was to fly behind the moon. An onboard transmitter was to send signals at a fixed frequency back to earth. Just before the probe passed behind the moon, signals would pass through the atmosphere of the moon, and changes in the frequency of signals would give an indication of the altitude and density of a lunar atmosphere, if there was one.

Dr. Ernst Stuhlinger, Director of Space Sciences Laboratory, was requested to arrange a short meeting to describe the experiment to Dr. von Braun. Dr. Stuhlinger and three other persons waited at the appointed time and place. Dr. von Braun arrived late, and had another meeting scheduled soon. A physicist who was most familiar with the experiment described it with great complexity, using mathematical formulae on a blackboard. Dr. von Braun listened patiently for several minutes before saying, "I just don't understand this." The physicist, without thinking, replied, "It's really very simple." Dr. von Braun, by then exasperated, said, "If it's so simple, why can't I understand it?" Dr. Stuhlinger then took over to salvage the awkward situation.

Electrical Connector
by Bill McPeak

I awoke from a deep sleep. It was the telephone. People from the Cape and Marshall, etc. were explaining that an open circuit had been isolated to an umbilical connector at the Instrument Unit (IU) to Launch Tower swing arm coupling during pre-launch check-out of the Apollo 15, Saturn V Launch Vehicle. I related that the problem seemed to be one that we were working in Huntsville. The prime contractor, IBM, had found the problem on an IU in their Huntsville facility. Smearing of adhesive isolated a spring member from the main body of a socket contact. When the mating connector pin lined up without any float and touched only the spring member, an open circuit could occur. IBM had a fix they were certifying in their lab. I understood what it was to be and had confidence that it would work although there were not yet approved procedures. The decision was made to fly me on the NASA plane to the Cape during the night to try to fix the problem and minimize the launch delay. I asked that an IBM guy be sent with me.

It was just daylight when we approached the Saturn V, with me looking and thinking of all the success of those great Saturns that I had studied in my entry level stint with the Marshall Center Reliability Office. We then rode the elevator of the launch gantry, going up past the Saturn S-IC stage, past the SII stage, and past the S-IVB stage. I thought of all our teamwork to assure the reliability and safety of the cryogenic feed through electrical connectors and wiring penetrating all these stage's fuel and oxygen tanks.

We reached the nosebleed level of the Instrument Unit, got off the elevator and I looked up. Somewhere above was the first Lunar Rover with the first space mission Zero-G astronaut operated electrical connectors on which so many of us had worked so long and hard all over the country.

We talked about the problem at hand. The decision was made to attempt the fix. The IBM guy with his hypodermic needles, alcohol, connector pin abrader adhesive cleaner and I were led across the swing arm to the access door of the Instrument Unit of which he and I were very familiar, except we were accustomed to it being only a few feet off the floor in Huntsville, not way up here. I had peered through the access door of an early Instrument Unit perched near the top of the smaller Saturn IB being prepared for its first launch at the Cape. I had traveled to the Cape to become familiar with the Saturn launch operations and to be on hand if any electrical connector issues or problems arose. We visited the Merritt Island Saturn V complex being built. We might need to know how to get out there someday. So now all the way until Apollo 15; I'm back. We had caught all the big connector problems before they got to the Cape until now.

The umbilical was not at the access door. A side gate opened to a "cat walk." We were shaky on this lofty perch. After a few cleaning cycles interrupted by quick retreats to the swing arm to calm our nerves, we were repeating the cleaning operation and I was using a dry air hose to blow liquid out of the socket contact cavity; until spray came out into my eyes. I jerked my head and my hard hat toppled off. I caught the hard hat. We looked at each other and agreed, "we're done." The umbilical connector was re-coupled, the fix and the launch were successful. We had flown back to Huntsville and I just missed my last chance to see a live Saturn launch at the Cape.

We used all the lessons learned with the Jupiter, Redstone and early Saturn connector problems along with much "can do" problem solving to inspire us to make sure for the future we always had well developed, qualified and fully coordinated published procedures and to continue to establish reliable families of launch vehicle and space rated standard electrical connectors for Skylab, Space Shuttle, Spacelab and throughout Marshall and NASA-wide programs. We supplemented them with special ordnance connectors with locksmith keying (no cross-mated cables) throughout the Space Shuttle stack, underwater re-useable high vibration connectors on Space Shuttle Solid Rocket Boosters and self-locking threaded cryogenic high vibration connectors on the Space Shuttle Main Engine (SSME), cold temperature umbilical connectors between the Space Shuttle External Tank and the Orbiter, and separation connectors between the External Tank and Solid Rocket Boosters.

We evaluated, tested and established insulated bus distribution and bussing modules and tools and certification techniques for launch vehicle and space rated interconnections inside electrical equipment and electronic components.

Marshall was especially successful with electrical connectors.

We Lived the Dream
by Clarence Jackson Driver

During my 25-year career, I was fortunate to work on several projects. The following accounts are sentimental to me, and they illustrate several fringe benefits of my employment.

I was a Radio Frequency Test and Checkout Specialist that performed Component and Systems Post Manufacturing Checkout of the Range Safety Command Receivers and Decoders as well as the Dovap Doppler Radar on Jupiter C, Jupiter and Juno series. One particular event stands out in my mind from those days. On August 7, 1957, I carried the Dovap ground receiving equipment to the fire tower on Monte Sano Mountain in Huntsville, Alabama. There I connected the Dovap receiver to the TV antenna, which was mounted on top of the fire tower. With this system in place, we could record the Doppler shift when Jupiter C number 40, which was carrying a scale model heat shield, was fired.

After liftoff of Jupiter C number 40, its height came into the line of sight of our receiving antenna, a signal came in loud and clear. Each time the second and third stages fired, there was a corresponding "Doppler shift" recorded on audio tape. These proved that the stages did indeed fire. This proof became extremely important given the fact that something happened with the tracking stations at the Cape. As it turned out, our audio tape was utilized to show that all three stages did fire and that proper velocity was reached. Later the nose cone of Jupiter C number 40 was recovered and shown on national TV by President Eisenhower.

This research effort was so successful that the remaining Jupiter C vehicles were deferred from the firing schedule, making them available for the Explorer series of satellites. Jupiter C number 29 was launched January 31, 1958 with Explorer I reaching orbit, thus, becoming the first US satellite. This event illustrates benefit number one—having a part in a historical event.

When I worked on the Saturn series, I performed checkout of a Digital Command System that was capable of updating the IU Digital Computer on the Saturn IB and V. Later, I was involved in the Saturn S-IC in-house Checkout at Marshall of the RF System.

I remember making several trips to Sacramento, California, Douglas Aircraft Corporation, regarding problems concerning the Saturn Command System. Because there were problems with the Secure Range Safety equipment that had been government furnished, it was our responsibility to determine the nature of the problem. On the first trip, it was discovered that the command ground transmitting antenna was mounted opposite the on-board receiving antenna that had cables entering the S-IVB in the line of sight of the transmitting antenna and onboard receiving antenna. This meant any movement or change of cables could drastically affect the signal level transmitted to the command receivers. Fortunately, that only required some repositioning.

When Douglas had problems again, I observed the Simulated Flight Test. Those test results were good. However, upon analyzing the test recordings, it was discovered that there was a drop in signal level shortly before the actual signals were transmitted from the ground command transmitter to the flight receivers. This was due to some movement in the checkout area. That was normal. Again, no serious problems were encountered.

Later, I was sent back to Sacramento, Douglas checkout area because they had another problem. This time it was more serious in that a signal could not get through at all. During the troubleshooting process, the source was located—not at the command encoder transmitter receivers or decoders. It was a relay between the command encoder input and the ground computers. When the signal was sent, this relay would send a coded message. However, it would also bounce, resulting in another message being sent before the first was processed. The Decoders were programmed to reject all spurious signals. Thus, the Decoders responded correctly. Upon replacing the defective relay, the test was completed successfully. That was the last time I went to Sacramento on a problem concerning the Saturn command system. Again, I had a small part in making history, but I also began receiving benefit number two—seeing parts of the country where I had never been.

These travels continued during an assignment on the Apollo Telescope Mount (ATM) Skylab program. I was a member of the Launch Planning Group. My primary duties were with the ATM Digital Command system that was able to completely reprogram the ATM computer as well as being able to perform others functions when astronauts were asleep or when Skylab was unattended. This duty required travel from Marshall Space Flight Center after Post Flight Checkout had been completed to Johnson Space Flight Center for further tests and finally to Kennedy Space Center for Pre-Flight Checkout and launch of Skylab I. After tests were completed there, I returned to the Marshall Space Flight Center and was assigned to the Gravity Probe A (Red-shift), which was a program that tested part of Einstein's theory concerning gravity affecting time. We carried the GPA capsule to Wallops Island where it was mounted on the Scout vehicle and was successfully fired. I was assigned to take the Space Shuttle Command Antenna, Directional Coupler, Command Receiver MCR-503, and a Radio Beacon Antenna through Qualification Tests. That brings me to the third benefit—satisfaction in being successful.

I was able to see first hand many launches: the Mercury rocket that carried Ham into space, Apollo 11, Apollo 17, and Space Shuttle Flight Number 1. These were all projects that I worked on. As I stood and watched those launches, I knew that each small part I had in them enabled the victory of the launch. It was then that I felt that all of my lost sleep and days of being away from home were well worth it. And when the final reports were completed without any anomalies being noted, I had great satisfaction in knowing that I had a small part. Although I didn't realize it at the time, I lived the dream!

Memories
by Lee James

A brief summary of my history with space programs is as follows: After graduation from West Point the Army sent me to the Army Guided Missile course at Fort Bliss, Texas. That was followed by a two year master's degree course at the University of Southern California and then by one guided missile assignment after another, and finally to the Marshall Space Flight Center at Huntsville. I had the honor to be the Director of the Saturn I, then Saturn IB, and then the Saturn V Moon Rocket during all of the Lunar flights.

It is difficult to separate out instances that stand out during those launches that were more stomach tightening than any other. During any countdown there were holds and the corresponding fixes, then the time came to consider all that had been done and say, "Go!" to the launch and try to think if you had missed anything. There were astronauts on board whose lives depended on everything being right. Fortunately we had great people working on this program and never had a failure of a manned rocket. It is nice now to look back on the entire Apollo program as a string of successes. My last position was to be the Deputy Director of the Apollo Program in Washington. What I remember was to look out over that network of Apollo and see such competent people in charge of each and every phase of the program. Everyone can probably visualize the feeling one would have in saying the final "GO" to a manned rocket to the Moon after having several holds and working out a number of problems where humans had to decide if enough had been done to correct all of the problems. It was a great program.

The Saturn V Breadboard
by Charles L. Riley

Aside from the scientific and engineering talent at MSFC, one of the major contributors to the Apollo program was the launch vehicle flight and ground equipment simulation provided by the Saturn V Systems Development Facility (SDF), commonly known as the breadboard, which I had the privilege of exercising overall supervision and direction. This facility, operated by the Boeing Company for MSFC, contained electrical simulation of the S-IC and S-II stages, actual fore and aft skirts of the S-IV B stage (including a J-2 engine capable of being gimbaled), and an Instrument Unit that was identical to the flight I.U. Included in the Breadboard was also a complete set of the required Ground Support Equipment (GSE).

All ground software, flight computer software hardware and all interfaces for both the flight vehicle and the launch site electrical and mechanical GSE were verified in the Breadboard for each Apollo vehicle beginning with SA-501 and going through the remainder of the program. Additionally, the facility was always powered up and completely staffed in support of every one of the Apollo vehicle launches.

One of the most satisfying events of my career occurred on the evening of December 6, 1972 when I was in the Breadboard and we were supporting the launch of SA-512 (Apollo 17). Lift-off was scheduled for 9:53 P.M. EST and all was proceeding smoothly until T minus 30 seconds when the countdown was stopped by a malfunction in the Terminal Countdown Sequencer, which indicated a problem with the S-IV B stage Liquid Oxygen Vent Valves. Engineers at the breadboard proposed a work-around that was developed and installed in the Breadboard and then several runs simulating the final few minutes of the countdown were made to verify the change. The verified hardware alteration was provided to KSC, thus allowing the countdown to be resumed. Lift-off occurred at 12:32 A.M. EST, still within the established launch window, thus avoiding a one-day slip of the launch.

Breadboard personnel, including myself, were recognized in a ceremony in Morris Auditorium with a Group Achievement Award, signed by James E. Fletcher, NASA Administrator.

Space Program Experiences
by Robert P. "Bob" Baker

When I came aboard in June 1962 the race was well underway. One could sense the pride and commitment to make our country first in a manned lunar expedition. For me, I knew when I joined the Marshall team, I had found a very promising and rewarding career.

The team I joined was the Airborne Electrical Systems Branch headed by Dick Smith and John Stroud. Our branch, the Ground Electrical Systems Branch, and the Power Systems Branch were part of Hans Fichtner's Electrical Systems Division. I cherish the memory of working with Dick and John along with Bill Shields, Al Woosley, Ethridge Pascal, Jim Stulting, Frank Matthews, Jerry Felch, Royce Mitchell, Ed Guerin, Dewey Greer, Bob Lewedag and Jack Gregg. I also recall the pleasurable working relationship with the Power and GSE guys: Gene Cagle, Charles Graff, Bill Glass, Walt Goodhue, Claude Baldwin, Gene Gallaher and Charles Mauldin to name a few.

Our task was fairly straightforward: we wired the launch vehicle. What that really entailed was integrating all the flight equipment with proper power, sequencing and signal flow. Our team designed and produced the electrical system schematics and hardware production drawings for the first stage boosters and instrument units of the early Saturn I, Saturn IB, and Saturn V flights. We also designed, integrated, and controlled all stage-to-stage, stage to GSE, and launch vehicle to payload electrical interface documentation for all the Saturn family launch vehicles. There was, perhaps, no better training ground for learning how all the elements fit together and functioned.

This experience and knowledge led to new assignments in the late '60s and early '70s. Our Center's role evolved from designing and building the Saturns to supporting the many flights leading to and including the Apollo Lunar Missions. Although many of my colleagues were busy designing and preparing for the Skylab Mission, my assignments were focused on the Saturn Launch Vehicle throughout the Apollo and Skylab Programs including the Apollo Soyuz Test Project. During this era, as a senior systems engineer, I was privileged to work under the leadership of Ludie Richards, Dick Smith, Ellery May, Harold Ledford and Jim Kingsbury. I particularly recall a couple of very interesting assignments; both involved teaming with Kennedy Space Center to solve.

It seems we were plagued with lightning strikes while on the launch pad, particular during the summer months. Typically lightning would strike the launch umbilical tower, resulting in current flow not only down the tower, but also across the umbilical arms and down the launch vehicle as well. The primary concern was the effects of induced transients on the umbilical cables connected to the vehicle. To assure that no damage had occurred major retesting was performed after each recorded strike. The task force for which I was the Marshall representative realized there had to be a better solution.

The team came up with the idea of routing the lightning surge away from the umbilical tower and the vehicle. The concept was a large fiberglass mast mounted on top of the tower with four equally spaced cables connected to lightning rods on top of the mast and running to the ground at 45 degrees to the tower. The concept was modeled, laboratory tested and proven to be effective. To my knowledge, after the system was implemented no current was ever measured on the tower or umbilical arms.

Another noteworthy assignment came prior to the Apollo Soyuz Test Project (ASTP) mission in 1975. Due to the Saturn IB/Apollo performance capability, NASA committed to perform the rendezvous and docking with an orbiting Soviet Soyuz. As I recall, we had a five consecutive day, less than 10 minutes per day, launch window. Needless to say, there was a lot of emphasis placed on launching on time for such an important international mission. Carroll Rouse of KSC, a senior engineer with years of experience in both ground and vehicle electrical systems checkout and launch, and I were charged with reviewing all the interlocks that could cause a prelaunch hold or shutdown. Interlocks are designed into the electrical system to prevent proceeding in the countdown with an out of configuration mission critical or unsafe condition. Eliminating any interlock requires a great deal of scrutiny. We did find particular interlocks that were thoroughly reviewed by all responsible parties and determined to be unnecessary. I'm proud to say I was honored with the NASA Exceptional Service Medal for the effort.

Following the final Saturn launch on the ASTP mission I joined the Space Shuttle Project. It was for me a very big change. I had been deeply involved in many of the technical aspects of Saturn and supported most of the launches. I remember many hours of teleconferencing, both day and night, working prelaunch problems. Now, the second half of my career was just beginning.

One of my earlier assignments turned out to be just what I needed to get up to speed on the Shuttle program. When John Yardley came onboard as the Associate Administrator of the Office of Space Flight he requested what was called a critical function assessment of the Shuttle Systems. Tom Milton of the Johnson Center Shuttle Systems Engineering Office was appointed to lead the JSC, MSFC, KSC tricenter effort. I was named the Marshall lead. I believe Bob Seick headed the Kennedy effort. We laid out a plan to organize a series of briefings essentially in chronological order of events of a typical mission. We were directed to have each responsible engineer present a detail description of how each critical element was designed, how redundancy or back-ups were implemented, and to identify any single failure points. It took several briefings and weeks to get through liftoff. The meetings were held at different centers depending on the topics addressed. The total review took about a year to complete. Although not many changes resulted from the effort, I think the process was very beneficial to all levels of engineering and management.

In the '80s I worked on Space Station Task Team, the Orbital Maneuvering Vehicle (OMV) program and the combined release and radiation effect satellite, retiring in May 1990.

In reflecting on my NASA career, I am very grateful for the many opportunities and take great pride in being on the team. Memories of working with some of mankind's finest are priceless.

Saturn V
by James T. Murphy

My background as an Air Force pilot and in design and development of ballistic missiles with the Air Force gave me experience directly usable with the development of the Saturn V launch vehicle. I had wanted to go into space work when I finished graduate school in 1961. I was interviewed for the deputy manager's job by Dr. von Braun, Dr. Rees and Dr. Rudolph and hired for the job in March 1965. NASA asked the Air Force to allow me to come to Huntsville in that position as soon as possible. I arrived in late March.

I was very aware of the many difficulties encountered in development of the Atlas and Titan Missile programs, where exact configuration had not been kept, including the impact of engineering changes on cost and schedule. The programs had become near cancellation and leadership was changed. Controls were established in engineering (called configuration control) in schedule, in cost and especially in safety. Both programs recovered.

Dr. Rudolph had a program control room that was a real asset. Bill Sneed was in charge. He also had started early to work problems related to the Saturn V and was starting to implement configuration control including costs and schedule impact. He was having problems getting information from the laboratory directors related to engineering changes, approval, accounting, and impact of those changes not only on the Saturn V but also from ground equipment at Kennedy and spacecraft at Johnson.

Dr. Rudolph asked me to see what we needed to do to get the problem solved on the Saturn V. There were three major items that had to be fixed. First, the laboratories were making their changes, drawings were being held in the labs and the contractors had no way to do their contracted responsibility. Second, there was little or no communications between the three centers on engineering changes or the impact on those changes. Third, the Saturn V had a good system established to record and track, but could not implement the system until control of changes was implemented totally.

We moved on the problem immediately, starting within MSFC. I was named as the engineering change maker on the launch vehicle. That included getting all of the drawings into the contractor's hands. Establishing a change board that met weekly to evaluate the requested change and make a decision that day with attendance from the program office, the requesting lab and the contractor. That form of control did work as soon as all participants realized the criticality of implementation to getting the Saturn V launched on the schedule, safely.

We then had total approval from the Apollo program director, General Sam Phillips. He had been commander of the Minuteman Launch Vehicle for the Air Force. He issued instructions to each center director (Marshall, Johnson, Kennedy) to immediately establish a three center configuration control team related to any interface changes requested that would impact the other centers. We met at one of the centers for discussion and approval/disapproval by the Apollo change manager. That brought the three centers together and did work very well. We were on our way.

Safety

After the astronauts lost their lives in simulation training at Kennedy, Dr. von Braun asked me to become director of safety for Marshall and to report directly to him. He wanted to have all of our potentially hazardous activities at the center inspected, classified, concerning actions required to negate possible damage or catastrophe. With the many tests being run with flammables, propellants, large structures, space vehicles our job sounded almost impossible.

We selected two categories to do our research, "manned" and "unmanned." We did a complete search of past problems, which had created risks or actual damage. We looked at measures taken to eliminate, reduce or resolve those problems. We arrived at a list of potentially new problems. Then through discussion and research we isolated the area and activity and named an individual to be responsible to initiate action and procedures, which could be tested and would demonstrate the elimination of the problem. This was by no means an easy task, but with the understanding of the potential hazards the results paid off. Many changes in procedures and tests were made.

In six months following this activity not one former potential or real problem occurred.

Dr. von Braun had in the meantime formed a new organization called program development. Dr. William Lucas had been named the director and he asked me to be a deputy. I took that job at once.

Program Development

Dr. von Braun was a pleasure to work for and associate with. He demonstrated over the many years a capability to search into the future. The results were amazing.

After the success of the Saturn V and long before the Apollo program was over, he created program development. He named Dr. William R. Lucas the director of that organization. He had often stated that our center would go out of business if we did not "spread our wings" into optics, Shuttle missions, transportation at much lower costs, space station and many other excellent opportunities. He gave Dr. Lucas an opportunity to bring some of the best and most diversified talent available from across the center. The job was to sell our plans for new space programs to NASA HQ by providing a well-studied concept, do design, arrive at costs and schedules and show how MSFC should be the lead center in that program.

The Apollo Team
by Gerald Wittenstein

During my nearly 25 years of working for NASA at the Marshall Space Flight Center,

50 Years of Rockets and Spacecraft

I was privileged to have been a part of the Apollo team. Although I have got to admit that I am not sure that I have all the facts straight nor all the people's names exactly correct, in thinking back over the Apollo program, one key meeting comes to mind. This particular meeting was something similar to a flight mechanics panel. Assembled in one of the bigger conference rooms at Marshall were contractors, employees of the Marshall Space Flight Center, and employees of Johnson Space Center (JSC).

Apollo 8 was the topic. A problem had occurred on that mission. The vehicle was ready; the payload, a lunar module, was not ready. The lunar module schedule had slipped. During the discussion about what that meant, a JSC employee, Carl Huss, stood up and asked a challenging question. The question was along the lines of "Can we do a translunar type orbit instead?" After some discussion, we concluded that there should not be a problem with doing a translunar type orbit. Why did he ask that question? An interesting discussion ensued among the team members: John Mayer, Gene Ricks, Ron Berry, Monroe Hardage, Jim Lindberg, Harold Scofield, as well as several contractors.

About two days later, Mr. Hardage, my immediate supervisor, says, "Mr. Wittenstein, we are going to take three men to the moon on Apollo 8. Keep it a secret. Do not even tell Jerry Weiler yet."

That momentous statement started the process in which I was involved, supporting JSC, before the launch, during launch and during the burn. Our tasks were to provide guidance information, performance information to the crew and to monitor the translunar injection. JSC asked our office to do this because of our familiarization with the guidance system. We started a team as a part of the flight mechanics branch, operations trajectory section, and in-flight trajectory team. Our counterparts at Johnson were Gran Polis, sometimes Jay Green as FIDO, and Charles Schultz, the retro burn officer. Those were the main JSC people with whom we interfaced. In the next year or so, our heads were down and our tails up. We worked crazy hours, and we had a ball. Thank God, we were successful. One of the descriptions later given about the work we did, was that this work was typical of MSFC. The work was designated as technically excellent. Our branch had that as part of its adjective. We actually ended up getting involved with the performing of a slingshot around the moon, and finally with lunar impact with the S-IVB stage.

Players involved in those projects were Paul Black, Terry Deaton, Hugh Brady, John Blackstone, Ann McNair, Jack Bradford, Jerry Weiler, Bobby Brothers, Pam Pack, Steve Tondura, Rocky Clark, Jim McMillan, Mack Henderson, G.E. Hall, Harold Scofield, Ray Bailey, John Wolfsburger, Ray Tanner, Jerry Mack, Ray Garcia, Jim Igou, Bill Cobb, George McKay and a cast of thousands. Apollo was a team effort. I am proud to have been a player on that wonderful, inspired team.

The Abort Dilemma
by Robert K. Wolf

During the Apollo program, Dr. von Braun placed a team of MSFC engineers at the Manned Spacecraft Center (MSC) in Houston to perform flight operations on the Saturn boosters. The MSFC team manned a console in the Mission Operations Control Room (MOCR) for each Apollo flight. I was selected as the lead Booster Systems Engineer for the unmanned launch of Apollo 6 (AS-502). Since this was an unmanned launch, abort was initiated from selected MOCR consoles, whereas abort was initiated by the astronaut from the Spacecraft on manned flights. AS-502 was launched on the morning of April 4, 1968. As I monitored the launch data from my console, the S-IC stage appeared to operate nominally. But during the S-11 stage flight, two of the J-2 engines prematurely shut down. This was difficult to immediately confirm due to erratic data being displayed on the console screens. The mission rule stated that the Booster Engineer should initiate the abort command at this time for the loss of two S-11 engines. I delayed taking action until I was assured of valid data. After confirming the loss of the two engines, the Saturn appeared to be under control and not tumbling at high rates as the prelaunch analysis had predicted. Therefore I elected not to send the abort command. The S-11 stage extended its burn with its three remaining engines and then the S-IVB stage operated nominally and achieved orbit. This memorable and exciting experience enabled the Apollo program to proceed without requiring another unmanned flight. George Low decided that the next launch of the Saturn V should carry men to the moon. This story is written in more detail in the book, *Apollo the Race to the Moon*, by Murray and Cox (Chapter 22).

The Saga of Apollo XI and the Freighter Vegasek
by R. J. Schwinghamer

The Apollo XI Saturn V Launch Vehicle left the launch pad at the Kennedy Space Center on July 16, 1969, headed for the first lunar landing. Separation of the first stage (the S-1C) occurred as planned at 150 seconds into the flight, and it came back to earth in the Atlantic from 38 miles altitude.

As we all know, the mission was a complete success, with the lunar landing on July 20, 1969 and then the successful return to earth. What is generally not known, however, is the fact that there was an international incident associated with the Saturn V, S-IC stage after separation when it plunged toward the Atlantic.

It turned out that at the same time the freighter Vegasek was passing a highly restricted area about 375 miles East-North-East of the Kennedy Space Center (the launch site). The captain was first frightened, then highly irritated as hot aluminum structure from the S-1C stage rained down on his vessel. He was taking the shortest way home to Hamburg, and at that point he was probably navigating on a great circle course.

Nevertheless, he continued on to Hamburg with his additional cargo of Saturn V S-1C debris. He immediately contacted the German government and complained bitterly that the Americans had dumped hot junk on his freighter.

When the German government had checked it out, they discovered that he was

MSFC Flight Controllers monitoring booster parameters in the Mission Operations Control Room in Houston during the launch of AS-501 (Apollo 4). AS-501 was the first Saturn V launch (unmanned), November 8, 1967. MSFC employees, from left: Bill Brady, Chuck Casey, Bob Wolf and Howell Hooper. Courtesy of Robert Wolf.

in a restricted area, so they confiscated the aluminum debris and subsequently delivered it to the U.S. Embassy. The Embassy flew the material to Washington, where we picked it up and brought it back to the materials laboratory at MSFC. Using wet chemistry analysis, we verified positively that the aluminum was, indeed, 2219 alloy from the Saturn V S-1C stage.

I had pieces of this aluminum mounted in clear plastic paper weights and sent one each to Armstrong, Aldrin and Collins (Apollo XI astronauts) along with a note explaining that, "here's some stuff you left behind on a freighter in the Atlantic on the way to the moon!" They were pleased.

To the Cape by Barge
by Billy C. Neal

It was recognized early in the space program that no equipment was available to transport the monstrous Saturn equipment, so a very large barge was built to move such equipment from Huntsville to Cape Canaveral via the Rivers system (Tennessee, Ohio, Mississippi) thence the Gulf of Mexico and through the intercoastal waterway. Since no propulsion or steering capability was included in the barge, the commercial towing contractor was faced with unheard of problems. Our barge was "rammed" at New Orleans by a Norwegian freighter (no damage to our cargo or crew) and ran aground several times in the varying depths of the waterway. Retiree Billy Neal rode the entire trip as the contractor liaison. He retired in November 1972.

History of the Apollo 11 Medallion
by Victor S. Grimes Sr.

From about 1963 to 1974, it was my job to create, layout, and have published manned flight awareness brochures; design, produce and distribute displays to contractors and government agencies; outfit, maintain and schedule the Craftsmanship Van; create, layout, and have posters printed; and create other communication motivational ideas related to employees goal of performing their jobs right the first time. In recognition for achieving outstanding work by employees, they were given Snoopy Pins awarded by astronauts, and were flown to Cape Kennedy to witness a space vehicle launch. While at the Cape, honorees attended an astronaut pre-launch party with VIPs such as I attended before the launch of Apollo 11, when Charles Lindbergh was invited by Dr. Wernher von Braun to speak to the Manned Flight Awareness Honorees. Other recognition given to Honorees, NASA, and Apollo/Saturn contractor employees, were certificates, decals, astronaut's crew photos and medallions such as the specially designed and produced Apollo 11 Medallion. The history of this medallion is as follows:

The issuance of the National Aeronautics and Space Administration (NASA) Apollo 11 Medallion is recognized as the second and the most unusual one ever produced in the form of medallic art or coinage and man's greatest technological achievement.

The 0.063 inch thick and 1.250 inches in diameter medallion contains metals that were structural members of Apollo 11 spacecraft, Eagle and Columbia. One structural member was a stainless steel bolt, one of several bolts used to secure the heat shield of the Command Module (CM) (Columbia). The CSM (Command and Service Modules together), piloted by Michael Collins, orbited the moon while Astronauts Neil Armstrong and Edwin "Buzz" Aldrin landed on the moon. One stainless steel bolt was removed from the heat shield of the CM after its return to earth. The other structural member was a composite metal clip used to secure the box of lunar rock samples to the Lunar Module (Eagle) ascent stage prior to lift off from the moon's surface. The clip and the moon rock box was transferred from the Lunar Module to the Command Module before leaving the orbit of the moon. The bolt and clip were requisitioned by Dr. Preston T. Farish through Dr. Christopher C. Kraft, Manned Spacecraft Center, Houston, Texas. Nickel silver alloy known as 1SC No. 268 with approximate composition of 70% copper, 18% nickel and 12% zinc, was added to the smelting process because of its Rockwell Hardness of B-24. It was chosen because of its ability to alloy with the stainless steel heat shield bolt. The planchets were smelted, rolled, and cut by the Meriden Rolling Mills, Inc., of Meriden, Connecticut, a subsidiary of the Internal Silver Company, during June 1970.

The dies used in making NASA's Apollo 11 Medallions along with two medallions have been encased by MSFC and officially transferred to the Smithsonian Institute, Washington, DC for display with other historic exhibits of the nation's money and commemorative coins.

The basic idea was initiated by Dr. Preston T. Farish, Chief, System Safety and Manned Flight Awareness, George C. Marshall Space Flight Center (MSFC), Huntsville, Alabama.

The original idea was to give each of the three astronauts a brass metal ticket 1/8 inch thick, about 2 inches high and 3 inches long to carry in their personal preference kit and later to be melted and produced into medallions. One side of the ticket was engraved with the words "Good for a successful flight to the moon in Apollo 11 aboard the finest booster that can be built - Saturn Spacelines Unlimited" and the signature of Dr. Wernher von Braun. On the other side of the metal ticket were the engraved initials of individual members of the MSFC Manned Flight Awareness Office. Just before lift off, the three metal tickets were

Apollo 11 medallion. Courtesy of Victor Grimes.

removed and replaced by a souvenir from a high authority in NASA's headquarters.

Subsequently a decision was made to use instead structural metal from the spacecraft Columbia and Eagle.

An unusual event happened regarding the artwork for the Apollo 11 Medallion. The original design and dummy layout concept of the Apollo 11 Medallion was created by Victor S. Grimes Sr., Visual Information Specialist, with the assistance of Technical Writer, Mitchell R. Sharpe, both of MSFC, Manned Flight Awareness Office and under the direction of Dr. Preston T. Farish, Chief. The mint that finally received the contract was to have a professional artist prepare the finished artwork. Because of a controversy over the selection of a sole source procurement, the contracting officer ruled that the bid would be competitive. Barco Mint of New Orleans, Louisiana, was the low bidder and they actually used the comprehensive dummy layout as the finished artwork. Procurement of 217,015 medallions was handled under the direction of Calvin L. Swindell by a contract through The Boeing Company, Huntsville, Alabama on May 20, 1970. The Purchase Order Number was L-19136951023, NAS8 Contract - 5608, with final delivery on August 18, 1970.

The Apollo 11 mission lifted off from Launch Complex 39A, Kennedy Space Center, Florida, July 16, 1969, on a Saturn V launch vehicle. The Lunar Module (LM) named the Eagle landed on the surface of the moon with Commander Neil Armstrong, who made mankind's first footprint on lunar surface soil, July 20, 1969 as pilot Edwin "Buzz" Aldrin watched from inside the Eagle. The Apollo 11 mission returned to Earth with a splashdown in the Pacific Ocean on July 24, 1969.

Other sources of facts regarding the history of the Apollo 11 Medallion: Man on the Moon - The Eagle has Landed, the Story of the NASA Apollo 11 Medallion by James C. Spilman, TAMS (Taken and Medal Society) #2116, Volume 11, Number 1, April 1971 and an article written by James C. Spilman, Numismatic News Weekly, October 22, 1974.

We certify that the above facts are true to the best of our knowledge.

Victor S. Grimes Sr.
Mitchell R. Sharpe
Dr. Preston T. Farish

F-1 Rocket Engine Research and Development Problems
by Frank M. Stewart

The F-1 with 1.5 million pounds thrust was nearly 10 times more powerful than any liquid rocket engine ever built and many difficulties were encountered before it could be flight certified, but two of these problems are remembered as major.

The first, called "combustion instability," was found by post firing inspection of the thrust chamber where localized overheating and erosion was discovered in the injector area. This was unforeseen, as it was not encountered on similar but smaller LOX RP-1 propellant engines used in the Jupiter missile and the Saturn I booster. A major effort was initiated almost immediately by a special task team at Rocketdyne and a parallel team with outside consultants at MFSC.

The difficulty of obtaining needed data in this region of the combustion chamber led to considerable effort and analysis and also testing at Rocketdyne's Test Facility. In several months, the Rocketdyne task team headed by Paul Castenholtz determined that a circumferential spinning mode near the injector plate was the root cause, and baffles were designed and incorporated on the injector base plate. This suppressed the lateral hot gas flow and solved the problem without impact on the Saturn-Apollo schedule. This was an outstanding engineering accomplishment by the task teams. At that time, the injector design and technical data were classified information and later when the Smithsonian requested an end-of-life F-1 engine for display, we hid the injector area from visibility and access by welding a thrust chamber throat plug prior to delivery.

Another serious problem arose on an early Saturn V-Apollo flight when "pogo" was encountered during the boost phase. Pogo was a word used to identify a potentially catastrophic closed loop oscillation involving a liquid rocket engine, the propellant, its feed system and the vehicle structure. Probably the term originated with the Air Force Titan vehicle when a propellant duct material change induced a serious osculation with pogo-stick like motion. This Saturn V pogo was determined to be triggered by pulses from the LOX pump inducer blades. (At the time, I was F-1 engine project manager, which was a project under the Engine Program Office headed by Leland Belew, a parallel but equal Apollo Program Office to the Saturn V Program Office with Arthur Rudolph as manager). The day that the problem was identified from telemetry data, Rudolph accosted me in the Marshall cafeteria at lunchtime and angrily charged, "Frank, your engine is shaking my vehicle." I responded, "Dr. Rudolph, it takes two to tango." He looked stunned and walked away. I assume he soon understood my comment, but he never discussed this with me again and our relations remained amiable throughout the Saturn/Apollo Program.

The explanation as to why the pogo was not found in the S-IC stage static firings at the Mississippi Test Facility is that the test stand structural ties suppressed the oscillation. Various solution options were studied and tested at Rocketdyne. These included drilling holes in the inducer blades to form cavitation bubbles, inert gas injection upstream of the pump, etc. A Marshall study committee considered several fixes and in a Center presentation to Dr. von Braun suggested the gas injection solution; however, there was not rigorous analysis support and reservations about the solution by Rocketdyne, Boeing and F-1 Project office resulted in von Braun directing further study even though a rapid solution was needed. Additional effort resulted in approval of a Boeing design suppressor installed in a LOX ducts in the S-IC stage, and the pogo problem did not reoccur in the remaining Saturn V Apollo flights.

How the Idea for a Crawler Launcher was Born
by G.D. Johnston

The first part of the year, 1962, we had just completed the second launch of a Saturn I, Block I vehicle. Everyone was feverishly working on future Saturn vehicle designs. We were evaluating the capabilities of the C-2, C-3, C-4, C-5 and Nova. The 'C' class vehicles later were all called Saturns. The C-5 design was settled on and it became known as the Saturn V. It was decided that we must make these boosters in the vertical attitude. A Vertical Assembly Building was constructed at the site of the existing manufacturing building, which could house two of the first stage boosters simultaneously. The term VAB was originated. Many years later the personnel at KSC began to ask themselves, "Why do we call this the Vertical Assembly Building?" When they realized it was a carry-over from the early influence of MSFC, they decided it had to be changed. The VAB was so ingrained in all the documentation that it seemed best to change the word 'Vertical' to 'Vehicle' and leave it at that.

"How do we launch the Saturn V?" A Task Team was formed, and Mr. Kroll, Chief of the Structures Design Branch, assigned Charles 'Chuck' Gassaway and myself to the Task Team as representatives from his design branch. There were many others from MSFC, KSC, HQS and other centers. It was an intense effort on the part of all of us to find the best way to launch the Saturn V.

Our first meeting as a Task Team took place in a conference room at KSC. R. P. Dodd opened the first meeting with a few introductions and a brief statement as to why we were assembled. When he had finished, a Corps Of Engineers Major stood up and announced, "My name is Rocco Petrone and I will be working with this task team to find the best way to launch the next family of Saturns." He then had each one in the room stand and tell his name, organization and a brief explanation of our fields of expertise. He had a Lieutenant with him taking detailed notes. When everyone had stood and given the information asked for by the Major, Petrone and the Lieutenant had a short session. Petrone took a page of notes that he had scribbled from that private session and stood up and said, "I want the following seven names that I call out to go across the hall into my office to have a splinter meeting." Five of those seven names were: Garland Johnston, Chuck Gassaway, Bill Bodie, Don Buchanan and Ed Davis. We seated ourselves around the small conference table in Petrone's office and left the seat at the head of the table vacant because it was obviously reserved for him. He came into the room and everyone quit talking immediately. Very formally, Petrone

Apollo 11 Launch Vehicle rollout to Pad 39A. Courtesy of Victor Grimes.

announced, "Well Gentlemen, you are the ones who are going to find a way to launch the Saturn V". He went on to explain what we would be expected to produce in the way of useful information in the short term. He further explained that we would listen to presentations by invited specialists from industry and government. These were people experienced in the missile business and had ideas of their own as to how this should be done. We would come in each morning at 8:00 am and begin listening to presentations from 9:00 am in the morning until 5:00 pm in the evening and sometimes as late as 7:00 pm.

My task was to produce the engineering data associated with vibration, acoustics, blast over-pressure, and community annoyance. I was given the name of a young German engineer who had been hired by General Electric to provide support to the USAF at the Patrick Air Force Base. Max was young, but he was an expert on the effects of an explosion. He was an over-pressure expert. He had hand-plotted charts on the acoustic pressure produced by dynamite, TNT and a lot of other explosives. He had TNT equivalent plots that were unique to his library of data. We worked together to calculate the blast effect of LOX/RP1 in the event of an on-pad structural failure that would cause the fuels to mix and explode. Assuming a first stage rupture from top to bottom and total mixing, we came up with a pad spacing requirement of 9000 feet. This was the distance that one launch platform would have to be spaced from the other no matter what kind of launcher was finally chosen. These criteria became the most important driver in the final decision.

The conference room sessions began, and I wish there were ways to go back and record on video tape all the ideas that we listened to in complete fascination. Some of the ideas were incredibly inventive and some of them were just incredible (the kind you don't want to remember).

One Monday, when we didn't go to the Cape, I went to my office. In the middle of signing the time cards, Bobby Erwin walked into my office with a cup of coffee in his hand. Bobby, a native of Kentucky, was a PFC in the US Army on loan to us from the days of ABMA. The week prior to this Monday Bobby had taken his new bride to Paradise, Kentucky, to meet his mother and father. As he came in and stood there he said, "Garland, you won't believe what I saw last week when I was up at my daddy's farm. Bucyrus-Erie shipped a strip-mining shovel down to my dad's farm in pieces and put it together there. It took 300 railroad cars to bring it all down there. It's eating its way over to the coalfields now and building its own road as it goes. It has a platform as big as a football field. It has tracks 8 feet high with diesel engines in each track. The platform has no vibration on the table and it maintains the table flat within one-quarter of an arc second".

As soon as Bobby said "—a platform as

Drafting specialists from the Propulsion and Vehicle Engineering Laboratory at work in the Huntsville Industrial Complex (HIC) building, temporary quarters for MSFC.

big as a football field", I made the connection with launching the Saturn V. We sketched the platform on a blue-lined scratch pad. He drew a flat bed and put large tracks at each corner. As he began to sketch the upper part of the big shovel, I said, "No, no, don't put anything on top of the platform." We then added a few notes that said 'Big as a football field' and put a 6 feet tall stick man by the tracks. We put the dimensions of 8 feet on the track height. I tore the sheet off the pad and ran down the fire escape to Don Buchanan's office directly below mine at the Huntsville Industrial Complex (HIC) building. Don was Chief of the design group for KSC but was still stationed in Huntsville to be near the design of the flight hardware. They had just been transferred on paper from MSFC to the new KSC and didn't have a budget yet. I ran into Don's office and told him, "Don, I have the way to launch the Saturn V". He leaned back in his chair and said, "Yeah, yeah, well let's hear your idea." I laid the sketch in from of him and repeated everything that Bobby had said. He knew immediately, as I did, that this was what we had been searching for. He called Ed Davis into the office and asked him what he thought. Ed concurred that it was what we needed. Joe Dunn put it on a 'C' size vellum, and Don called the Redstone Airfield and got the NASA plane reserved to fly to the Cape immediately. Don flew to the Cape on that Monday and presented it to Debus, Dodd and Petrone. Don called me and said they have bought the idea. Get Bucyrus-Erie into Huntsville so we can tell them what we want." KSC had no funds to pay Bucyrus-Erie for travel and drafting a set of specifications. Francis Tucker, Mr. Kroll's administrator, hurriedly found $10,000 to cover the cost of our effort with Bucyrus-Erie. They wrote a set of specifications as thick as a Sears catalog and sent us 30 copies. We decided to make a few changes and put a NASA cover on it and put it out on the street. We were sure that no one except Bucyrus-Erie could bid the job. They lost the bid to Marion Power Shovel. Marion Power Shovel did an outstanding job in producing the first crawlers.

That was how the idea for the Crawler came into being. We didn't care who received credit for it because it was so urgent just to get the job done. I am certain, however, that if any one person deserves credit for the idea it has to be Bobby Erwin. If he had not come into my office at the precise time he did and tell me about the "Big Hog" as Bucyrus-Erie called it, the idea would never have occurred to me through my own faculties. I don't believe that Don Buchanan would have ever come up with such an idea as that if left to his own thoughts. We probably would have been using some technique like the Russians use. The most important thing of note to me so far as credit goes is that the Marshall Space Flight Center deserves credit for the idea or concept. We gave it to the KSC personnel to help resolve one of our most pressing problems at the time. To their credit, they took a good idea and did an outstanding job of design and fabrication.

Static Power Inverter
by Carl E. Winkler

During the Saturn era in the 1960s, work on the development of a static inverter for changing direct current power to three-phase alternating current power was assigned to the MSFC (Marshall Space Flight Center)

Astrionics Laboratory's Applied Research Branch headed by James C. "Tex" Taylor. At that time Dorrance L. Anderson and I of Taylor's organization had the opportunity to attend in-house classes on Boolean Algebra taught by Hugh Zeanah of MSFC's Computation Laboratory. It was realized that the Boolean principles had immediate application to the static inverter development work. It enabled us to systematically design the logic to control the switching of power transistors that had recently become available. The goal was to replace rotary inverters with lightweight, more reliable and more efficient devices having no moving parts. A scheme for simulating sine waves via a sequence of steps was designed, built and tested. The results were reported and discussed at an AIEE (American Institute of Electrical Engineers) conference. Subsequently, a patent application was submitted with the help of George Porter, MSFC's patent attorney. The names of Albert E. Willis and John M. Gould were added as inventors along with Anderson and Winkler since all had contributed significantly to the static inverter development. A conflict, however, emerged when Westinghouse engineers attempted to patent a similar device. To resolve the situation, a meeting was held at MSFC before a Federal Court reporter. All inventors gave depositions describing their contributions, including pertinent documentation with dates showing when the work was witnessed. Engineers in private industry tend to be more diligent along these lines, always aware that this information could be critical if a patent becomes a possibility. Their records were better than ours, but under cross-examination by the lawyers we were able to support winning arguments. Eventually both parties were awarded patents covering specific claims. MSFC's patent attorney said that this case was the most interesting he had been involved with in his nearly 40 years of service.

The Flight Vehicle Thrust Vector Control System
by Zack Thompson

When the Saturn I was static fired in the Huntsville test area, the clustered engines, which were attached to a large beam structure, caused the thrust vector control system to oscillate. The structure was beefed up to increase the spring rate, which caused the resonant frequency to move out of the bandwidth of the thrust vector control system. For all other stages and vehicles, a mechanical filter was designed for the thrust vector control system. The beams were designed for structural strength only, after the first stage of Saturn I. All that was required, by the thrust vector control designers was the spring rate of the structure. At that time any hydraulic designer, for thrust vector control, could get a contract to build a mechanical shaping network to solve the differential equations. This mechanical shaping network was to be placed between the servo valve and the hydraulic actuator. All sizes and shapes were delivered. Some could be placed in the palm of your hand and some were the size of the hydraulic actuator. They had various names: some were called dynamic pressure feedback (DPF), some were called dynamic pressure flow (DPQ), and another one was known as a dynamic load damper (DLD). The DLD flew on Saturn V and the space shuttle.

A hydraulic laboratory was developed to dynamically test all actuators. Load simulators were designed to simulate the inertia and spring rates of all the MSFC engines and some that we never built. The engines included the H-1, M-1, F-1, RL-10, J-2, and the NERVA. The simulators would simulate the loading of the actuators, such as friction on the gimbal bearings, forces on the engine when moved in the air stream, thrust vector offset, and the dynamic thrust vector. These simulators are still in use today.

The Space Shuttle Main Engine (SSME) Hardware Simulation Laboratory (HSL) was located in the eastern part of building 4487. The purpose of the laboratory was to design, validate, debug, and checkout the SSME computer software before static test or flight of the engine. The equations were arranged so that as the engine controls hardware became available the simulated hardware could be replaced with actual flight hardware. The hardware consists of all the sensors, the actuators, the computer and a hydraulic system. The SSME flight computer is redundant and adjusts the thrust level and the mixture ratio. This is done by five redundant hydraulic actuators, which move the adaptive control system. When the thrust is changed from one level to another the mixture ratio will also change. Therefore the control system must adjust its parameters automatically in order to hold a desired mixture ratio. It is the only adaptive control system that MSFC has flown to date.

SA-5 Assembly Underway
by Vincent P. Caruso

(Excerpts from an interview with Vincent Caruso which appeared in the *Marshall Star*, August 1, 1962.)

The fifth Saturn flight booster, SA-5, is moving through the assembly phases at the Marshall Center. It began taking shape at the Manufacturing Engineering Division when the Fabrication shop delivered parts in the Assembly Engineering Branch on June 27. Assembly began the next day.

Already the big vehicle that is paving man's way to the moon is taking shape.

Boosters for SA-5 and subsequent vehicles are called "Block II" of the S-I design. The re-design was made by the Propulsion and Vehicle Engineering Division, with other divisions participating.

The changes, including addition of fins at the tail section, were generally to improve the design and to provide increased assurance of vehicle stability which is desired for manned flight-to follow later.

Vincent Caruso, MED project engineer for the tail section SA-5, who is keeping a meticulous logbook on the fifth Saturn, predicted it will be ready for static tests about Dec. 1, 1962.

Looking at his log, he unfolded the history of SA-5. First, he pointed out a partial retooling program transpired because of design changes. Many of these changes were in the tail section. The outriggers were completely re-designed. Thrust rings and the barrel assembly were also completely re-designed.

The second stage adapter had a considerable amount of re-design. It now incorporates a honeycomb seal plate.

The clustering fixture had to be modified extensively to accommodate the longer tanks. Caruso said the new tooling, at the completion of this program, will be used to supplement the tooling program at Michoud, which is a future assemble site for Saturn S-1.

When the rocket finishes this phase of assembly, it will follow the others to Quality, then to the Test Division, back to Quality and finally to Cape Canaveral.

Astrionics Laboratory, Instrumentation & Communications Division
by Warren Harper, James Rorex, Ted Paludan, and Tom Barr

The successful development of a launch vehicle or spacecraft requires the acquisition and transmission of a large amount of data regarding the flight performance of many systems and subsystems, as well as the transmission of uplink commands from the ground. There is also a need for tracking the craft as it moves along its path. These requirements rested on the shoulders of what would become the Instrumentation and Communication Division of the Astrionics Laboratory. These people designed, developed and tested the hardware and procedures by means of which the requirements were met.

The work performed by the I&C Division had its historical roots in the projects conducted at Fort Bliss, TX and later at Redstone Arsenal. Those who came from Fort Bliss were Otto Hoberg, Edward Fischel, Walter Schwidetzky, Helmut Zoike, Hugo Woerdemann, Antonio Beltran, Thomas Barr and Donald French. Others joined the group after the move to Huntsville.

Significant among the efforts of the group during the pre-NASA period was the work associated with the development and launch of our nation's first satellite, Explorer I. These efforts led to the design, fabrication and testing of other launch vehicles and a series of scientific satellites. The communication and tracking requirements for these missions were challenging, but were always satisfactorily met.

Mr. Otto Hoberg served as director of the group until 1967. He was followed in that

NASA - Marshall Flight Space Center

Members of the I & C Division with Explorer VII, 1958-59. From left: Ted Paludan, Jesse Seal, James Howell, Harlan Burke, Olin King, Bill Threlkeld, Walter Haeussermann, unidentified, Otto Hoberg, Grady Saunders, Frank Emens, Heinz Kampmeier, Tom Barr, Leslie Thomas, Paul Swindall, Bill Edens, Ray Lowery.

Ellington Pitts, Ross Evans, Billy Reed, Ed Gleason, R. Edens, William Shields, George Burson, Harold Schultz, Ray Roberts, Frazier Williams, Rudolph Decher, Dave Lowery, Pat Lakey, Leon Bell, Dave Harris, Grover Tucker, Ed Martin, George Harris, Buford Freeman, John Chase, Randy Clinton, Bill Case, Joe Kerr, James Howell, George Lambertson, Mary Stowe, and Norma Oberlies.

Tracking

Adherence of the rocket vehicle to its programmed trajectory was accurately measured by several tracking systems. The tracking responsibility involved the use of radar transponders and other on-board tracking equipment. It also included analysis and performance evaluation of various ground-based tracking systems.

This work was performed by John Gregory, Andrew Bratcher and Carl Huggins.

Television

The transmission of video data played an important part in the assessment of system performance in several flight tests. Cameras on the vehicle captured scenes that revealed information that would otherwise not be feasible to transmit. A notable example was a video record of the behavior of the hydrogen fuel in the S-IVB stage of the Saturn V vehicle under zero-g condition during the parking orbit.

The television work was performed by Carroll Bordelon, Tom Barnes, Al Kosis, P.D. Nicase and Daryl Craig.

Engineering Planning Office

This office was staffed by John Cox, Heinz Kampmeier, Charles Chambers, Harvey Golden and Antonio Beltran. Its function was to plan and coordinate the program of work that was carried out by the Instrumentation and Communication Division.

Theoretical Studies and Analysis Office

Grady Saunders, Elbert Peters and Bob Mixon studied advanced technology and recommended improvements in operational methods and equipment.

Utilizing Space Garbage to Acquire Lunar Science
"The Moon Rang like a Bell"
by Bob Beaman

During Apollo 11 and 12 missions, dumping of unused propellant tanks thru the J-2 engine and burning the Auxiliary Propulsion System was used to achieve a safe separation from the Apollo Spacecraft. The resulting retrograde velocity, along with the influence of the moon's gravity, prevented Lunar or Earth impact of the spent three stage of the Saturn V, S-IVB and Instrument Unit (IU) hardware by placing it in a solar orbit.

A Lunar Seismic Profiling Experiment was flown on Apollo 12-17 lunar landing missions and placed on the lunar surface by the crews.

---

office by Lucian Bell, J.T. Powell and others. Willa Eslick served as secretary. The Division consisted of several functional groups, each performing a part of the task of gathering, processing, transmitting and recording the required data. These groups were—

Measuring

In the instrumentation of a rocket vehicle, a large number of parameters, such as temperature, pressure, vibration, stress and fuel flow must be continuously measured, transmitted to ground and recorded for later analysis. The Redstone missile, for instance, required several hundred such measurements. The Saturn V had 904 measurements on the first (S-IC) stage, 975 in the second (S-II), 590 in the third (S-IVB) and 322 in the Instrument Unit; a total of 2,791 measurements, exclusive of the Command and Service modules. The responsibility of the Measuring Branch personnel was to convert each of these parameters, by the use of "transducers", to an electrical form that would satisfactorily represent the parameter to be measured and which would be acceptable to the telemetry equipment that would generate the radio signal that carried the data to the ground. The instrumentation was monitored during ground static firings, during factory checkout, during launch checkout and finally during flight.

The personnel that made up this group were C.T. Paludan, W. Escue, A. Touchstone, Harlan Burke, Joe Zimmerman, Troy Ponder, Alonza Davis, W. Sutherland, Bill Lewter, James Derington, James Power, Ray Holder, Pat Layke, Sanford Downs, John Avery, Orville Smith, Drayton Talley, and John Hamlet.

Telemetry

The set of electrical signals provided by the measurement equipment was organized by the telemetry group into a single data stream, taking into account data rates of the various measurements. For example, temperatures usually changed very slowly, while vibration produced a rapidly changing value. This data stream was used to modulate a radio carrier signal, which was amplified to a level of power sufficient to be transmitted over the miles of distance between the vehicle in flight and the ground stations. At the ground site, a suitable configuration of equipment had to be provided to receive, demodulate, separate and record the transmitted data. Records were made on strip chart recorders for rapid post-flight evaluation and on magnetic tape for later, more exhaustive evaluation.

The Telemetry Branch consisted of the following personnel: James Rorex, W.O. Frost, William Threlkeld, William Pittman, Frank Emens, R. Williams, T. Lawson, Gabriel Wallace, Leo Arsement, Charles Morris, K. Stephens, M. Teal, Robert Eichelberger, Barnes Beasley, R. Coffee, and B. Adair.

R.F. Systems

This branch was responsible for radio transmission of the down-link and up-link signals associated with flight data and radio tracking. The RF transmission task involved the design of antennas and other components for transmission and reception of R.F. signals, the design of cable arrangements for power division and phasing to provide maximum RF power transfer between vehicle and ground receivers. It also included the analysis of communication link performance, taking into account the effects of any rocket vehicle that might go out of control and impact any ground area. The implementation of this system, including the use of security codes, was a responsibility of this branch.

The R.F. Systems Branch personnel were: O.T. Duggan, Thomas A. Barr, Warren Harper, Olin Ely, Paul Swindall, Donald Stone, Leslie Thomas, John Price, Ursula Mrazek Vann, John Williams, Lee Malone, Ray Lowery,

165

There was a desire by the experimenter to calibrate the seismic instruments with a force much closer to that produced by a meteor impact or moon quake than by the small explosive carried with the experiment. An impact on the surface of the moon caused by the spent S-IVB/IU weighing over 30,000 pounds and impacting the surface at over 2500 meters per second (approximately 5600 miles per hour) would equal to the explosion of 21,000 pounds of TNT.

Marshall Space Flight Center was asked to look at the vehicle changes that would be necessary to impact the S-IVB/IU on the lunar surface. The only significant change required was to add a battery to the S-IVB stage in order to retain command and control capability of the S-IVB long enough to accomplish the necessary propulsion system burns.

The battery change was implemented and a new objective was added to the remaining missions beginning with Apollo 13. This new objective was to impact the S-IVB/IU on the lunar surface within 350km of the targeted impact point to excite the seismic profiling experiments to determine the actual impact point within 5km and the time of impact within 1 second.

MSFC supported this objective using the In Flight Trajectory Team in the Huntsville Operations Support Center. This team provided the preflight analysis, as well as the analysis during the mission to define the necessary impulse durations to accomplish the lunar impact task. The MSFC booster systems engineer provided commands to the S-IVB/IU.

The Seismic Experiment used very sensitive instruments to measure very small vibrations of the moon's surface. Using the small explosions that went with the experiment and the S-IVB/IU impact data, valuable scientific information was obtained about the lunar crust and its core. The seismic properties of the lunar crust were a great surprise. The moon rang like a bell for several hours after the S-IVB/IU impact.

The In Flight Trajectory Team was headed by Monroe Hardage now deceased of the Astrodynamics laboratory. He was assisted by Jerry Weiler, James Craft, Gerry Whittenstein, Don MacFadden, Hugh Brady, Clark Owen, Robert Beaman, Jim Hutchins, Frank VanRensselaer, Don Townsen and Bill Brady.

Divine Guidance
by Harrison K Brown

It was 1962 and NASA was still in its infancy. We were moving along in the Ranger Program to put an instrumentation package on the Moon. We had five flights planned and each of theses resulted in failure.

At that time the planet Venus reached optimum position where a flyby mission became possible. So flights Mariner I and Mariner II were planned. The launch vehicle was to be the Atlas-Agena. The Atlas vehicle would take the Agena and the spacecraft to its designed optimum orbital altitude. The Agena upper stage with its Mariner spacecraft would then do a burn to placed the Agena and the Mariner in another orbit to await the right orbital position for a restart of the stage thus launching the spacecraft toward Venus.

The Atlas was controlled from the ground and had an internal gyro to maintain roll so that a communication antenna was always pointed toward ground control. Mariner I was launched, but shortly in the flight a problem developed. Apparently the gyros had not worked as expected and the Atlas lost ground control and became erratic. At this point Range Safety sent the command to destroy the vehicle.

The launch had occurred at 2:00 AM, so the rest of the night was spent with the NASA people beating on the poor Air Force Captain for blowing up their vehicle. Of course he was only doing his job but we were mad!

About three weeks later, on August 27, 1962 the Marnier II spacecraft was launched. During the launch I saw the Atlas shutdown, the Agena burn and watched as the spacecraft was placed in its required orbit and attitude. I later went to the JPL Center in Pasadena, California and there we watched the Agena restart and then saw the spacecraft as it headed for Venus. JPL Mission controllers said that everything was looking great.

I then went to our NASA tracking station and was told that the Agena with the Marnier II had started rotation and had revolved about 30 times, going as fast as one revolution per second. Then it stopped exactly at a point where the communication antenna was pointing directly at the ground station but the Atlas was not pointed at a 2-degrees up angle as programmed but was pointed to the earth. I told them that they were crazy. Once the Atlas rotated out of the line of sight of ground control it would have gone erratic,

On March 16, 1966, an Atlas booster launched an Agena Target Vehicle for the Gemini 8 mission. The flight crew for the 3-day mission, astronauts Neil A. Armstrong and David R. Scott, achieved the first rendezvous and docking to Atlas/Agena in Earth orbit.

NASA - Marshall Flight Space Center

and chances that it would stop its rotation at the right spot was impossible. Also, if it had not gone into a parallel attitude to the earth the Agena would turn down at -2-degrees, crashing to the earth. To check on their telemetry, I went over to the Air Force Tracking station and they confirmed the NASA data. All of this telemetry and other tracking data indicated that we had problems, but we didn't based on the JPL tracking data that we had just observed. So we said that we had "Divine Guidance " for this mission so far.

Several weeks later the Air Force decided since it was their mistake and that we should get all the engineers involved together in a meeting there in Los Angeles and go through the data (point to point). The Air Force Colonel in charge of the meeting said that anybody who mentioned "Divine Guidance" would be thrown out of the meeting. The conclusion:

Why did the Agena start to rotate? (Don't know). Why did the Agena stop its rotation with the communication antenna pointed toward the ground command station? (Don't know). Why did the Agena tip over just parallel with the earth (Don't know). Why didn't the Agena tip to the programmed 2-degree attitude up? (Don't know). The Colonel's final statement was, "Let's Pray! We decided we had had a successful mission so we would all keep our mouths shut about "Divine Guidance" and see what happens as the mission progressed.

But it wasn't the end of this story. To assure that the high gain antenna was directed to face the earth, the spacecraft had to be positioned and maintained that position if we were to get any data back. To do this the spacecraft was programmed to go into a search mode and lock on the sunlit earth. It did that, somehow it locked on another light source, which nobody knows yet the source. I remember the program manager telling me if we could give it a kick it would break the lock on mode and go back in the search mode. Several weeks later during the mission a micrometeorite hit the spacecraft, caused a break lock mode and the spacecraft went to go into the search mode, but this time it locked back on the earth. "Divine Guidance" again?

Everything was going fine after that problem until about three weeks before the intercept of Venus, and then the JPL controllers informed us that the batteries on the spacecraft was losing their charge and it was felt that the solar cells had maybe failed. If this was true we would not get any spacecraft data, but within about two days of the encounter the batteries came back up. "Divine Guidance" again?

From this mission we learned that Venus is a very hot girl, about 700-800 degrees on the surface. It is covered with a heavy atmosphere of carbon dioxide gas, enough to crush a man to death. It has strong winds that sweep its surface features. This thick atmosphere causes Venus to have a greenhouse effect, thus causing it to be become hot because most of the incoming thermal energy cannot radiate back to space.

We have sent additional spacecraft to Venus many times since this mission, and the Russians have put a probe on the surface, but you'll never convince me that we didn't have a little help from the man upstairs on the Mariner II mission.

Millisadic
by Earl Choate

I started at MSFC in 1960, as the Millisadic engineer. The Millisadic was one of the first digital data acquisition systems. Each Millisadic contained about 800 vacuum tubes, and therefore, required constant maintenance. The failure rate was about once per day. This system however, was a beginning of the era that requires no manual data reduction. The random error in the data was such that six data points were usually averaged together in order to produce a truer reading. This system was used to measure temperature, pressure, flow rate, position, rpm, and other

A large group of ME Lab officials with a S-IC-T booster at building 4705, March 1, 1965. Courtesy of Max Sharpe.

parameters. The pressure transducer used at this time was a variable reluctance type. The output drift of this type transducer was such that it was recalibrated after each test.

During the mid-sixties, the all-solid state "Beckman 210" digital data acquisition systems came into being. These systems had less than one tenth the random error, or the uncertainty of the vacuum tube-type system. At about the same time, the temperature compensated strain gauge pressure transducer came into use. This type transducer was considerably more stable than previous types. With these systems and transducers a new era of data acquisition was born.

These systems were computer controlled, and could monitor red line parameters, and print out data in engineering units. The systems did not, however, have enough data storage capacity to print out all the data in a final format in the required time period. This made it necessary to send an engineering unit magnetic tape to computation laboratory for final data requirements.

Saturn V First Stage
by Orville Driver

When President Kennedy made the announcement in early 1960 that we would send a man to the moon and return him safely to the earth in this decade, there were no test facilities available for testing the vehicle stages. Neither were there any design physical performance parameters known. MSFC was given the responsibilities for the vehicle propulsion systems and the first stage development and flight acceptance testing.

In early 1961 the test facility design proposals were received and evaluated. Aetron, a division of Aerojet General, was selected in November 1960. Due to critical lead time of the facility construction, Aetron was given the best real time vehicle design information for the Saturn V first stage. Since time was of the essence, the facility design proceeded with some concern that we may have undersized the stage dimensions and the engine thrust. This concern later proved to be true when MSFC structures and engineering validated the vehicle length would be 145 feet, diameter 33 feet and engine total thrust of 7.5 million pounds. The test stand design had been based on a slightly smaller vehicle. This caused the test stand foundation work in progress to be terminated by the Corps of Engineers (Mobile) on September 28, 1961. Following the test stand redesign change orders, construction changes implemented, the construction began anew on March 30, 1962. The test stand blockhouse, and booster services were designed concurrently, but construction bid packages were phased to ensure competitive bidding of potential contractors with competitive specialized skills.

In late 1964, when the facility activation and checkout was in process, the S-IC-T was delivered from MSFC manufacturing laboratory and was installed in the test stand on March 1, 1965. The S-IC-T stage was delivered less the F-1 engines as planned so that a load test of the stage thrust structure could be made prior to test firing of the engines. Following the structural load tests which simulated flight conditions to certify the stage structural integrity, the test hardware was removed and one of the 5 F-1 engines was installed. While the facility activation continued, the stage checkouts began and consummated in the first static firing (S-IC-01) of the S-IC-T on April 9, 1965. The first static firing was inadvertently terminated by an observer at 40% full thrust. The test data was quickly evaluated and the second static firing (S-IC-02) was conducted two and one half hours later. Static firing (S-1C-03) conducted on April 10, 1965, completed the planned one engine test program. Four other engines were installed in positions 1-4 for the subsequent five engine static firings (S-IC-04 thru S-IC-15). Static firing S-1C-04 was successfully conducted on April 16, 1965, using a timer for engine shutdown at 6.5 seconds.

The S-IC-T development test program was highly successful in accomplishment of all the test objectives with only a few minor problems and ahead of NASA program schedule. In summary for the 15 development static firings; nine tests achieved the planned firing duration, three tests were inadvertently terminated by an observer, two tests were terminated by instrumentation red-line observers, and one test terminated due to a failed broken wire in a safety circuit. The development testing was completed with static firing (S-IC-15) on December 16, 1965. The removal of the S-IC-T stage from the test stand on January 16, 1966 ended the development testing of the Saturn V first stage, which was accomplished in only eight and one half months.

There were several eventful occurrences during the course of the program: (1) The low bidder for the LOX pumps could not produce an acceptable pump, which required Test Laboratory to justify sole source procurement from a reputable contractor. (2) During the excavation of soil down to bedrock (40 feet below grade) for the test stand foundation, an eight-inch diameter water stream was uncovered. This required a de-watering pumping system be installed. (3) The vice president of the U.S. Steel Company voiced a complaint that the design of the high-pressure bottles specified dimensions greater than the company could fabricate. This was voiced as discrimination against the company. However, the rationale for the bottle size was later explained and resolved. (4) On November 24, 1965, following completion of static firing S-IC-13, the earth was vibrating in Birmingham, causing a radio station announcement that the city was experiencing an earthquake.

The acceptance testing of the flight stages S-IC-1, S-IC-2, and S-IC-3 that were fabricated and assembled at MSFC, were successfully tested and completed on November 15, 1966. This completed a highly successful Saturn V first stage development and acceptance test program by a dedicated, competent, and professional propulsion test team in MSFC Test Laboratory.

Short Stories
by J. N. Foster

The following is a series of short stories about the Space Program at MSFC and Redstone Arsenal.

I arrived in Huntsville in September 1955 and retired from the Government in 1988 after serving with the Army prior to 1960 and later with NASA. I subsequently worked with aerospace industry and the Canadian Space Agency as a consultant.

First Integrated Lunar Graphic

President Kennedy's announcement of the Apollo Program in May 1961 required many changes in NASA. The Space Task Group [STG] based at Langley Research Center was assigned the development of the Apollo Spacecraft and the development of the Astronaut Corps. The STG was later transferred to Houston and became the nucleus of the Johnson Space Center. The MSFC was assigned the responsibility of providing the rockets to carry the Apollo Spacecraft to the Moon based on the Saturn family of vehicles, which were largely paper-planned rockets at the time. Initially, the STG and MSFC had to learn to work together and this was a challenge. For several months, much effort was devoted to determining the method of going to the moon. Three methods were considered: Earth Orbit rendezvous, Lunar Orbit rendezvous, and direct Earth to Moon. The Lunar Orbit was selected and for many months the STG and MSFC went their own way in developing schedules and plans for the initial Lunar landing. The problem was that the vehicles and the Spacecraft were planned for launch on different days with somewhat varying sequences of operations. This caused some confusion and the MSFC Lunar Planning Office under Hans Maus put together the first illustrated concept of how all the major elements of the initial Lunar Landing would work together. The primary individuals involved were Hans Maus, Bud Abbott, and Jay Foster.

Shuttle All-up Ignition Sequence Test

After the Shuttle Challenger accident, the President's appointed accident investigation board required that a series of four full-scale Solid Rocket Motor ignition tests of 0.6 seconds be conducted prior to resumption of flight. The design and construction of the required test facility and an associated refurbishment facility were on the critical path for resumption of the Shuttle program. This testing at full scale was an engineering challenge considering the shock load for the ignition and also for the rebound after the 0.6 second shutdown. The test stand cost was within MSFC funding authority, and design and construction was initiated. For the refurbishment facility, a bit of bureaucratic slight of hand was required. It

Unloading and installing the CRAY computer in room 160, building 4663. Courtesy of Jay Foster.

should be known that NASA and MSFC had authority to proceed with "Minor Construction" of new stand-alone facilities up to $500,000 without specific approval of the Congress, while the Agency had authority to modify existing facilities up to $750,000 without specific congressional authority. The refurbishment facility was estimated to cost $600,000 and we did not want to spend the time necessary to go back to the Congress for this urgent task. Upon careful review of the proposed site for the refurbishment facility, it was discovered that a low reinforced concrete wall remained from a decade's earlier U. S. Army test facility, which had been demolished. The wall remained because of the high demolition cost due to the steel reinforcement. Well, this was what was needed. A major search was initiated of NASA and Army records, and we found the original title and facility number for the building that included the wall. Consequently, we built a $600,000 facility modification onto that old remnant of a building.

MSFC's 1st CRAY Supercomputer

All of the NASA Research Centers, AMES Research Center, Langley Research Center and Lewis [now Glenn] Research Center had been able to justify and acquire CRAY supercomputers by the early 1980's. The belief in NASA Headquarters was that the development centers did not need the large computational capability provided by the CRAY. MSFC embarked on a major effort to convince Headquarters that the development tasks did indeed require this capability. The MSFC Information System Office jointly with the Science and Engineering Directorate worked for several months putting together the justification required by Headquarters and the Office of Management and Budget. The justification was briefed at several levels at Headquarters with negative results. Finally, with continuing pressure from MSFC Management, the NASA Administrator, Dr. Fletcher, agreed to hear the case. The result was that he could not make a firm negative decision, and he turned to the Deputy Administrator, Dr. Hans Mark, and asked him to visit MSFC and see for himself if a Supercomputer was really required.

A few weeks later, Dr. Mark visited MSFC. We arranged 20 twenty-minute interviews in the senior engineer's office for each discipline including: structures, optics, thermal, aerodynamics, electrical networks, computational fluid dynamics, etc. After visiting the first eight offices Dr. Mark threw up his hands and said we could have the dammed supercomputer.

The 1st Saturn I Fabrication and Assembly

Back in the late 1950's and early 1960's, the 1st stage for the first flight Saturn I was being assembled by the Fabrication Laboratory. Money was tight and Hans Maus, then the Laboratory Director, went ahead with the necessary steps to complete the fabrication and assembly of the stage. This stage had a center core of a Jupiter tank surrounded by eight Redstone tanks. In part, this vehicle was to demonstrate that tanks and engines could be clustered. Low and behold there was no money for the 8th Redstone tank. Hans Maus had out-maneuvered the others trying to justify Dr. von Braun's budget reserve. He simply stated to Dr. von Braun that if you want to fly, we need funds for the eighth tank!

Two Interesting Things Happened on a Trip to the West Coast

Dr. von Braun made periodic trips to the West Coast to check on the status of several contractors developing elements of the Saturn vehicles. I accompanied him on one of these trips in August 1968.

We flew in the NASA 3, a Gulfstream I 12-passenger airplane. We had arranged for a fuel stop at a SAC [Strategic Air Command] Air Base. Dr. von Braun was flying the aircraft and he was starting his descent in the vicinity of Mt. Rushmore, and he decided that we needed to see the Great Stone Faces. He tilted the aircraft on its side so the passengers could get a good look. It was August and there were a large number of tourists visiting Mt. Rushmore. I don't remember any of the stone faces, because I was looking at the tourists looking up at the crazy airplane flying sideways.

Later on the same trip, we visited Mt. Palomar to see how the best ground based telescopes performed. We were bussed up to the Observatory in a Marine bus operated out of Camp Pendelton. The bus broke down half way up the mountain.

Fortunately, the Marines had a premonition and had sent along a pick-up truck with spare fuel tanks. Undaunted, Dr. von Braun and the rest of our group of 12 climbed aboard the pick-up truck. Dr. von Braun sat on the first fuel tank he could find. We had a great tour of the facility, had a catered dinner, and after dark were able to look thru the telescope and had a lecture by Dr. Jessie Greenstein, an expert on White Dwarf Stars. At about 1 AM the next morning, as we were leaving the observatory, Dr. von Braun said he was all keyed up by what we had seen and he invited the group to his room in the Camp Pendelton Officers barracks for an astronomy lecture. We arrived at the barracks about 2 AM and he lectured until 5AM, when we had to depart for the airport. This was the most outstanding lecture.

Formality of the von Braun Team

When Dr. Seamans was the Deputy Administrator of NASA he and Dr. Eberhard Rees, Dr. von Braun's Deputy, were looking south from the MSFC Headquarters building 4200 at the test stands in the distance. Dr. Seamans commented on the formality of the German group. He asked Eberhard if he had always called Dr. von Braun, "Dr. von Braun." Eberhard looked into Dr. Seamans eyes and said, with a straight face, that he used to call him "Herr Dr. von Braun."

Dr. Eberhard Rees's trip to Washington

One day Dr. Rees was preparing to go to a Management Council Meeting at NASA Headquarters. I was called to his office to bring a set of briefing papers that he needed.

I entered his office and sat at the table with several others, both men and women, who were assisting on other subjects. Subsequently, Dr. Rees was called to the telephone at his desk. He rose from his chair and walked across the room to the phone as if nothing were amiss, but he was not wearing pants. I later discovered that his zipper was broken and he had whipped off his trousers and handed them to his secretary to get them repaired in time for his flight. However, nothing could get in the way of his preparation.

The Courageous All-up Decision
by Rein Ise

In late 1963 the Saturn V development was rapidly gaining momentum. The system design was maturing and decisions needed to be made on how to approach flight testing and man-rating of the Saturn V for it's destined role in the manned lunar landing program. The conventional Marshall approach, influenced greatly by the conservative German rocket team members, was to proceed with such testing carefully one step at a time, starting with the most basic system and then adding complexity as satisfactory test results were obtained.

I was a young engineer/manager during those days, heading the Saturn Systems Office systems engineering section, when I was asked to prepare a presentation to NASA Headquarters management outlining the Marshall proposed flight test program. After many meetings with Dr. von Braun and the Marshall management team, I summarized the approach with its supporting rationale in about a dozen charts. The Marshall team felt strongly that the first launch configuration should include only a live first stage. Then, after the design of the first stage had been proven, the second stage would be added, and when the team was satisfied with that configuration, the third and final stage would be flown. It was estimated that this approach would require anywhere from a minimum of three launches to possibly as many as nine launches before the Saturn V could be considered fully qualified to carry a crew into space. The recommendation was strongly supported with reliability analyses and data from previous experiences.

The NASA management contingent from Washington headed by the Program Director Dr. George Mueller, and directors from other NASA Centers, arrived at Marshall for a full day Program Review. As such meetings go, the agenda stretched considerably, and my topic finally came up at the end of a very long day. After I had turned to the second chart, Dr. Mueller asked me if we had considered any other options. I told him we had, but that this was the only recommendation that Marshall could forward, and began delving into the rationale and justification. At that point Dr. Mueller, in his gentle but determined manner, asked me to sit down and not waste everyone's time as the hour was getting late. I was abashed as I felt I had not represented our case very well. But Dr. Mueller got up to the blackboard and said, "Here is what we are going to do," and proceeded to lay out a plan which required flying all three stages live on the very first launch, and placing the crew on the third one. Everyone in the room, and particularly the Marshall team, was stunned. No one knew how to react, or take up the arguments we had developed in our recommendation. Finally, some concerns and reservations were voiced around the table, but no one could strongly argue that Marshall was not ready to tackle that challenge. I felt somewhat let down, but it turned out to be a good and pivotal decision, as no serious problems were encountered, and any other approach would probably not have achieved the nation's goal of landing man on the moon before the end of the decade. Of course it required an even more focused and diligent engineering effort from all of Marshall and it's many contractors to make it happen.

After reflecting on that night's events, I came out of that experience with deep new insights into several management principles including: the strong leadership of one person can make a huge difference in any undertaking; challenging goals are achievable, provided they are clearly stated and vigorously implemented; and, when dealing with higher management, one should maintain a certain degree of flexibility.

Redstone
by Jim Pearson and Charlie Gillespie

During the Cold War and Korean War, the U.S. Army Ordnance at Redstone Arsenal was charged with developing a missile with a 75 mile range. Von Braun, with the Ordnance Guided Missile Development Division, immediately began a design based on the Alcohol/LOX propellant V-2 technology. An engine test stand for the 78,000 lb thrust Redstone engine was built by Rocketdyne. A need for a test stand to acceptance fire the flight Redstone vehicle was urgently needed, but funds were not available. The test lab found old chemical tanks from WWII chemical arsenal scrapyard and the ingenious engineers and technicians used $25,000 to pour a concrete foundation and scrapped together the Interim T.S. (IITS), which used a carbon block dry deflector, Navy bottles, with the ends cut off, for the steel columns of the T.S. with scrap iron beams. The interim name inferred that the test stand would be replaced as money became available and justified the one million dollars for the East Test Area, but was never used for Redstone.

In 1953 the first flight missile was static fired (20 Sec) by the test crew, who then took it to Cape Canaveral and launched it. The crew returned to the IITS to static fire #2. This continued until a firing lab was formed in 1955.

Bill Grafton, mechanical leader of the test and launch crew, remained as chief of IITS until 1959. When NASA came in, he moved with ABMA to a new location. The IITS continued as the only acceptance firing for Redstone, Jupiter C, and Mercury-Redstone flight vehicles. (Total of 364 tests) The carbon block deflector failed so a water-cooled deflector was put in. A surplus tugboat engine (12 cylinder Waukesha) drove the water pumps for the deflector operated by IITS.

In the mid '50s, a Redstone firing was the best and only attraction at the arsenal. After the water cooled deflector was added, spectators were told not to park cars along the road downstream of the deflector, as they would be coated with a red mud spray. The firing schedule was determined by arrival of a commercial LOX truck from Birmingham.

Many development firings for tactical Redstone were conducted, including -25 degree F alcohol and engine compartment, which found need for heaters for peroxide tank, regulators, valves, and TMRV.

On the Redstone cold temperature test with a full load of propellants for firing, a box covering the nozzle exit had been provided for liquid nitrogen to chill to minus 25 degrees. Then we ran a sequence test to verify all valve operation. On completion of this sequence Frank Kopera, the test stand talker, said, "Cut off the LN2 to the nozzle cover, it is overflowing." The test conductor who had cutoff the LN2 earlier, realized that the main LOX valve had opened during the sequence and filled the chamber with LOX. On hearing this, Frank dropped the headphones and jumped off the test stand, fearing a LOX explosion.

The engine for the Redstone rocket.

NASA - Marshall Flight Space Center

Dr. Wernher von Braun is shown briefing the seven original Mercury astronauts in ABMA's Fabrication Laboratory. From left: Gus Grissom, Walter Schirra, Alan Shepard, John Glenn, Scott Carpenter, Gordon Cooper, Donald Slayton, and Dr. von Braun.

The Jupiter C added 10 feet of propellant tanks for added burn time. Vibration measurements and spin tests of upper stages, including Explorer I, were tested. The Mercury capsule was mounted in the modified stand for test firing, thoroughly instrumented to assess vibration level and effect on astronauts during firing.

Astronauts came to observe Mercury Redstone firings. They all wanted to be inside the capsule, but our safety accommodations were inadequate. Gordon Cooper (Mercury 7) stood outside the bunker to feel the vibrations during a firing. When he left in his T-80 Jet, he approached the test stand at tree top level, then abruptly climbed straight up as to simulate a launch and as a salute to the test crew.

Prevalves had been added earlier as a requirement on Jupiter C using hydrazine fuel. The prevalves isolated toxic hydrazine from the engine during pretest, but was kept as an added safety feature to isolate turbopumps from tankage and was used on all subsequent vehicles.

On Redstone flight vehicles the access to the engine compartment was a 2 foot by 2 foot door. To squeeze in, one would hold the test stand beam of the floor above, lift and swing the feet inside requiring some athletic ability; then climb up to set the regulators and check for leaks etc. LOX leaks were common at the manhole cover at the bottom of the LOX tank requiring torquing until the leakage stopped. Also, checking the peroxide tank heater, which was a tight squeeze. Bentley Irwin, at 130 pounds was the prime candidate for the job, which sometimes occurred late in the sequence with the propellants loaded. These guys earned their pay.

The Army was asked to supply Redstone missiles to launch the atomic bomb to explode at 25 mile height. IITS static fired several tactical vehicles (Hardtack or Willow code names) which were sent to the S. Pacific (Kwajalein Atoll) for launch-test. Some of the IITS crew assisted in the launch. Results of the test included loss of communications after the blast blew a hole in the atmosphere.

RL10–Liquid Hydrogen

As the plans for Moon missions evolved, the requirement for liquid hydrogen for fuel was seen. A familiarization program for hydrogen was originated, as no previous large-scale use was made. Pratt & Whitney (P&W) developed the RL10 (the first H2 rocket engine) from an attempt to develop LH2 turbine for jet engines. MSFC built a stand for the RL10 as a part of the H2 familiarization project. We used P&W design for horizontal diffuser, which we had local contractors build. We purchased a steam Jenny and diffuser from a Birmingham welding shop and had Fritz Vandersee's test shop assemble parts into a test stand.

The LH2 tank was insulated by the test crew using a bare 2,000 gallon SS tank. We put a steel pipe through top and bottom port so as to devise a rotisserie capability. Then, with a spray gun purchased from a company in Atlanta, we applied spray-on polyurethane as insulation for the LH2 tank; a procedure used later by companies making the S-II and tanks for the space Shuttle. Later, we purchased a vacuum-jacketed 2,000 gal LH2 tank for repeated use as the spray-on foam cracked after a few cycles using LH2. Our first attempt at filling the tank with LH2 resulted in loss of 1/2 the LH2 on venting the tank pressure to the burn stack. We had omitted the baffles in tank vent, allowing the liquid H2 to go out the burn stack and setting a big grass fire. This allowed us to try out the off-stand remote controlled firex. A 'water gun' from the gold mining use was modified with remote controls. Charlie Gillespie got a U.S. patent for this device.

Many other items, i.e. H2 sensors for leak detection, H2 flowmeters, purge and inerting procedures, ignition systems for burn stacks, were devised during this early program. Leading this study under Dan Driscoll were: Warren Johnson, Rufus Collins, Lou Scarbrough, Jim Pearson, Frank Rutledge, Paul Artis, Charlie Gillespie and crew. This group also monitored contractor work on S-IV, Centaur, S-IVB, S-II and Shuttle.

The RL10 program provided good experience to the government who not only monitored, but directed test programs on new stages. Many untried conditions were tested such as: Zero NPSP on LOX pump inlet–no cavitation. First LH2 tests on the LH2 Pesco pumps used on S-IV, S-IVB, S-II etc as pre-chill conditioning for the turbopump & helium heater on S-IV & S-IVB. Tests were made with no steam system to evacuate nozzle prior to firing into a diffuser in an attempt to simplify facility operation and lower costs.

Memories
by Ellery May, Jr.

I came to work with the Army's Missile Command in July 1951. I was hired to do wind tunnel testing on the Redstone Launch Vehicle, but over the span of the next 28 years, I was assigned to work in several assignments including Wind Tunnel Development, Mission Planning, Systems Engineering and Program Management. The focus of my experience was centered on the Apollo Saturn Launch Vehicle. This program became a national priority in 1961 when President John F. Kennedy announced the goal to send men to the moon and return them safely to earth by the end of the decade. The incredible success of the Saturn Launch Vehicles in reaching this goal is a fitting tribute to the outstanding team of men and women, government and contractor, who designed and developed them. Thirty-two Saturns were launched without a failure. All the launches were memorable, but three especially stand out in my mind.

Apollo 8 was launched on Dec. 21, 1968. This was only the third Saturn V to be launched and the first to be manned. That in itself was incredible, but what made it so memorable was that its primary objective was to put the Apollo spacecraft into lunar orbit. The hairs still stand up on the back of my neck when

I remember that Christmas Eve and heard Frank Borman read from the book of Genesis as the Apollo spacecraft orbited the moon. The first earthlings had escaped the earth's gravity and we got our first ever picture of our home planet taken from space.

Apollo 11, launched just seven months later, has to be included. Man's first landing on the moon occurred on July 20, 1969. The Presidential goal was accomplished when the spacecraft returned safely to earth and was recovered in the Pacific Ocean on July 24, 1969.

The U.S. Apollo Spacecraft, our half of the joint US/USSR Apollo Soyuz Test Project, was launched from the Kennedy Space Center on July 15, 1975 on top of a Saturn 1B Launch Vehicle. This was the 32nd and last Saturn to be launched. It did not occur to me at the time, but someone reminded me years later that at a time when the Cold War was in high gear and the world was just one phone call away from nuclear destruction, we used our last Saturn launch vehicle to send three American Astronauts into space to shake hands with the Russians. Twenty-six years later, astronauts from the United States, Russia and several other nations are continuously living and working in the International Space Station. I am privileged to have had the opportunity to participate in the accomplishments of this remarkable rocket development team, which helped make all this possible.

The Saturn Parts Catalog
courtesy of Stephen J. Dobbs

In 1956, a failure occurred in an electronic part used on the Atlas being built by the Air Force that would have resulted in a mission failure if flown. Soon after, Dr. Walter Haeussermann, Director of Astrionics Laboratory, called on me and inquired whether we had any such part in the Saturn. I was able to name four assemblies that contained the bad part. Dr. Haeussermann expressed the need of a parts program and felt I would be the one the head this program. At that time I supervised a group of 50 civil service and contractor personnel performing design and documentation of electronic assemblies for the Saturn. I prepared specifications for electrical parts not covered by any documentation. Therefore, the Design Guide Lines Parts and Materials was born as well as Design Guide Line Procedures outlining coatings for circuit boards as well as anodizing of aluminum, etc.

Letter to a Friend
by Jesse N. Bradley

Dear Harry and Dimple,

I was so pleased to see you again, Harry, and also to meet Dimple.

You are part of my history, and I of yours. Seeing you stirred up old memories, good and bad (mostly good) of those long-ago days we soldiered together in Test Lab. It really was a special place and a special group of people, and I am proud to have been a part of that.

When I retired, I didn't miss the schedule pressures, the constant budget battles, the reductions in force and reorganizations, or the chicken —, but I sorely missed the day-to-day association and fellowship of the guys I worked with. Sadly, but inevitably, many of those are now gone, and at our age, no one can tell who will be next. That's life, however, and I don't dwell on it.

I reminded Bob Saidla that on September 11, 1962, President Kennedy visited Test Lab to watch us fire SA-4, Saturn IB, on the east side of the Static Test Tower. Although Dan Driscoll really was not needed in the blockhouse because Saidla was running the test, Dan was hanging out in the control room anyhow, and was wearing an intercom headset and mouthpiece. As usual, when we had important visitors, the countdown was on hold at X minus 10 minutes to allow everyone to get settled in the bunker, but this time, the discussion between Kennedy, Webb, von Braun, and Heimburg dragged on for several minutes longer than usual.

What Dan did not know or had forgotten was that at X minus 10, one of our guys always plugged in a jumper that tied the operational intercom channel onto the public address system so that everyone in the whole test area could listen to the final minutes of the countdown. Sure enough, the predictable happened. Dan grew impatient at the delay, and in his usual impulsive, no-nonsense fashion, called to Saidla and asked loudly, "What in the hell are those guys up on the hill doing?"

However, Karl Heimburg had asked several of us including Jack Balch, who he thought would not be intimidated by the press or the important visitors, to circulate among the crowd of several hundred people to answer questions about the test. So there we were, dressed up in coat and tie and white hard hat, eager to share our vast store of technical knowledge, when Dan's peevish question came booming out of the loud-speakers. We were, of course, shocked, and I'm sure von Braun and Heimburg were terribly embarrassed, but Kennedy and Webb and the senators, generals, reporters, and other visitors all seemed merely amused. Dan later told me that nobody ever chastised him about the remark.

Incidentally, I was asked only one question. A senator whose name I've forgotten asked, "How tall is that test stand?" Although I had no idea, without hesitation I quickly answered, "A hundred and seventy five feet." He never knew the difference.

I heard you discussing with someone the effort we in Test Lab made to bring the first few S-II's to MSFC and fire them in the S-IC stand. There is a Dan and Jesse tale about that, also. What you may not remember is that a group of us flew up to Washington to present our plan to Dr. Mueller. Dr. Rees, Col. Yarkin (S-II manager), Charlie Brooks (Quality Lab), me, and two or three others whose names I can't remember, flew up on the Viscount from the old airport, and Jack Balch came up from Mississippi Test Faculty (MTF). Dan and Karl, however, decided to fly in the MSFC Queen Air with Gen. O'Connor who needed some air time to keep qualified for flight pay

The next morning at Headquarters, we met with Dr. Mueller and several other HQ types. After a few remarks by Col. Yarkin, Jack Balch explained the situation at MTF, and then Dan and I made the Test Lab pitch. After Dan expounded eloquently on the mechanical, tanking, piping, structural, and scheduling plans, I talked about how we could lash-up the S-IC electrical GSE to fire the S-II. This was the major problem at MTF; North American was having severe problems fabricating and delivering their electrical gear on time.

Dr. Mueller seemed interested and asked a few questions, and then he and Rees went into an inner office to discuss it some more in private, while we twiddled our thumbs. After about an hour Dr. Rees came out, looking rather subdued, but not saying anything about their discussion. On the way to the airport, I was on the front seat of the cab with Dr. Rees; Heimburg, Dan, and Col. Yarkin were on the back seat. Dan and Karl had enough of flying in the Queen Air, and were going back with us on the Viscount.

We were all waiting eagerly for Dr. Rees to give us the lowdown on Dr. Mueller's comments, but he told us nothing concrete, evidently wanting to wait until he'd talked to von Braun. He did say something like this, however, "Dr. Mueller said, 'If those guys Driscoll and Bradley know so damn much about the S-II, why aren't they down at Mississippi Test Facility (MTF) getting the job done?'" Dan and I immediately began to stutter and protest and offer excuses, and Dr. Rees just sat there chuckling at us. We later learned that our trip was a waste of time; that the decision had already been made to leave the work at MTF.

That's the kind of stories I have. While they might be interesting to old Test Lab hands, they would not be suitable for an official history book. As I told you at lunch, for us in the Electrical Branch and for John K., all tests were about the same. You told us to open or close a valve at a certain time, or to pressurize or vent a tank, or to measure pressure, thrust, temperature, or flow, and we did. Your bunch, with the intimate knowledge of the item being tested, and the responsibility for getting it right, are the ones who can produce an accurate, definitive account of the tests we ran.

T-Bird
by Howard Hester

In the spring of 1965, Test Laboratory was preparing to start testing of the S-IC test stage, better know as the "T Bird", of the Saturn V booster, often called the Moon rocket. At one time we were working 12-hour days so much that when I got off on a Saturday after putting in only 8 hours, it was like I got home in the middle of the day with nothing to do. I also remember receiving a paycheck with over 60 hours of overtime, at time and a half, in addition to the 80 hours of regular time on it.

The first test came, as I remember on a Friday. We made an attempt at a 10 second single engine burn, but had a premature cutoff, automatically as I remember. We recycled that day and made another attempt that evening, but again cutoff came prematurely. Ron Tepool was given credit for that cutoff that came just as the fire flashed in the center engine. We instrumentation types at first thought that he might have been startled as he watched thru the scope. However, he said he was reaching for the cutoff button and he bumped it, and that was accepted. It really did not matter why the test was terminated. We then recycled the next day and had a successful 10 second burn.

Test Laboratory skipped the three engine tests that were never done and went straight to all up five engines. After the five engine burn it seemed pointless to go back and do a three engine burn. Before long we had worked up to full duration of 155 seconds.

After a while we were doing a five engine S-IC firing every week, or we would do two single F-1 engine tests every other day depending on what was required. Bud Sallo spent most of his time getting the digital system setup for the next test, and I would spend most of mine recovering the data from the last test and putting it into final format. Then it would be time to run over and help conduct the next test.

At the instrumentation peaking, which occurred early in the testing program, the four Beckman Systems, each an analog to digital data acquisition system with 192 analog channels were fully loaded. They also had 8 channels of digital signals each, but these were not put to full use until later when we started to record events.

Each channel was sampled 25 times a second, so a lot of data was being generated especially with a 155 second firing. This data was recorded on magnetic tape in a raw format during the firing and then converted to engineering units after the firing was over. The engineering unit tapes were then taken to Computation Laboratory where over night the data was printed, plotted and microfilmed. I picked it up early in the morning and got ready for the data review.

The only exception to this occurred after someone got on an energy saving kick. The maintenance personal came in during a period when we had most of the equipment turned off and adjusted the temperature to what had been set of offices. At the next test, which happened on a hot summer day, when the equipment was turned on the air conditioner could not keep up. The digital crew was having a real problem in that the computers were overheating even with their own air conditioners, since they dumped their heat into the room. At one point we got cardboard and placed it on one of the computer air conditioner outputs and packed it with ice, to improve the cooling just to get the test off. As soon as the test was completed we cut the systems off and went home. The next morning we made the engineering tape and took it down for processing. This caused a one-day delay in reviewing the data, which did not set well with the mechanical types and the bosses. The next day the maintenance people had returned and it was so cold in the recorder room that I think you could have processed meat in it. Anyway our technicians got in the habit of coming in and turning on all of the equipment so the room would warm while they got their morning coffee.

Another abnormal event was when the timing system failed just before test time. Ray Proffitt got the idea to push the timing reset at about ten second intervals. This put a mark on the strip chart records, oscillographs, etc., and gave a reset to the Beckman System. Thus, there was something to match the records from one system to another with. However, the Beckman saw it as 15 ten-second tests strung together. We must have fought this problem for a week or two. In the end we had to have Computation Laboratory count records to recover the time on the digital systems. While this seems straight forward, we did have to account for some drop records that normally would not have given us any problem.

Test Lab Retiree's Mementoes
by Frank Rutledge

Background

In June 1956, the only test stand that was operational was the "Interim Ignition Test Stand (IITS). The Test Laboratory was part of the Army Ballistic Missile Agency (ABMA). Walter D. McNabb was Chief of Instrumentation Section. When the IITS became part of the National Aeronautics and

Walt Disney toured the West Test Area during his visit to the Marshall Space Flight Center on April 13, 1965. The three in center foreground are Karl Heimburg, Director, Test Division; Dr. von Braun, Director, MSFC; and Walt Disney. The S-IC Booster Test Stand with the S-1C stage being installed is in the background.

Space Administration (NASA); McNabb stayed with ABMA and I became Chief of the IITS Instrumentation Section. Although the name "Interim Ignition" was envisioned, the facility became permanent even to the extent of becoming the only Historical Test Facility. At the time the IITS was named; the Test Lab was under the organization of the Army Ballistic Missile Agency (ABMA) later to be come part of NASA. In 1956, the only booster that was available for test was the Redstone. The operation of the stand during test was basically three elements of organization; Mechanical, Control, and Instrumentation. All three performed their tasks independently in preparing for test with the Mechanical group preparing the missile/booster itself and being responsible for the overall facility operation and the test conductor. All three functions had to be ready before the test could be performed. These different functions are somewhat obvious; the mechanical and control were required for the operation during the test, the instrumentation was required only for the "Red Lined" values during test but was essential for evaluating the parameters as to the satisfactory performance of the test booster. This basic information is required for understanding and appreciating some of the following.

Lesson Learned

In 1956, vibration was measured using velocity sensitive type transducers (pick-ups). I found that piezoelectric sensitive type pick-ups were directly proportional to acceleration and were available. This would save the designers the tasks of manually approximating and converting velocity to acceleration, which was required. The designers required that the accelerometers be used along with the old velocity pickups, which were big and heavy and were powered by external sources. This was required for comparison purposes. The small accelerometers required power for the conditioning electronics close to the pickup mounted inside the boattail of the Redstone missile. Unfortunately, a fire occurred on the first test when the accelerometers were used. A battery box was immediately suspect but later, it was determined that some oily rags were found to be the cause of the fire. However, Bill Grafton, the engineer in charge of the test facility and the test article, informed Frank in no uncertain terms (commonly called a Chewing Out) that he was to know of any thing that was put into the test article. A lesson learned by me.

"Hop" Stand Tests

The "Hop" stand tests were to fire a Jupiter and allow it to move up to attain an acceleration of one(1) g for a duration of one hundred (100) milli-seconds (these are approximate values) and stop it not allowing it to fall back down (catch it so to speak). This meant that cables would have to move with the test article (Jupiter). An acceleration measurement would have to be used to measure D.C. i.e. not just vibration (A.C.). A strain gauge accelerometer was selected. After the test while evaluating the data, the test conductor (Guy Perry) asked, "do we average out the trace on the oscillograph?" I said, " Hell no! That is exactly what the Jupiter did." We had always looked at pure vibration traces not actual movement of the test article, therefore we had always taken an eye ball average of the peaks and valleys of the trace during the quick look data session.

Test Laboratory Memories
by Bob Saidla

I went to work for the Army right out of college in June 1956. At that time, the East test area was nearly complete. I was assigned to the Static Test Tower East (STTE) to work for Mr. Erich Kaschig. I had requested that assignment because the test tower itself was a fascinating structure. Price Clanton was the shop foreman for Fritz Vandersee, who was doing the structural modifications. The technician crew from the Interim Test Stand were divided as foremen among the East Area test positions. Bill Grafton, Chief of the Interim Test Stand, stayed there a while longer. Others who accepted new leadership positions included Bill McMahan, Instrumentation; Gardner Maples, Network Control; Ronald Chittam, Mechanical, Static Test Tower East (STTE); (Toddy) McClendon, Power Plant Test Stand (PTTS); Ben Curtis, Blockhouse; Floyd (Pat) Patterson, Components Test Facility (CTF); and Gene McLemore, Cold Calibration Test Stand (CCTS). Paul Kennedy was the general foreman over all the Test Lab mechanical technicians. Our test crew were mostly veterans that had also worked on the ammunition lines on Redstone Arsenal after WW II and the Korean War.

We were visited by a famed TV news anchor, Walter Cronkite. For a more impressive photo opportunity, a live static firing was scheduled. Our leaders determined that we would further impress him if we could fire on the STTE, and after it shut down, could switch to Static Test Tower West (STTW), and fire there also. To accomplish this, the flame deflector water had to be rerouted from the East side to the West, as they shared common plumbing. We ran a water flow test to work out the details. Gordon Artley, our chief, insisted on switching under full flow. After the first test was run, and over our objections, we began to change valve positions to switch the water flow. One by one, the resulting water hammers broke the 36" water line at each elbow in the test area. This caused a series of major geyser-like eruptions. I'll never forget the look on Mr. Artley's face. I wish I could have been present when he reported to Mr. Heimburg. Naturally, the pipe fitters got lots of overtime. We put orifices in the control valves for the deflector valves and worked out a safe, smooth transition procedure, allowing the dual feat to be conducted with a slight pause between firings. The irony was that the live TV show gave viewers a brief look at the STTE firing, then switched to downtown Huntsville at the courthouse square where they interviewed an elderly man as to what he thought about the rocket business coming to Huntsville.

A day in my life I'll never forget was the day the original seven astronauts came to watch their first Jupiter static firing on the STTW, similar to the beast they were going to ride. To give the astronauts some confidence in the controlling of this awesome power of a rocket engine firing, which was in reality a controlled explosion, Gordon Artley had gotten approval for several volunteers to walk out of the safety of the STTW onto the top of the Jupiter vehicle during the static firing. From seven thick viewing windows in the observation room of the block house, the seven astronauts watched, in safety and disbelief. The Test Laboratory volunteers from the test crew and Gordon Artley marched out of the seventh level back door, up, over and around the nose cone platform on the STTW, out into the deafening roar of this powerful engine and bravely walked on top of this raging beast and back in the front door. They then changed jackets to look like different people and repeated their walk around so that it looked like a lot of people were involved. This they continued for most of the static firing. The noise level was painful for them. The plan was to repeat the test with the astronauts participating, however Gens. Medaris and Barclay were not willing to take the risk and wisely cancelled the repeat performance, but we were disappointed.

One day we will all remember was when we put on a Saturn I test firing for President Kennedy in Huntsville at the East Stand. After the test stand was secured, he was to meet the bunker crew at the front base of the static test tower, then depart the area. When everyone was in place with the dignitaries watching from the observation bunker on "Heimburg Hill", the static firing went very well. After briefly securing the test article, the Blockhouse crew made an exit to see the President drive by on the way to the test stand. I joined them on an upper outside stairway balcony. Unplanned, the President, riding with Dr. von Braun in an open top car, asked to stop at the back of the blockhouse. He said to Dr. von Braun, "Who is in charge of the test we just witnessed?" Dr. von Braun asked the crowd, "Who is in charge?" "Saidla," answered the crowd. The word was passed up to me on the back balcony, "The President wants to see you, Bob." Down I trotted and sprinted to the back of the blockhouse. Not wanting to keep the President waiting longer, I hastened to get to the side of the car, almost propelled by the crowd. Dr. von Braun was on my side of the car, so I reached over him to shake hands with President Kennedy. After receiving their congratulations, etc., I thanked them and backed away from the car and then the motorcade continued. Von Braun looked very relaxed. The President was tall, well-tanned and very athletic looking. Unknown to me, the Secret Service people in a following car did not understand this unplanned event. All they saw was the President stop, a murmur in the crowd and someone running up toward his car. Fortunately, they didn't shoot first and ask questions later.

I think most of us in the test stand suffered hearing loss. It was our own fault that we were careless with our ears. First, we loved to stand out in the open and feel the rocket engine's power. When in a rear observation bunker, we would stand outside where we could not be seen from the control room. When the sound level got too painful, we would put our hands over our ears. We continued this practice when we were observers and not in the control room. And the other cause of hearing loss was the loud noise in the test stands. Sometimes the 5000 psi high pressure valves would trip and we would have to go to the top of the stand and manually close the shut off valve. We had no ear protection and it was so painful we would hold a hand over one ear and turn the cutoff valve with the other.

The first installation of the S-IC-T into the west test stand was not a good day. It was bitter cold and all was in readiness. The S-IC-T came on the special road from the Fabrication Laboratory on its transporter. All preparations were made for lifting, however, we were told by Mr. Heimburg that we had to wait for Dr. von Braun and others to be present. They wanted a photo with the S-IC-T. Von Braun stood beside it and wonderful pictures resulted.

When we were finally told to proceed into the test stand, the wind had increased to a real stormy force. The S-IC-T had to be raised up and over the load platform to set down on the hold down arms. There was very little clearance. The S-IC was so large the wind had a big sail to work with. We were not properly prepared. The S-IC was banging from side to side in the load platform as it was lowered. We had crew all around it at every platform level to guide it in with our hands and feet as was our custom. We could not control it and we were literally using our bodies as fenders to keep it from damage. It was a very tense hour, but we did get it in and set down with no major damage. Mr. Heimburg was not a bit happy about this. He appointed Julian Hamilton, a third party, to investigate. Next time we had cushion feeders and a maximum wind velocity criteria. We used this system both here and in Mississippi for the S-IC stage test stand insertion or removal. After a successful static test program of the S-IC "T-Bird", the first three flight S-IC Booster Stages, were static fired with no major problems.

We performed captive static firings for many notables like President Eisenhower, Lyndon B. Johnson, President Kennedy, Lady Bird Johnson and Walt Disney. I think seeing and feeling our Saturn I static firing gave President Kennedy the confidence that the national goal of a lunar landing in that decade could be accomplished. All in all, the entire test years were very stressful. There were long hours and extreme weather exposure on open steel towers. At the same time, it was an exciting wonderful adventure to be an important part of man's greatest scientific achievement, to put a man on the Moon and the subsequent space travel.

JFK's Visit to Marshall
by Ed Buckbee

We had our share of VIP's visits at the Marshall Space Flight Center in the 60s—Eisenhower, Humphrey, LBJ, Ladybird Johnson, Ford—and a slew of senators and congressmen. The most prominent visitor was President John F. Kennedy. This was probably the biggest VIP event in the decade of the 60s for Marshall. It was big for Wernher von Braun. It was big for all of us.

I was working in the Public Affairs Office one afternoon when we got the call that the White House was considering a visit to Marshall. The caller didn't say it was the president, but he left little doubt. We began discussing what we could show him that was in the Manufacturing Laboratory, which we referred to as Fab Lab. We also believed a full duration Saturn booster static test firing was a must. My job was to organize the people and hardware in the Fab Lab where the President would tour. Next was working out the route he would take and where he would watch a test firing from. We wanted to take him to four stops and the Army, the landlord on Redstone Arsenal, wanted three. Soon we found out we were a little ahead of ourselves with the plan. The White House staff and Secret Service visited us, which was my first such experience but not the last for me. We showed them our planned route and activities, and soon learned they had other ideas. The Secret Service told us where we could take the President and the White House staff told us what we could do with him. Even though we were on a military installation, Redstone Arsenal, the Secret Service combed the place looking for hazardous things, places that could be used by a sniper especially on top of buildings. We had to provide a list of the names, date of birth and Social Security number of all persons who would be in contact with the President's party.

After several visits and many meetings, we finally got approval for his walk through of Fab Lab and to observe a test firing. The Army got one stop at their Guidance and Control Lab. For us the test firing approval was big. At first the Secret Service said, "no way" because of safety concerns. But after we convinced the staff that this would be the event of his visit, we got approval for the President to watch it from Heimberg Hill, 2,000 feet from the test stand. Hard hats were a big discussion point. I remember the safety people announcing that everyone would be required to don hard hats as soon as they departed the plane, including the President. That's when I learned Jack Kennedy didn't like hats, which was made very clear by his staff. I don't believe Kennedy wore a hard hat the entire visit, although an aide stuck one in his hand while in the Test Lab.

It was the President's first exposure to a NASA Center involved in building something for the Apollo Moon landing program, which he had set the goal for the U.S. to achieve. It was his program and he flew here on Air Force One accompanied by the Vice-President, administrator of NASA, Secretary of Defense and all the ranking space people on the Hill. The White House press corps arrived in a separate plane. Nearly 200 press people—representing all major networks and news services in the country—were on hand.

Kennedy came to see the Moon rocket and meet the man and the team who were expected to deliver it. It was a fascinating experience to observe these two men from such different worlds get to know each other. Von Braun—the rocket man—did more than anyone else to convince President Kennedy and the leaders of Congress that this Nation was capable of undertaking a complex and dangerous mission like landing man-on-the-Moon and returning him safely to Earth. He was literally a salesman of the idea and

President John F. Kennedy and Dr. Wernher von Braun during JFK's visit to MSFC. Courtesy of Farley Vaughn.

invited anyone who would listen, including the President to come to Huntsville. Dr. von Braun met the President and his party at the Redstone Airfield with Army officials and full military honors, including a 21-gun salute. After a speech by President Kennedy to foreign military students, the party traveled to the Army's Guidance and Control Lab for a short tour and demonstration of various classified missile systems. From the time they climbed in the open top limo and rode through the Redstone Arsenal gate, Kennedy and von Braun hit it off.

They proceeded to Marshall's Fab Lab where the 152-foot-long Saturn I was displayed along with a F-1 engine that would power the Saturn moon rocket booster and many other components of the booster. Several hundred employees were there waiting, and as the President and his party stepped on the floor of this giant rocket hanger there was thunderous applause. Then it was very quiet as von Braun began his enthusiastic explanation of the rocket hardware for the president and his party. You could see the chemistry working, each respecting the others abilities and understanding the vision they shared together—sending an American to the Moon. It was important to Kennedy to personally be reassured by von Braun that we could meet the commitment Kennedy had made to the people of this country. He asked that very question of von Braun and got a firm answer, "Yes, Mr. President we are going to meet your commitment of landing a man on the Moon and we're going to do it within the time you set."

Last stop on the tour was the Test Division, where a Saturn I booster was being prepared for a test firing. We had planned on a full duration test—two and a half minutes—but it was changed to thirty seconds for reasons never made clear.

Kennedy watched from Heimburg Hill, 2,000 feet away, a Saturn booster ignite, shooting smoke and flames skyward, generating 160 million-horse power. After the firing, Kennedy was excited. He grabbed von Braun's hand, shook it, congratulating him and beaming with that Kennedy smile. As Kennedy described the firing later, he said the ground and structures nearby shook. He could feel the vibration in his chest and he felt the pressure and heat hit his face as he looked through the slit in the front of the bunker. He pointed down to his feet and said he felt heat coming up his pant's legs. Kennedy was truly moved by this experience. He was now convinced more then ever that his goal of landing an American astronaut on the Moon first, before the Russians, was going to happen with the help of von Braun's team.

Hundreds of Marshall employees lined the road as the Presidential party returned to the airfield for a mid-day departure. He was so impressed with the enthusiastic team and the work he saw going on at Marshall and the Saturn booster test that day, he invited von Braun to accompany him on Air Force One to visit other NASA centers. Von Braun said he continued asking questions about the Saturn and how the trip to the moon would to be accomplished. They got to know each other very well. We sent von Braun's briefcase and toothbrush on another plane.

Shortly after that visit, President Kennedy and First Lady Jackie Kennedy extended an invitation to Dr. and Mrs. von Braun to attend dinner at the White House. Unfortunately, the date of that dinner was one week after President John F. Kennedy was assassinated. The President who challenged the United States to land man on the Moon within the decade never had the opportunity to see the Saturn Moon rocket fly.

Moonlight Requisition of the Saturn V
by Ed Buckbee

When Wernher von Braun sold the idea of a space museum being built in Huntsville, he promised the people of Alabama that a complete Saturn V moon rocket would be on display. In 1967, I was given the job by Dave Newby, who got his marching orders from von Braun, to go find the Saturn V. After a few days of checking the labs and talking to Saturn V project people, I learned our Saturn V was not to be found. Stages, parts and pieces of Saturn V's were scattered coast to coast with some in Louisiana and Mississippi. After reporting this information to Newby, I was charged with the responsibility of conducting a moonlight requisition--NASA-style--for a Saturn V. I had received training for this endeavor during my U.S. Army tour at Redstone Arsenal.

Having no program funds or approval for such an expensive project, we came up with the idea of moving the Saturn V components as a manned flight awareness training exercise. Further, it was determined no approval had been granted from NASA Headquarters to move the government-owned Saturn V off NASA property and 'store' it at a local museum. When von Braun was informed of this situation, he granted approval for the training exercise and stated, 'it is much easier to ask for forgiveness then request permission,' from NASA Headquarters.

I was authorized to use the full resources of the Center as long as it was considered a training exercise. I became known around Marshall as the "space junk man." I had access to all forms of transportation, which included ocean-going barges like the Poseidon and the Super Guppy--Marshall's flying freighter--and special rail and ground transporters. Rocket stages, engines, interstages and spacecraft were located and prepared for shipment. The booster, the S-IC-T, the "Shop Queen" was at Mississippi Test Facility. Marshall Test Lab helped me have the crown jewel of the Saturn V test program, returned for "additional testing and training" at Marshall. The other stages, engines, spacecraft and components came from Cape Kennedy and contractor plants from across the country. All in all, there were seventeen shipments of hardware to Huntsville by air, land and water.

Once the stages and components were assembled at Marshall, preparations were made to transport the hardware to the Alabama Space & Rocket Center. On June 27, 1969, five rocket stages--the Saturn V, S-IC, S-II and S-IVB stages and the Saturn I, S-I, and S-IV stages with Command and Service modules--were moved the five mile trek within a day. This was the largest move of space hardware in the history of NASA. The S-IC-T booster was moved without the five F-1 engines to lessen the weight of the booster in order to move across a cotton field onto Center property.

Ed Buckbee (left), first director of the Space and Rocket Center and Wernher von Braun view the Saturn V booster, S-IC-T on display at the opening of the space museum in March 1970. The vehicle has been declared a National Historic Landmark and is on display at the Center's new Davidson Center for Space Exploration. Courtesy of Marshall Space Flight Center.

NASA - Marshall Flight Space Center

Once the vehicles were assembled in place on site, excavation and shaping of the Rocket Park was completed.

During final assembly, von Braun visited to see the first complete Saturn V on public exhibit. The rocket he had promised Alabamians became the focal point of the earth's largest space museum, opening on March 17, 1970. Today, it is proudly displayed in the new Davidson Center for Space Exploration.

CTL
by W. L. Shippey

The Components Test Lab (CTL) performed a vital function in developing critical hardware for NASA space programs. It was built and activated along with other major test stands and their supporting facilities.

One of its primary capabilities was a 2,500 horsepower "prime mover", which consisted of a huge generator/motor set for testing missile engine turbopumps. It was built by Westinghouse and put into service in late 1956. Although it was not used operationally for that purpose, it provided the muscle for a system for solving one of the most critical problems ever encountered in the Jupiter flight test program. We received "orders" to put together a test setup and have it operational within a two-week period. Since Willie Shippey had been intimately involved in the installation and checkout of this and all other electrical equipment at the CTL, he was assigned the lead task of integrating the Prime Mover into a system for driving a test fixture intended to solve the problem caused by resonance of sloshing liquid propellants with the thrust vector control system. In addition to the Prime Mover generator, a surplus streetcar traction motor was obtained along with an aircraft generator and a salvage yard speed reduction gearbox. All instrumentation and electrical control panels and associated wiring were fabricated and installed by in-house technicians. The technicians were led by Bo Cloud, our resident control systems lead foreman. Without the hard work, perseverance and ingenuity of these dedicated people, the tight schedule imposed on us would have never been met.

Tests were run, data collected and analyzed permitting the flight control people to adjust the TVC systems to overcome the resonance/instability problem.

The CTL expanded over the years to solve many other flight hardware problems. A small sample of these projects included: fuel disconnects, heat exchangers, hot-gas generators, model engines, propellant characterizations, flight engine gearbox development, and dozens of others.

Memories
by John Shirey

One of the strong points of Test Lab was often overlooked until there was a crisis in the program. The ability of the lab to analyze a situation, develop requirements for testing, design the test facility and hardware, fabricate the hardware, install it at a test location, provide test procedures and test personnel, perform the test, evaluate the test, and make program and hardware recommendations. All of these items were accomplished in the minimum of time, with a minimum of program impact, utilizing the Test Lab personnel. This was made possible by the way our in-house personnel were utilized to provide total support for all phases of the test requirements.

A good example of the capability of our Lab was when the time came for the flight data of the first Jupiter to be analyzed, it was determined to be very unstable. The instability was believed to be sloshing in the tanks. We designed a test facility, procured the hardware, and had tests being performed within seven days. To do this required borrowing a railroad engine electric traction motor, analyzing the motor requirements for power, using a prime mover power supply and two aircraft energizers for power to replace the normal power source, procuring a reduction gearbox from a salvage yard in St. Louis, MO, in addition to providing all fabrication and the instrumentation system for test analysis.

Another example was the time the Saturn transport barge was docked at the arsenal docks for loading for the trip to the Cape and the locks at Wheeler Dam were damaged beyond use. Test lab outfitted another barge (below the dam), provided for unloading above the dam, roadway, loading below the dam and support for making the transfer. This was again done without any impact on the schedule for shipping.

Playboy and Other Stories
by Richard (Dick) N. Stone

From October 1964 to March 1970, tests were conducted on the turbopump and propellant feed system for the F-1 engine. The F-1 was the main engine for the S-1C, the first stage of the Saturn V. During one of the tests, cutoff was initiated by an observer who was monitoring one of the critical red-line pressures, which was being recorded on a strip-chart recorder.

Investigation of the cut-off revealed that a jokester had attached the centerfold of a *Playboy* magazine in a roll of strip-chart recorder paper. When the picture of the Playmate became visible on the recorder as the test was being conducted, the red-line observer was so shocked, or maybe so interested, that he failed to realize that the critical pressure had started to exceed its limit. An observer monitoring a redundant pressure on another recorder cut the test before any damage was done. This "record" was studied very carefully by all the engineers and technicians. Even management showed a great deal of interest in it.

In 1966, Test Laboratory conducted tests to study how liquid hydrogen in the S-IVB (Saturn V third stage) would react to flight conditions. The boil-gas of the liquid hydrogen in the test tank was vented through a large pipe to a vent stack located several hundred feet from the test stand. An open flame at the top of the vent stack was used to ignite the gaseous hydrogen boil-off for safety reasons. Sometimes wind would blow the hydrogen away from the flame and the gas would not immediately ignite. When the hydrogen would ignite, a considerable amount had accumulated and a loud but harmless explosion would result.

At the beginning of one test, the test engineer, Jim Wise, had cleared the area around the stand of all personnel and started tanking liquid hydrogen. The vented liquid hydrogen had not ignited when a man on a tractor-mower rode into the area. Before Jim could inform the man to leave the area, the hydrogen ignited with a loud explosion. The man jumped off of the tractor with it still moving and started running. All at once, he realized that the tractor was still moving. He ran back, turned off the tractor and ran away again. We thought he was more frightened by what his boss would do than what would happen to him by the explosion.

This story relates a time when MSFC proved that the U.S. Patent Office could be wrong. In September 1964, Test Laboratory received a request to design, fabricate, install and run a zero gravity test facility. The facility was to be able to accommodate test packages of about 200 pounds and would investigate how instruments and liquid in containers and under flow conditions would react in zero and low gravity conditions. This package was to be released from a height of 300 or more feet, allowed to free-fall for 4-5 seconds and then caught without damage. Test Laboratory designed a facility to be installed in the Saturn V Dynamic Test Stand. The test package was to be dropped while inside an aerodynamically-shaped shield, which was also to be dropped. The shield, which was 24 feet long and 8 feet in diameter, would protect the package from drag forces. The shield was dropped from the 336 foot level and rode along two rails straight down to the bottom of the test stand when released. The shield was stopped by entering a 36-foot-high cylinder located at the bottom of the stand. The shield upon entering the cylinder compressed the air in the cylinder, building pressure sufficient to stop the shield. Orifices in the side of the cylinder controlled the rate of deceleration of the shield and prevented it from excessive "G" forces.

The shield, with a total weight (including the package) of 2,000 pounds and traveling at approximately 100 miles per hour when entering the catch cylinder, would be slowed to a stop and brought gently to the ground.

The Zero Gravity Drop Tower was operational in October 1965. Since the method of stopping the shield with the package was quite unique, a patent was applied for. The U.S. Patent Office replied that the request was denied because it was impossible for this method to

177

work. Of course, we replied that the method had been working for several tests. We were given the patent.

Dr. von Braun enjoyed bringing visitors to the facility to witness tests. On several occasions, he would place the guests in line with the cylinder orifices. When the shield would enter the cylinder, a strong gust of air from the orifices would result. Von Braun would be greatly amused when the gusts blew the visitors' hard hats off. This facility operated well into the eighties and produced a great amount of data over the years.

Short Stories
by Elmer Ward

When I First Met Heimburg

Harry: "Karl Heimburg this is Elm..."

Karl: "Harry, how is the activation of the Test Stand for the Jupiter missile coming along?"

Harry: "Well, everything is coming..."

Karl: (Looking At Me) "Harry, just who is he?"

Harry: "This is an engineer I hired."

Karl: "Take him out of here and put him to work."

I don't think Harry ever did get to tell him my name during this meeting, and I don't think I ever got one word in "edge ways".

Incident at the Power Plant Test Stand

We were conducting a test of the H-1 engine at the Power Plant Test Stand (PPTS); the area had been checked and the command, 'Okay To Fire' had been given. The test was successful and the Pill Box Crew was cleared to the PPTS. The 'All Clear' was given, then we found a very pale and trembling, shaking gentleman sitting in a corner of the PPTS. I asked if he had been sitting in that corner during the firing? Didn't you hear the siren and the warning to take cover? He said, "Yes, but I thought it was another test stand firing." Boy, did he get the surprise of his life. He had only a concrete wall between himself and a firing engine about ten feet away.

Another Incident

Our boss, Gordon Artley, called Harry Johnstone, Jim Williams and myself into his office. Something had happened and none of us recalled what it was about, but we knew we were in for a "Chewing Out".

Gordon had sat down and was just about to start when I knocked over a quart container of hot coffee in his lap. He jumped up and said, "I wasn't going to fuss at you that bad." He left to change clothes and he never called us back to his office.

Memories
by Jim Williams

Early in the Saturn I program, Test Lab was required to perform a series of static firing of the H-1 Engine at the Power Plant Test Stand, using flight configuration propellant suction ducts. This was not a problem for the outboard engines, since they could be gimbaled to the correct alignment to be compatible with the test stand's water-cooled flame deflectors. The inboard engines, however, were a different story. They were installed in a fired, center position, not at all in alignment with the flame deflector.

Test Lab personnel solved the problem by designing and constructing a mounting block for the engines on the thrust structure that was slanted in the opposite direction and in the same number of degrees that the engine was canted. This solved the problem, and a successful series of tests resulted.

On many occasions Dr. von Braun would come to the Test Lab control center to witness static firings. Sometimes he came alone and sometimes he brought some visiting VIP that he wanted to impress. On one occasion, during an important test program, I was acting as the test conductor and Dr. von Braun was in the control center. After a successful test and a safe shutdown were accomplished, Dr. von Braun approached me and said that he wanted to give me something. He then handed me a cigar that noted the birth of a son. I prized that cigar and kept it for many years until it succumbed to age. I still cherish the memory of that moment when Dr. von Braun took time from his busy schedule to honor me with a memento of an important event in his life.

In 1961 I was detailed to the Los Alamos Scientific Laboratory. My job at the lab was to assist in the development of a flight type hydrogen pump to be used in a nuclear rocket engine being developed. During my time at Los Alamos, I was present for several tests at Jackass Flats, Nevada. For one of the tests, President John Kennedy was in attendance. When he arrived at the control center, Keith Boyer, who was the chief of the test organization, greeted him as follows, "Welcome to Flatass, Jack."

Jimsphere Detail Wind Profile Measuring System
by William W. Vaughan, Ph.D.

In order to ensure the maximum launch opportunities for the Saturn vehicle, and later the Shuttle, we needed a much better measurement of the pre-launch wind profile structure than could be provided by the conventional measurement systems used by the operational weather services. After considerable work and interactions on the subject with the range support groups, which within itself makes an interesting story, we developed in the mid-1960s a two-meter diameter, super pressure, Mylar aluminized balloon with protuberances. It was tracked by a very high precision X band range radar. I named it the Jimsphere after my Deputy, Dr. Jim Scoggins, the person assigned to lead the development. During the development period we tested many of the Jimsphere configurations at MSFC for aerodynamic characteristics and design details.

There was a period when we received several news photos and reports about farmers and others in the outlying areas seeing them fall or finding them draped half-deflated over fences, power lines, trees, etc., thinking they were from outer space or even put there by the Russians. One photo published in a local newspaper showed a farmer pointing his gun at the Jimsphere that had fallen on his farm. We finally put out a notice plus attached a tag on the experimental Jimsphere to tell people what they were.

The development and deployment of the Jimsphere Detail Wind Measuring System significantly minimizes the risk of vehicle failure due to in-flight wind loads when properly integrated into a flight simulation program. The resolution of the wind profile structure permits calculation of structural loads associated with the first bending mode and generally the second mode for vehicles during the critical, high-dynamic pressure phase of flight. The figure on operation of the system illustrates the utility of the Jimsphere Detail Wind Measuring System data during pre-launch monitoring of wind loads. Although developed during the 1960s, the Jimsphere is still in use today for prelaunch monitoring of the Space Shuttle in-flight structural loads.

Memories
by Sylvia Balch Thomas

In 1952 Dr. Gerhard Heller hired me as an engineering draftsman in the Propulsion and Fuels Design Section of the Guided Missile Development Division, a strange choice for a recent graduate of Smith College with a French major! He needed a female to hand letter and draw India ink charts, a task engineers absolutely hated to do. The pay ($3,100) was fabulous and far exceeded that of a high school teacher. The Design Unit under Hans Palaora was my work place and I supported Krafft Ehricke, Chief, Gas Dynamics, Ramjet Project. While I had no idea what ramjet or gas dynamics meant, I believed I could become an instant India ink expert! I still have a copy of his article with my graphs called, "A Method of Using Small Orbital Carriers for Establishing Satellites." On one occasion, Ericke asked me to enlarge a sketch of his to fit a 30" by 40" poster board. His sketch depicted pipes and ductwork for a missile profile. He took my chart to the meeting. I thought it looked great. When he returned he said my chart saved the day by breaking tension at the meeting and bringing peels of laughter. When I learned the cause of the laughter, I was mortified. Not being an engineer, I copied the labels from his sketch writing "duck" instead of "duct!" To me, the German "t" written in longhand looked like a lower case English "k!"

A year later, I was a technical illustrator on Dr. von Braun's staff and as you all know government job descriptions conclude with "other duties as assigned." Well my extra duty

NASA - Marshall Flight Space Center

Seated, from left: Martha W. Smith, OML; Ethel R. Ethridge, OML; Nancy R. Wright, Comptroller; Rayford H. Latham, PM; James L. Brewer, OML; Jesse E. Long, Jr., PM; Joe Covington, OML; Sylvia B. Warren, OML; Frances Greenwood, FSD. Standing: Paul Kennedy, OML; Jim Hill, Comptroller; A.R. Crutcher, OML; Earl M. Wysong, OML; Gus H. Grace, OML; Forrest S. Brazzell, Comptroller; William F. Pylant, Redstone Depot; Alfred M. Jones, OML; Leonard C. Duke, OML. Courtesy of Sylvia B. Thomas.

was to translate Dr. von Braun's hand written answers to letters received from young French space enthusiasts. When I applied for a night teaching position at UAH, he wrote in his recommendation: "Judging from my limited knowledge of French (which used to be a lot better, but has become pretty rusty) I can certify that Mrs. Warren (my name then) has handled these translation assignments very efficiently and well. Her mastery of the French language, even in difficult scientific matters, appears to be such that she should well qualify for a teaching position in French." I got the job!

I have also fulfilled a prophecy Dr. von Braun wrote in my copy of his co-authored book *The Exploration of MARS* published in 1956. "To Mrs. Sylvia Balch Warren (now Thomas) who rendered such invaluable help during the fledgling five years of exploration of space, and who, we hope, will still be around when we start exploring Mars." I'm still around!

In 1970, many believed an MSFC Alcohol Program unnecessary. When NASA first took steps to implement the federally mandated program, little was known about alcoholism and its potential to adversely affect job performance. Training became a first priority. Jim Hiers of S&E became the first MSFC program administrator and was reassigned to Community Relations under Jim Johnson where I worked. Sadly he died just before assuming full responsibility for the newly created position. Jim Hiers was a recovering alcoholic active in AA and his expertise was a loss to the Center. However, he left excellent material as to what an industrial program should include. Jim's death led to my appointment as program administrator. My first task was to write an alcoholism management instruction. In doing this I worked closely and shared ideas with NASA Headquarters personnel. I consulted with alcohol program managers at the Army Missile Command and local industries. To meet the Center's immediate training requirements, I developed an introductory course on alcoholism and taught the program fundamentals to over 900 Marshall managers/supervisors. Teaching was a challenge. It was not something supervisors wanted to hear. I passed state licensing requirements, but had to use annual leave to take the examinations! My role as administrator was limited to a few hours; it was not a key space program. However, this tiny pioneer effort, mandated by law, evolved over the years into the broader based MSFC Employee Assistance Program that exists today. The original MSFC Alcohol Program never made front-page news, nor was it ever viewed as an important program, but in the scheme of things it had the potential to save lives. I am proud I had a role in its development.

Planning the MSFC Management Development Program (MDP) was a major challenge for Jim Johnson, my supervisor, and for me due to what seemed impossible time constraints. NASA Headquarters asked each Center to develop a management seminar modeled on the Agency's successful Management Education Program (MEP) held at Wallops Island, Virginia. The goal was to offer managers a chance away from their jobs to interact with each other, hear management philosophy from our top leaders, and gain knowledge from nationally known scholars in management training field.

MSFC's initial reply to Headquarters was a flat "no" based on critical mission priorities. NASA Headquarters issued a new directive telling all Centers to proceed at once with the development phase. Don Bean, our Personnel Director, asked Jim and me to work the action. We went into high alert as though in a count down to launch! Credit goes to all dedicated people at MSFC whose teamwork made it possible to cross what seemed uncharted territory in the areas of procurement and contracting. We worked with Dr. Constance Dees of Alabama A&M University to contract for the facility and provide nationally recognized management speakers. The five-day off Center live-in program finally became a successful reality.

The first MDP took place November 1980 at Wheeler Refuge with an observer from NASA Headquarters. He wrote many tiny notes to us each day! I don't believe I slept a wink that week. The remaining sessions were at Guntersville Lodge. After 29 sessions, MPD ended in 1994. More than 700 MSFC managers attended. Dr. William R. Lucas listed MDP as a successful first-of-a-kind MSFC program for managers.

Communications at NASA/MSFC
by E.D. Hildreth

I was general manager of the DeKalb Telephone Cooperative at Alexandria, Tennessee when Robert O. Sparks, Army, later MSFC employee, called me to come to Redstone Arsenal in July 1960 for a job interview. I was not impressed with the Army offer, but Sparks said they are forming NASA and I said what's that? I was introduced to the Director of Personnel and was employed in October 1960 as a communications engineer section chief with eight employees. It was a challenge to create a communication system that would support the research and development that was taking place at MSFC. The Central Communications building 4207 was built in 1962 thanks to Chauncey Huth. Thanks to Dr. Eberhard Rees for approving the center's underground cable and duct system with support from Erich Neubert and Victor Sorensen that was constructed in 1964-65. These actions and the installation of MSFC's own telephone exchange in 1968 provided for an improved communication system. I visited with Mr. Neubert frequently never advising my supervisors and we discussed communication improvements that would eventually be approved.

Ed Buckbee of the MSFC Public Affairs Office prepared scripts for 32 public television programs that required use of the Communications Office television studio. Dr. Wernher von Braun narrated most of the programs. He did not have time to rehearse, so we prepared the information with large print on a teleprompter mounted on the camera for him to read. He usually did the programs without mishap but one day we almost completed a program when he suddenly stopped and said that word is misspelled and we had to start over - never again did we misspell a word.

Personnel of the Communication Office were motivated like other Center elements to provide communication support in order that we might achieve the moon landing within the time frame President Kennedy had established. Some of the support was the installation of radios on the Saturn barge within one week including procurement and delivery, developing communications for the Michoud Assembly Facility, New Orleans, Louisiana; Slidell Computer Facility, Slidell Louisiana; Mississippi Test Facility near Bay St. Louis, Mississippi; movement of the Hubble telescope mirror from New England to California, off-loading, loading, moving and testing the Shuttle Enterprise at MSFC, support at Pass Christian for Hurricane Camille, obtaining and coordinating the use of radio frequencies for MSFC administrative and launch vehicles operations, installing and maintaining communication equipment for the Huntsville Operations Support Center (HOSC) including Skylab simulation and training, maintaining conference room and medical equipment and other communications requirements for the various program offices and their contractors.

The Communication Office was responsible for the MSFC telephone directory.

Misspellings will happen as we misspelled General O'Conner's name the head of Industrial Operations. I had to go and assure General O'Conner it would not happen again, but it did in the next publication. Classified and non-classified teletype service was provided to the center by a group of dedicated teletype operators. I proposed an emergency warning system in the Communication budget until my boss said I don't want to see it again; that is until the 1974 tornado hit the center and I was asked where is that warning system? The emergency warning system (EWS) was installed very quickly and is in use today (2002).

The technical control facility was established in MSFC building 4207 to improve support to the Center and its programs. This facility enabled center management personnel to be contacted and included in a voice conference call, regardless of location, 24 hours a day. This facility was used for all NASA flight readiness review conference calls prior to vehicle launch that included various centers and their contractors. The technical control also had a secure voice capability.

Information transfer capability continued to improve as MSFC was selected to develop and manage a communication network for the entire agency. The Communication Office developed and implemented this network called the Program Support Communication Network (PSCN), later called PSC, to provide switching and transmission of voice, high speed and packet data and facsimile between NASA centers and NASA contractors via landlines and satellites.

The center's private automatic branch telephone exchange (PABX) was upgraded in 1985 to provide digital switching and voice mail for center employees. Computer terminals were also installed and connected to the communication network throughout the center to expedite response to action items and improve schedules for employees.

Breakfast was prepared in building 4207 break room for civil service and contractor employees supporting each launch by the ladies who worked in communications. Retirees such as former Center Director Dr. Rees, my buddy Erich Neubert, Karl Heimburg and other retirees would meet for breakfast in 4207's conference room, enjoy breakfast, view the launch and discuss their decisions effecting Shuttle etc. until one day in 1984 I was advised to cease and desist as I was not in the food service business. Another incident occurred in the picnic area early one Saturday morning as I observed a man with a beard sitting at a table alone; as I got closer it was Dr. von Braun who had arrived from Washington for the picnic. We talked but in a few minutes a crowd gathered to say hello to our former Center Director.

I wish to pay tribute to all civil service and contractor employees who were a part of the Communication Office effort, especially my branch chiefs Charles Bryant and Robert Carlin (both deceased) and engineer Irv Sainker (deceased) to whom I assigned the most difficult engineering tasks. The Communication Office utilized various support contractors but one contractor manager named George Ward who came with RCA Service Company in 1963 was the most outstanding manager as he switched from one contractor to another until he retired at Huntsville. My secretary, Carol Busby, kept everyone straight.

A Young Engineer's Amazing Moments
by Hermon H. Hight

After completing my 26-month obligation in the US Army in May of 1962, I reported to my new employer, Marshall Space Flight Center (MSFC), to start my new career with NASA. I was assigned to the Astrionics Laboratory in the Guidance and Control Division under F. Brooks Moore. My immediate supervisor was James E. "Jim" Blanton. Our Unit performed Saturn Vehicle control system tests using an actual Saturn Instrument Unit flight computer in a simulation that included flight type actuators and flight type vehicle engines, such as the Saturn first stage H-1 engine. Other hardware components and vehicle dynamics were simulated on analog computers due to the need for "real-time" simulation requirements. Digital computers were too slow for that function in those days.

My father, A. Hill Hight Sr. of Jackson, TN, was so proud of his son (me) not only for working for the National Aeronautics and Space Agency at MSFC, but especially for me working for Dr. Wernher von Braun, the genius and world famous rocket expert! He bragged about that to his fellow railroad employees, friends, and family all the time. I felt the very same way; also, very fortunate to have such an opportunity. My Dad's pride in me and this accomplishment is better understood in light of him not having the opportunity for a college education himself and for me being the first member of our family to go to college thanks to Dad's support. He could see a very happy result of his and my efforts. Making it to Huntsville working for Dr. von Braun was my first amazing moment!

One evening in the Astrionics Simulation Laboratory, testing the Saturn Control System for an upcoming flight, I and my companion, Donald R. "Don" Scott, were working in the "Hole," what everyone called the test facility in the east end of Astrionics Laboratory in Building 4487. Everyone worked a lot of overtime in those days, trying to meet schedules and make sure everything worked right. We kept testing even when we had done everything planned and necessary. We tested right up to launch time! (Saturn flight software never failed!) On this particular night, just the two of us, Don and I, were working diligently when two unexpected visitors walked into our laboratory! We were both amazed. Dr. von Braun was escorting one of his best friends on a night tour of MSFC. His friend was Walt Disney! So here we were, in the presence of the two most famous and popular persons in the world! We introduced ourselves, shook hands, and explained what we were doing! They only stayed a few minutes, but we felt so honored and on top of the world to have such visitors, and especially to know that Dr. von Braun knew where the "Hole" was located and thought that our efforts were worthy of a demonstration to Walt Disney! How important we felt! My second amazing moment!

To this day, I consider these events to be the personal highlights of my space career. These personal events were topped, of course, by the mega-event of the Apollo 11 landing on the Moon, with my wife Pearl and two daughters Anita and Vicky watching the culmination of the Saturn/Apollo effort on our den TV. I have been so blessed and honored and have so enjoyed my MSFC career!

Looking Back
by Ruth von Saurma

After the launch of the first U.S. satellite, Explorer 1, January 31, 1958, the eyes of the world began to focus on the Army's successful rocket team under the inspiring leadership of Dr. Wernher von Braun. Their achievements became the major headline of all Western media reports. An avalanche of congratulatory telegrams and messages reached the desk of von Braun. Through speeches to various national organizations and publications in scientific and popular magazines, he had become known as "Prophet of the Space Age" (*New York Times Magazine*, Oct. 20, 1957). After the devastating shock of Soviet Russia's Sputnik I and II, it was he who had assured many reporters and audiences that the U.S. did not lack the genius required to compete in the area of rocket technology. Explorer I proved his point and provided a sense of relief in the Western world combined with a growing interest in the new frontier of space.

It was at that time that I joined the von Braun team as a translator. My initially temporary job was to review and reply to the stacks of congratulatory mail and requests for information and interviews from abroad. In the course of years, my job expanded into an intriguing permanent position as international relations and public information specialist with many diversified and unique assignments. I learned a tremendous amount from my journalist colleagues, supervisors, and directly from von Braun. He was a master in written and oral communications and a brilliant and skillful manager. With his broad range of knowledge and interests, keen curiosity, awareness of the past and visionary outlook on the future, he was the modern version of a Renaissance man.

On my job I got a unique insight into the complexity of rocket technology and in particular the vast array of public relations behind the U.S. rocket and space program. Gordon Harris, ABMA's public information officer, was fully aware that increased public understanding and congressional support were necessary to maintain a viable space flight program. With untiring efforts and insight he represented ABMA's commander, Maj. Gen. John B. Medaris and von Braun, its technical director, in many negotiations with government officials and the media. His office released a great amount of visual and printed material, illustrating the technical and scientific advances made by the team's creative scientists and engineers, working at the forefront of new technology.

After the transfer of the von Braun team to the National Aeronautics and Space Administration and the establishment of the Marshall Space Flight Center, it was Bart Slattery who skillfully and astutely directed the Center's public affairs. Briefing for VIPs, special events programs, television and radio broadcasts had to be coordinated and a steady flow of news releases for the media prepared to give a full account of the Center's progressing activities.

Slattery's office also had to support von Braun's numerous public appearances with speech material geared to the scientific community or the public at large. As the most effective spokesperson for the need and benefits of an active space program, von Braun was on the lecture circuit often once per week. In addition, stacks of incoming mail and telephone requests had to be dealt with expediently. They came from everywhere in the nation, but also from abroad. The U.S. space race with Soviet Russia had all Western nations truly concerned. Their media representatives, scientific and technical experts, representatives of the International Astronautical Federation from all parts of the world, space fans and interested individuals flooded the office with special requests.

It was fascinating for me to screen the variety of foreign requests addressed to von Braun and in important cases compose replies for his signature or prepare drafts of comments or articles for foreign publications. If time permitted, he gladly shared his vision of future space ventures with visiting journalists and dignitaries and his contagious enthusiasm was most effective. He also had a special knack of establishing rapport with his visitors. Before an interview with a reporter from a German aviation magazine, he told him of his great joy of flying private planes and doing takeoffs or landings when aboard the Center's Gulfstream. Having recently received his license for multi-engine transport aircraft, von Braun joked that he could at least get another job as pilot.

In July 1965, a production team of the French-Canadian Broadcasting Company spent several days at the Center to gather film footage and get an interview with von Braun. They were delighted when he agreed to address their audience in French. With inputs from our news branch and the interviewers, I had prepared the French text for von Brauns' comments and had it put on the teleprompter in the TV studio of our communications branch. Von Braun's French was remarkably good, having attended in his youth the French Gymnasium in Berlin. When the producer and the interviewer thanked him for the interview and complimented him on his good French, he smilingly admitted that a French girlfriend had helped a lot to improve his language fluency.

Unforgettable for me will always be the day of July 16, 1969. At 5:00 a.m., Wernher von Braun, Cornelius Ryan, the noted author, and I climbed into a helicopter waiting behind the Holiday Inn in Cocoa Beach to fly to the Melbourne Airport. It was still pitch dark. At the launch pad, the huge Saturn V was illuminated by a series of searchlights forming a brilliant star against the dark sky, a truly awesome sight. At the airport we met a French jet chartered by the renowned *Paris Match* magazine. The passengers included Sergeant Shriver, U.S. Ambassador to France, and 80 high-ranking European VIPs. We joined them

French-Canadian TV interview with Wernher von Braun, July 1965. Pictured from left are: Canadian producer of the space series, Canadian interviewer, MSFC News Branch Chief Joe Jones, MSFC International Relations Specialist Ruth von Saurma, and Wernher von Braun. Courtesy of Ruth von Saurma.

for breakfast at the airport restaurant, where von Braun officially welcomed the guests in French and English. Afterwards, von Braun and Ryan left by helicopter to the Launch Control Center, while I escorted the European group to the VIP viewing site.

Watching the giant Saturn V-Apollo vehicle take off for man's first landing on the Moon was an ecstatic, once-in-a-lifetime experience.

Looking back on those days in Huntsville and the Cape - I can only say that I am so glad I was there, too.

NASA Wives Week
by Hamp Wilson

In May 1964 with Secretaries Day just around the corner, my wife Peggy and several other wives of NASA employees, primarily from the Procurement Office, were discussing the amount of attention that NASA secretaries received and how little the wives received.

My wife approached Huntsville's Mayor, R.B. "Spec" Searcy, to see if there was anything that could be done for NASA wives. He suggested a Proclamation that he could sign, which she developed and is reproduced below.

STATE OF ALABAMA)
COUNTY OF MADISON)
PROCLAMATION

WHEREAS, a group of citizens of the City of Huntsville, State of Alabama, who did solemnly swear to be the wives of the male employees of Mr. JIM MORRISON'S BRANCH at the George C. Marshall Space Flight Center, came before me to state their cause, and;

WHEREAS, the above WIVES did state that they have dutifully served their husbands for their entire marriage, and;

WHEREAS, said WIVES reared the children, cooked the meals, cleaned the house, washed and ironed the clothes, Listened patiently to the trials of the office work, Listened smilingly to the measurements of the new secretaries, Listened dutifully to the office gossip, Listened disgustedly to how much harder HIS work is than HERS, Listened sympathetically to the aches and pains, fed the ulcers and chilled the martinis for their husbands, and:

WHEREAS, said WIVES may not be the "girl Friday," but they are of great value and worth on Saturday nights, and;

WHEREAS, said WIVES have kept themselves charming, beautiful, and enticing while their husbands have become increasingly fat, gray, bald, and senile, and;

WHEREAS, I, R.B. SEARCY, Mayor of the City of Huntsville, recognizing the outstanding virtues, hard work and extreme sacrifices of this group, do hereby proclaim to each and every spouse of said WIVES the week of May 17, 1964, to May 24, 1964, as NASA WIVES WEEK.

GIVEN under my hand and seal this 15th day of May, 1964

R.B. Searcy, Mayor (SEAL)

Footnote: Jim Morrison was James A. and his branch was really the Mississippi Test Branch and the proclamation was personally delivered to his office by my wife. In addition to the author, some of the other NASA male employees affected by this endeavor were Ollie Hirsch, Manley Williams, Stu Jones, Jim Morrison, Hal Minney, Jack Hyer, Jack Jones, Wally Foxworth, Paul McCutcheon, Ancil Kent, Ross Hunter, Ron Backer, Don Cornelius, Jim Goldsmith, Milt Story, Bob Mick, Wayne Nichols, Jim Locke, Bill Goodrich, Cecil Law, Harold McMillan, Earl Eubanks, Morgan Moore, Al Jolliff, Milt Story, Ken Sowell, Craig Mitchell, Bob Woodham, Gordon Dison, Jim Maxwell, Bud Lively, Jack Coldiron, and Max Mallernee, just to name a few. The original of that proclamation resides on a wall in our home here in Tallahassee, and I think of those early days at Marshall every time I look at it. For folks who were not at MSFC during those days, Mayor R.B. Searcy was one of the first to hold that Office and from what we learned later on, that proclamation was his last official act before leaving office.

A 1968 Trip to Norwalk, Connecticut, with Gerhard Heller
by Jim Harrison

It was 1968 and my idea was to fly to Bridgeport, Connecticut, or to an airport near Norwalk, Connecticut, where we were to meet with Perking-Elmer (an optics company) about the thermal design problems on Princeton University's Stratoscope II balloon borne telescope (Perking-Elmer was the manufacturer). Gerhard Heller, my boss, thought it best to fly to Newark, New Jersey, and then drive fifty miles north directly through the middle of New York City. Since he was the boss, his plan was adopted. Mechanical problems caused our flight out of Huntsville to be about two hours late and we therefore missed all our connections, not arriving in Newark until very late. We got our rental car and headed for Norwalk. Gerhard drove through New York City in the middle of the night (the reduced traffic was a fortunate consequence) as if he did it every day, cheerful and talking all the while and totally oblivious to the lateness of the hour or our misfortunes of the day. I had worked for Gerhard for eight years, but it was not until that night that I got to really know him. He was an extremely brilliant engineer, scientist, and manager. In 1969 he became the director of the Space Science Laboratory, where we were working at the time of this trip. That night we were thrown together for many hours, and I enjoyed listening to his stories about Peenemunde and the assignments he had in the development of the German V-1 and V-2 rockets. As I remember it, his task was to design the rocket nozzle and to determine how it could be kept cool enough not to melt from the hot exhaust gasses. At Peenemunde, the engineers often did not have time to prepare complete drawings, so Gerhard had to work hand-in-hand with the machinist to describe to them using crude sketches what he wanted them to build. He would sometimes go out in the shop and actually run the metal lathes himself to make the parts that were needed. I recall that Gerhard's ideas on the history of Europe before World War II and especially the root causes leading up to that momentous conflict were extremely interesting. We finally arrived in Norwalk at about 2:00 AM.

The Stratoscope II balloon borne telescope made its maiden flight the next year (1969) with more successful flights accomplished during the next several years.

A Remembrance
by Jack Conner

As the Saturn V Swing Arm Czar for MSFC, I had the pleasure of flying to Kennedy Space Center (KSC) to attend status reviews of testing, modifying, retesting and manufacturing three sets of changes for three sets of swing arms. On many occasions I was accompanied by Karl Heimburg, who always carried a plastic attaché case but never revealed its contents. We normally flew the twin engine Beechcraft and more often than not, Air Force General Edward O'Connor was our pilot.

In a separate action unrelated to swing arms, MSFC issued a directive that no government sedan would be seen parked in front of the ABC stores in Florida and should a traveler bring back some of that Florida liquor, it would be so recognized by proper authorities at MSFC airport for Alabama taxation if taken off the arsenal.

On one of these trips, after a very successful meeting with the KSC hierarchy (success was measured by a private meeting with Kurt Debus, Karl Heimburg and myself), we left the complex to start our journey back to Huntsville. We first traveled in a grey GSA car to get to Patrick Air Force Base. When we got to Cocoa Beach, Karl requested with authority, that we stop at the ABC store on Highway A-1A. Who was I to deny him his wish?

After rummaging through the store for 30 minutes or so, Karl had decided, "A case of Jim Beam bourbon whiskey." The goods were brought forth and immediately the cash register rang out $130.00 or thereabouts. Karl opened his wallet only to find it empty. Ruth had not given him his travel allotment and if she did, it did not include a case of booze. Without hesitation, Karl turned to me and said, "Pay the man!" For some unknown reason I had enough "green backs" to get him out of the hole.

At this point, I learned the contents of his plastic attaché case: a small coil of clothesline rope and a metallic handle. As I was paying, he was wrapping the rope around the case, twice, in two directions. After tying a satisfactory knot, he attached the handle so he could carry the case like normal luggage.

When we arrived at the Arsenal airstrip, I had the pleasure of providing transportation to our abodes, via a 1961 VW Beetle. The Jim Beam case was proudly displayed in the middle of the back seat. We ignored the MSFC directive and started home. As we approached the Martin Road checkpoint, the Military Police stopped us, and politely informed me that it was vehicle inspection time. Needless to say, his flashlight caught the contents on the back seat.

After pulling off to the side and waiting for the MP with his pad, pencil and flashlight; good ole Karl came through again. He reached in his wallet and passed a card to the MP. The card, signed by the Commanding General, excused Karl Heimburg and the vehicle he was traveling in from search. The MP snapped to attention, apologized, returned the card and gave us a very dignified salute.

When we got to Karl's house, I was invited in, not for a drink, but reimbursement from Ruth and her apology for not properly preparing Karl for the day's activity.

Moral—When working with Karl, expect the unexpected!!!

Remote Sensing of Earth Resources at MSFC
by Ted Paludan

MSFC has been directly involved in remote sensing of Earth resources in three different and distinct periods. The first was during 1965-66 when MSFC was in competition to be Lead Center (versus the Manned Spacecraft Center). The second was from 1970 to 1979 when the Environmental Applications Office and the Earth Resources Office in the Data Systems Lab were in operation. I don't know when the third activity began as part of "Mission to Planet Earth"—maybe 1991 This is still on-going with major activity as part of the Global Hydrology and Climate Center.

In 1965-66 MSFC made efforts to get assignments from the Office of Space Science and Applications (OSSA) under the Apollo Applications Program. By early 1966 we had convinced OSSA leaders that MSFC should be the Lead Center for Earth Resources. Others at MSFC seemed to be successful in getting Lead Center assignments for an orbital workshop (later the Skylab space station, the Voyager Mars probe (which involved remote sensing) and a large space telescope (later the Hubble Space Telescope).

There was just one more detail: approval by the Office of Manned Space Flight, headed by Associate Administrator Dr. George Mueller. There was to be a meeting of the Directors of MSFC and MSC, a few engineers from each, and Dr. Mueller. On June 5, 1966 I and some orbital workshop personnel flew to Long Beach, California with Dr. von Braun. The next day we met with Dr. Mueller. I made a pitch for Earth Resources. After a brief consideration, Dr. Mueller said he was deciding to assign the Lead Center roles for the orbital workshop to MSFC and for Earth Resources to MSC. The official decisions were prepared at an OMSF "Hide-Away" meeting at Lake Logan, North Carolina in August 1966. I visited the MSC people in Houston December 12-15, 1966, and the final close-out was accomplished by letters exchanged in January and February 1967.

In 1970 OSSA invited all ten NASA Centers to help manage the many experiments expected to grow out of the data from Skylab's Earth-viewing instruments and the planned Earth Resources Technology Satellite (ERTS, later named Landsat). This "Regional Center" program began officially by a letter from Acting Administrator George Lowe on November 19, 1970. MSFC began participation immediately with temporary assignments. Dr. George McDonough was assigned as Director, Environmental Applications Office on September 17, 1972, and I (Ted Paludan) was assigned as Deputy Director on October 1, 1972.

In 1971 and 1972 we began the "Applications" by meeting with state and local officials in Alabama, Georgia and Tennessee. These included the Alabama Development Office, the Top of Alabama Regional Council of Governments (TARCOG), the Alabama Marine Environmental Consortium, the Georgia Science and Technology Commission, the Geological Survey of Alabama, the Tennessee State Planning Office, the Tennessee Valley Authority and several universities.

Landsat-1 was launched July 23, 1972 and Skylab on May 14, 1973. Landsat and its follow-on satellites have provided continuous global data. The last data from Skylab was returned by astronauts February 8, 1974.

By use of "Task Work Packages" in Fiscal Year 1973 about 82 people in various S&E Labs were working on Environmental Applications projects. Our own offices and laboratory spaces were consolidated into the old cafeteria area on the first floor of Building 4201. In addition to satellite data, we had high altitude data from NASA's U-2 and RB-57F, and had a C-45 based at MSFC; with an array of instruments.

On March 16, 1973 a flood hit Huntsville. We used the C-45 to collect Infrared photographs of its extent. These were used by the Geological Survey of Alabama to produce flood-zone maps. We also provided aerial photographs of the damage done by the "Super Tornadoes" of April 3, 1974 to state and federal disaster relief agencies.

With little effect on our activity, a reorganization in June 1974 abolished the Environmental Applications Office and relocated us to the Earth Resources Office within Data Systems Lab in Building 4708.

We maintained about 22 contracts for specific applications during each year (about $500,000/ year), primarily with universities, which in turn often supported application needs of state and local governments. Each of us were also conducting in-house projects in mapping, hydrology, geology, etc. Our small group was responsible for publication of 25 to 30 percent of all of MSFC's technical papers during the 1974-1978 period. We were trying to help our region by providing information on resources and environmental problems—and our sister organizations were doing the same in other regions. Publicity on this helped with grass-roots support for NASA—something in short supply since the last lunar landing.

Lunar Roving Vehicle (LRV)
by Saverio "Sonny" Morea

The Manned Lunar Roving Vehicle at times became affectionately known or referred to as the "moon buggy" or the "moon car." Both these titles engendered an image of simplicity. Nothing could be further from the truth. A more apt description that would more accurately capture the essence of this vehicle would be how we in the project referred to it as the "Spacecraft on Wheels." Consider its features.

It had redundant batteries, woven wire mesh tires, four independently driven and braked wheels, double Ackerman steering that enabled the vehicle to turn completely around within its own body length, independent front and rear wheel steering – a feature that had to be used on the first attempted EVA, when the astronaut may not have appropriately flipped the right switch – a navigational system capable of allowing a return to the Lunar Module within 600 meters after being driven 5 kilometers away from the Lunar Module (actual performance fell well within this requirement), a thermal control system, and last but not least, a semiautomatic deployment system used to unfold the LRV from a small package, the volume of a Volkswagen into a large sized American automobile.

Another issue, one that focused a great deal of outside attention on the project, was the realization early in the program by the LRV team that costs were exceeding accomplishments in a substantial manner. Within the first six months into the contract, steps had to be taken to replace and strengthen the contractor management of the LRV project. As is so often the case in contractor personnel assignments to project teams, the winning contractor proposal team is rewarded by the contractor with the task of managing the project, should they win the bid. Although an argument can be made for this practice, it can and in the LRV case, would have been disastrous to the final outcome, had the contractor not been strongly encouraged to correct the overall situation. Although distasteful to me to replace a contractor manager, my many years of management experience told him a correction needed to be made. Thus in the spring of 1970, the Boeing project management was strengthened with the assignment of a more senior and experienced project manager supported by the Boeing Plant manager of the Huntsville Space facility. In addition, a portion of the work was transferred to a Boeing manufacturing facility in Kent, Washington. Final assemble and thermal vacuum testing was accomplished at the Kent facility.

Concurrent with the management fixes being made was the realization that the fears the government team had during the competitive negotiations with Boeing regarding whether the cost they bid could be attained or not. Boeing bid approximately $19.6 million, whereas the government estimate approached $40 million. Due in part to this large difference the use of the "overriding multiple incentive contract," and the use of the "performance type specification appeared as the prudent way to protect the governments interest.

As the summer of 1970 was reached, contractor and government "run out" cost estimates began to clearly confirm that the run out project cost overrun was approaching 100%. The technical and schedule milestones were being met, but cost was clearly beyond what the contractor had proposed in bidding the project.

It was during this time that our team experienced some of the most stressful and difficult days. Congress had picked up on the potentially large percentage overrun of a NASA project, and initiated action for the GAO to run a full audit of the program. I was asked to put together a briefing to NASA headquarters personnel.

Some members of NASA headquarters were critical of the use of a "performance Spec." approach, indicating that it would have been more "politically" acceptable to have a 100% "cost growth" rather than a 100% "cost overrun." The former putting the blame on the government for directed changes and allowing the contractor to earn more profit, while the latter approach punishes the contractor for perhaps coming in with an undoable low bid just to win the contract, or for mismanaging the project, or both.

Much to my surprise, I learned that the congressional staffers were so impressed with incentive contract negotiated and the performance Spec approach used by the MSFC team that they arranged to call off the GAO audit. To our knowledge, never before had a GAO audit ever been called off, once such a request had been made. Ben Milwitski indicated that the congressional staffers were well pleased with how the governments interests were protected and expressed surprise that a contractor such as Boeing would sign such a contract. Needless to say, the LRV team at MSFC felt vindicated in their judgments.

A final issue deals not with the LRV story, but rather its legacy. Each spring high schools and colleges throughout the United States are invited to participate in a "moon buggy" design and race competition. The race is conducted in Huntsville at the Space and Rocket Center. High schools compete against high schools and colleges against other colleges. Designs are judged by some the original LRV team members. Students are required to design and build a human powered vehicle, that can be folded into a box 4' by 4' by 4', unfolded quickly and raced on a specially prepared obstacle course simulating the lunar surface and obstacles.

It is a marvelous sight to behold to see the best of America's youth applying their innovativeness and creativity and mechanical to solving a design problem as complex as this.

In addition to the moon buggy legacy comes a more pragmatic legacy of the LRV, and that is the all aluminum car currently being sold by the Audi Company.

Lunar Roving Vehicle (LRV) Navigation and Control
by Pete Broussard

When my branch started on the problem of the navigation system for the LRV, we knew very little about surface navigation. Everything we had done in fact, it seemed that everything anybody had done had been in the area of three-dimensional space navigation. Boeing proposed using a strapdown system. This system was more complicated than required; moreover, it had not even been designed. This meant we were looking at—and paying for—the design, fabrication and testing of a brand new system. We thought there had to be an adequately accurate system that was simpler and cheaper.

The team I formed for this project started considering the problem of the simplest way lunar surface navigation could be done. We considered a dead reckoning system suggested by one of the team members, Jon Knickrehm, which did not differ in theory from the Boy Scout and Army technique of navigation by knowing a fixed direction (e.g. north) and knowing distances traveled in two directions "Northings and Eastings." Our confidence was bolstered after I went through a purely theoretical document on the differential geometry of surfaces as it may be applied to navigation, although surface navigation wasn't addressed directly.

At this time, Astrionics Lab was well equipped to do in-house work. To do the analog of Boy Scout navigation we would need a directional gyro which, once started, maintains a fixed direction in space - except for the ever-present random drift for direction, and would need to count wheel rotations for distance traveled. We had an old Bendix directional gyro in the lab, which while noisy because its bearings were about shot, was deemed to be suitably accurate. The navigation was simulated by running the output of the directional gyro as well as simulated outputs of the odometer through a processor. The results were very good. We next mounted the gyro on a vehicle that had small magnets attached to the rims of the wheels of the vehicle. As the wheel rotated, the magnets passed near a magnetic reed switch, which gave a pulse each time a passing magnet tripped it. This gave us incremental distances.

We were convinced it could be done with this simple system, but Boeing wasn't convinced. After much arguing, they agreed to use our system on the LRV with the understanding that it was MSFC's choice, not their choice.

In the spring of 1970, we started full-scale field trials at a site a few miles from Flagstaff, AZ, a site chosen because of the presence of the United States Geological Survey in the area. They had very accurate maps of the area. The area was near ideal for what we wanted: sandy, with some rocks. In closed sorties exceeding 20 kilometers, we were well within the error budget, proving that this simple inexpensive system was adequate.

The team tested a hand-held device, designed and fabricated at MSFC, that used the sun's shadow falling on a scale; this would allow the astronauts to update the gyro setting on the LRV, if required. It turned out to be an amazingly accurate device whose cost was negligible. The gyro, because its drift rate was so low, did not have to be updated during lunar sorties.

I was told to give a presentation on the navigation subsystem to the then-director of MSFC, Dr. Eberhard Rees. At the time, the astronaut's display panel hadn't been completely designed. As I started explaining how the system worked, I showed a picture of the display panel and where the navigation switches were. Dr. Rees interrupted me by asking if it was possible for one of the astronauts to accidentally hit a certain switch. I explained what would happen, but added that precautions were being taken – guard rings over the switches – that this didn't happen. I continued the presentation. Dr. Rees interrupted me again and asked the same question. I gave the same explanation. I continued. Dr. Rees interrupted me again with the question. By this time, I was exasperated and told him, "It's possible that one of the astronauts could fall out of the vehicle and get run over, but I don't think it will happen." This was a short reply, but I guess Dr. Rees realized that I had told him all I could, and had given him all the assurances I could. He laughed. He and I always got along well.

While we were confident the navigation system would do the job (and it performed superbly on the lunar missions) there still remained the question of the drivability of the LRV. While this area didn't directly fall within the purview of the lab, it was concerned with control, so we undertook a study of the dynamics of the LRV under 1/6 gravity. We used a 1/8 mile track, which was located east and slightly north of the Redstone Air Field. We put sand on the track to a good depth. This sand consisted of grains that closely duplicated the distribution of grain size in the samples that had been brought back from previous Apollo missions except for the finer mesh sizes. No sand and gravel company had grains that size. We also placed small boulders along the track.

To simulate a vehicle moving on a surface in a 1/6g gravitational field, it is necessary that 5/6 of the weight of the vehicle be removed while retaining the same mass. This was done by placing a large diameter aluminum pipe on a 5-ton truck in such a way that it extended over the side of the truck. There was a cylinder on the boom, which assured that a constant tension was exerted on a cable that was attached to the center of mass of the LRV, an LRV assembled by the branch from leftover parts. The simulation was done by driving the truck and LRV at the same speed while maintaining a constant lateral separation between the two. Complete 1/6g of the vehicle was achieved by suspending each of the four wheel assemblies from the boom by constant tension potentiometers. These potentiometers are suitable for relatively light weights only, and aren't capable of lifting 5/6 of the weight of an LRV plus operators. During these valuable tests, the driver of the LRV was principally John Burch of my branch, an indispensable member of the team. Results from these tests foretold some of the things experienced in the lunar driving, e.g., skids and wheel spin.

After the success of the LRV on Apollo 15, my branch was asked to investigate the possibility of controlling the LRV remotely after the astronauts had returned to Earth. This was because they had learned that over half of the LRV's battery life was left when the astronauts left it on the moon. Remote driving with a time delay appeared to be an area nobody knew anything about. Certainly, we didn't.

We had an old aluminum LRV-size chassis from another lunar project in the lab that we used to make a remotely controlled vehicle. Before assembling it, we decided to get some experience in actually driving a totally enclosed vehicle while seated in the vehicle. We outfitted a commercial Jeep with an LRV navigation system and a video system. The pan-and-tilt camera was mounted on top of the vehicle and its output was sent to the driver who could watch the terrain on TV while he drove in his completely enclosed cab. Additionally, he had access to topographic maps in his cab. Before he started, the driver was given a description of the landmarks he was to find as well as their coordinates. Thus, he drove and navigated. A chase vehicle, equipped with a back-up navigation system, was used to prevent the driver of the Jeep from getting into any serious trouble, such as driving into a ravine. None of the drivers who participated in these trials out at Flagstaff had any trouble driving and navigating to the targets. For various reasons, NASA decided not to equip the LRV for remote control.

I was very lucky to get in on the beginning of space exploration. There were no precedents to follow. This phase lasted about 15 or 20 years after 1960. After that, meetings tended to take the place of work, in-house capability faded and less engineering was done.

It is interesting that a person like myself, and others like me, products of the Depression who had no exposure to science and technology growing up, were able to score notable successes. This was made possible by hard work and common sense. Most of the people involved were pretty average people, mostly GI Bill-educated engineers and technicians from Alabama and neighboring states, who had grown up in a relatively poor era and area.

Lunar Roving Vehicle (LRV) Program Systems Technology Development
by William W. Wales, Jr.

The Systems Technology Development process spans the period starting with the commitment by President John F. Kennedy to land a man on the Moon in 1961 until the Lunar Roving Vehicle (LRV) Project Development contract was awarded to Boeing and General Motors in 1966. The development and operation of the Lunar Roving Vehicle ranks as one of the Center's most successful programs in the history of the Marshall Space Flight Center.

The idea of putting a man on the Moon raised a number of concerns such as how do we provide a shelter for the astronauts while on the lunar surface and how will the men be able to explore the lunar surface. The first thing we looked at was a lunar shelter which would provide long term life support, however this concept required another Saturn V launch vehicle to deliver the shelter and did not address the lunar exploration concern. Next we looked at a combined shelter/exploration vehicle called the Mobile Laboratory (MoLab). Here again this concept required a different launch, from the astronaut delivery vehicle system, since the weight was approximately 12,000 pounds. Once it was decided that the astronauts would have their life support system in a backpack we evolved the exploration system into a Local Scientific Survey Module (LSSM), which could carry two men up to 10-12 miles and return them to the Lunar Lander. This vehicle would weigh approximately 3,000 pounds and could be carried aboard the Lunar Lander or with a separate launch of a smaller launch vehicle system such as the Saturn 1B. With the decision that the mobility system must be capable of being carried to the lunar surface with the astronauts required that the vehicle be much smaller and storable into a compartment on the Lunar Module's landing stage. This concept also allowed for the later landing missions to go to different lunar locations. The ultimate decision to go to the smaller LRV was driven by the overall program cost and LRV weight and system complexity.

The development of the LRV involved numerous organizations developing technology to assure the successful operation and safety of the astronauts while on the lunar surface. Understand that when this project was initiated NASA knew nothing as to the condition of the surface of the moon. We knew that there were many large and small craters on the surface and there was probably a soft sandy-type material that covered the complete surface. Having these concerns required that we investigate various combined wheel and drive designs for the surface rover plus determine the feasibility of using a flying vehicle to perform the exploration. The surface mobility approach required that we look at not only four or six wheel vehicles but track vehicles and even the possibility of driving the vehicle remotely from earth to traverse from one lunar landing site to another. All these factors were evaluated and surface models built along with full scale functional vehicle mockups to evaluate the vehicle design and performance,

MSFC Director Dr. Wernher von Braun test drives the Mobility Test Article (MTA), a developmental vehicle built by the Bendix Corporation to test lunar mobility concepts, June 1966. The data provided by the MTA helped in designing the Lunar Roving Vehicle.

knowing we had to deal with one sixth earth gravity, a very soft surface, and a six second time delay for communications.

Early on in the program we discarded the idea of a remotely driven vehicle, because this concept had more complexities. Also we ruled the need for a multi-wheel (6 or 8) for similar reasons. The Lunar Flying remained as a viable option and contracts were awarded to Bell Helicopter in Niagara Falls, NY to develop designs, which could transport either one or two men to certain points of interest and return them to the Lunar Module. The flying vehicle option was eliminated in the 1966-1967 time period after we received data on surface conditions from the Lunar Orbiter.

General Motors, Santa Barbara, CA and The Bendix Corporation, Ann Arbor, MI manufactured two Mobility Test Articles (MTA) which were full-scale mobility systems. Bendix built a four-wheel vehicle, which had a flat footprint with suspension arms, which could be controlled to clear boulders and better negotiate the Lunar craters. General Motors built a six-wheel concept of the woven wire design and incorporated the articulated chassis. One MTA concept incorporated a Neutater drive and the other had a Harmonic drive. Both vehicles were transported to Yuma, AZ and underwent four months of vigorous testing on different surface conditions. Yuma Proving Ground had about seven different surface textures ranging from boulders to very fine sand.

In conjunction with the MTA tests, smaller (wheel-alone) test articles were built of the LRV size wheels and tested on the various soil conditions at Waterway Experiment Station at Vicksburg, MS. Different soil conditions were simulated in approx. 50 ft. long vats and the various wheel concepts were driven the entire length of the vat. The footprints and surface tension were measured at different speeds and various simulated vehicle weights. These tests were strictly to investigate the wheel to lunar surface characteristics.

While all these tests were being conducted, similar activities were being worked in-house at the MSFC using a vehicle constructed by Teledyne Brown Engineering to look at the man-machine interface to develop some guidelines as to the operational limitations of this type vehicle. A test bed/driving range was built south of Building 4610 and was lined with various obstacles, which closely related to the lunar terrain. The vehicle operator would drive at different speeds with and without the pressurized suit.

These developments all supported and had significant influence of the final design of the Lunar Roving Vehicle as we know it today. These activities occurred over a period of about four years (1963-67) until the program went to flight hardware and the project office was established. A few of the folks involved in this activity are: Lyn Bradford, Harold Trexler, Herb Schafer, Bill Wales, Richard Love, Jerry Peoples and others.

A Real Handyman
by E.L. Schneider

During the hectic days when I was privileged to attend one of the top management weekly meetings in the ninth floor conference room of Building 4200, one of the topics was the costly overrun of the Lunar Rover Project. I can vividly remember Dr. Eberhard Rees' comment, "Give me a Sears Roebuck catalogue, a Volkswagen, and some local off-the-shelf hardware, and I will build you the Lunar Rover."

Saving Project HEAO
by Dr. Fred Speer

In 1971 there was a great deal of joy when Marshall received word that a major new science program was awarded to the Center. This culminated a long period of hard work by many of our scientists, engineers and program planners. Two huge satellites were to be built to explore the universe outside the earth's atmosphere by searching for X-rays, gamma rays and cosmic ray particles and opening a new window to the far reaches of the universe. They were called High Energy Astronomy Observatories or HEAO in short. Each of the satellites would weigh more than 20,000 lbs. Such heavy payloads required a launch initially by Titan rockets and later the Space Shuttle.

This event was far more important than just getting another job for the Center. NASA was in a major transition from the Apollo era to more austere times as we were to find out shortly. The various field centers had to establish their respective roles and missions and became rather competitive with each other. The Marshall Center, well-known and respected for developing the Saturn family of launch vehicles and Skylab, and very busy with the development of the Space Shuttle, saw itself approaching a time without any new launch vehicle on the horizon. New missions had to be found that would be right for our Center, and space science missions such as HEAO were obviously prime candidates.

The first steps were to select the prime contractor, establish a project office at Marshall, and begin the spacecraft design. The schedule was very ambitious, but everything looked good for about two years. Then, in early 1973 we were suddenly and unexpectedly hit by a bombshell. NASA Administrator Fletcher, facing major budget problems for the agency, had decided to cancel our embryonic project. Eberhard Rees called me into his office, told me the bad news and what we had to do following the direction from Headquarters.

Having learned from other setbacks in the past, we were not willing to give up. What developed was a hectic but well-orchestrated game plan to turn this decision around into something less drastic than a cancellation yet responsive to the tight money situation. We began immediately an all out effort that involved all project organizations inside and outside our Center. It paid off, and within a single year we were ready to present a brand new project plan to the administrator that he was willing to approve.

The single most important step was the change from the Titan launch vehicle to the smaller Atlas-Centaur. This brought the satellite weight down to one third and at the same time imposed severe limitations on both its length and diameter. Mission planners know that spacecraft weight translates almost directly into cost; so this was a huge cost reduction.

At the same time NASA Headquarters took on the difficult job of uninviting several of the scientists whose instruments had suddenly become too large and heavy to be carried on the smaller satellite. In a Washington meeting Jesse Mitchell, our boss at Headquarters, compared the situation with a shipwreck. The captain had to toss some of the most valuable cargo overboard in order to save the ship and what was left in it. The elimination of instruments from a science project that had already been formally approved was a very drastic step indeed and might have jeopardized the mission; however, after many painful sessions we finally got the full support of all the HEAO scientists to proceed with this much smaller payload. They launched an effective and wide-ranging campaign to convince Congressional members and staff that high energy astronomy was an endeavor of the highest scientific priority and that the new HEAO program was the best way to accomplish that.

The spacecraft contractor, TRW, had to go through a complete redesign. Rather than developing new tailor-made components for HEAO, they agreed to proceed largely with already existing designs, sometimes called off-the-shelf items, in order to shorten the development time and reduce the cost. Of course, everybody liked one positive aspect of the program change, which was the increase from two to three missions.

At Marshall we started our own soul searching by taking a hard look at the need for the prototype. A prototype is an early version of the spacecraft built for verifying overall systems compatibility and performance before the final flight hardware is built. Since the prototype will not be flown, it can be subjected to harsh tests even exceeding the levels the flight unit will actually see during launch. Although manufactured early, the prototype would be nearly identical to the flight unit. The Marshall engineers had used the prototype concept on all previous programs and considered it one of the cornerstones of their successful design philosophy. Therefore, it was an integral part of the original HEAO Project as well.

Now we had to convince ourselves that giving up the prototype for HEAO would produce the required savings and, more importantly, was an acceptable risk that would not jeopardize mission success. The reasoning was simple; the cost saved would translate into more instruments to be flown and consequently produce more and better science. A few smaller projects in other centers had already demonstrated the viability of the proposed elimination. On the other hand, there was the question of adequate testing. Without test results from a protoflight unit, how could we be confident that the flight unit would survive the actual flight? After a long debate we agreed on appropriate test levels for the flight unit and to go ahead without the prototype. We argued that this was not a manned mission where the life of the crew was the paramount concern, and that new rules for unmanned science missions were appropriate if Marshall wanted to stay in this new and promising business.

This intense and multifaceted restructuring effort saved HEAO. It became a much smaller spacecraft that was essentially common to all three missions. Existing components largely determined the spacecraft design, and non-essential changes were not allowed. A total of 12 science instruments were carried into orbit between 1977 and 1979. After successful completion of all three HEAO missions the Marshall Space Flight Center continued with more and much larger space science missions.

Test Lab at MTF
by Jim Pearson

After acceptance firing the Redstone, Jupiter, Saturn I & Saturn IB, and S-IC's in-house, MSFC Test Lab had the job of building and activating the Mississippi Test Site for S-II and S-IC flight stages. The design started in the early 60's of brick and mortar, transforming a mosquito and snake infested swamp to a test site in four years. Gordon Artley headed a group nicknamed "Alligators" to prepare the facility.

The first test stand activation was the S-II A-1/ one of 2 S-II/ in 1965, which required all facility systems ready: fuel, HP water, air, nitrogen, helium, hydraulics, etc. Propellants, which were supplied by barge from the manufacturing plant in New Orleans by canal, were positioned at the dock for fill and drain. Blast and firex protection were required as barges remained in place during tests. Two barges each were required for a test. Tugboats were used to transport propellant barges from New Orleans as well as the S-II stages which were transported from California thru Panama Canal by ship.

All supporting facilities, shops, offices, warehouses, etc. had to be ready for test operations by North American Aviation builder of the S-II stages. A MSFC Test Lab team for activation was formed to work plans for activation and testing by North American. Bill Grafton headed this team composed of engineers and techs from Test Lab with support from other labs.

North American had prior experience with the S-II battleship firings at Santa Susanna, California using the old Rocketdyne Test Stand blockhouse. MSFC also had a team of experienced people who worked this job.

The S-II-T was the first to arrive at MTF and was installed in the test stand from the barge by contractor teams and MSFC. Test procedures, check out of the engines and systems proceeded using "lessons learned" from Test Lab and S-II battleship - J-2 stands at Canoga.

Many problems found were corrected, which paved the way for flight stages to follow; i.e., thrust beam deflection hit the pre valve, slam arm drop device at start for gimbal. Fill and drain valve tension on a tank later destroyed S-II-T when a vent valve closed (blocking valve) and over pressurized the tank. This occurred near the end of the tests on the S-II-T bird and did not delay the flight test program.

Test Lab in the 1950s
by Jack Waite

I worked in Test Lab in the 1950s. Karl Heimburg was the lab director. Karl was a very dynamic leader and took great pride in meeting all challenges and schedules that appeared, to most people, impossible.

I was special assistant to Heimburg, and shared an office with Bernard Tessman, the Deputy Director, and Fritz Kramer, another special assistant to Heimburg. We worked together to solve many challenging problems that arose during developmental test activities. Some examples are: need for fast response load and pressure devices and jet deflector to withstand full duration missile static test firing (the carbon deflector for Redstone firing lasted 7-11 seconds before burning through). We developed a water film cooling system for the deflector, using large quantities of the high pressure water, a pressure higher than the static pressure of the exhaust gasses, to cool the exhaust deflector during captive testing (only slight discoloring occurred on the steel deflector paint during a full duration test), and a dynamic test stand with a "soft mount" for the vehicle to study the vibration spectrum of the assembled vehicle.

We had in-house capacities to do most anything required to support all component testing, major sub-system and system testing, and full rocket vehicle testing. The Test Lab design group was under Frank Looney. The machine shop, welding and fabrication shop was under the guidance of Fritz Vandersee. The science and engineering labs supported these activities.

I was assigned to coordinate required support activities leading to the captive testing of the Saturn test vehicle. I will give an example highlighting one of these critical tasks:

A new thrust measuring system, which we designed and developed in-house for the modified Saturn test stand, was one of the long lead-time items in our schedule. Three, 50,000 lb. tension/compression load cells were needed. There were only two companies, Cox-Stevens and Baldwin who both supported the railroad companies, that made large load cells, but they made only compression types. Baldwin would not bid on the tension compression load cell.

We had a contract with Cox-Stevens to manufacture three, 50,000 lb. tension compression load cells, with delivery in 18 months. We made several visits to Cox-Stevens during the contract. About two weeks prior to the delivery date, Gene Cataldo and I went to their plant for the final acceptance test. They were testing at 120% of the rated load, and during the fifty cycle the load cell shattered like glass at 60% of the rated load. They had used high-strength brittle tool steel.

We immediately set up a conference call at 1:30 am with Marshall design personnel to start working this critical problem. The team included: Jim Verbal (S&E vehicle design); Gene Cataldo (S&E metallurgy); Frank Looney (test lab design); Fritz Vandersee (test lab machine & fabrication shops); and Jack Waite (test lab, lead for the overall activity). We selected the proper material, 4340 chrome molly steel, chose a tube in order to aid in the heat treatment, did the stress analysis, and completed the design. Harlan Harmon worked the strain gage selection and load cell design. We ended the telecon listing all action items, assigned responsibilities, and established a schedule compatible with the Saturn test schedule. The assignments were as follows:

- Locate the 4340 tubing (Fritz Vandersee)
- Prepare for fabrication (Fritz Vandersee)
- Prepare for heat treatment (Gene Cataldo)
- Final grinding (Fritz Vandersee)
- Install strain gages (Harlan Harmon)
- Locate a facility capable of testing and calibration of the three, 50,000 lb. load cells (Jack Waite)

Everything was started immediately.

The only location found for load testing and calibration was at Georgia Tech University in Atlanta, Georgia. The load cell was completed in four or five days. Harlan Harmon and Jack Waite loaded the load cell in the car, went to Georgia Tech University, completed load testing and calibration successfully, and returned to Huntsville three or four days later. The thrust measuring system was back on schedule.

The Program Development Organization and Activities
by Charles Darwin

The Program Development Core organization included an advanced projects group called the Mission and Payload Planning Office to provide early project leadership and science interface; the Preliminary Design Office for conceptual and preliminary design; the Program Planning Office for engineering cost, project schedules, and program plans; and an Operation Support staff for personnel staffing and accommodation needs. The personnel staffing and dedication was generally excellent and strong emphasis was placed on establishing and maintaining very interactive interfaces with the breadth of Center organizations, with industry, with NASA Headquarters and with other NASA Centers. The staffing has varied from 150 to 250 for the core organization plus 5 to 50 people in focused project groups called Task Teams (the precursor of hardware program offices).

To mention some of the core staff without mentioning all is difficult since many individuals contributed so effectively, but a few are noted here. Directors included Dr. William R. Lucas, Col. James T. Murphy (USAF Ret.), Bob Marshall, Charles Darwin, Jim McMillion, Axel Roth and Jack Bullman. Deputies included O.C. Jean, Jim Downey, Bill Snoddy, and others. Some additional very senior contributors were Bill Sneed, Erich Goerner, Herman Gierow, Dr. Herman Thomason, Bill Rutledge, Jim Jackson, Don Bishop, Joe Hamaker, Ken Fikes, Cecil Gregg, Ron Harris, Gene Austin, Pete Priest, Carmine De Sanctis, and Fred Digesu plus many others. The other senior and junior staff were also very sharp, effective and dedicated. The activity and organizations were supported in many operational aspects through the years with a two-to-three person staff led early on by Al Flynn and in the last many years by Wayne Parks.

Project work included the conceptual and preliminary design focus for many, many projects including: the Space Shuttle, a very major early activity including many challenging and interesting aspects; early Phase A and Phase B Space Station definition studies which were evolved and iterated for years until the approval of the International Space Station; the strategies, conceptual, and design approach to a large observatory program which led to the High Energy Astrophysical Observatory, the Hubble Space Telescope, the Chandra X-Ray Observatory, and future large optical programs; a significant effort in the progression of energy programs such as coal extraction and gasification; solar energy systems, fuel cells, and space based power systems; extensive new launch vehicle options, designs, and technologies including Space Shuttle replacement options, the Shuttle derived heavy lift vehicle options, new heavy lift and reusable vehicle options–all done working with multiple Centers, NASA Headquarters, and many jointly with the Air Force; the effective integration of technical and operational requirements for scientific payloads into the Spacelab, the Space Shuttle and the Space Station; the concept and preliminary definition of the elements leading to Spacelab systems; upper stage program concepts, designs, and technologies leading to the Inertial Upper Stage and multiple upper stage designs and technologies; a program definition and system approaches for microgravity as a productive science for materials and processes; and many manned MARS mission definitions, plans, concepts including the late 1980's definition work for the President Bush administration. This last item started as a small two to three week clandestine activity in a remote building at JSC with one MSFC participant, followed by President Bush's MARS initiative announcement at the Lunar Landing 20th Anniversary Celebration and a picnic on the White House lawn attended by five to six members of MSFC management and then moved on to a three to four month Skunk-Works type definition effort in a dedicated building at MSFC with significant effort also at JSC and other Centers (just one example of the many interesting facets throughout all of the above activities). Many other activities were carried on by Program Development on other payloads, science, transportation systems, spacecraft and related advanced development or technology efforts. The organization was a major contributor to Agency-wide planning on many new programs, Agency and MSFC strategic planning, and approaches for new ways of doing business.

This organization was unprecedented in scope, effectiveness, responsiveness, and products for the resources invested. Comparing with other NASA organizations and government agencies as well as industry organizations of the '70s, '80s and '90s this group was probably the most effective systems organization (very attuned to needs, closely located together, and highly interactive) in government and industry (except for industry 'Skunk-Works' which focused on a single project objective for a few months). This era has passed and the organization was dissolved in 1998 after thirty years. Perhaps change and time or administration objectives and approaches had overtaken this unique capability. There is no evidence of a more dedicated and unique systems preliminary planning and definition organization in or beyond NASA.

The Early History of Program Development
by Wayne Parks

Near the end of the Apollo Program in the late 1960s, attention was turning to "What is next for NASA and especially the Marshall Space Flight Center?" The Marshall Center was know primarily for it's technical expertise in building large scale rocket boosters like the Moon Rocket, and much concern was generated within the Marshall Center and perhaps even the local community over what the future might hold and what this Center would be doing to further the Space Program. There was a strong, growing interest and need for the Center to diversify in terms of future projects and activities. After many meetings and much discussion by Center management over this potential dilemma, Dr. von Braun announced that a new and major organization would be established at Marshall to initiate in a more serious and formal manner planning for the future. Thus, the Program Development organization was initiated December 16, 1968.

Following the announcement of the new organization, over the next several months efforts were concentrated on staffing this new organization. First, the Advanced Systems Office (containing 97 people), in the Science and Engineering Directorate (S&E) was the first major organization consolidated to become the nucleus of this new organization. Next, several groups that were co-located from S&E with the Advanced Systems Office were converted over permanently to the new organization. So, Program Development received a jump-start with staffing from these former organizations and was soon well on it's it way to becoming a premier organization at the Center. Program Development was staffed with a wide variety of technical, scientific, and business professional personnel. The idea behind this diversity in staffing was that this new organization would have the capability to do well-balanced technical and business management planning for new projects and activities without disturbing or taking away technical resources and support from other already ongoing projects and activities at the Center.

The Program Development Directorate has enjoyed a long and rich history of operation including very few organizational changes and has had a total of seven directors during its organizational life. Program Development has the distinction of being one of, if not the, longest living organization at MSFC (almost thirty-one years). The first Director appointed by Dr. von Braun was Dr. William R. Lucas, who was a former Director of the Propulsion & Vehicle Engineering laboratory in S&E. His successor was the Deputy Director, Program Development, James T. Murphy. After several years service as Director, PD, Mr. Murphy was selected to head the Institutional and Program Support Office at MSFC. Mr. Murphy was followed by Bob Marshall, a former Program Development employee who had been Director, Preliminary Design Office. Following the Challenger tragedy in 1986, Mr. Marshall was selected as Director of the Shuttle Projects Office. Another former Program Development employee, Charles R. Darwin became Director, PD. Upon Mr. Darwin's retirement late in 1993, Jim McMillion became Director, PD. Upon his transfer to Director, S&E, his Deputy, Axel Roth became Director PD. With recent retirements and changes at Marshall, Mr. Roth was selected to replace Richard Marmon as Director, Flight Projects Office. Following this assignment Jack Bullman was appointed Acting Director, Program Development with responsibility for bringing the organization to appropriate closure during the current major reorganization of MSFC.

Program Development's primary mission was to create new business opportunities for NASA and the Marshall Center. Early in Program Development's history, Project Task Teams were established to guide projects efforts during the early formulation, preliminary design, and definition phases. Technical and scientific personnel were drawn from S&E, Program Development, and other Center organizations as appropriate to staff these teams and to carry the program development definition for project and activities to maturity. At the end of the Preliminary Design and Definition Phase (Phase B), when a project or activity received approval as a new start, the effort was then transitioned to either a new Project Office or an existing one as appropriate. The cadre of personnel serving on the Task Team became the nucleus of the new Project Office or the new organization responsible for the Hardware Design and development phases (Phase C-D). Looking back over the extensive period of time that Program Development was in existence, it has been a highly successful organization. Over the years, Program Development has given birth to many current and past projects and activities for the Center. Some of the most significant ones include the following: the Spacelab, the Space Shuttle, High Energy Astrophysical Observatory (HEAO), the Hubble Telescope, the Microgravity Science Program, the design of the International Space Station, and many other similar significant projects and activities.

A considerable number of people at Marshall have served in various capacities in Program Development on different project activities.

Program Development had played a major role in the Center's history and made many valuable contributions to NASA's efforts in the continued exploration of Space.

Program Development claims unique functions within NASA...the Engineering Cost Group is known Agency wide for it's expertise in costing small and large-scale space projects.

The Engineering Cost Office in Program Development is largely responsible for providing the economic justification for initiating the Space Shuttle as a major transportation system for the Agency.

The Preliminary Design Office has laid the technical foundation for many new projects and challenging space system over the years.

One bit of trivia...it was never intended that Program Development design, build and fly flight hardware. However, in the history of PD one such occasion did occur. The Laser-ranging Geodetic Survey (LAGEOS) Satellite was designed, built and flown as a flight project while never leaving the Program Development organization.

The Tank and Skylab
by Charles D. Stocks

I can remember when I joined the German team under Dr. Wernher von Braun in 1956. One of our problems in those early days was how to determine the true take-off weight of the old Redstone rocket. My first task was to calibrate the best load cells we could find on the market. We developed a system of dead weights and ran tests until we could come up with a system that would accomplish this. It was a fine program but the Redstone rocket was never fired without full tanks. I even remember the time when the Army had just been restricted to a 250-mile range because the Redstone was declared to be a tactical missile. It was an accident that the destruct system malfunctioned one time and the Redstone flew a little over nine thousand miles. It came down somewhere on the other side of Africa.

In 1960 when NASA was formed, the Saturn was already on the drawing board but because of the time constraints and engineering requirements it was decided to start with tanks that could be made from a proven design and with the welding jigs already on hand. This allowed the existing hardware to be used while the development of the new engines and tanks could proceed as a parallel effort. During this time we could use the Redstone rocket for the early flights. Soon after this the Russians orbited the Sputnik and we had to break out our Explorer satellite. We would have been able to orbit it much earlier except the missile people didn't want the USSR to know we had that capability. The Explorer sat in a warehouse under guard for about four years before we got permission to launch it. We did launch it in 90 days, while at the same time were working on a manned flight vehicle. The Mercury flight system was the beginning.

Of course in those days we didn't know what would happen when man ventured beyond our atmosphere. To prove that man could survive in this unknown environment, monkeys were trained in a centrifuge to push a button whenever a red light flashed. The idea was to find out if the monkey was still alive at various points during the sub-orbital flight. It was not necessary to stimulate the monkey at all. When the rocket lit off and started to climb towards space, he began to punch the button rapidly and continuously for the whole flight.

After the first flights into this new environment, NASA asked for suggestions for things that could be done in space. Shortly after the first Gemini walk in space, Charlie Cooper and myself examined the tapes of the walk because Charlie thought that we could develop a training program for spacewalks by going

Astronaut Ed Gibson in MSFC's Neutral Buoyancy Simulator during Skylab Extra Vehicular Activity training. This view shows a mockup of the Apollo Telescope Mount Transfer Work Station.

underwater in scuba gear. He told me that he would make up a list of what we would need if I would order it. I sent in the order for the first two sets of scuba diving equipment that NASA had on hand. I made a mistake though, I forgot to order boots and gloves. Our first tank was eight feet in diameter and eight feet deep that had been used for exploding bridge wire forming of metal parts underwater. It was cold. Charlie was teaching me to use the scuba equipment and he had a good idea for a restraint that the astronauts could use. We tried it out and it worked pretty well, and it was a lot better than the tripod arrangement that was being proposed at that time. We built one but could not get it accepted. Years later I saw the same idea used on a Sears clamping tool. Just too early I guess.

We had been working with the gear for several months, when Bob Schwinghamer noticed the suits drying out one day and commented that it was nice that we brought our own gear out to MSFC to try out an idea. We told him it was NASA gear and he became concerned as he thought that the lab director should have been told about what we proposed before going out and buying it. Bob took about six months before the right opportunity presented itself for him to tell the director what we had started. The upshot was that the director approved and we took over the explosive forming tank and made it our space tool development facility.

Money was tight in those days so we found a Saturn corrugated section that had been damaged by a forklift at the cape. We had it shipped back up here and put about eight men on it with saber saws to cut it down to ten feet high. Then we had the shop fabricate a pipe frame and the fabric shop to make us a fiberglass cover for the top. The section was 33 feet in diameter, which just fit over the 25 foot diameter by 15 foot deep in-ground tank we took over. We bought a Sears swimming pool filter and tapped into an adjacent steam line for heat. It took the chemists almost a year to get the correct balance of chlorine and acid to keep the algae at an acceptable level. We had the control room set up in a trailer next to the pool.

We got qualified to use a high altitude suit underwater and set out to prove our concepts of operations for the space environment underwater. We developed safety procedures to get breathable air to the man in the suits first and then to bring him back to the surface. It had to be this way as it took a 150 pounds of lead weights to make the suit neutral in the water. In those first days, the S-IVB stage could not be used as a station in space because it had to carry propellants and could only be set up as a habitat after the fuel had been expended. The Saturn was set up for Moon missions, but without the LM on board there was space for material and supplies in the forward nose section. A tunnel could be placed above the hatch in the bulkhead of the LOX tank. This concept required the astronaut to make up a rubber boot with 120 bolts to be removed and the bellows like boot then made up to the bulkhead hatch. This makeshift airlock was 10 feet long and about four feet in diameter.

We picked up a tube of the right size and length placed it in the water with a hatch at the end of it. We used several types of tools, setting up time lines. We found that the best tool was an air tool we used to simulate an electric multipurpose tool system we were thinking about. Had to use air, as electrical tools don't work too well in water. It was by far the best way to remove all of those bolts. We built our own underwater movie camera system and set up some underwater speakers for communication with the divers and picked up a video recorder for some TV cameras we had. All in all we had a pretty safe system with qualified people to run it. The only problem was that Houston wanted to have the Neutral Buoyancy training facility in Texas.

We had one of the astronauts, Alan Bean, Apollo 12 moonwalker, try our facility. He got into the only suit we had that was small enough for him, and the first thing he had trouble with was a hole in one of his gloves. We got him out and replaced the glove and put him back into the tube, where the suit ripped open under his arm. I was controller that time and while I was hitting the claxon for the safety divers, I was seeing in my mind's eye the headlines of the *Huntsville Times*, "Astronaut drowns at MSFC." Alan was very cool and asked if the divers could hear him. I told him that they could so he asked them to keep his head up as he had air in the helmet. The divers kept his head up and got him out OK. We didn't have any more suits for him, so he went underwater in regular scuba gear and worked the various tools for his own evaluation. The beginnings we made here led to the much larger neutral buoyancy facility with a tank 40 feet deep and 75 feet in diameter.

It was interesting how the large tank was acquired. All we had was Directors Discretionary funds available at the time. We were not allowed to construct a new building for the tank, so a leak was discovered under the model and prototype building. This leak required about eight feet of concrete under the floor to repair. Then we were able to fund the purchase of the steel for the tank and we used all of the welders we had to join the steel sections together. The last of the available money went for facility module and the water conditioning equipment. We made a first class facility for a small amount of money and lots of hard work. This facility was the best until it was decommissioned. Houston had built a similar facility and was more politically savvy than we were. We were the first and the best and the facility should have been made a historical landmark.

We had a lot of fun setting up the recovery team that went to Australia to recover the parts of Skylab when it re-entered the atmosphere. The team was in place before it got permission to seek pieces of the vehicle. Liability was a paramount issue and the State Department had to be on top of everything so the team had to sit in Sydney for two days before beginning the search. NASA people were very concerned about the film safe as it was made of lead, which made it quite heavy. It was not found. The team flew to a central site and advertised for people to search the area. About 40 people from the outback responded. One of them found the only casualty reported from the re-entry pieces. When one of the glass wrapped tanks was turned over they found a flattened out jackrabbit underneath. It probably never knew what hit him.

One of the local residents drove in and told how she had heard this swishing sound in the middle of the night. When she got up the next morning she found this 20 foot long jagged piece of metal in her farmyard. It was a part of Skylab. Then there was the fisherman who started to raise some cain because a large chunk of something had just barely missed his boat. It sent up a huge cloud of steam and splashed water all over him and his crew. This man became rather forgetful though when he was asked just exactly where he was at the time. He had noticed that there was a policeman with the team and it appeared that he could have been in a no fishing area.

The bunch of people searching left the job suddenly after three days. It seems that one of them had hit on some gold somewhere, so everyone hightailed it to the strike. Then there was the junkyard dealer that had bought up a whole trailer full of old missile parts and bits. Our team offered to check out the bits for him but he would have no part of that. He moved the trailer inside of his fence and chased everyone away. While the team was negotiating with the junk dealer, they got some good pictures of a string of one-room huts across the street. It seems that each room had a girl installed in it and was available for pleasure. I never did understand why they got so many shots of these girls standing in front of this building. We did discover quite a few bits of Skylab, which we were able to make a quick check on with a radiation detector. Because Skylab had been in space so long, it had a much higher level of radioactivity than the other regular missile parts from the test ranges in the desert. Then if it passed this first test it was then examined to determine where it had been installed on Skylab.

I have been a sort of cartoonist over the years and have quite a collection of goofs that have happened through the years. The crew on the accelerator in 4605, the Light Gas Gun in 4612, the Ukrainian Electron Beam tool, the light weight external tank, the support activities by various contractors, the traffic habits of some of our people, lots of retirement roasting and many of our key people with wise sayings. It has been fun to have been a part of history for all of the years. I did not want to retire when I did but was made to feel that I was no longer useful, so after 47 years I hung up my tools and took on other projects.

Memories
by James W. Thomas Jr.

I started my NASA career in June 1962 with the Marshall Space Flight Center (MSFC) as a Project Engineer in the Engine Program Office on the 1.5 million pound thrust M-1 LOX/Hydrogen Liquid Propellant Rocket Engine. I was located in the WWII Cement Block Building 4610, and 40 years later that building is still in good shape and being used by the Structures and Propulsion Division of the Science and Engineering Directorate.

There were six engineers and two secretaries that shared one not too large room on the first floor of 4610. It was hard to hold a telephone conversation with one of our aerospace contractors or another NASA Center due to the crowded conditions, but we did not let that interfere with our daily work-load and activities. We had the "Can Do" attitude and the desire, determination and commitment to the Apollo Lunar Landing Program, which was very evident in the daily activities of our group of engineers and secretaries. We were proud to have been selected to be a part of NASA-MSFC and the new and elite group of young aerospace engineers that would develop the Saturn I, IB and V launch vehicles capable of taking the United States to the Moon. We accepted President John F. Kennedy's challenge of developing a space launch vehicle capable of landing men to the Moon and return them safely to Earth in this decade.

The Air Force's Rocket Propulsion Laboratory at Edwards Air Force Base had started the development of the F-1 Engine in 1958.

When NASA received the approval to be the government agency to develop the Launch Vehicle for the Lunar Landing Program, NASA-MSFC inherited the F-1 engine.

I supported the Program Managers in the development of the RL-10 and J-2 LOX/Hydrogen Engine Systems that would be used as upper stages of the Saturn I, IB. NASA Headquarters established a Technology Propulsion Team to study advanced propulsion systems for potential upgrade of the Saturn V Vehicle and future Space Vehicle Propulsion Systems. The test and analytical data derived from the F-1, M-1, J-2 and RL-10 Engine Programs convinced the Propulsion Technology Team that a Large Thrust and Higher Performance LOX/Hydrogen Engine would be needed for a first-stage application for large payloads in Low Earth Orbit with a two-stage vehicle and be cost effective. Through the joint efforts of the NASA Propulsion Technology Team, MSFC and RPL, the XLR-129 was initiated as an experimental LOX/Hydrogen Engine System. The XLR-129 Engine Program, even with its development problems, was very successful and provided excellent technology and data. It demonstrated that a new High Performance LOX/Hydrogen Engine System was doable and paved the way for the development of the High Performance Space Shuttle Main Engine (SSME) currently flying on the Orbiter.

After the successful landing of Apollo 11 on the Moon in July 1969, the U.S. announced that it would develop a new Space Shuttle that could be used by both the Air Force for its military missions and NASA for its space missions. The Air Force and NASA were in competition for the responsibility to develop the next generation vehicle that was required to be reusable, low cost and capable of putting 50,000 pounds of payload in Low-Earth Orbit.

After several years of studying many different vehicle configurations, NASA received the direction to develop the Reusable Space Shuttle Vehicle. NASA Headquarters selected JSC to be the Center responsible for the program management of the new Space Shuttle vehicle. MSFC had become the Structures and Propulsion Center for Space Vehicles during the Apollo Program, and MSFC engineers were assigned the responsibility to support JSC in the design of the structures and propulsion systems for the new Shuttle. In late 1969, I became the MSFC propulsion member assigned to support JSC in the engineering and cost studies for the new shuttle. The selected configuration was a two-stage vehicle with a vertical take-off and horizontal-landing. The Delta-Wing Booster was a huge 12-Engine vehicle and the Delta-Wing Orbiter was a somewhat smaller 2-engine vehicle. Both would use a new high performance LOX/Hydrogen Engine. The payload requirement remained at 50,000 pounds with a payload bay of 15 feet in diameter and 60 feet long. The team would rely heavily on the experience of reuse, maintenance and turnaround from the aircraft industry, to lower the operational costs. The large size of the two-stage vehicle, the complexity of the propulsion system with a 12 engine cluster of rocket engines, pushed the costs for the Design, Development, Test and Engineering (DDT&E) and first six flights, far in excess of the budget guide lines given to NASA from the Office of Management and Budget (OMB).

Based on the final budget guidelines from OMB in 1970-71, the Space Shuttle vehicle configuration changed extensively from those we started out with in 1969. After three years of studying many different configurations and the budget guideline, it was determined that the most cost effective reusable space shuttle vehicle would be a one-and-one-half-stage vehicle. A 3-engine orbiter would use the new SSME LOX/Hydrogen Engine that had already been started by MSFC, and the SSME engines would burn from lift-off to orbit eight and one-half minutes. To reduce costs and lower the Orbiter weight, the jet engines were removed giving the Orbiter a one-time capability to land. The size of the Orbiter with a 15 feet by 60 feet Payload Bay, dictated that the LOX/hydrogen propellant tanks be removed from the Orbiter and the Expendable External Tank Configuration was born. The structure and recovery system required to return the External Tank from Orbit altitude with its weight and configuration and land it in the Ocean with minimal to no damage, would produce a tank so heavy, that the lift-off loads would far exceed the thrust capability of a three-engine Orbiter and two strap-on SRMs. The boost-stage became a two strap-on SRM configuration and each SRM would produce 3.3 million pounds of thrust and burn for a little over two minutes, from lift-off to separation, creating a one-and-one-half stage vehicle.

Using the technology from the Army's parachute system used to land the large battle field tanks on land, a new and larger parachute system was developed to land each SRM with an inert weight of approximately 155,000 lbs. into the ocean with minimal to no damage, so they would be reusable.

There were five SRM companies bidding on the Space Shuttle SRMs. Three were located in California and two in Utah. At the time, the Shuttle was to be launched from KSC, so the transportation study evaluated rail transportation from California and Utah to Florida. The study showed that California had the limiting rail overpass height and in order to be fair to all of the SRM companies bidding on the SRM Program the motor case diameter was base-lined at 146 inches. MSFC's preliminary studies showed that the aft segment with the nozzle, less aft exit cone, would weigh between 300,000 and 350,000 pounds. A loaded 350,000-pound segment would require special handling rings and supporting chocks during the handling and processing of each motor segment at the SRM Plant. The handling rings would also be used to support the loaded segment during rail shipment. A protective cover was required to be placed over the handling rings, the loaded motor segment and attached to the bed of the rail car. Since the motor case diameter was the variable in the equation, its diameter was determined by the height of the lowest overpass the train would be required to go under from California or Utah to Florida.

Following three Shuttle flights and the successful Return-to-Flight, I retired from NASA in April 1989. I was proud to be a part of the Saturn V/Apollo Lunar Landing and Space Shuttle programs. In 1958, I was fortunate to be in the right place at the right time, to be given the opportunity to start my aerospace career on the U.S.'s first Ballistic Missile System that would prepare me for my aerospace propulsion career with NASA. I would like to believe that my efforts and contributions to America's space program has helped provide a better Earth for people to live on.

Early MSFC Recollections
by Dale Johnson

I came to MSFC in May of 1966 directly out of college from the University of Michigan, where I majored in meteorology. O.E. Smith was looking for a fresh-out atmospheric type to replace retiring climatologist J.W. Smith, and I turned out to be the one. I was told that I would basically work in the terrestrial environment area and deal with modeling the thermodynamic properties of the atmosphere. The first day on the job I reported to a processing-in office on the first floor of building 4202. There, I was given a paper that listed my new job responsibilities. To my amazement, the job category and responsibilities did not match at all what I was told I would be doing here. Listed were various tasks dealing with the space environment; i.e., magnetosphere, solar, thermosphere, etc. They were subjects that I knew little about (at the time). I was horrified! However, after I reported to Bill Vaughan and O.E. Smith, I did end up doing atmospheric thermodynamic modeling work. I was later told that I had to be hired-in under the space environment branch, as a veteran (George West) had priority over me in hiring and he had to be hired in first within this job category with me to follow. My job category did change back to being terrestrial after being here a few months; but I was a little apprehensive initially!

It was a pleasure working for the von Braun team in those early days in Dr Geissler's Aero-Astro Laboratory. I even became a substitute weather forecaster for the Saturn engine static test firings at Test Lab, whenever our regular forecaster Joe Sloan was off. We had to send up radiosonde balloons the day prior to the firing to measure the vertical temperature and wind profiles over the arsenal. Then I would brief Dr. Heimburg and Dr. Tessman as to what I thought the temperature and wind profiles would be on the following day. As they didn't want the maximum sound propagation from the engine test to be reflected down into the Huntsville residential and commercial areas. They would then make a decision to test or not, based on my forecast. It was quite exciting to apply what I learned at college directly to my job here at Marshall.

I was also initially involved in Saturn Flight Evaluation work as well. After each Saturn I or V launch I would obtain the surface and aloft weather measurements taken at the Cape (provided through Comp Lab's Paul Harness and Wade Batts), and would then summarize the weather at L-0 for the MSFC flight evaluation team. I would construct the atmospheric and wind L-0 final vertical profile for engineering use both here at MSFC and JSC in trajectory and other analyses. Things were very active in those days. It was truly a blessing and a dream come true for me to be a small part of the space team here at Marshall.

Von Braun Moves to NASA HQ
By J.N. Foster

After Apollo 11, the 1st Manned Lunar Landing, Dr. von Braun took time off and traveled to the Caribbean. Tom Paine, NASA Administrator, came to MSFC and confirmed the rumor that Dr. von Braun was transferring to NASA Headquarters as the Deputy Associate Administrator. There was great shock and awe around the Marshall Center!

Wernher returned with a full beard and confirmed to all hands that he was moving to Headquarters. He wished to get out from under the major workload of managing the large Marshall Center, 7,000 employees plus 100,000 contractors, and he wanted time to think about the future. Wernher asked Frank Williams and me to help him form a new Headquarters Planning Element.

Tom Paine requested Wernher's office define the agency's program for the year 2000, 31 years hence. Wernher's efforts focused on two elements: 1) Defining shuttle configuration, cost benefit studies, and testimony before Congress. This effort helped in obtaining 1st funding for Shuttle hardware development. 2) Developing plan inputs from Senior Headquarters personnel, a round robin of the NASA Centers, and discussions with industry and government leaders. These meetings produced a plethora of ideas for a tapestry of future space flight.

Administrator Paine called for a conference to help pull the information together. The conference took place over a 4-day weekend (June 11-14, 1970) at Wallops Flight Center. Tom Paine and Wernher developed an attendance list of the top NASA futurists and Tom Paine invited Arthur Clark, the science fiction writer, to be the keynote speaker. Dr. Clark provided a challenging kick off to the conference, but he was quiet during most of the meeting. He was in awe in the presence of the most renowned, real (vs. imaginary) space pioneers.

The attendees made presentations about the way they saw the future developing.

Prior to MSFC Director, Dr. von Braun's transfer to NASA Headquarters where he had been appointed Deputy Associate Administrator for Planning, he was honored during a series of events recognizing his contribution to the space effort during his career in Huntsville, Alabama. In this photo at the Madison County Courthouse, Dr. von Braun is shown seated next to his wife, Maria, as U.S. Senator John Sparkman comments on his career in Huntsville, where he worked for both the Army and NASA. February 24, 1970.

Dr. von Braun presented the "Orion" vehicle, a vehicle capable of interplanetary flight powered by nuclear propulsion. A pressure plate at the rear had an orifice that expelled aspirin-sized pellets in a continuous stream. These pellets passed a laser beam at a precise distance from the pressure plate, igniting a continuing stream of nuclear explosions that propelled the vehicle to extreme velocities.

After the conference, a plan was produced and Tom Paine took it to President Nixon, who was up to his eyeballs in Watergate. I paraphrase Tom Paine: "NASA's marching orders are to continue working on assigned programs and to not bother him with future programs". All copies of the plan were sealed.

A few weeks later Tom Paine announced his departure from NASA. (Twenty-plus years later, first President Bush appointed Tom Paine to chair a Blue Ribbon Committee to develop a long range Space Plan. This committee published a plan that looks exactly like I remember the Paine/von Braun plan of 1970!)

Wernher announced his retirement a few months later, and Eberhard Rees offered me a job back at MSFC as Assistant Deputy Director Management.

The Headquarters period allowed me to become better acquainted with a remarkable individual, Wernher, the man, the world's greatest space pioneer. Many people who were not in Headquarters comment on the success/failure of the Washington years. My view is that Washington was a success for Wernher since the Shuttle was initiated, and a plan was developed that still resonates for the future, even though it is under the Paine committee's aegis.

My time in Washington with Wernher and the extended period at MSFC, where I had the opportunity to put my knowledge and experience to use, were clearly the highlights of my career. *How I loved that man!*

Skylab Mission Adversities Overcome and MSFC's Drop Tower
by Charles C. Wood

Skylab can be considered the first United States space station. Extensive use of hardware and technology developed in the Mercury, Gemini and Apollo programs was made in its development. The Skylab consisted of the main living and work areas called the Orbital Workshop; an S-IVB stage with propulsion hardware replaced with living quarters, experimental equipment, food lockers, ventilation and passive thermal control system, solar arrays, etc.; a docking adapter; an air lock module with life support and environmental control systems; the Apollo Telescope Mount Solar Astronomy experiments; and a Saturn V Instrument Unit. Skylab, launched from Cape Kennedy by the Saturn V rocket, was placed in approximately a 200 nautical mile circular earth orbit.

During the journey between earth and earth orbit a cable tray cover on the Orbital Workshop exterior surface dislodged and resulted in irreparable damage to the electric power generating capability and to the passive thermal control system on the Orbital Workshop. On arrival of Skylab in orbit the extent of hardware damage became apparent, as did the future capability of the Skylab unless immediate actions were initiated.

The most pressing requirement was to prevent, if possible, the environment within the station and the temperature of special hardware from reaching unacceptable extremes—both hot and cold—which would destroy food, experimental hardware, and other equipment. The longer term requirement was to redesign the thermal control system considering hardware which could be quickly designed, constructed, tested on earth and carried into space aboard subsequent planned manned flights and then to be installed external to the Orbital Workshop by astronauts during EVA operations. Also it was necessary that electrical power budgets be revised for compatibility with power generating capabilities. The various operations were required 24 hours per day for multiple days. Portable trailers with sleeping provisions were delivered to the Center to accommodate personnel as exhaustion overtook them.

By monitoring temperatures on the Skylab and frequently changing orientation of various locations of Skylab with respect to the sun and earth, it was possible to maintain acceptable environments. Design, construction and testing of hardware, which could be installed on the workshop by astronauts, continued in parallel. Subsequently, the planned Skylab repair kits and astronauts were delivered to orbit and the repair kits installed during EVA operations. Skylab became habitable and was a productive home for astronaut crews for several months.

To acquire critically needed information to support on-going reduced gravity fluid mechanics analyses for the Saturn V program, Marshall Space Flight Center personnel designed, constructed and operated a 400-foot high drop tower. At this time in the mid-1960s a limited understanding of fluid behavior in a low gravity environment existed and only two significant drop towers existed within the United States which could contribute to the virtual data void. Those two facilities were both 100 feet or less in height, the actual test time was between two and three seconds in duration and the model sizes were quite small.

The MSFC facility was located in one corner of the 400-foot tall dynamic test stand. Vertical rails guided the drop package from the 400-foot elevation level to a deceleration chamber located at ground level. The deceleration chamber was a large round tube approximately 30 feet in length, and had adjustable orifices to control the rate of escaping compressed air as the drop package entered the deceleration chamber. The deceleration device is a unique feature of this facility since the other two facilities at this time experienced significant dollar cost per drop for drop package deceleration.

The actual test time was between four and five seconds and model sizes were significantly larger than models for the two other facilities, although they were not large. The model experiments were located internal to a large drop shield and the combined weight was between 2,100 and 3,000 pounds.

The drop tower was used extensively and contributed much to understanding zero and reduced gravity fluid mechanics problems to be experienced by the Saturn V Apollo program. The facility has been used subsequently for other research programs.

The Sky is Falling
by Dr. Ronald K. Olsen

No, Chicken Little, the sky is not falling. Skylab is falling. One of NASA's greatest scientific endeavors was coming to an end.

Skylab was 86 feet long and weighed 85 tons. It was lifted into orbit by a Saturn V moon rocket. The Marshall Space Flight Center made major contributions to both the rocket, and Skylab. It was launched on May 14, 1973, but had immediate serious problems. The solar panels did not deploy correctly, resulting in a substantial loss of power. The further result was that the interior of Skylab was too hot for human endurance.

The first manned crew, launched May 25, erected a sunshade, which dropped the

NASA Conference on the Year 2000 – June 11-14, 1970 – Wallops Station, Virginia. Front row: Edward Schmidt, Special Assistant to the Administrator; Dale D. Myers, Associate Administrator, Manned Space Flight; Jerry Truszinsky, Associate Administrator, Tracking & Data; D.D. Wyatt, Assistant Administrator, Planning; George M. Low, Deputy Administrator; Wernher von Braun, Deputy Associate Administrator; Thomas Paine, Administrator; John E. Naugle, Associate Administrator, Science & Applications; Oran W. Nicks, Acting Associate Administrator, Research; Arthur C. Clark, Keynoter & Science Fiction Writer; Willis Shapley, Associate Deputy Administrator; Homer Newell, Associate Administrator; Clare Farley, Manager, Headquarters Administrator. Back row: Neil Armstrong, Astronaut, Robert L. Kreiger, Director, Wallops Station; Robert Jastrow, Director, Goddard Space Science Center; Bruce Lundin, Director, Lewis Research Center; Charles W. Mathews, Manager, Gemini Program; Robert R. Gilruth, Director, Johnson Space Center; Abraham Spinak, Deputy Director, Wallops Station; J.N. Foster, Deputy Assistant Administrator, Planning; Ray Kline, Assistant Administrator, Administration; William H. Pickering, Director, Jet Propulsion Laboratory; Julian W. Scheer, Director, Public Affairs.

temperature to reasonable levels. A spacewalk deployed the solar panels and this crew spent 28 days in orbit. They returned June 22. The second crew launched July 28 exceeded their objectives by 150%, spent 59 days in orbit and returned September 25, The third crew, launched November 16, spent 84 days in orbit and returned February 8, 1975. The last command to shut down was sent to Skylab on February 9, 1975.

Many months later, the Skylab orbit was decaying. There was no longer any control of the gigantic laboratory, and it was so large that large pieces of it were expected to strike the earth's surface. There was much speculation as to where it would come down.

Richard Smith, a former deputy director of the Marshall Space Flight Center, was now an associate administrator at NASA Headquarters in Washington. He was given the impossible assignment of making sure no harm came from the return of Skylab.

During this period, an organization of space fans, calling themselves the Orion Group, claimed to have discovered pieces of Skylab that had fallen to earth. They brought these articles to the Marshall Space Flight Center. They were directed to my office, the Regional Inspector General.

I recognized the material as chaff, the metallic substance released by combat aircraft to confuse enemy radar. I tried to explain this to the Orion Group, but they would not be convinced that they had not found part of Skylab. To further convince them, I sent them to James Kingsbury, the Director of Science and Engineering at the Marshall Space Flight Center. He confirmed that the material was chaff, and the Orion Group left.

The Orion Group was still not satisfied. Even though Skylab was still in orbit, they claimed that their material may have been brought down by space aliens and that this fact was being covered up by Kingsbury and me. They reported this to two television stations in Nashville.

As news of Skylab was pretty hot at the time, both stations sent news crews to the Marshall Space flight center. One crew, which had come in a van, came to interview me, and the other crew, which had come in a helicopter, planned to interview Kingsbury. The van parked in the north parking lot of the administration building, and later the helicopter came down in a vacant part of the same lot.

Prior to any interviews, Kingsbury and I conferred on the telephone as to how to handle this matter. We did not want to appear to be covering up anything, but space aliens were too ridiculous to even discuss with the media.

After explaining that the material was chaff, I pointed out the helicopter, which could be seen from my window, denied that there were any space aliens in it, but said that people who got out of it were interested in the debris. This was all it took and the TV crew, which had come in the van, unknowing that another TV crew was interviewing Kingsbury, was out the door heading for the helicopter.

Meanwhile, Kingsbury told the TV crew, which had interviewed him, that while there were no space aliens, some people who were interested in the debris were taking an unusual interest in their helicopter. They were out the door heading for the parking lot.

Kingsbury and I were greatly amused watching the two different crews taping and interviewing each other. For some reason, there was no publicity of the incident by either station.

Skylab came down in the outback of Australia and Richard Smith was promoted and named the Director of the Kennedy Space Center.

Nice Launch, Marshall!
by George Harsh

In the mid-1970s the Test lab conducted several Space Shuttle related tests, one of, which was an acoustical energy test of a scale model of the Shuttle on the Kennedy Space Center launch pad to be fired at Test Stand 116. Conducting a test on time was not easy working with many red line cutoffs resulting in test schedule changes by the hour. The NASA Center Directors had scheduled a meeting at Marshall, and as part of the agenda they planned to witness a firing of the model of the Solid Rocket Booster on the simulate KSC launch pad. The firing was scheduled for 5:00 pm. However, around 3:00 pm we were informed by Center director Rocco Petrone, who was familiar with our lack of on-time testing, had decided that the Center Directors would pass on the test and close the day at the Redstone officers Club. But we also received word that the Deputy Director of KSC and two of his engineers declined the Officers Club invitation and decided to watch the test no matter what time it occurred, since it was a test of their launch pad. When this word was received at Test Lab, some of the older technicians became interested because they had observed that if "company" was coming, there was a good chance something interesting might happen.

Amazingly enough, we did fire on time at about 5:00 pm. Upon ignition of the Solid Rocket Motor, the head end of one of the motors blew out, severing the only connection to Test Stand 116. The motor danced off the launch pad and rose majestically into the late afternoon sky, flying end over end, doing loops and heading toward Huntsville. Since it was burning from both ends, I knew it wouldn't go far. It went downrange and or up range for about one mile at a maximum altitude of 300 to 400 feet and fell harmlessly among some pine trees. That's when we heard the KSC guys say, "Nice Launch, Marshall!"

This resulted in the first and last rocket launch from Marshall. That event plus all of the other strange happenings at TS 116 caused Earl Choate, an instrumentation engineer, to write Ode to 116.

Ode to One Sixteen
by Earl Choate

Keep your eyes wide open and your ears tuned keen
If you plan to venture near 116.
A spirit lives there, perhaps an Indian chief,
With his mind set firm on dealing out grief.
With his C.T.F. or (C.F.M.)
Your life may be at the old chief's whim.
The basement leaks, the hydrogen too
And fire balls burn your wires clean thru.
The lightning streaks, the transducers fail.
You'll need to wait, for more in the mail.
The sequencer skipped the whole darn test,
Hydrogen embrittlement ruins the rest.
There never was nothing like 116
With hydrogen and oxygen that burn so clean
And a phantom spirit that lurks in the night
Setting solid motors on unscheduled flight
If you plan to go near 116
Keep your eyes wide open
And your ears tuned keen.

Measurement Acquisition Post 1976
by Houston Hammack

In 1972, the need for a distributed measurement acquisition system to support Space Shuttle testing was obvious. Dedicated systems for each test location would not be budgetarily feasible. I was appointed chief engineer to develop requirements and specifications for the system, which was presented to an evaluation board. In 1973, NASA Headquarters and Congress approved $5M for the system. A contract was awarded to AVCO Electronics in 1974 to fabricate the system from commercially available components and develop the software programs for quasi real-time numerical and graphical measurement presentation to analysts. The system was delivered in early 1977 and operationally accepted in mid-1977. The control and processing programs calibrated the system, acquired measurements, and presented the data graphically during test conduct.

The system was a new concept and consisted of three digitally based tiers, a Central Facility (CF), four Data Selector Units (DSU), and 24 Signal Input Units (SIU). Each SIU comprised 250 channels, totaling 6,000 physical channels. Each SIU sampled measurements at 20,000 samples per second with system accuracy better than $\pm 0.15\%$ of full scale range. Although computers in the CF and DSUs were acceptable, putting computers in the SIUs exceeded acceptable quantities for one entity. The computer domain was an empire that forbade generalized and unregistered use. Consequently, another session with Congress developed to justify use of the computers versus hard-wired controllers regardless of the higher cost and lack of versatility of the latter. However, common sense and economics prevailed and the computers remained as the controllers in the SIUs. The DSUs and SIUs were transportable and configurable to match the test location requirements up to 3 miles distant from the CF. The CF supported two separate tests simultaneously, or combined all of the measurements as one test. The system was paramount to success of Space Shuttle testing at the MSFC. It was designed for 10 years service, but remained in service until 1995, almost 18 years, before being retired by technological advances in electronics and demise of the computer empire. The replacement system is similar, but more accurate, more easily transportable, and faster due to technology enhancements.

Solid Rocket Booster STA1 Structural Test
by Robert E. Johnson

The first structural test of the Space Shuttle Solid Rocket Booster was conducted at the East Area Test Facility on the west side of the static test tower that had been modified for this type test. The test was actually conducted in the horizontal position in the flame pit of the test facility.

A new data acquisition that had never been used before, with a capability of recording as many as 4000 measurements at the same time was used for this test. This system was designed and assembled by AVCO engineering using ModComp hardware. All of the cables for these measurements had to be fabricated and routed from the test article to the acquisition system.

There were several types of measurements that were used on this test. Most of them were strain gages used to measure stress and strain throughout the test article, which consisted of the aft skirt, three solid rocket booster segments, and a forward skirt that was used to house flight instrumentation. Other measurements types were load cells for measuring applied load, transducers for measuring displacement, pressure, and liquid level.

Different series of tests were performed such as lift-off loads on the whole test article, aft skirt loads, splash down loads, parachute loads and external tank attack point loads. The most measurements that were recorded on any one test were approximately 3,850. For all tests there were more than 4,500 measurements.

All of the above activity for the instrumentation portion of these test was planned and coordinated by Robert (Bob) Williams, who was the lead engineer. He was assisted by lead technician Robert E. (Bob) Johnson and technician Kenneth Dodd.

Interesting Visitors
by Woody Bethay

Soon after relations between the U.S. and Red China began to open up, a group of Chinese aerospace scientists made a visit to this country and included the Marshall Space Flight Center in their itinerary. A group of about 10 came, of course with the official escorts from Washington. They were given appropriate and permissible presentations about the Center and our programs as well as a tour to carefully selected sites of the Center. From the outset all communications were through interpreters. Throughout the morning the entire Chinese delegation acted as if they understood no English. We adjourned to the Officers Club for a lunch, and I happened to be seated by a distinguished looking Chinese gentleman. I decided to explore the potential for some direct communication so I asked the gentleman, whose name I believe was Dr. Pao, if he understood any English. His response was that he did, so we began to converse. I complimented his English and asked if he had visited the U.S. before. Imagine my surprise when he said that he had earned his PhD from Cornell University. The next surprise was his asking if I happened to know a Dr. Walter Hauessermann (a leading member of the original von Braun team and retired director of the Astrionics Laboratory). It seems that they had met at some time in the past at a technical conference. I gave him Walter's phone number and Dr. Pao quietly slipped out for a few minutes to give him call.

Every day was full of surprises at Marshall. Especially while Dr. von Braun was Director, many interesting and often famous folks came to visit. One was Dr. Christian Barnard, the South African surgeon who performed the first heart transplant. Needless to say, those of us who were privileged to be at Staff Luncheon on that day were fascinated to meet and hear such a noted person. We were also fascinated by his beautiful young wife. Dr Barnard was much more interested in hearing about our development projects than in talking about his history making medical advance.

When preparations were underway for the filming of Apollo 13, an entourage of Hollywood personalities visited the Space and Rocket Center and Marshall. They spent a day touring Marshall to gather basic information for the film. It fell my lot to welcome the group to the Center. Since the group was traveling "incognito," I was not aware of just who was in the group. However, I did recognize "Opie" (Ron Howard) from the Andy Griffith show. When I returned to the office, I was talking to a coworker and mentioned where I had been. I was asked about who was in the group and responded that "Opie" was there and a very nice fellow named Tom, but not being an avid movie goer I did not know his last name. From that day forward I was branded by my secretary and my family for being ignorant for being in the presence of a great movie actor, Tom Hanks, and not knowing it.

When Neil Armstrong came in 1985 to be the keynote speaker for the 25th Anniversary of the Center's founding, he flew his own plane. He was late because he had to do some repairs on the plane before he left for Huntsville. He was a most pleasant guest with a great message for the Center. Another most delightful visitor was John Denver, who came to be the principle speaker at a von Braun Dinner sponsored by the National Space Club. On at least one occasion he brought his two adopted children. He was a dedicated flier and supporter of both the space program and environmental matters.

Two notable Apollo Soyuz Test Program (ASTP) cosmonauts (Leonov and Kubasov) visited the Center in 1990 along with their escort Tom Stafford, an ASTP astronaut. We hosted them for briefings and tours, then the city fathers entertained them for a lovely dinner that evening. I remember that they were shocked that we had not already closed the loop on air or water on a manned space vehicle. The teen age son of Leonov, after visiting Marshall and the Space and Rocket Center found the whole visit to be "boring".

Throughout its history the Center hosted presidents, vice presidents, senators, congressmen, captains of industry, academia and entertainment, European royalty, astronauts, and many others of notoriety. A year or so after the *Challenger* tragedy the finalists for the "Teacher In Space" competition visited Marshall. After speaking to them, I was impressed that to a person they all said that given the chance they would fly "in a New York minute" in spite of the tragedy. This speaks volumes, I believe, about the place that the space endeavor has in the hearts of the American people.

Memories
by Charles E. Meadows

My NASA/MSFC career journey spanned the time frame from February 1962 until September 1987. My work assignments included electrical engineer, program management, systems engineer, Space Shuttle systems analysis and integration, Space Shuttle SRB structures, materials and processing in space, MSFC representative assigned to the U.S. Air Force Space Systems Division, program development for advance launch systems and future program definition.

The programs I was assigned to were: Centaur, Saturn Apollo S-IV and S-IV-B Stages, Skylab, Space Shuttle, Space Shuttle/Space Lab Micro-Gravity Materials Processing Experiments, Inertial Upper Stage (IUS), NASA/Air Force Advanced Launch Vehicles.

I am humbled and privileged to have been associated with such a historic, monumental venture as the "Peaceful Exploration of Space." To have witnessed the visionaries and architects of space exploration and to have been a part of the team that translated those visions into the reality of space travel was awe-inspiring. Those team members were characterized by their resolve, their dedication to perfection, attention to detail, and wrought with the skill and competence to meet the goal, with outstanding success. To those Marshall Space Flight Center teammates, I give my thanks, appreciation and gratitude for the professional and personal guidance, council and friendship I received along the way.

In closing, I recall a comment by Astronaut Wally Schirra, when President Kennedy established the NASA goal for the Manned Lunar Program and said, "We will send Men to the Moon, and return them safely back to Earth during this decade." Astronaut Schirra said, "And That's one Word," and by the Grace of God, that's the way it was.

NASA's "Man Under the Sea"
by Chester B. May

While a fascinated world watched the Apollo 11 moon expedition in July 1969, another very important scientific voyage was taking place. With similar goals of exploring unknown places, seeking answers to previously unanswered questions, and recording scientific data, Jacques Piccard was leading an undersea drift mission in the Gulf Stream off the East Coast of North America.

Both the immensity of the oceans of earth and the vastness of the space surrounding earth have captured the curiosity of men since the dawning of knowledge. One can, from the surface, peer down into the depths of the ocean and also upward at the heavens. But to explore fully either of these frontiers requires extensive new technology. And, in some ways, the technology requirements are startlingly similar.

Dr. Piccard designed the vessel, the *Ben Franklin*, which, when submerged, provided a livable environment for its crew, as a spacecraft does outside the earth's atmosphere. During long voyages, the crew must live and work together in a situation such as might be encountered during long stays in an earth-orbiting space station.

Because of these similarities, a young NASA engineer, Chester B. May, was allowed to accompany the drift mission to observe and evaluate the crew at work, rest, and play and to relate the experience to the development of future NASA space stations. NASA was also concerned with the Tektite Program, in which scientists spent long periods in habitats on the ocean floor. Several NASA employees were involved in that operation, studying habitability and observing the crews from both a physiological and psychological standpoint.

I think there is a common bond - or a common dream, if you wish - between the men who explore these two vast unknowns - the oceans and space. It is a fervent hope that nationalistic aggressions will cease and fade into a new age of knowledge and cooperation, that the very immensity of the ocean depths and the uncharted avenues of space will show that, to extract their benefits, the nations of man must work as one.

Snowy Review
by Bill Hallisey

One of the many Shuttle program budget reviews in Washington, DC, requiring the presence of project and program managers that I often reflect upon took place on a very snowy day.

The meeting went long into the early evening of what was a grueling budget justification for all of the Shuttle elements. Representatives from MSFC, JSC and Kennedy Space Center participated over a period of several days. All participants' suitcases were neatly stored in the room during that final day, ready for everyone's departure to their homes. JSC and KSC participants had concluded their briefings, picked-up their suitcases and left for the airport. It just happened that my suitcase was identical to JSC's Bob Thompson. You guessed it, mine was headed to Houston.

When I found out my case was missing and recognized who had it, I called the dispatch in the General Aviation office at National Airport. Since my plan was to fly to Boston, I did need my suitcase. Fortunately, they were able to retrieve my bag before the Houston aircraft took off.

Well on leaving the Headquarters building and noticing that many inches of snow had fallen, taxicabs were nowhere to be found, and time was running out to catch my flight. The Metro, DC's new subway system, had recently became operational and, alas, one of our project managers indicated we should take the Metro, as he was familiar with the new operation.

So off we went, down into the caverns of the subway to catch our train to National Airport. As we were waiting, there was an announcement over the address system that our train was arriving on the opposite tract. There we were, with suitcases and briefcases full of financial viewgraphs, climbing up and over the bridge to the other side. We had just settled in and as the train to National was on approach, another announcement was made that the train we want is arriving on the opposite side of the station, where we were in the first place!

Here we went, this illustrious group of NASA managers, fully loaded with overcoats, briefcases and suitcases, hustling "up" the DOWN ESCALATOR. It was quite a sight, as some had to rest, trying to go up the "down" escalator. *In slow motion it's even funnier.*

We were all successful in getting to our respective aircraft, as my commercial flight to Boston was delayed due to the heavy snowstorm. Thanks to Jim Odom, George Hardy, J.R. Thompson, Bob Lindstrom, and others for a very exciting, entertaining and memorable evening.

Shuttle Main Propulsion Test Article
by Harry M. Johnstone

I was designated as the Main Propulsion Test (MPT) Manager for Marshall Space Flight Center by Roy Godfrey, the Shuttle Projects Manager, on January 25, 1974. However, I initially turned down the job because there appeared to be no way that you could keep seven organizations happy on one test program. There were Headquarters, Johnson Space Center (JSC), Marshall Space Flight Center (MSFC), Kennedy Space Center (KSC), Rockwell, Martin and Rocketdyne, all to agree on the test requirements. However Bob Lindstrom applied his persuasive talent and said, "I will give you my total support", which he did. I signed up and was delegated the authority and responsibility to manage and direct the program by Robert F. Thompson, Shuttle Program Manager, JSC. I was given the responsibility for managing and directing the MPT activities to assure adequate, timely, cost effective and safe accomplishment of the program. I was to assure that the in-house MSFC expertise in the propulsion testing area was utilized to support the program. In addition, I was to keep the Manager for System Integration, Space Shuttle Program Office, JSC, Owen Morris and Space Shuttle Program Office, MSFC, Bob Lindstrom and the element Project Managers External Tank (ET), Jim Odom and Space Shuttle Main Engine (SSME), J.R. Thompson apprised of the MPT activities, status and progress.

The Integration Contractor manager, Jerry Wilson, North American Rockwell, managed the Overall Vehicle Integration System and Operations, the Main Propulsion Test Article (MPTA) and the Orbiter Systems. Rockwell would prepare the test requirements and test objectives under my guidance and direction. These test requirements and test objectives would be prepared in the form of the Test Requirements And Specification Document (TRSD) and the necessary criteria to certify and verify that the propulsion system met these requirements. Our primary purpose was to test all these new systems and components and verify, certify that each met all test requirement prior to the first Flight Vehicle Launch.

Al Reeser, Rockwell, was Chief of Operations and Tom Baggett, John Plowden and Tom Lydon were the test conductors for the MPTA. Colin Harrison of Martin Marietta managed the ET Systems and Operations. Rosco Nicholson of Rocketdyne managed the SSME Systems and Operations.

Test Team and Responsibility

I transferred to Stennis Space Center (SSC) in February 1976. The SSC B-2 Position was selected to static fire the MPTA. Again, I selected a dream team that established and encompassed all associated test activities such as schedules, pretest engineering, test requirements, all test planning documentation, coordinating facility design, construction, modification and activation. This team included Orville E. Driver, Facility and Program Requirements; Bob Saidla, Facility Construction and Modification; Bill Lindsey,

Electronics and Instrumentation; who was responsible for the data requirements documents, data acquisition and data evaluation. Jim Williams, Ground Support Equipment Design, Manufacturing and Installation. Cecil Houston, Budget Management and Control. Once again, I had praise and admiration for this team. Our initial static firing was scheduled October 1977, however it did not occur until April 1978 due to delays of the arrival of all three systems, the Orbiter, the ET and the SSME. We had no major delays on any facility hardware and once again, our team met all its requirements.

Test Team Growing Pains

The MPTA consisted of three elements, a propellant tank (ET) for supplying Liquid Oxygen (LOX) and Liquid Hydrogen (LH2) to the SSME's, an open bridge structure called a "strongback," simulated the fuselage. An aft compartment is attached to the "strongback" and the aft compartment houses the three SSME's, the LOX and LH2 propellant feed lines with prevalves in each line. Prevalves are upstream of the engine main valves and the turbopumps. If an engine explodes destroying the engine main valves and there are no prevalves, thousands of gallons of LOX and LH2 would be dumped into the aft compartment, feeding the explosion and causing an extreme temperature fire. This would destroy the test vehicle and do heavy damage to our test facility. Therefore prevalves are a mandatory requirement of the test organization.

One of our big battles with JSC and Headquarters was our requirement for prevalves. It was all about money for management, but it was about protecting the test vehicle and our expensive facility. We were asked to make our presentation to JSC management to justify our requirement. We presented to Robert (Bob) Thompson, Space Shuttle Manager, JSC. Bob decided to send us to Space Shuttle Headquarters where we made our presentation again. We were turned down again and sent back to Bob Thompson, JSC. We were beginning to get aggravated and in the middle of our presentation, Bob Thompson asked, "Harry, what would you do if I do not approve these prevalves"? I thought a moment that I had been chosen MPT Manager because of my test knowledge and my experience. In a moment of frustration, I said to Bob, "You can get yourself another boy". A long silence, and I thought, here it comes, "You are fired". Instead Bob said, "If you feel that strong about these prevalves I am approving them now". It sounded insubordinate, but I had personal experience with LOX/Fuel fires/explosions and I knew in my heart what was required for test stand/vehicle safety. I was very fond of Bob and enjoyed every minute that I worked with him. He was a great boss.

First Four MPTA Firings Successful

Although our first test ran only 1.5 seconds, it was long enough to be evaluated as a successful test. We ran three more successful test, 20 seconds, 42 seconds, and 104 seconds which made a total of four successful tests. By the way, we had prayer over the nation wide intercom system just prior to tanking propellants before each firing. At the end of the fourth test, I transferred to Kennedy Space Center as the Senior Science and Engineering Representative in the Marshall Space Flight Center Resident Office. I worked all problems on the Shuttle Launch Vehicle directly with Science and Engineering engineers at MSFC until I retired in December 1979.

Rockets and Wind
by Michael Susko

As NASA and Marshall Space Flight Center celebrated the launch of the first Space Shuttle some 25-plus years ago, I remembered what it was like to stand on "Astronaut Hill" waiting for STS-1 to launch on April 12, 1981.

As an engineer in Marshall's Space Sciences Laboratory, I used electrets—sensors—to measure environmental contaminants from the Shuttle on the first four Shuttle flights. I used the weather knowledge I gained while assigned to the 20th Weather Squadron of the Far Eastern Air Force of the US Army Air Corps in the Philippines during World War II.

I was stationed in the Philippines with the 20th Weather Squadron when the United States dropped the atomic bombs on Hiroshima and Nagasaki. We helped to calculate the wind and direction for that drop, so we knew the thing was going off.

The electrets used at NASA picked up the specific environmental effects in the cargo bay of the Space Shuttle and at the launch pad and downwind from the pad. In the case of contaminants, the sensors identified the type and amount. The information was then evaluated by x-ray spectroscopy for the contamination rate. For the first launch, the electrets were part of the Passive Optical System Array (POSA). The Environmental Contamination Monitor (IECM) was not part of the payload because of the nearly 2000 pounds of extra weight that would have been added. We needed a lighter Space Shuttle in case we had to turn around after launch. Turning around if a problem was detected would have been much harder because of the added weight. We had to measure the contaminants for the first four shuttle launches to make sure we met the Environmental Protection Agency guide lines. The electrets and the IECM were flown on the next three Shuttle flights.

We used the electrets to measure the exhaust cloud effluents from solid rocket motors (SRM) during the demonstration model static test firings at Salt Lake City. Also, during the 6.4% model testing at MSFC by the Tomahawk missile for acoustical data vital to the design of the Shuttle, the electrets obtained pollutant measurements. For the mission to Mars at Launch Complex 51, Kennedy Space Center, Florida for Titan III Viking, I was the principal investigator in performing air quality experiments at the launch.

One of the most significant contributions in determining the atmospheric wind loads on a space vehicle was the development of the roughened Jimsphere wind sensor because of its accuracy. It replaced the traditional smooth balloon, which behaved erratically with spurious motions in the atmosphere.

The wind loads for all the Saturn launches and over 100 Space Shuttle flights have been computed from the roughened Jimsphere tracked by an FPS-16 radar. The Jimsphere with a 100 gram mass attached provides wind profile measurements averaged over 25 to 50 meter altitude intervals with an accuracy of 0.5 m/sec or less to an altitude of 18 kilometers tracked by an FPS-16 radar.

During the Saturn V Apollo 11 launch to the Moon, I manned the meteorological tower, which was located a mile and a half away from Launch Complex 31 at Kennedy Space Center. A series of Jimspheres, one hour apart for 12 hours, were released and evaluated for wind loads of the Saturn.

The roughened Jimsphere was instrumental in obtaining accurate winds in the successful launch of the Saturn V to the moon and the successful Mercury, Apollo and Space Shuttle programs.

During my 47 years of Federal service, I was a member of Marshall Space Flight Center's team for over 34 years.

Small Steps to a Space Station
by Wilbur E. Thompson

Skylab had an exciting beginning unknown to most people. It involved Dr. Wernher von Braun's continuing dream for an orbital laboratory. Skylab achieved von Braun's quests for more astronaut involvement at MSFC and development of manned payloads at the center. Our first briefing to von Braun on a project that would become Skylab was a difficult start on a long path.

In the early 1960s, MSFC blossomed with concepts of future programs to follow the lunar landing initiative. The agency was seeking ways for expanded exploration of space and viable program(s) to continue after Apollo/Saturn was fully developed. During this early period, an Orbital Base Section was formed in the Propulsion and Vehicle Engineering (P&VE) Laboratory. We interpreted this title to include earth orbiting stations, and space stations in lunar and planetary orbit. We were a skeletal organization with myself assigned as chief.

My initial assignment included travel to other NASA centers to keep abreast of on-going space station studies. With budget concerns, MSFC focused on hardware developed for the Apollo/Saturn program. We stepped through various concepts such as experiment canisters mounted external to the crew cabin; to a small experiment module mounted inside the spacecraft adaptor; then to using the empty hydrogen tank of the S-IVB stage.

At this point, our office learned von Braun

was aware of our spent stage activity and wanted a meeting. We understood he wanted a work session with his participation in the brainstorming. He had recently returned from a Management Retreat Meeting where he had discussed use of the expended S-IVB stage for manned occupancy in earth orbit. Others there concurred in his further investigation.

With viewgraphs, I reviewed the mission profile, including tank venting, CSM separation, transposition and docking with the expended S-IVB. The briefing included concepts and requirements for docking fixture and modifications for crew entry into the empty tank. Key issues were the size of the hatch and desired rapid entry for crew. Also reviewed was initial assessment of requirements for crew protection including venting high-pressure gas bottles inside the tank, protective covers for potentially sharp edges and other safety issues. The capability for conducting EVA experiments and performing other crew tasks was discussed, as was options for an airlock and pressurized S-IVB stage in the design.

Von Braun asked probing and insightful questions as the briefing progressed. It seemed to go well, but von Braun's words were guarded. He stressed the center needed to have a better focus on advanced studies. Later that day, Jim Shepherd, von Braun's assistant, met with us to explain what von Braun wanted and had expected an elaborate configuration he called the S-IVB Workshop. His scheme envisioned an equipment module mounted in the payload adapter area above the S-IVB stage. The mechanism would be designed for telescoping down into the empty hydrogen tank following docking and hydrogen tank evacuation. The module would provide crew passage and include electrical power, environmental control system (ECS), crew and experiment equipment, etc. Based on the description conveyed to us, what von Braun really wanted was the empty stage converted on-orbit into a primitive space station. We really had missed the mark!

With guidance received from Shepherd on von Braun's "ideas" and contractor hours made available, our task on the workshop was clear. Jay Laue and I worked non-stop for two days and straight through an entire night to prepare a new briefing. This workshop concept incorporated all, or most, of the desired features expressed by von Braun. Laue presented the results at the MSFC dry run, and von Braun was very pleased with the briefing. Frank Williams, Director of the Advanced Systems Office, presented the concept at the Management Council at NASA Headquarters. MSFC received approval for continued study.

Management of Scientific Payloads on Spacelab Missions
by Robert E. Pace, Jr.

In the time before Spacelab missions, for NASA-sponsored scientific investigations the Principal Investigator (PI) was selected, a contract issued for the development of the investigative hardware, the hardware was flown, data collected and delivered to the investigator, analysis made and a report of the results issued. NASA assumed responsibility for the performance of the investigative hardware by monitoring development, accepting delivery, integrating into the mission, training of the astronaut crew, operation of the experiment during the mission, and delivery of the investigative results to the Principal Investigator subsequent to the flight.

With Spacelab a number of changes were implemented. NASA adopted an approach of contracting with a scientist for a complete investigation. The Principal Investigator was responsible for the development of his own scientific hardware, its performance, its operation during the mission, and the publication of the results. NASA was not any longer responsible for assuring the experiment hardware performed properly. Informing the Principal Investigators they would not be selected for flight again if their investigation were not successful because of their poor performance mitigated concerns over failure.

The Spacelab, with its seven different configuration options, had the capacity for a number of different experiments in various disciplines. To assure maximum return from these flights a Mission Manager, and a Mission Scientist, was selected for each mission. The Mission Manager was responsible for integrating the entire payload into an efficient operating package for maximum scientific return, and was the primary interface to the Spacelab and Shuttle projects. The Mission Scientist was the primary interface with the scientists selected to participate in the flight.

Prior to Spacelab, the astronauts operated all flight experiments requiring onboard crew participation. Spacelab missions introduced the Payload Specialist; an individual from the scientific community selected to represent the Principal Investigators onboard. The selected Payload Specialist, as a member of the flight crew, had to undergo a certain level of astronaut training to become familiar with the operation of the Shuttle systems.

The first Investigator Working Group (IWG) was chartered for the first Spacelab mission. The IWG was comprised of the Principal Investigators selected for the mission, or their representative. The IWG was responsible for the allocation of onboard resources available for use in the scientific investigations. These resources include crew time, electrical power, environmental control, data access and storage, spacecraft pointing, operating time, and other items that might limit the conduct of an experiment. These allocations by the IWG went into the mission planning that resulted in the pre-mission timeline. During the mission period these allocations were adjusted in real time with inputs from the IWG.

An important function of the IWG was the selection of the Payload Specialist for the mission. The objective was to have one or more of the Principal Investigators, or co-investigators, selected to represent the entire scientific payload onboard the Spacelab. A backup Payload Specialist was selected for contingency purposes.

In early manned space flights all communication with the onboard crew was by the capsule communicator (CAPCOM). An astronaut occupied the CAPCOM position resulting in all communications being from astronaut to astronaut. All commands to the onboard hardware, including scientific experiments and the returning data stream was from the Mission Control Center at the Johnson Space Center. During the Skylab mission, as sort of an experiment, some of the Principal Investigators were permitted to talk directly to the astronaut crew during operation of the scientific hardware.

With the beginning of the Spacelab missions and the increase in scientific investigations in space, NASA recognized the need for more direct participation of the Principal Investigator, and team, both onboard and directly from the ground. The use of Payload Specialists discussed above resulted from the need for onboard participation. The Payload Operation Control Center (POCC) was established at MSFC to permit more direct participation from the ground.

The POCC was the central place from where the entire payload was operated and controlled. Each Principal Investigator, or team, could communicate with and control his or her own experiment in accordance with the approved mission plan. If required, the Principal Investigator could bring to the POCC his or her own equipment used to operate with the equipment onboard. Direct communication to the flight crew operating the scientific experiments, Payload Specialists or Astronauts, was made from the POCC.

Origin of Spacelab Concept
by James A. Downey, III

I was assigned to Program Development Directorate in 1969 as a charter member to manage the Mission and Payload Planning Office. This Office served as the scientific arm of Program Development.

Both the Space Shuttle and the Space Station were in Phase B definition in 1969. One of our first initiatives was to plan possible scientific payloads for the Space Station. Numerous sources were contacted from academia, other NASA Centers, industry, and other government agencies. Literally hundreds of hypothetical experiments and payloads for the Space Station were assembled into a large document, referred to as the "Blue Book" because of the color of its cover. The first approach considered was that the Space Station experiments would be assembled in Experiment Modules which would be delivered to orbit by the Shuttle and docked to the Space Station to stay in orbit a long time.

Upon a detailed study of the hypothetical Space Station experiments and payloads,

Jean Olivier and Max Nein of my office, while interacting with the scientific community, made the observation that a large majority of the experiments could be completed in a few days on-orbit, using power and other resources which should be available from the Space Shuttle. The obvious followed: why attach the Experiment Module onto the Space Station? It would be much simpler operationally to leave the Experiment Module in the Shuttle and operate in the Shuttle cargo bay for a week or two, and then return the Experiment Module to Earth with the Shuttle and change it out.

Although this new Experiment Module concept certainly did not help to sell the Space Station, it caught on rapidly. In-house studies of Experiment Modules were done by the Mission and Payload Planning Office and the Preliminary Design Office of Program Development. Also, free flying Experiment Modules which could be deployed and picked up by the Space Shuttle were considered for astronomy payloads. The Hubble Telescope concept involving repair and up-dating of the telescope on orbit during Shuttle visits was derived indirectly from the free-flying Experiment Modules concepts.

Bill Armstrong, who worked advanced missions in NASA Headquarters' Office of Manned Space Flight, strongly supported the concept of leaving the Experiment Module in the Shuttle Cargo bay. He promoted an analogy with the on-going CV-990 scientific aircraft program at the Ames Center. (Much later Stan Reinartz managed for Marshall an extensive series of simulations for Spacelab Payloads using the CV-990 scientific aircraft program at Ames.) Funding was made available from NASA Headquarters to us in 1970 for concept studies of the Experiment Module. The first contracted study of Experiment Modules was done by the Convair Division of General Dynamics. This study was managed by Max Nein and Jean Olivier and resulted in the definition of free-flying and Shuttle-attached Experiment Modules.

In 1971 the name Experiment Modules was changed to Research and Applications Modules. Phase B studies of the RAM concept were accomplished by General Dynamics over a period of some fifteen months in 1971 and 1972.

Earth Science research at Marshall Space Flight Center had its origin in the research accomplished within the Aeroballistics Laboratory's Aero-Geophysics Branch of the Army Ballistic Missile Agency. At this time, in the 1950s, the primary activity was in support of definitions of the natural environment for use in the design and development of ballistic missile systems. In 1960 the group became part of the newly formed NASA Marshall Space Flight Center's Aero-Astroballistics Laboratory. It became the Aerospace Environment Division and in latter part of the 1970s the Atmospheric Science Division of the Space Sciences Laboratory. The Division's Earth Science research provided the basis for development and interpretation of natural environment design, development, and operational requirements for the Saturn, Space Shuttle, and International Space Station plus a variety of smaller NASA flight projects at MSFC and other NASA Centers.

In 1995, the Earth Science Department joined its long-standing informal partners of the University of Alabama in Huntsville and the Universities Space Research Organization at an off-site location in the Huntsville Industrial Park. The collective effort, cemented under a NASA Cooperative Agreement, adopted the name of the Global Hydrology and Climate Center (GHCC) as its umbrella title. Within the GHCC, each member institution retained its line connection with their parent institution. For the Earth Science Department, its reporting line remained through the Space Science Laboratory of the Marshall Space Flight Center. In 1996, the Department's reporting line changed to that through the Hydrologic Project Office and in 1999, to the newly formed Science Directorate of the Marshall Space Flight Center. In 2000, the Department and it's Global Hydrology and Climate Center cooperative agreement members, joined the newly formed National Space Science and Technology Center (NSSTC) as a core-member resident at the new Center's location on the campus of the University of Alabama in Huntsville.

Since the mid-nineties, the focus of the research within the Department has remained in the mainstream of NASA's Earth Science Enterprise. Research emphasis is on it's leading role in atmospheric electricity using space-based systems, urban heat island and impacts health of the urban system including air and water quality, improvements short term weather prediction, precipitation studies using passive microwave capabilities and the occurrence of extreme weather events. On a longer time scale, studies of climate variability, including using remote sensing techniques to monitor as well as to better understand the dynamics of the climate system, are conducted as well as climate related studies using global observing systems. Links between the past and the present are accomplished through a unique effort using remote sensing techniques to gain archeological insight on impacts of climate variability and change on development and decay of social systems.

The department today is a by-product of its long history in support of the aerospace environment as well as the science and space flight accomplishments of recent years. Many of the products of the early accomplishments as part of the MSFC's Aerospace Environment and Atmospheric Science Division's, such as the Global Reference Atmosphere Model, 150-m Ground Winds Tower at KSC, and Jimsphere Detailed Wind Sensor, are still in use today for support of Space Shuttle operations and upgrades, the International Space Station operations, and the Space Launch Initiative. More recent accomplishments such as the Optical Transient Detector and the Lightning Imaging Sensor on polar orbiting platforms, aircraft flight programs to study land falling hurricanes, and use of un-inhabited aircraft as well as accomplishments in research, lead national efforts in gaining a better understanding of the world we live in using technology and science unique to NASA. The most recent efforts to successively link NASA's accomplishments in observing and science to the improvement of NOAA's operational weather forecast is an example of interagency partnering to bring capabilities of two agencies together to address the recognized goal of improving the short term forecasting skill at by combining capabilities of different federal agencies.

Space Shuttle History
by Max Akridge

In September 1961, Deiter Fellenz and myself drafted, edited and integrated a final statement of work for what became NASA/MSFC's first efforts in manned reusable orbital transportation systems. In 1966 I was Technical Manager of a Martin/Denver contract to investigate horizontal versus vertical take-off, and with this we had our first set of common agreed-to requirements for a system.

The Apollo fire took any funds we might have had designs on for 1967-68, but it gave us a chance to pull together NASA and DOD efforts to proceed with Phase A contracts. A statement of work was written, meetings were held with all centers chaired by Dr. George Mueller in 1967, funds were requested but only approved in late 1968. An RFP was issued and responses evaluated, but awaited Dr. Mueller's "go-ahead". Dr. Mueller gave this approval in January 1969, forwarded money and directions for awarding four contracts, two to be directed by MSFC, one by Langley, and one by Manned Spacecraft Center.

A group to refine requirements, configurations and operations was established in Washington, DC, composed of all NASA and DOD elements. At that time, every contractor and center had several names for the vehicles.

On May 5, 1969, we had a status meeting with Dr. Mueller in which we were discussing progress in which Dr. Mueller said he wanted to assure that the vehicle would not be dedicated to any one or set of payloads, but deliver men and payloads to earth orbit and return "like a shuttle bus". I wrote Shuttle, put Space in front, and hence, the name "Space Shuttle" was born. I was the technical manager of two Phase A contracts and had a meeting with one that afternoon and one the next morning, and directed that they use the name "Space Shuttle" exclusively. Even though NASA went through a name selection process, the name never changed.

One Persistent SOB
by Ronald L. Young

During the 34 years of my service at MSFC as Manager of Project Control and Business

Offices for several Projects within major programs, such as Shuttle, Hubble Space Telescope, and Space Station, I had many interesting and challenging tasks. One of the earliest ones involved negotiations with the contractor for providing Solid Rocket Booster hardware. We had just completed negotiating a Fixed Priced Incentive contract for around $15M when we issued our first change order. The contractor's proposal was for $5M, my estimate for the work was $300k. Since this was a fixed price contract, my contract negotiator and I knew that how these negotiations turned out would set the tone for the remainder of the contract. After over a year of negotiations and the threat of arbitration, we settled for under $500K. Afterwards, one of the contractor's managers told me, "You are one persistent SOB." I replied, "Just make sure my boss knows you think that." Throughout my career I found that the challenges of managing the project effort (whether in contract management, schedule assessment, or budget development) were as difficult and important as the engineering challenges.

Early Space Mistake Due to Use of Both Metric and English Units
by Roger Chassay

In 1985, I was the NASA/MSFC project manager for a $50M class payload called the Fluid Experiment Facility (FES). We had had great difficulty in developing this highly complex apparatus for use in microgravity, but with numerous "patches" or quick fixes to work around design problems, we had successfully gotten it aboard the Spacelab 3 flight of the Space Shuttle. The experiments were controlled by the MSFC FES team, located at, NASA JSC for this mission. The control room assigned to us was very crowded, and noisy, but we took what we got and worked two shifts per day, 12-18 hours per shift for some of us, for the seven day mission.

During one night shift about midway through the mission, I heard one of our team members yell out a question above the din in the room, wanting to know a particular temperature reading. Another team member across the room looked up the parameter, and yelled back the temperature reading. The first team member updated his software table with the temperature input and also updated the associated stored commands for the FES to execute; during the next communications pass, he uplinked these FES commands to the Spacelab. Then I heard another question yelled, "Did you give me that temperature in Fahrenheit or in Centigrade?" The answer was yelled back, "Fahrenheit." Unfortunately, the uplinked commands had assumed "Centigrade," so the engineer had to hastily revise the commands and uplink the corrected commands, with only seconds remaining in the communications pass.

Having been advocate of the use of the Metric System for years, I knew at that moment that this "close call" could have been averted if NASA and the U.S. Congress had fully addressed this issue, and earlier provided the needed funding to abolish the use within NASA of the old English system of units. However, I also knew that for years this issue had been placed "on the back burner" as a low priority action, due to the reluctance to divert funding from "more important" projects. Also I knew this lack of action was likely to continue, so I concluded that it would take more than my "Close Call" on FES to rid NASA of this hidden risk.

Fast forward to the year 1999, when the NASA Mars Climate Orbiter $250M mission was lost due to a mislabeled data table using the English system of units, and the ground software using the Metric system!

Unlikely Commendation

Around 1965, the Chrysler Corporation's Space Division (CCSD) at NASA/MSFC's Michoud Assembly Facility was heavily challenged during the development of the booster stage (S-IB) of the Saturn IB rocket. It was a time when mistakes were frowned upon, since each mistake was costly in time, money, and credibility for CCSD.

One day a large tear was discovered in the outer skin section of one of the fuel tanks of an S-IB stage, which was needed very soon at the Kennedy Space Center for an upcoming launch. The technician who discovered the hole reported it to management. As a new NASA employee in the resident office at Michoud, I was sure that this incident would cause the top management of CCSD to rant and rave, and severely discipline the workers on the production floor.

To my surprise, the president of CCSD made a profound announcement that the reporting of this damage to the fuel was an outstanding and exemplary event. He then announced that anyone caught not reporting such events would be promptly fired. His rationale was that CCSD already had the most experienced, best trained, and highly motivated employees on the job; if one of them made a mistake, it would be foolish of management to fire that employee and replace him or her with a "second string" employee.

Instead, he insisted that for each mistake made at CCSD/Michoud, their management team would be responsible for finding the "root cause" of the mistake, e.g., poor lighting, insufficient training, improper tool availability, etc. In summary, he put his trust in his employees, and knew that the "system" is the root cause of virtually all mistakes made by an excellent team of space workers.

Challenger Recollections
by Charles R. Mauldin

As the Solid Rocket Booster (SRB) Integration Branch Chief at MSFC, and with a background in launch vehicle electrical systems, it was logical for me to be designated as the SRB Senior Test Representative to serve on site in the Launch Control Center (LCC) at KSC for all shuttle launches. I coordinated the efforts of a support team of engineers at the Huntsville Operations Support Center (HOSC) along with several on site USBI contractor and NASA KSC Resident Office personnel. Our task was to assure that all SRB systems were functioning properly and to advise the SRB Project Manager of the launch status of all active systems where measurement data was available.

I had served in this capacity for the 24 previous launches leading up to STS-51 L, which was the mission designation for the next launch of the Space Shuttle Orbiter named Challenger. This launch was to be attempted on January 28, 1986 after a delay caused by a forecast of unfavorable weather on January 26 and high crosswinds at the KSC return-to-launch-site runway on the 27th that came up during the launch countdown. The news media was highly critical of the missed launch window on the 26th because of bad weather, which did not materialize. There was much more media coverage for this launch because of the presence of the first school teacher to be launched into space aboard the Shuttle as a crew member.

On launch morning, I had assumed my position at the SRB console during the launch countdown hold that is scheduled after the completion of loading the liquid oxygen and hydrogen aboard the External Tank (ET). All SRB systems were normal for that point in the countdown. At this time the so-called Ice Team that goes to the launch pad and surveys the vehicle for ice in forbidden areas on the launch vehicle was on the mobile launch platform (MLP). This ice is usually present as frost on the ET and, if in a forbidden area, could present a debris hazard to the fragile Orbiter heat shield tiles when jarred loose by vibrations during Shuttle main engine ignition, SRB ignition and liftoff. I could see on the TV monitors that it wasn't the normal icing on the ET that was of concern, but the ice on the MLP from the freezing overnight temperatures. There were icicles hanging from structure, and teams were sweeping loose ice off the MLP with push brooms! I listened to the assessments being made by the engineers at the pad and their managers as to the risks being posed by the free and hanging ice on the MLP and surrounding structure. I also heard discussions on whether any launch commit criteria were being violated. Even if there were no criteria to cover this highly unusual situation, it was obvious to me and many others on the teleconference loops that there was no way to analyze the potential impact areas on the Orbiter from this massive amount of ice in the short period of time left before the prescribed launch time. I was so convinced that we were in a launch scrub mode, that I put my headset in my briefcase and was about to head for the firing room door. One of the USBI engineers present in the row behind my console and who was monitoring some of the management loops said, "Charlie, you'd better come back here. They're talking about launching!"

When I got back on the intercom the tone had changed radically. The NASA Shuttle Program Manager, Arnold Aldrich, was asking for reasons why we should not launch. And, more specifically, where was it documented that the conditions that we were in represented an unsafe condition. The burden of proof had shifted from "prove to me it is safe to fly" to "prove to me it is unsafe to fly." There were no documented constraining criteria to cover the unforeseen situation we were in. After much discussion and voiced reservations, no one made an absolute statement that they were "no go" for launch. There was talk of not having 100% assurance that ice will not hit the Orbiter, but no positive recommendation not to launch. Not having a "no go," Mr. Aldrich and his Mission Management Team decided to proceed with the launch. The teams at the pad cleared the loose ice the best they could, left the launch pad, the countdown clock was started and the rest of the countdown was nominal. At SRB ignition and liftoff, all the SRB data I was monitoring looked normal. As the Shuttle cleared the launch tower, the USBI engineer behind me, Larry Clark, exclaimed, "That lucky school teacher!"

I didn't see anything abnormal until more than 60 seconds after liftoff. Then the SRB thrust vector control (TVC) data showed that the systems were providing vehicle flight control authority beyond that normally required. The gimbal angles on the SRM nozzles exceeded what I was accustomed to seeing. They had gone outside the pre-launch gimbal limits and flashed red. Then my data screen stopped updating. I said, "What happened." Then I looked at the TV monitor and saw that the vehicle had come apart. I said, "My God, it exploded!" The USBI Chief Engineer, Ike Rigell, sitting beside me said, "Charlie, I never thought it would happen."

We were kept in the firing room for hours. Even when we were able to leave, all data relative to the launch had to be left behind for investigation purposes. We went and viewed TV tape of the launch and explosion on monitors at the MSFC Resident Office. I had been unaware of all the discussion and teleconferences about cold temperature effects on the SRM O-rings and the safety concern that Morton-Thiokol engineers had expressed the night before. Innocently, I remarked to Al McDonald, the senior Thiokol SRM representative who was sitting beside me during the tape viewing, that it looked like an SRM case burn-through. He looked chagrined and crestfallen and just nodded.

Later, I drove to Patrick AFB to board the MSFC plane, NASA 3, for the return to Huntsville. It was dark, but as we taxied out for takeoff, we passed the T-38 aircraft that the Challenger crew had flow in. The trauma of the moment and the realization that those people were not coming back set in. When he saw the planes, Alex McCool, a Science & Engineering Laboratory Director, said, "Oh no. Look at that. That's awful!" Indeed it was.

After the investigation of the Challenger accident, in which I participated, I was selected by the new MSFC Director, J.R. Thompson, to be the director of a newly formed Systems Safety and Reliability Office. Our first major task was the total re-evaluation of the failure modes and effects analyses, hazards analyses, and critical items of all the MSFC Space Shuttle hardware. We then put into place specific control measures and a tracking system to assure that proper countermeasures were taken to prevent catastrophic failures. All during this activity, the image of those T-38s waiting for their crews was with me and is still today.

A Laughing Moment During a Very Difficult Time
by Ted Carey

Late winter of 1986 was a very difficult time for us. The Challenger accident had just happened. It seemed as though the whole world was after NASA, especially MSFC since it was our rocket that was at fault.

While most of my memories of that time are painful, I cannot help but remember a laughing experience I had with Neil Armstrong. Immediately after the Challenger accident, President Reagan appointed a Presidential Commission to investigate the cause of the accident and to make recommendations. Neil Armstrong and Air Force General Don Kutyna were the two members of the Commission that were to lead the part of the investigation that had to do with MSFC.

We were notified of their arrival time at Huntsville airport and were making plans for their visit to the Center. I was Director of Executive Staff at the time. A large number of national press members were in Huntsville doing everything they could to "get the latest scoop." When we received the flight number and time of arrival for Neil and Don, I decided that the best way to get them to the Center was to personally pick them up in my Honda instead of a NASA staff car so as to draw as little attention as possible.

It was either late January or early February and was very cold. While we were driving from the airport to the Center, the heater in my Honda was not putting out enough hot air to warm us up. I asked Neil, who was sitting next to me in the front seat, if he would try to adjust it so as to get warmer air. He fidgeted with it for a few moments, then said he couldn't get it to work any better. I looked over at him and said, "You, who landed that Lunar Module on the moon under the most difficult of situations, can't get warm air out of my Honda heater? We both laughed and Gen. Kutyna, who was in the back seat, laughed so hard I thought he was going to fall over.

My Experiences
by Otha H. Vaughan Jr.

I was honorably discharged from the USAF as 1st lieutenant in 1953, and returned to graduate school at Clemson A&M College to get my master's degree in mechanical engineering. After a few years of going to graduate school and working with my father in his Radio and TV Business, I then decided to go to Huntsville, AL to work for the Army Ballistic Missile Agency (ABMA). My experiences with the von Braun team there were what a young engineer could only hope for. My Division Chief was Hans Paul, a von Braun Rocket Team member, and while working for him I did cooling system design for the Redstone, Jupiter C, Juno and Jupiter. I also did Aerodynamic Heating Studies and Flight Test Evaluation for these missile systems during the period of 1957 through 1960. One of my big thrills was to see Jupiter C/Explorer I launched successfully into orbit and also the launch, ballistic flight, and recovery of the Monkeys Able and Baker and later to see the launch of the Saturn IB. During January 1959 I received my master's degree in mechanical engineering from Clemson A&M College.

On July 1, 1960, I transferred from ABMA to the newly formed Marshall Space Flight Center and really began working in developing the vehicles and hardware for the Apollo Program. Some of my best memories were

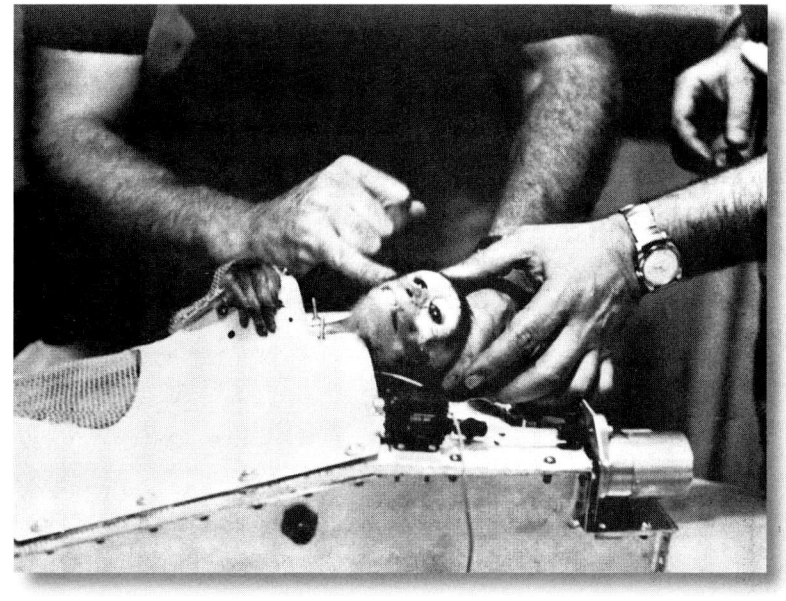

On May 28, 1959, a Jupiter Intermediate Range Ballistic Missile provided by a U.S. Army team in Redstone Arsenal, Alabama, launched a nose cone carrying Baker, A South American squirrel monkey and Able, An American-born rhesus monkey. This photograph shows Able after recovery of the nose cone of the Jupiter rocket by U.S.S. Kiowa.

in 1968, even before we had landed a man on the moon, was of the times that I had many discussions with Dr. von Braun about the many lunar surface features that could be seen from the photo's taken by the Surveyor Lander and Orbiter spacecraft's and what interesting places there were to be explored on the moon once we had successfully landed a man on the moon.

I was a member of the Scientific Advisory Team for the Apollo 8 mission. Its purpose was to do a fly-by around the moon and to photograph the terrain on the back side of the moon and to photograph the potential landing sites for Apollo 11.

I also provided Lunar Surface and Environmental Design Criteria for the design of the Lunar Rover, developed a Lunar Driving Simulator, and developed possible Lunar Traverses. I participated in the MSFC Missions Operations Control Center during each of the Apollo 15, 16, and 17 Lunar surface landings in supporting the Rover operations which explored the lunar surface. It was quite a thrill to see our Rover operate as it did and the great feeling of satisfaction that my design criteria were correct.

During the SkyLab Program I was fortunate to be able to propose a number of simple Science Demos that I conducted in the NASA Zero-G-aircraft (Vomit Comet) that illustrated how fluids react in zero gravity and then to have them performed by the SkyLab crewmembers in orbit. I participated as a test subject and experimenter on many of the Zero G flights in the NASA KC-135 Zero G aircraft.

Prior to the flight of the Space Shuttle, I was a test subject in the Component Verification Program (CVT). Its purpose was to develop experiments for future space flights and to see how well the experiments could be conducted with limited resources. This was also an exciting experience for me.

During the first series of shuttle flights, STS-2, 4, and 6, I was fortunate to have the crewmembers conduct an observational program which I and my University colleagues proposed called the Night-Time/Daytime Observational Experiment of Lightning (NOSL). Its purpose was to observe thunderstorms and their lightning from space and to collect video images and electronic signatures of the lightning for analysis later. This first observational program led me later to propose in 1986, the Mesoscale Lightning Experiment (MLE) to be conducted from the space shuttle and from this program in 1989 we were the first to observe and obtain video data on the Red Sprites and Blue Jets, a new atmospheric phenomena that had been reported earlier by pilots but had not been captured on film or video. Observations of these phenomena were a great satisfaction to me and the data that was collected was used as design criteria for a lightning mapper satellite called the Optical Transient Detector (OTD) and for another satellite called the Lightning Imaging Sensor (LIS). Both of these satellites are in orbit. This was one of the most interesting highlights in my career.

During the shuttle program I served as Mission Support Scientist for SpaceLab 3 and was assistant mission scientist for the SpaceLab J, a joint Japanese National Space and Development Agency (NASDA) and a NASA Space Shuttle mission.

On January 2, 1999, I retired from the Global Hydrology and Climate Center, NASA Marshall Space Flight Center here in Huntsville, AL. During my 45 years with NASA and the US Army Ballistic Missile Agency as a employee with the Federal Government I had so much fun working over these years it has not been a job because of the many interesting and challenging things that I had the privilege of working on throughout my career.

Memories
by Grady Sherman Jobe

I had the incredible good fortune to be assigned to a group of people who, I later learned, were some of the "best in the world" at what they did. There was a pink Lincoln and a red and black Buick who's owners were on a fast-track to enjoying life (which included work). Their actions demonstrated early that you can work hard, very hard, and still laugh and have a good time. The owners of those automobiles were Gerry Turner and Bill Burdine. Working in the lab with Webb, Brothers, Cochran, Albright, Yost, Belvedier and Kuznicki, while working for Burdine and Turner was a beginning which guaranteed future successes. Klan, Pinkerton, Wilson and Dupree next door were contributors of sage advice on the virtues of questioning everything and everybody. Stories told, on a continuous basis as if stored on a loop recorder, dealt with fishing, girls, hunting, girls, golfing, girls, launching missiles, girls, and what the next project was all about, called Saturn. Most of the leaders (maybe all) were ex-military people from WWII; either German or American. That structure provided a very direct path for decision making (not a lot of time spent on consensus building or voting) but it put enormous pressure on those leaders to make the "right" decisions; von Braun plus a dozen others, max. I believe this decision making process in the early years of launch vehicle development is frequently overlooked. That approach would never work in today's environment and maybe that is why people say we could never go to the moon again.

Today is different and better in a lot of ways, but with the work ethic coming from a world-wide depression and discipline coming from a world-wide war, during the midst of a cold-war race, there were incredible, history making, accomplishments and it all started here in Huntsville, Alabama (affectionately called Hunts-patch by those from afar). The floors shook and the windows rattled and the only thing known for sure by the local community was those rocket people are at it again; they really are going to the moon. Orville Underwood, who lived at New Market, never really believed that it happened when it happened. I believe he later believed.

Huntsville's Rocket City Astronomical Association and Space Journal
by Jim Daniels

The Rocket City Astronomical Association (RCAA) had its beginnings in 1954-55. After a series of informal meetings on astronomy at the home of Dr. Martin Schilling, scientist at Redstone Arsenal, RCAA was organized and officers elected on April 4, 1955. Officers were President, Dr. Wernher von Braun; Vice President, B. Spencer Isbell; Secretary, George A. Ferrell; Treasurer, Erwin Priddy. Board members were Dr. Ernst Stuhlinger, Gerhard Heller, Wilhelm Angele, Conrad Swanson, and Quincy Love.

Inspired by Dr. von Braun's leadership, the group raised $800.00 to purchase a telescope. Plans were adopted for an observatory, designed by Mr. Angele. In March 1955, Dr. von Braun purchased a 16.5-inch telescope (with a $2,000 mirror) for $600. RCAA accelerated plans for its observatory and needed to house the telescope. A 13.5-acre tract in Monte Sano State Park was leased from the State of Alabama for 20 years for one dollar. After site clearing, construction of a concrete-block observatory began in April 1956. At the same time, a dome for the building was begun under the supervision of Mr. Angele. Shortly afterward, the dome was installed on the observatory. In September the facility was used for observation of an historic event, the planet Mars on its closet approach to Earth in 17 years. Official opening of the observatory was finally celebrated on June 12-13, 1959.

More than $3,000 cash and materials was contributed by Huntsville businesses and individuals for the observatory. But more money was needed. RCAA Vice President Isbell proposed an alliance between the RCAA and a group of writers, illustrators and engineers that he had assembled to produce a magazine devoted to space exploration and the associated sciences. Isbell and the other members of the magazine, called *Space Journal*, agreed to contribute profits from public sales and subscriptions to RCAA for use in completing and equipping the observatory. Some RCAA members were skeptical of the venture, but Dr. von Braun was convinced by Isbell's assurance that *Space Journal* would function as a completely separate entity from the Association, and that the staff of the *Journal* would assume all financial responsibilities for the publication.

In the spring of 1957 the staff found a publisher to print the first *Journal* issue on credit. The *Space Journal* hit the newsstands in 12 cities within two weeks after the Russian Sputnik was launched. The *Journal* sales were phenomenal, but the poor business arrangement with the publisher resulted in a net loss for the issue. Good press and media notices

NASA - Marshall Flight Space Center

Dr. Hermann Oberth (center) examining the first issue of the Space Journal, summer 1957. Journal founders and staff members pictured include, seated, from left: Ralph E. Jennings, Associate Editor; James L. Daniels, Associate Editor; B. Spencer Isbell, Editor in Chief. Standing: Gordon D. Willhite, Art Director; Yewell Lybrand, Business Manager; Harold Price, Composition-Layout Director; George A. Ferrell, Distribution Director; David Christensen, Advertising Manager, Lew Cimijotti, Production Manager. Courtesy of George A. Ferrell.

of the new magazine brought the *Journal* to the attention of several potential investors. The staff concluded a one-year contract with Space Enterprises, Inc. of Nashville for four issues, guaranteeing all production expenses plus a minimum of $5,000 for RCAA.

The magazine experienced continued growth in sales and subscriptions throughout the year, but the staff was unable to reach agreement with Space Enterprises on renegotiation and terminated the relationship and the magazine. Although the *Journal* ceased publication after the one-year contract, it contributed more than $10,000 to RCAA.

The Wernher von Braun Research Hall at the University of Alabama in Huntsville

by Charles A. Lundquist, Ph.D.

The University of Alabama in Huntsville and the Army-NASA rocket programs have followed parallel development paths. In October 1949, officials of the U. S. Army reached the decision to transfer the Army rocket activities from Fort Bliss, Texas to Army facilities at Redstone Arsenal adjacent to Huntsville, Alabama. Meanwhile, in an unrelated 1949 effort, the community leaders of Huntsville convinced the University of Alabama to establish a center for instruction in Huntsville. The first classes at the new Center began in January 1950. Shortly thereafter, in April 1950, personnel from the Army rocket activity in Fort Bliss began moving into Huntsville. The new residents included Dr. Wernher von Braun and his rocket team transplanted from Germany after World War II.

Army management quickly realized that it must provide advanced technical courses for its military and civilian personnel. By September, 1951, the Redstone Arsenal Institute of Graduate Studies began offering graduate courses through a contract with the Graduate School of the University of Alabama. By 1957 this graduate program had grown to the extent that the Army initiated an Army Ballistic Missile Agency/Redstone Arsenal Graduate Study Steering Committee to oversee the program.

A necessary feature of a viable graduate program is an opportunity for the students to conduct research to satisfy thesis and dissertation requirements. Independently, the development objectives of the Army organizations required research support of a scale beyond that possible with only in-house Army personnel. Academic organizations traditionally conduct such research under contracts from the government. Again the interests of the Huntsville components of the University of Alabama and the Army Ballistic Missile Agency were on parallel paths. The University activity in Huntsville was motivated to establish a research capability as needed by its graduate students and also as appropriate to satisfy Army requirements.

The preferred mode to attain the desired research capability was formation of a research institute or research centers at the University of Alabama Huntsville operation. This was voiced by Major General John Bruce Medaris in a talk to the Alabama Legislature on June 23, 1959. He stated that to support the Army functions in Huntsville, "a fundamental is creation of... research centers." To further promote this concept, the Graduate Study Steering Committee created a Research Institute Subcommittee. On May 4, 1960, they initiated a meeting with President Rose to inquire what position the University would take on the proposition to establish a Research

203

Institute in Huntsville in conjunction with the University's functions there. This meeting was followed with a formal letter to President Rose on the Institute proposal signed by Dr. J. C. McCall (assistant to Dr. von Braun) acting for the Joint Redstone Graduate Study Steering Committee. In response, a delegation from the University of Alabama met to discuss the proposed Research Institute with representatives of the Army Ordnance Missile Command and the Marshall Space Flight Center (MSFC came into existence on July 1, 1960). By October 1960 the concept had been accepted and University of Alabama Professor Ferdinand H. Mitchell had been appointed interim director of the Research Institute.

The first building on the Huntsville campus, Morton Hall, was dedicated in May 1961, primarily for instructional use. It was at once overcrowded by the rapidly expanding class schedules, and there was no room for the fledgling Research Institute. To remedy the situation, on June 20, 1961 Dr. von Braun, director of MSFC, addressed the Alabama Legislature to request a three million dollar general revenue bond issue to build and equip a Research Institute building in Huntsville. Both houses of the Legislature passed a bill to implement the bond issue and the Alabama voters gave 3-1 approval in December 1961. Immediately thereafter, the City of Huntsville and Madison County pledged funds to buy land on which the Institute could build

In January, 1962, Dr. Rudolph Hermann accepted the permanent Directorship of the Research Institute. He came from the University of Minnesota where he was a professor in the Department of Aeronautical Engineering. There, he was also concurrently Director of the Rosemont Aeronautical Laboratories and Director of the Hypersonic Laboratory. He joined the University of Minnesota in 1950. Dr. Hermann came from Germany to the United States at the end of World War II. He spent 1945 to 1950 at the Wright-Patterson Air Force Base in Dayton, Ohio as a technical consultant. In Germany, from 1937 to 1943 he was at Peenemunde as director of the Supersonic Wind Tunnel Laboratory and was in charge of the aerodynamic development of the V-2 missile.

During 1962 - 1964, while their building was underway, Dr. Hermann and his growing staff operated out of space loaned by Brown Engineering. The Research Institute Building was completed and occupied in fall 1964. It was the second building on the Huntsville campus. After some eight years of productive research, Dr. Hermann retired from the RI Directorship in 1970, but continued to teach. He died May 17, 1991.

Recognizing the steady expansion of instruction and research in Huntsville, the president of the University of Alabama designated Huntsville as a branch campus in March 1966. In 1969, the Trustees created the University of Alabama System with a campus in Tuscaloosa, Birmingham and Huntsville, each with its own president.

In the following years, the University of Alabama in Huntsville, UAH, has added several additional research centers and buildings, but the RI Building has continued to house its central research promotion and administration functions. From its formation in 1960 until 1972, the Research Institute coordinated campus-wide research. Subsequently, the Dean of the Graduate School performed the coordination function. In 1982, the position Director of Research was established (with Dr. C. A. Lundquist filling the position). As UAH continued to grow, a Vice President for Research and Graduate Studies was created in 1986 (filled by Dr. L. Guy Donaruma), again with offices in the RI Building. The offices of the current Vice President for Research (Dr. L. Ron Greenwood) and the RI Director (Dr.

Members of the original rocket team from Germany at the von Braun Research Hall dedication. They are standing in front of a large mural exhibiting the V-2 to Saturn V rocket family. From left: Ernil Lange, Johann Tschinkel, Dieter Grau, Konrad Dannenberg, Werner Dahm, Gerhard Reisig, Fritz Mueller, Otto Hirschler, Ernst Stuhlinger, W.A. Schulze, Rudi Schlidt, Walter Jacobi, Hans Fichtner.

Richard G. Rhoades) remain there today. Drs. Greenwood, Rhoades, Lundquist, and other UAH-RI officers are Army or NASA alumni.

With the strong rocket and space tradition at UAH, many noteworthy accomplishments have resulted from research in these disciplines. In cooperation with MSFC, UAH scientists in 1973-4, provided successful materials experiments that were performed on Skylab, the first space station. These experiments were followed by many others carried on the Space Shuttle and planned eventually for the International Space Station. Working with the U. S. Army, Research Institute personnel have developed missile system simulations, particularly with actual hardware components in the simulation loop. With NASA funding, UAH promoted commercial operations in space by contracting with private industry for eight suborbital rocket flights with several-hundred-pound payloads for investigations in microgravity. In fundamental science, the first material that is superconducting above the temperature of liquid nitrogen was discovered at UAH.

During the decades while UAH was evolving, the Army and MSFC organizations also evolved to adapt to new missions. For example, MSFC had an essential role in the Apollo Program that flew humans to the Moon in 1969. Thereafter in 1970, Dr. von Braun moved to the Washington headquarters of the National Aeronautics and Space Administration. Subsequently he took a position in private industry. He died of cancer on June 16, 1977.

The year 2000 marked 50 years since the first University of Alabama class in Huntsville and also since the arrival of the Army rocket personnel. As part of the celebration of these events, the Research Institute Building, whose existence owes so much to Dr. von Braun, was rededicated as the Wernher von Braun Research Hall. The dedication ceremony on May 13, 2000 brought many of the surviving members of the original von Braun team to the Huntsville campus. An extensive permanent exhibit in von Braun Hall has a 3 by 5 ft display board for each of the years from 1950 through 1970, when Dr. von Braun was in Huntsville. Each year's board displays highlights of Huntsville rocketry events of the year. The exhibit also has several other elements, including a 25 foot high illustration of the rocket family from the V-2 through the Saturn-V, all reproduced to the same scale.

It is accurate to observe that the University of Alabama in Huntsville grew in a synergistic relationship with the Army-NASA rocket and space organizations in Huntsville. The size and scope of UAH owes much to the adjacent federal organizations. The dedication of von Braun Research Hall with its public exhibits is one recognition of that fact.

Michoud Assembly Facility – NASA Mission Essential Infrastructure
by Robert McBrayer

A big facility is necessary to manufacture and assemble big rockets. The Michoud Assembly Facility just outside New Orleans, LA is among the few places in the world where large space structures such as the Saturn 1-B and Saturn V stages could be built and assembled.

The size and complexity of the 832 acre Michoud Assembly Facility is difficult to grasp, especially Building 103, which has 42 acres under one roof and provides 2.2 million square feet of space for manuvering large space structures. MAF can be compared to a city that has its own waste processing; satellite communications system; 27 major utility systems; interstate access; deep water access port and harbor facilities, specialized pneumatic, structural load, and hydrostatic testing; and advanced and state-of-the-art machines for metal and non-metal manufacturing. A layout of MAF is provided below.

The Michoud Assembly Facility (MAF) was originally constructed in 1940 at the village of Michoud, Louisiana by the United States government for production of World War II plywood cargo planes and landing craft. During the Korean War, it manufactured engines for Sherman and Patton tanks.

Major Buildings at MAF

101 Admin Bldg	203 Reclamation Storage	303 X-Ray Mod & Staging	404 Structures Test
102 Engr Bldg	207 Boiler House	318 Component Ablator Fac	420 Acceptance & Prep
103 MFG Bldg	220 Vehicle Component Supply Bldg	320 Facilities Ops Bldg	450 Main Pumping Station
110 Vert Assy Bldg	221 Hazard Mats Storage	350 Office & Engr Bldg	451 LH2 Pneumatic Test

MAF came under the management of NASA in 1961 and was used for the construction of the S-1C first stage of the Saturn V rocket and the S-1B first stage of the Saturn 1B rocket. In 1975 NASA began construction of the Shuttle External Tank, which continues to this day. In January 2006, the NASA Administrator announced that Constellation will utilize MAF for manufacturing and assembly of ARES I, Orion and ARES V elements. This strategic decision dictated significant changes in the management of MAF.

The Michoud Assembly Facility had been used solely for the production of the Shuttle External Tank since 1975, under the direction of the External Tank Project Office at the Marshall Space Flight Center. Lockheed Martin Space Systems was contracted in a dual role of producing the External Tanks for Shuttle and operating the facility. The decision to also produce Constellation hardware at MAF precipitated a NASA management change to an independent MSFC organization, more NASA involvement in the management of MAF, and the selection of an independent contractor to perform manufacturing operations and facility support for all the Non-NASA tenants (Department of Agriculture, United States Coast Guard, and the University of New Orleans, among the largest), Programs, and prime contractors that would utilize MAF. In addition, all the processes for multi-program production management and support would be required to ensure a smooth transformation and long term stability for MAF.

The Orion structure, Ares I upper stage and instrument unit production areas are currently being placed inside Building 103. Out year plans have space allocated for production of Ares V.

The Michoud Assembly Facility is NASA Mission Essential Infrastructure for the production of America's space-faring vehicles of the future.

Artist's Role In MSFC's Space Program (Before Computers)
by Victor S. Grimes, Sr.

Like many administrative personnel, Illustrators and Visual Information Specialists were seldom recognized by either the media or public affairs. Artistic trained personnel played an important role in the development of space vehicles. Managers and executive personnel needed daily, weekly and monthly presentation materials to keep higher officials informed of various project achievements. It was up to a small group of artists to provide them with illustrations, publications, flip charts, wall charts, vu-graphs, slides, and displays for reviews. Most of the time, their requests were needed in "rush job" speed. The Graphics and Engineering and Models Branch under Gerd deBeek and later Floyd Duke provided all of these services. Often managers needed 3-dimensional illustrations to confirm what an engineer's scaled blueprint would look like as a finished product. Also, it allowed the Model Shop to produce scaled models showing what the vehicle would look like before construction. This branch was first under ABMA and transferred over to the newly activated Marshall Space Flight Center, along with Dr. von Braun's team of engineers and scientists in June 1960. In the early days, most charts were done freehand or by Leroy Lettering in India black or color inks; therefore, fellow artists were called Leroy Jockeys. Contractor artists were later assigned to deBeek's Branch in 1961.

At the peak of the Apollo/Saturn program in the '70s, space technology brought forth a mechanical artist, now known as the computer. Any person with artistic talent can now, after learning the computer's operational skills, become a technical Illustrator or a Visual Information Specialist. In the future,

Astronaut Walter Cunningham (Apollo 7) presenting the Snoopy Pin Award to Victor Grimes, with Col. Lee James of the Saturn program present, 1970.

One of the Saturn Apollo vehicles.

NASA - Marshall Flight Space Center

skylab 3

JACK R. LOUSMA
Command Module Pilot

ALAN L. BEAN
Commander

OWEN K. GARRIOTT
Science Pilot

THEY WILL DO THEIR JOB...WILL YOU?

MANNED FLIGHT AWARENESS
MARSHALL SPACE FLIGHT CENTER / HUNTSVILLE, ALABAMA

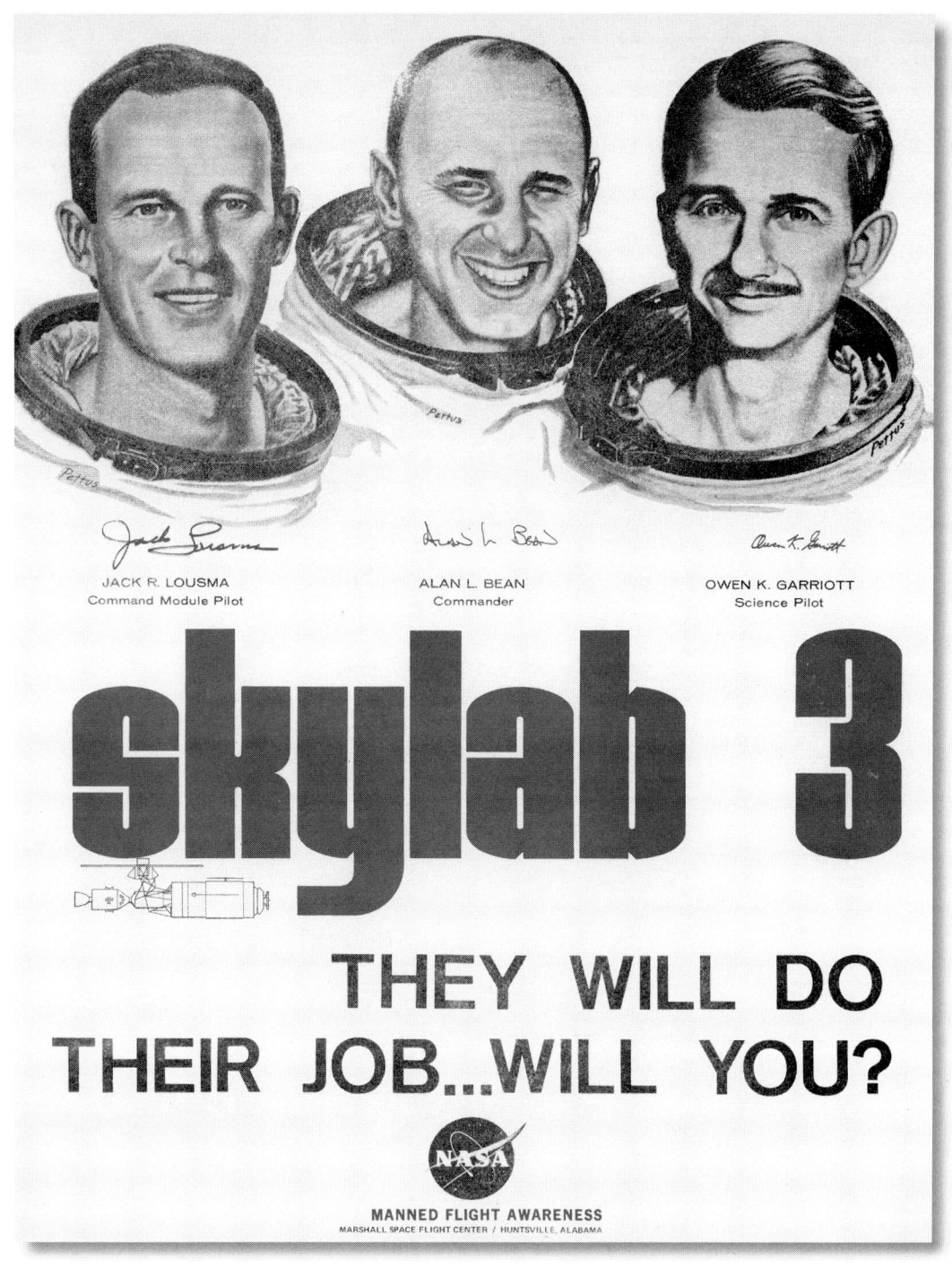

old-fashioned trained professional artists will be gone unless they learn how to use a computer. If Norman Rockwell were alive today, he could not qualify as a civil service illustrator GS-5.

I joined Gerd deBeek's branch in November 1958 and moved with it over to MSFC in June 1960. One of my first MSFC jobs was to design and freehand letter a masthead for the new "Marshall Star". As an example, the following will describe some of the duties I performed (before computers) in the space program, such as the Apollo/Saturn, Skylab and Shuttle missions.

In addition to performing general illustrator duties, I was contacted by Dr. Preston Farish, MSFC, to paint a rocket and space scene to be used on a post card advertising the building of a "Research Institute" in Huntsville. Dr. von Braun had directed Dr. Farish to develop and promote a plan to get the local community, as well as people across Alabama, to "Vote Yes" on Amendment Number 2. The $3 million bond was needed to build the institute to support America's new space program. I designed and prepared camera-ready art for both sides of the post card and engaged a local printer to print thousands of cards to be distributed throughout the state of Alabama. The people of Alabama approved the bond issue and the Institute was built. It is now known as the University of Alabama in Huntsville (UAH).

In July 1965, as a Visual Information Specialist, I was assigned to the newly Manned Flight Awareness (MFA) Program, headed by Dr. Preston Farish, to do the layout and design of brochures, vugraphs, charts, slides, booklets, displays, and to regularly update the exterior designs which included a display-type painted of lettering on a traveling NASA van.

Dr. von Braun was very concerned about the production of thousands of parts in construction of the Apollo/Saturn V vehicles. It was MFA's responsibility to motivate government and contractor employees to do the best job possible in their area of service because the lives of the astronauts in space were at stake. Employees working on the Apollo/Saturn V vehicles were constantly reminded to perform good work through safety posters, films, displays and NASA's traveling van. The successful launch of all Apollo/Saturn vehicles shows that the awareness program was effective.

The MFA Office also interfaced with the Astronauts when visiting with space workers. They presented to outstanding employees their individual Snoopy Pin Awards. Dr. Farish was unable to assist in making this report, since he is a victim of Alzheimer's.

I designed the Apollo 8 and 11 Medallions for MSFC, and our office furnished every Apollo/Saturn worker throughout the country (government agencies and private contractors) a medallion in their excellent performance. I also designed the NASA/MSFC Retiree Association's logo after retirement in 1976.

I was honored as a "Manned Flight Awareness Honoree" and invited to Cape Kennedy to observe the launch of Apollo 11 (first moon landing) in 1969. A celebration was held at the motel the night before the launch for all MFA Honorees, with several astronauts, the Governor of Florida and Dr. von Braun present. The word was out that the famous flyer, Charles Lindbergh, was at the Cape to witness the launch. Dr. von Braun sent a representative to locate and invite him to the party. First, Lindbergh was reluctant, but finally said okay. Lindberg arrived late that night and Dr. von Braun introduced the famous flyer to the awareness group. He stepped up to the podium and praised the honorees for their good work and selection to witness this great event. Lindbergh's speech lasted about five minutes, and then he left.

In 1970, Astronaut Walter Cunningham (Apollo 7), with Col. Lee James of the Saturn program present, visited MSFC and presented me their Snoopy Pin Award. A picture was made during the presentation, and I later received a copy of the print. Thirty-seven years later, I met Cunningham again and showed him the photo, which he immediately autographed for me. This event occurred during the Fourth Annual Saturn/Apollo Reunion, U.S. Space and Rocket Center, Huntsville, Alabama, in July 2007.

207

The Computer Center at Slidell
by Charles Bradshaw

Computers of all descriptions played an important role in the success of the first moon landing. The large computer centers at the Marshall Space Flight Center, the Manned Spacecraft Center and in particular the computers used in the Mission Control Center were essential to the success of the mission. The computers employed were the earliest of what is generally referred to as the second generation (transistors). A computer center at Slidell, Louisiana was brought together in record time and made important contributions to the mission at Michoud (Louisiana) and the Mississippi Test Facility at Bay St. Louis, Mississippi.

The MSFC had taken over the large WWII facility named the Michoud Facility (some twenty miles from New Orleans) with plans to use the facility as the site to construct large fuel tanks for the Saturn launch vehicle. Chrysler and the Boeing Company were selected as prime contractors for the facility and both presented plans for a large computer facility to be located at either end of the large building. This seemed to be duplication to George Constan and he asked that I come to a meeting to review these proposals. I soon met with George, his deputy, Jim Stamey, and "Ab" Abernathy, contract specialist for Michoud. It was obvious to all present that a joint use facility would save one and a half to two million dollars on the start-up and several hundred thousand a year thereafter. It was decided in this meeting there would be a single computer center, designed and built by MSFC, with computers and operating staff provided by MSFC. This facility would provide the computing resources needed by Boeing, Chrysler and other contractors at Michoud.

As plans were being made for this joint use facility to be located in the Michoud building, a just completed building in Slidell (some 15 miles from Michoud) was declared surplus by the Federal Aviation Authority and was acquired by MSFC. It was decided to locate the joint use computer facility in this new building, which was well configured for a computer center and would require only minor modifications. Specifications for two computer systems, one for engineering computations and one for administrative systems, were developed by MSFC and reviewed by a committee of the contractors formed for the purpose of coordinating the joint use facility. Several changes were made and there continued to be strong opposition from Boeing and Chrysler to the concept of a joint use facility.

In early July 1962 Bob Reeves was transferred from the Computation Laboratory in Huntsville to head the developing center at Slidell. He faced the awesome task of bringing together the computers, the personnel and other resources to have an operating computer center in just six months. The computers had been selected (IBM 7090 for engineering and Honeywell 800 for administrative) and the next task was to select and contract with a firm to provide computer personnel. This, along with dozens of other tasks, was completed and the Slidell Center became fully operational in January 1963.

Maintaining an Even Strain
by Robert Saidla

One way we survived the extreme stress and very long hours of our R & D static firing test program was to find humor in the worst of situations. Our first rocket engine had just arrived at the Static Test Tower East. It was sitting on the pavement in front of the tower. Ronald Chittam, our mechanical technician foreman had the crew preparing to connect the gantry crane hook to the engine to lift it to the tower. Our German bosses and the Rocketdyne engine representatives had emphasized to us how careful we must be to avoid damaging the exposed thrust chamber tubes, which formed the nozzle. We were all standing around the engine getting familiar with it. Ace Perry, a mechanical technician approached the engine with a large wrench in his hand to tighten the lifting sling, tripped and the wrench slammed into the nozzle, making a loud clang. There was a deadly silence. We were horrified. Then another of our technicians, perhaps Emmitt Powers or Edward Ball, said, "Man Ace, you really lost a lot of points then!" Then Erich Kaschig, our German chief of the test tower, said, "No! Ace didn't lose any points. You have to have points to lose them." That broke the tension and we all laughed. In checking the nozzle there was no damage.

Skylab, America's First Station in Space
by Ernst Stuhlinger

Skylab was launched by a Saturn V rocket in May 1973. It had the size of a house and weighed 90 tons. Orbiting the Earth at an altitude of 230 miles in a plane inclined by 50 degrees toward the plane of the equator, its trajectory covered 75 percent of the Earth's surface, 80 percent of its food-producing area, and 90 percent of its populated regions. During its active lifetime, Skylab cameras took 35,000 photos of the Earth; about 140 different groups of scientists from 21 nations had developed and built, for operation by the astronauts, scientific or technological experiments for Earth observations, and for physical, chemical, metallurgical, astronomical, biological, and medical studies under the vacuum and near-weightlessness conditions in orbit. Three groups of astronauts, each with three men, spent a total of 184 days on Skylab. It was by far the largest and most complex spacecraft of its time. After its life in orbit had ended, more than 1,000 scientific and technical papers were published about the results of its experiments.

Around 1963, support for the still very tentative and fragile space station idea came from an other place. It began with a modest inquiry by Dr. Nancy Roman, in charge of astronomy projects at NASA Headquarters. "Could you mount an astronomical telescope on a Command and Service Module, as you use it on your Moon rocket, take pictures of stars and their spectra, and have the astronauts bring the films back to Earth?" she asked me. This simple question, coming from Headquarters, enabled us to legitimize our quest for an active space station project. Over the years, it resulted in a landslide of activities that led to the very impressive and successful Skylab Project.

I first talked to Jerry McCall, special assistant to Dr. von Braun, trying to obtain his help in establishing a planning activity for an orbiting research laboratory. His reaction was very negative. "This does not make any sense to me," he said. "We should keep our hands off a project like this." Then I talked to Walter Haeussermann, Director of the Astrionics Laboratory. He was immediately enthusiastic about the prospects of a research station in orbit. We went to von Braun together, and from that moment on the space station project was wide alive. Although deeply entrenched at that time in the huge Saturn-Apollo program, von Braun took the decisive steps to activate the project. He joined forces with Dr. Gilruth, Director of the Johnson Center in Houston; the two of them convinced George Mueller, Director of Manned Space Flight Projects at NASA Headquarters, of the great value of such a project for NASA. He promptly became a strong supporter of the project and took part in the planning of its technical details.

Von Braun had always been very eager to have scientists use the potential of his rockets to carry their instruments into space where they could make observations not possible from the Earth's surface. One of the duties he assigned to me soon after we had joined NASA was an effort to develop contacts with scientists at NASA Headquarters and at other places in the country, and to make them aware of the ability of our rockets to provide opportunities for scientific research in and from outer space.

Instead of one astronomical telescope on the small Command and Service Module of a Saturn-Apollo rocket that Dr. Roman had requested earlier, we could offer her the prospect of a large orbiting observatory and workshop, tended by astronauts, including a collection of telescopes for a variety of astronomical studies.

The hydrogen tank of one Saturn IB would be equipped with an access hole, an air lock, and a multiple docking adapter, and also with a diaphragm dividing the interior of the tank into two compartments. The other Saturn IB would carry a Command and Service module, three astronauts, and all the instruments that would make up a manned orbiting laboratory. The second Saturn IB would be launched one day after the first. After arrival in orbit, the first hydrogen tank would be completely emptied, cleaned, and set-up by the astronauts.

NASA - Marshall Flight Space Center

The wet workshop scheme sounded simple, and it could be offered as a very cost effective project to the planners in Washington. In reality, however, it was a nightmare for the project engineers at the Marshall and Johnson Centers. So, they played, although inconspicuously, with the idea of a 'dry workshop' scheme: an empty hydrogen tank of an S IVB stage would be fully converted and equipped on Earth as an orbiting laboratory, complete with telescopes, solar panels for electric power, crew quarters and accommodations, air, water, and food supply, and then put on a two-stage version of the big Saturn V rocket. It would be launched unmanned, and checked out remotely. Only when everything worked properly in orbit, the three-man crew would be launched with a Saturn IB. This dry-workshop scheme was very satisfactory from the standpoint of the project planners, but it was considerably more expensive than the wet-workshop scheme. During the late 1960s, considerable planning for the dry workshop was underway at Marshall and Johnson, but very quietly, and no effort was made to obtain a go-ahead for that scheme from Washington.

George Mueller at NASA Headquarters, eager to obtain a go-ahead for the dry workshop project from Congress, knew that he could hope for an approval only if he could submit his request at a moment of exceptionally favorable circumstances. Fortunately, such a moment occurred in July 1969, after Apollo 11 had enabled Neil Armstrong and Buzz Aldrin to land on the Moon. Mueller, talking to Congress, and shining in the glory of this wonderful achievement, concluded his presentation: "By the way, we have decided that we want to equip our first longtime space station with a dry workshop. This will make a far better project, simpler to build, and far more efficient as a research station." Congress approved the project immediately, and work in Huntsville, in Houston, and at a large number of research laboratories all over the country went into high gear. The project received the official name Skylab in 1970.

For most of the members of the Marshall and Johnson Centers it was a novel experience to adjust their engineering work to the wishes and needs of scientists that were dictated primarily by the uniqueness of the scientific experiments. Fortunately, the promoters of the Skylab Project, who had to act as mediators between the scientists and the engineers, enjoyed the strong support of the two senior members of the administrative staff at Marshall, Harry Gorman and Richard Cook.

The rest of Skylab's history has been told many times. Von Braun left Marshall early in 1970 and joined NASA Headquarters as Administrator Tom Paine's Chief Planner. Eberhard Rees, and then Rocco Petrone succeeded him as Marshall Directors. As the Skylab Project proceeded, some staff members became concerned that Skylab might be hit by a meteoroid, and suggested that it be protected against that danger. A meteoroid shield -- a thin sheet metal layer about an inch above the wall of the tank -- was installed, against the advice of some of the older members. When Skylab was launched on May 14, 1973, on top of a Saturn V rocket, atmospheric forces ripped off the shield and with it one of the large solar panels, leaving the workshop and crew quarters with reduced electric energy, and with much increased solar heating of the workshop's interior. A very intense damage control and repair effort started immediately under the command of Center Director Rocco Petrone. The first crew, launched belatedly (on May 25) with a Saturn IB, with a sun shield that could be installed like an opening umbrella, and with other repair tools, was successful in putting the ailing Skylab back into a state that allowed the astronauts to live and to work on Skylab for 28 days. The second crew left the Earth in July and stayed for 59 days. The third crew, leaving in November, stayed for 84 days.

Skylab I was still in orbit when Skylabs II and III were in preparation. The members of the Skylab Project at Marshall and Johnson would have been eager to apply their newly-won experience with scientific space instruments and their creators in further Skylab launchings, making these projects smoother, and more efficient and fruitful, than Skylab I. However, all-mighty Washington had different ideas. Skylab II and III were not authorized for flight; they ended up as museum pieces. MIR, launched in Soviet Russia on February 20, 1986, became mankind's next space station. It orbited the Earth for 13 active and very successful years, sometimes encountering dramatic incidents, and it established a new endurance record that may be broken once the International Space Station has become operational in space.

The completed Saturn I S-I stage (booster) during the checkout in the Marshall Space Flight Center, building 4705, January 23, 1961. The Saturn I S-I stage had eight H-1 engines clustered, using liquid oxygen/kerosene-1 (LOX/RP-1) propellants capable of producing a total of 1,500,000 pounds of thrust.

Retiree Roster

Abernathy, Agnes E.
Abernathy, William J.
Abernethy, Hosea B.
Able, Robert M.
Abraham, Ronald W.
Absher, Archie C.
Acker, Richard M.
Ackerman, Donald J.
Ackerman, Grace M.
Ackerman, John E.
Acuff, James B.
Adair, Billy M.
Adair, Fred T.
Adair, Jessie B.
Adair, Lynda P.
Adams, Albert
Adams, Arlis C.
Adams, Arvard K.
Adams, Charles L.
Adams, Clarence C.
Adams, Elizabeth J.
Adams, George D.
Adams, Ira A.
Adams, John Q.
Adams, Leslie F.
Adams, Opal L.
Adams, Ralph G.
Adams, Robert L.
Adams, Sallie B.
Adams, Sammie G.
Adams, Warren A. Jr.
Adams, William R.
Adams, William T.
Adamson, James Z.
Adcock, Janes H.
Adcock, John C.
Adcock, Marvin E.
Adcock, Mavis A.
Aden, Robert M.
Aderhold, Harlin H.
Aderholt, Leroy
Adkins, Allison J.
Adkins, Maxine M.
Adkins, William D.
Admire, John R.
Agar, Robert C.
Agee, Kay C.
Ahlander, Barbara A.
Ahlander, Dan H.
Aichele, David G.
Aiello, David E.
Ainsworth, Claude C.
Akbay, Ismail
Akens, David S.
Akins, James P.
Akridge, Charles M.
Albright, Eddie R.
Alcott, Russell J.
Aldrich, Billy R.
Aldridge, Betty K.
Aldridge, Burch H.
Aldridge, Carlton W.
Alexander, Carolyn J.
Aldridge, Charlotte G.
Alexander, Douglas W.
Alexis, Everett C.
Alford, Ben
Allbritain, Robert W.
Allday, Leslie J.
Allen, Albert R.
Allen, Bruce B.
Allen, Charles S.
Allen, Clarence Y.
Allen, David W.
Allen, Dona A.

Allen, Genenne S.
Allen, Gilbert V.
Allen, Hazle M.
Allen, Henry D.
Allen, Howard W.
Allen, Hurley W.
Allen, James G.
Allen, James K.
Allen, James M.
Allen, Jesse B.
Allen, Joseph H.
Allen, Lawrence H.
Allen, Leon B.
Allen, Phillip W.
Allen, Robert C.
Allen, Thomas W.
Allen, Vern P.
Allen, Waldo A.
Allen, William N.
Allen, William R.
Allfrey, Barbara G.
Allison, Paul R.
Allred, Paul J.
Allums, Steve L.
Almon, Winston C.
Alston, Earl H.
Alsup, Brice R.
Alter, John D.
Alverson, Eva M.
Alverson, Russell E.
Alward, Paul B.
Amalavage, Albin J.
Ambrose, Colin E.
Amburn, Herbert G.
Amos, Carl E.
Amos, Edna F.
Anders, Hoyd J.
Anderson, Alvin W.
Anderson, Audie E.
Anderson, Bruce Henry S.
Anderson, Dorrance L.
Anderson, Louis M.
Anderson, Matt W.
Anderson, Nehemiah
Anderson, Oakley S.
Anderson, Otis C.
Anderson, Paul E.
Anderson, Philip N.
Anderson, Samuel D.
Anderson, Sherwood H.
Anderson, Sylvia B.
Anderson, Wannie H.
Anderson, William L.
Anderson, Wilson A.
Andres, Merle S.
Andressen, Christian E.
Andrews, Bill
Andrews, Clayton D.
Andrews, Donald R.
Andrews, Douglas E.
Andrews, Henry C.
Andrews, John W.
Andrews, Ronald M.
Andrews, Thomas L.
Angele, Wilhelm
Anglin, William T.
Anthony, Kenneth G.
Apkarian, Sadie
Apostolos, Ethel S.
Applegate, James F.
Appling, Marvin
Archambault, Amos D.
Arcilesi, Camillo J.
Arden, Mary D.
Arden, Oliver W.

Ardoin, Joseph E.
Arender, Barney C.
Arias, Carlos M.
Armistead, Reginald R.
Armistead, William C.
Armstrong, Frank W.
Armstrong, George B.
Armstrong, Murphy D.
Armstrong, Rex W.
Arnett, Carl D.
Arnold, David A.
Arnold, James E.
Arnold, Ray J.
Arnold, Thomas H.
Arnold, Willie M.
Arnts, Gerald W.
Arrington, Fred D.
Arsement, Leo A.
Artis, Paul T.
Ashley, Houston W.
Askew, Charles R.
Askew, William C.
Askins, Carl C.
Asquith, Robert C.
Aston, Charles M.
Atchley, Hazel T.
Atherton, James R.
Atkins, Harry L.
Atkins, John H.
Atnip, Pink A.
Attaya, Henry E.
Auer, Ann C.
Austell, Gilbert H.
Austin, John G.
Austin, Robert E.
Austin, Robert W.
Austin, Thelmer M.
Avery, Alvin H.
Avery, John D.
Avery, Patricia E.
Avery, Robert M.
Avery, Samuel H.
Aycock, Claude R.
Ayers, Mary Y.
Ayers, Verda R.
Babakitis, Nicholas
Babb, Lloyd C.
Babb, Mary W.
Babington, Herbert R.
Bacchus, David L.
Bachtel, Frederick D.
Backer, Ronald D.
Baddley, Rusell E.
Baerg, Arthur
Baggett, Alvin
Baggs, Ellis T.
Bagley, Ben D.
Bailes, Claude W.
Bailes, Imogene F.
Bailess, Lewis E.
Bailey, Brenda K.
Bailey, Calvin V.
Bailey, Curtis R.
Bailey, Donald
Bailey, Doyce R.
Bailey, Doyle J.
Bailey, George A.
Bailey, John C.
Bailey, Judson O.
Bailey, Wilbur M.
Bailey, William M.
Bailey, William R. Jr.
Bain, Billy J.
Baites, John M.
Baker, Billie B.

Baker, Cleamon O.
Baker, Clyde D.
Baker, Edward N.
Baker, Edward R.
Baker, Frank F.
Baker, Fred A.
Baker, John
Baker, Joseph W.
Baker, Leonard D.
Baker, Richard
Baker, Robert P.
Baker, William P.
Balance, James O.
Balch, Charles W.
Balch, Harry A.
Baldwin, Abel H.
Baldwin, Claude M.
Baldwin, Dalas B.
Baldwin, Jerry B.
Baldwin, Lorine S.
Baldwin, William P.
Balentine, Robert C.
Ball, Edward W.
Ball, Erich K.
Ball, James A.
Ball, Leslie W.
Ball, Otto C.
Ball, Sanford G.
Ballard, Carl L.
Ballew, Calvin
Banasiak, John S.
Bandgren, Harry J.
Banholzer, Gordon S.
Banks, Coy C.
Bankston, Benjamin F.
Bannister, Tommy C.
Barber, John A.
Barden, Winfred M.
Barefield, Alexander
Barfield, Thomas K.
Barisa, Bartholomew B.
Barkley, Billy
Barkley, Hillard R.
Barkley, Homer
Barksdale, Thomas A.
Barley, Billy R.
Barnes, John W.
Barnes, Ralph W.
Barnes, Robert O.
Barnes, Scottie P.
Barnes, Wayne L.
Barnett, Darthy M.
Barnett, Harry B.
Barnett, Thomas A.
Barnett, Zack Jr.
Barr, Glenn P.
Barr, Thomas A.
Barraza, Robert O.
Barraza, Rodolfo M.
Barret, Chris
Barrett, Thomas W.
Barrett, William R.
Beltron, Antonio
Barron, Virginia A.
Barron, William R.
Bartlett, Elmer O.
Bartlett, Frank K.
Bartley, Stephen C.
Basel, Leonard C.
Bass, Billy B.
Bass, Carl H.
Bass, James W.
Bates, David K.
Bates, James P.
Batey, Johontas G.
Batey, Zack T.

Batson, Clarence A.
Batson, Norman M.
Batson, Obed E.
Batte, James O.
Battle, James L.
Batty, Frank R.
Bauer, Dee V.
Baxter, Albert F.
Bayless, Ernest O.
Bayless, Syble C.
Beal, John D.
Beale, Stephen P.
Beall, Frank E.
Beam, Alicia C.
Beam, Everette E.
Beam, Hoyle W.
Beaman, Robert E.
Bean, Clyde D.
Bean, Coy
Bean, Jack R.
Bearden, General B.
Bearden, Loyce J.
Bearskin, Alvin W.
Beasley, Barnes
Beasley, John L. Jr.
Beasley, Travis E.
Beason, Fred H.
Beatty, Elwin J.
Beaty, James B.
Beavers, Robert R.
Bechtel, Robert T.
Bechtel, Thomas D.
Beck, Elwood M.
Beck, Marvin D.
Beck, Richard
Beckham, Richard H.
Becks, Robert
Beddingfield, James W.
Bedell, Thomas G.
Bedsole, Hurley H.
Beduerftig, Herman F.
Beecham, Martha S.
Beecherl, George R.
Belanger, Karin L.
Belcher, Edwin J.
Belcher, Jewell Green Jr.
Belcher, John T.
Belew, Claude M.
Belew, Herschel W.
Belew, James A.
Belew, Leland F.
Belew, Robert R.
Bell, John R.
Bell, Joseph L.
Bell, Lucian B.
Bell, Robert A.
Bell, Watt R.
Bell, William W.
Bellingrath, Albert F. III
Belsom, Edgar L.
Belton, Paul D.
Barrett, William R.
Beltron, Antonio
Bencaz, Harold R.
Bendall, Sharon T.
Bendel, Vincent P.
Benefield, Abraham
Benefield, Dairrel M.
Benefield, James C.
Benefield, William L.
Bennett, Harry K.
Bennett, Mary J.
Bennett, Russell H.
Bennich, Jerry B.
Bennight, J. D.
Bensko, John

Bentley, Agnes C.
Bentley, Foster L.
Bentley, Fred
Bentley, Margarette Q.
Bentley, Mark W.
Bentley, Robert M.
Benton, Ethel G.
Benton, John T.
Beranek, Richard G.
Berg, Gerald H.
Berge, Leroy H.
Bergeler, Herbert R.
Berger, Edwin A.
Berisford, Edward B.
Berkebile, Marlin J.
Berridge, Charles A.
Berry, Charles E.
Berry, Harry H.
Berry, James W.
Besherse, John B.
Besler, Julius H.
Bess, Herman L.
Best, Cecil A.
Best, Robert G.
Bethay, J. A.
Betsill, Herbert L.
Betterton, Garland G.
Beumer, Joseph H.
Beutjer, Jayne L.
Beutjer, William H.
Bevels, Charlie F.
Bevels, Peggy G.
Bevis, Angus L.
Bey, Franklin A.
Beyer, Rudolf E.
Beyerly, Frederick J.
Bianca, Carmelo J. Jr.
Bibb, John B.
Bice, Kenneth H.
Biggerstaff, Charles R.
Biggs, Patrick M.
Bilbro, James W.
Billinghurst, Edward E. Jr.
Billings, Emma A.
Billmayer, Hanns L.
Binkley, Robbye M.
Birdsong, Joe C.
Birdsong, Sam A.
Birdwell, Boyce R.
Birdwell, Frances L.
Birdwell, Lorene
Birney, Herman H.
Bishop, Donald A.
Bishop, Donald F.
Bishop, Gerald G.
Bishop, Guy R.
Bishop, James R.
Bishop, Leonard M.
Bishop, Maylond T.
Biss, Carl A.
Black, Amos C.
Black, Claude E.
Black, Dolphus H.
Black, Lillion F.
Black, Norris M.
Black, Paul E.
Black, Ray
Black, Richard E.
Black, William H.
Blackmon, Patricia B.
Blackstone, John H.
Blackwell, Douglas L.
Blackwell, Patricia A.
Blackwood, Barbara P.
Blackwood, David L.

NASA - Marshall Flight Space Center

Blaine, Dana M.
Blair, Howard W.
Blair, James C.
Blair, Roy W.
Blair, Winfred E.
Blaise, Herman S.
Blaise, Herman T.
Blake, Bernard S.
Blake, Guy D.
Blake, Virgil H.
Blanchard, Mary S.
Blanche, James F.
Blankenship, Arlee M.
Blankenship, Betty B.
Blankenship, Jack G.
Blankenship, Jimmie R.
Blankenship, Stuart M.
Blanks, Elizabeth B.
Blanteno, James S.
Blanton, James E.
Blasczyk, Erwin W.
Blaylock, Gordon P.
Blaylock, Leonard E.
Bledsoe, Elgan F.
Bledsoe, Ronald B.
Bledsoe, Theodore T.
Blenis, Dan R.
Blevins, Calvin B.
Blevins, Frederick H.
Blevins, Harold R.
Blevins, Kathy L.
Blocker, James R.
Blomberg, Carl G.
Bloom, Kenneth W.
Blount, Dale H.
Blue, Matthew P.
Boardman, Frank S.
Boatwright, Carroll W.
Boatwright, Doyle C.
Bobo, David E.
Bobo, James E.
Bobo, Mildred B.
Bobo, Virginia Y.
Boddie, Thomas T.
Bodie, William G.
Boehm, Loudene O.
Boehm, Richard J.
Boffola, Frank P.
Boger, John C.
Boggs, Alma R.
Boglio, William L.
Bolding, Herbert A.
Bolding, J. W.
Bolin, Betty M.
Bolin, Curtis W.
Bolton, Charles H.
Bombara, Elwood L.
Bomgaars, Merlin J.
Bond, Alpha R.
Bond, Jean B.
Bond, William M.
Bonney, Henry M.
Book, George W.
Boone, Charles T.
Booth, June A.
Boothe, Jewel M.
Boothman, Irvine S.
Borcherding, Paul W.
Bordelon, Carroll S.
Bordelon, Terence T.
Borelli, Michael T.
Bosch, George W.
Boshers, Telitha W.
Boshers, William A.
Bost, Johnnie
Bostwick, Leonard C.
Boudra, Paul W.
Bouldin, Lamuel C.
Boutell, Richard R.
Bowen, Adrian E.
Bowen, Arthur L.
Bowen, Claude J.
Bowen, Robert O.
Bowen, William R.
Bowers, Robert B.

Bowling, Aubrie L.
Bowling, Bernice J.
Bowling, Cullan O.
Bowling, Thomas E.
Bowman, Harry F.
Boyanton, Arthur O. Jr.
Boyanton, Arthur O. Sr.
Boyd, Green E.
Boyd, John W.
Boyden, Richard
Boyer, Esther T.
Boyer, Robert G.
Boyett, Marvin E.
Boykin, Claude C.
Boykin, Clifford H.
Boykin, Nell B.
Boyle, James H.
Boyle, Stanley F.
Boylen, Corrine B.
Braam, Frederick W.
Braddy, Sidney H.
Bradford, Alma W.
Bradford, Dorcas M.
Bradford, Joel H.
Bradford, Robert L.
Bradford, Robert N.
Bradford, William C.
Bradley, Andrew L.
Bradley, Jesse N.
Brady, Hugh J.
Brady, Jewell W.
Brady, William L.
Bragg, John L.
Bramlet, James B.
Brandner, Friedrich W.
Brandon, Larry B.
Brandon, Patricia D.
Brandon, Walter W.
Brannan, Henry C.
Brannon, Julian R.
Brantley, Lott W. Jr.
Brasell, Richard P.
Brasfield, Buel T.
Brasher, Charles H.
Bratcher, Andrew L.
Bray, James R.
Brazelton, Annie T.
Brazelton, Hazel M.
Brazelton, John W.
Brazelton, Silas A.
Breazeale, William L.
Breece, Cyrial J.
Breedlove, Theresa L.
Breithaupt, William G.
Breland, Charles G.
Breland, Norris O.
Brennecke, Margaret W.
Bressette, Edwina J.
Brewer, Bradley
Brewer, Edwin B.
Brewer, Henry R.
Brewer, James B.
Brewer, Lee H.
Brick, Robert W.
Bridges, Chester C.
Bridges, Doris S.
Bridges, Jack K.
Bridges, James P.
Bridges, Robert A.
Bridges, Winston L.
Bridwell, Gene P.
Brien, William L.
Bright, Cora S.
Bright, Covington L.
Bright, Frazier E.
Bright, Lillian S.
Brightwell, Wanda T.
Bringman, Ronald R.
Brinkley, Dorothy W.
Britz, William J.
Brizendine, Charles E.
Broad, William M.
Broadway, J. B.
Broadway, James R.
Brock, Benjamin F.

Brock, James L.
Brock, Jimmy W.
Brock, Martha G.
Brock, Paul H.
Brock, Ronald C.
Brock, Warner H.
Brockwell, Thurman E.
Brodowski, John E.
Brolliar, Richard H.
Brooks, Alvis C.
Brooks, Charles
Brooks, Charles G.
Brooks, Columbus
Brooks, Eddie L.
Brooks, Eugeal S.
Brooks, Floyd L.
Brooks, George R.
Brooks, Joe D.
Brooks, Melvin
Brooks, Otis L.
Brooks, Thomas A.
Brooksbank, William A.
Brookshire, Josef O.
Brosemer, Charles A.
Brothers, Bobby G.
Brothers, Johnie C.
Brouillette, Everett A.
Broussard, Peter H.
Brown, Alvis C.
Brown, Ben L.
Brown, Boyce C.
Brown, Carole A.
Brown, Claude Kelly
Brown, Coy J.
Brown, Edwin L.
Brown, Elmer L.
Brown, Francis C.
Brown, Fred O.
Brown, Harold E.
Brown, Harrison K.
Brown, Herbert L.
Brown, James L.
Brown, James V.
Brown, Joe H.
Brown, John A.
Brown, John D.
Brown, Kenneth D.
Brown, Marvin H.
Brown, Mary E.
Brown, Richard L.
Brown, Robbie
Brown, Robert W.
Brown, Sayre C.
Brown, Wallace F.
Brown, William D.
Brown, William D.
Bruce, Ann S.
Bruce, Foch P.
Bruce, Henry W.
Brumback, Byron M.
Brumley, James M.
Brunell, Richard C. Jr.
Bryan, Comer H.
Bryan, Edward J.
Bryant, Charles H.
Bryant, Charles W.
Bryant, Leslie E.
Bryant, Melvin A., III
Bryson, Grady L.
Bryson, Joe E.
Bubke, William T.
Buchanan, Harry J.
Buchanan, Jack B.
Buchanan, R. A.
Buchanan, Walter H.
Bucher, George C.
Buck, John F.
Buck, Theodarit M.
Buckelew, Denvil C.
Buckelew, Volis L.
Buckelew, William P.
Buckner, Garland G.
Bufkin, John N.
Buford, Harry W.
Buford, Jewel M.

Bugbee, Gary D.
Buhmann, Eugene J.
Buist, Charles A.
Bulette, Floyd E.
Bull, James T.
Bullard, Thomas
Bullington, Ernestine
Bullington, Marvin
Bullock, Tulon
Bunce, Earl G.
Bunn, Wiley C.
Bunnell, Alice S.
Bunnell, Charles R.
Burch, John L.
Burdett, Hazel C.
Burdett, Lewis H.
Burdine, William M.
Burgess, Gloria R.
Burgett, Maude C.
Burgreen, Jerre H.
Burke, Clyde L.
Burke, Harlan D.
Burkett, Audey S.
Burks, Alfred J.
Burner, Beverly J.
Burney, Sam T.
Burns, Annie R.
Burns, David H.
Burns, Deborah D.
Burns, Howard
Burns, John W.
Burns, Milan D.
Burns, Ralph A.
Burns, Rowland E.
Burrell, Martin O.
Burrough, Roy A.
Burrow, Patricia B.
Burrows, Dale L.
Burrows, Roger R.
Burrus, Ed R.
Burruss, Hattie M.
Burruss, James L.
Burruss, James W.
Burson, Joe H.
Burton, Donald E.
Burton, Edward H.
Burton, John W.
Burton, Thomas L.
Burton, Von L.
Busbin, Homer D.
Busby, Carol D.
Busby, George M.
Bush, Billie J.
Bush, Donald R.
Bush, Edward T.
Bush, Herbert T.
Bush, James C.
Bush, Robert A.
Bush, William C.
Butcher, Shirley A.
Butler, Albert L.
Butler, Charles S.
Butler, Connie H.
Butler, Earl M.
Butler, Floyd H.
Butler, George
Butler, John M.
Butler, Lydia H.
Butler, Ralph Q.
Butler, Raymond D.
Butler, Robert E.
Butler, Shirley S.
Butler, Teddy L.
Buttram, William H.
Buzbee, Claude Q.
Byerline, Charles R.
Byers, James E.
Byler, Elizabeth S.
Bynum, Bobby G.
Bynum, Frank W.
Bynum, Julian E.
Bynum, Watson
Byram, Kenneth C.
Byrd, Ambrose W.
Byrd, Clarence A.

Byrd, Katherine W.
Cabler, James F.
Caddy, Barbara H.
Caddy, Larry A.
Cadle, Sam A.
Caffey, John B.
Cagle, Billy R.
Cagle, Euel D.
Cagle, Eugene H.
Cagle, Floyd E.
Cagle, Fred A.
Cahalan, Paul F.
Cahall, Robert B.
Caillouet, Vernon G.
Cain, Sandra G.
Caldwell, Edward L.
Caldwell, Roy E.
Caldwell, Walter D.
Calhoun, James D.
Callaway, Richard C.
Calsoni, Arthur
Calvert, John A.
Cameron, Frank M.
Camp, Dennis W.
Camp, Howard C.
Campbell, Edward C.
Campbell, Elbert C.
Campbell, Evelyn L.
Campbell, Glenn
Campbell, Hugh M.
Campbell, James L.
Campbell, Jeffrey L.
Campbell, John D.
Campbell, Laurence B.
Campbell, Oscar S.
Campbell, Robert N.
Campbell, Wallace G.
Campbell, Willa J.
Campbell, William L.
Campbell, William M.
Camper, Jack L.
Caneer, Rosanna B.
Cannon, Robert J.
Cantrell, Eddie R.
Cantrell, Joseph P.
Cantrell, Roy O.
Cantrell, Thomas D.
Cantrell, Will A.
Capley, Fred M.
Capley, Hilda I.
Capps, John D.
Caraccioli, Philip M.
Carden, Ernest
Carden, James R.
Cardin, Kelsie T.
Carey, William T.
Carlile, Chalmus D.
Carlin, Robert S.
Carlisle, James L.
Carlisle, John G.
Carlson, William A.
Carlton, Joseph M.
Carmack, Sybil L.
Carmody, Robert J.
Carmosino, Vito N.
Carpenter, Gene T.
Carr, Arthur M.
Carr, Billy W.
Carr, Judith M.
Carrington, John B.
Carroll, Jack M.
Carroll, Joe D.
Carroll, Marvin E.
Carroll, Stanley N.
Carroll, Tommie R.
Carter, Calvin E.
Carter, Charles H.
Carter, Dewey E.
Carter, Evelyn H.
Carter, James A.
Carter, James H.
Carter, Jimmy H.
Carter, Kathryn F.
Carter, Nell E.
Carter, Roy L.

Carter, Ruby B.
Carter, Thomas J.
Cartwright, Bradley P.
Cartwright, Johnnie M.
Caruso, Salvadore V.
Caruso, Vincent P.
Carver, Robert R.
Carver, Tillmon
Carwile, James A.
Case, Lee R.
Casey, Charles W.
Casey, Roy T.
Cash, James D.
Cash, Mitchell
Cason, George R.
Cassidy, James G.
Cassimus, George D.
Casteel, Mable L.
Castleberry, William V.
Castleman, Murray W.
Cataldo, Charles E.
Cater, Albert S.
Cates, Paul E.
Cather, John C.
Cathey, John M.
Caudle, Billy G.
Caudle, Alexander G.
Caudle, John M.
Caudle, Ronald J.
Causey, Wilton E.
Cauthen, Joseph M.
Cavalaris, James G.
Cavaliere, Benjamin A.
Cavanaugh, Francis J.
Cavanaugh, John W.
Cavnar, Sandra C.
Cawthon, William P.
Cerny, Otto F.
Chafin, Carolyn M.
Chafin, James D.
Chamberlain, James D.
Chambers, Charles H.
Chambers, Charles M.
Chambers, John A.
Chambers, Roy L.
Chamness, Mary O.
Champion, Peggy J.
Champion, Ray H.
Chandler, Carl C.
Chandler, Cleo W.
Chandler, Doris C.
Chandler, Harry L.
Chandler, Keith B.
Chandler, O. W.
Chandler, Rebecca J.
Chandler, Thomas A.
Chandler, William A.
Channell, Dewey B.
Chapman, George M.
Chapman, Harvey J.
Chapman, Howard L.
Chapman, John D.
Chapman, Judy S.
Chapman, Richard A.
Chappell, Charles R.
Chase, John B.
Chassay, Roger P.
Chassay, Roger P. Jr.
Chatham, Roy L.
Chaviers, Vernon O.
Chenault, Alba P.
Chereek, Benjamin
Chesley, Albert E.
Chesser, Charles M.
Chesteen, James H.
Chevaillier, Lewis B.
Childers, Brady C.
Childers, Chalmer J.
Childers, Delbert E.
Childers, Harry K.
Childers, John H.
Childress, John H.
Chisam, Charles L.
Chisholm, William L.
Chittam, Bobby L.

211

50 Years of Rockets and Spacecraft

Chittam, James R.
Chittam, Jefferson C.
Chittam, Ronald C.
Choate, Bernice K.
Choate, Earl S.
Chou, Lynn C.
Chovan, William
Chow, Alan S.
Christian, Clarence R.
Christian, David C.
Christian, Hugh J. Jr.
Christner, Amos P.
Christopher, Ashford H.
Christopher, James A.
Christopher, Jennie C.
Christy, James O.
Chubb, William B.
Chumley, Herman G.
Chumley, James F.
Churchey, Jackie E.
Cicero, Florence S.
Cizek, Richard J.
Clanton, Esly A.
Clanton, Price
Clark, Adrian V.
Clark, Arthur E.
Clark, Beryl M.
Clark, Billy D.
Clark, Charles G.
Clark, Charles H.
Clark, David C.
Clark, Egbert E.
Clark, Evelyn Q.
Clark, Floyd M.
Clark, Herman H.
Clark, James B.
Clark, James E.
Clark, James H.
Clark, James W.
Clark, Johnny C.
Clark, Keith H.
Clark, Marvin R.
Clark, Max H.
Clark, William D.
Clark, William O.
Clarke, Walton G.
Clarke, William A.
Clarke, William D.
Clatworthy, Ralph W.
Claunch, William C.
Clay, Lura P.
Clay, Mary B.
Clay, Richard W.
Clay, Ruby I.
Clay, Russell W.
Clayton, Archie B.
Clayton, Judith T.
Clem, Lottie B.
Clem, Marguerite C.
Clem, Richard G.
Clem, Robert V.
Clem, Thomas J.
Clements, Maelon H.
Clements, Walter
Clemmons, Charles B.
Clemons, James A.
Clemons, Michael C.
Cleveland, Bill
Cleveland, Linda N.
Clever, William W. II
Clifford, Carolyn M.
Clift, J. R.
Clifton, Kenneth S.
Cline, Doris S.
Cline, Starke T.
Clines, Oren L.
Clonts, Samuel E. Jr.
Clotfelter, Wayman N.
Cloud, James W.
Cloud, Osceola
Clough, Daniel R.
Clough, William W.
Coates, Keith D.
Cobb, Benjamin C.
Cobb, George H.

Cobb, Oleta B.
Cobb, Robert J.
Cobb, William A.
Cobb, William E.
Cobb, William O.
Cochran, John R.
Cochran, Joseph M.
Cochran, Roy J.
Cockrell, Daniel R.
Cody, Joseph C.
Coffey, Sherry T.
Cofield, Kester L.
Cofield, Melvin B.
Cogbill, Beodric
Coggin, James M.
Coiner, Warren G.
Coker, Doris W.
Colbert, Mary B.
Colburn, Kenneth P.
Coldiron, Henry C.
Coldwater, Harold R.
Cole, Bethel A.
Cole, Billy B.
Cole, John D.
Cole, John W.
Cole, Orbin
Cole, Ralph E.
Cole, William J. Jr.
Coleman, Archie D.
Coleman, Harvey
Coleman, Ruth L.
Coleman, Sandra C.
Coleman, Weldon G.
Coller, Louis J.
Colley, Carl E.
Collier, Eugene B.
Collier, Glyn
Collier, Imogene J.
Collier, Shelley D.
Collins, Charles L.
Collins, Dewey S.
Collins, Elizabeth R.
Collins, Lester R.
Collins, Oliver O.
Collins, Rufus D.
Collinsworth, R. I.
Collum, Jepthy B.
Colson, Elizabeth S.
Comer, Gene E.
Compton, Eugene T.
Compton, Milford B.
Conard, Doria H.
Conard, Gertrude K.
Condo, William L.
Cone, Robert F.
Congo, Grady
Conklin, Albert J.
Conley, Bernard L.
Conley, David W.
Conn, Jake C.
Connally, John W.
Connell, Harvey A.
Connelly, James M.
Conner, John B.
Connor, Edward J.
Connor, Herschel W.
Conrad, James F.
Constan, George N.
Constant, Carl
Conway, Alason A.
Cook, Arthur E.
Cook, Donald B.
Cook, Jimmie D.
Cook, John E.
Cook, Leon A.
Cook, Lewis J.
Cook, Richard W.
Cook, Warden H.
Cooley, Charles I.
Cooley, Daniel J.
Cooley, Marvin G.
Coons, Harry J.
Cooper, Allen B.
Cooper, Charles R.
Cooper, Doris M.

Cooper, Durwood B.
Cooper, Elbert M.
Cooper, Garvis O.
Cooper, James L.
Cooper, John J.
Cooper, Milburn
Cooper, Orval R.
Cooper, Robert L.
Cooper, Roy M.
Cooper, Sallie B.
Cooper, Sharon R.
Cooper, Thirby W.
Cooter, Cloyd R.
Cope, Dudley O.
Copeland, Hubert G.
Copeland, James A.
Coppock, Huey W.
Corbin, Charles M.
Corbitt, James L.
Corder, Billie H.
Corder, Bobby J.
Corder, Wilson P.
Cordes, Hoyt
Cornelius, Charles S.
Cornelius, Donald W.
Cornelius, Gladyce V.
Cornelius, Harvey R.
Cornell, Arthur J.
Corners, William M.
Cornett, Darrel D.
Cornett, Varge
Corp, Lynn J.
Corry, Coleman R.
Cortner, William M.
Corum, Clarence C.
Coskrey, Henry N.
Costes, Nicholas C.
Cothran, Charles A.
Cothran, Ernestine K.
Cothran, Philip C.
Cothren, Bobby E.
Cothren, James M.
Cotton, Thelbert R.
Cottrell, Nellie H.
Couch, Ferrell D.
Couch, J. D.
Couey, Arvle J.
Coulter, Robert M.
Counter, Duane N.
Counts, Parker V.
Counts, Peggy M.
Counts, Richard H.
Courtenay, St. John
Cove, Jean S.
Covington, Arnold L.
Covington, Clydeas H.
Covington, Joseph L.
Covington, William J.
Cowart, Walter R.
Cowley, Walter W.
Cox, Emma J.
Cox, Glenn
Cox, Jack A.
Cox, Jerry W.
Cox, John B.
Cox, Manuel L.
Cox, Phyllis U.
Cox, Robert N.
Cozac, Glenn L.
Cozelos, Charles L.
Crabtree, William L.
Craddock, Julius Q.
Craft, Harry G.
Craft, Ray H.
Craft, Richard C.
Crafton, Edmund W.
Crafts, James L.
Craig, Carl G.
Craig, Clinton A.
Craig, Elbert B.
Craig, Glenn D.
Craig, Hazle A.
Craig, Lawrence D. II
Craighead, Paul T.
Craighead, Willie B.

Cramblit, David C.
Cramer, David E.
Crane, Charles G. Jr.
Cranford, Clara M.
Cranford, James M.
Cranford, Louis M.
Cranford, Olen B.
Crawford, James W.
Crawford, Janie T.
Crawford, Victoria J.
Creek, Harold B.
Creekmore, Terrell M.
Creel, Cecil D.
Cremin, Joseph W.
Cresap, John
Crews, Wilson E.
Crider, Byron
Crider, Duron
Crisler, Johnny P.
Crisp, Amos C.
Crockett, Charles D.
Croft, Freddie M.
Croft, Robert M.
Crooks, Frank E.
Cropp, Norman L.
Cross, Ernest B.
Cross, Jesse C.
Cross, Judy C.
Crossman, Robert L.
Crouch, Charles E.
Crouch, Johnie R.
Crouch, Louis C.
Crouch, Ray M.
Crow, George C.
Crowden, Mildred M.
Crowe, Delmar N.
Crowell, Clifford S.
Crowell, Clifford S. Jr.
Cruise, William B.
Crumbley, Andrew B.
Crumbley, Robert
Crumbley, William F.
Crumpton, Walter G.
Cruse, Clarence T.
Cruse, Mami N.
Crutcher, Andy R.
Cryar, Newell
Cryer, Norman V.
Cucarola, Gerald A.
Cucarola, Lana H.
Culbreth, Ronald C.
Culver, Wilbur M.
Culver, William N.
Cumings, Nesbitt P.
Cummings, David A.
Cummings, Robert E.
Cummins, Verl G.
Cunningham, Freddie L.
Cunningham, Roy E.
Cunningham, William C.
Cupp, Hollis A.
Curet, Gerald J.
Curran, Lyle C.
Currie, James R.
Currie, Roy E.
Curry, James E.
Curry, John D.
Curtis, Benjamin F.
Curtis, James C.
Cushman, William L.
Cutter, Lester H.
Dabbs, Joseph R.
Dabney, Richard W.
D'agostino, Alex J.
Dahm, Werner K.
Dailey, Carroll C.
Dale, Duvone
Daley, Daniel M.
Dalins, Ilmars
Dalrymple, Emmet O.
Dalton, Charles C.
Daly, Arthur V.
Daly, John A.
Damron, Alwayne
Damron, Claudine S.

Damron, James W.
Damron, Norman G.
Danford, Merlin D.
Daniel, Charles C.
Daniel, Gene E.
Daniels, Glenn E.
Daniels, Harry J.
Daniels, James L.
Daniels, Jan C.
Daniels, Norman C.
Dannenberg, Konrad K.
Darden, James H.
Darmer, John C.
Darnell, Roy R.
Darsey, Joe H.
Darwin, Charles R.
Daugherty, Leonard J.
Daugherty, Richard A.
Dauro, Vincent A.
Daussman, Grover F.
Davey, Frederick W. Jr.
Davidson, Bobby L.
Davidson, Frances.
Davidson, Thomas O.
Davidson, William E.
Davis, Arnold L.
Davis, Bervil D.
Davis, Betty J.
Davis, Billie
Davis, Billy G.
Davis, Billy K.
Davis, Carl M.
Davis, Charles H.
Davis, Charles M.
Davis, Desmond L.
Davis, Donald G.
Davis, Dorothy J.
Davis, Earl
Davis, Earnest H.
Davis, Eugene M.
Davis, Gerald A.
Davis, Hilda G.
Davis, Hubert E.
Davis, Joe B.
Davis, John C.
Davis, Juanita J.
Davis, Karen D.
Davis, Larkin D.
Davis, N. Jan
Davis, Paul A.
Davis, Paul R.
Davis, Riley L.
Davis, Robert E.
Davis, Samuel L.
Davis, Sandra T.
Davis, Thomas A.
Davis, Virginia H.
Davis, Wallace
Davis, Wanda R.
Davis, Wilbur S.
Davis, Wilda B.
Dawes, Richard P.
Dawkins, Donald D.
Dawson, Edna W.
Dawson, James K.
Day, Edwin M.
Day, Mary J.
Day, Maybella H.
De Armond, James G.
De Castra, Joseph E.
De Sanctis, Carmine E.
Deal, Walter J.
Dean, Charles A.
Dearing, Waldo H.
Dearman, Commodore C.
Deason, Neal
Deaton, Alva W.
Deaton, Emsley T.
Debeau, Jay W.
Debeek, Gerd W.
Decher, Rudolf
Dedrick, Richard L.
Defreez, Wilbur T.
Degroff, Audrey C.
Dehaye, Robert F.

Delabar, Claude T.
Delaney, Albert M.
Delaney, Robert
Delany, John
Delinoback, Leon M.
Delinski, Bernard F.
Dellinger, Thomas P.
Deloach, Anthony C.
Demarest, John R.
Dement, Samuel B.
Demirjian, Edward
Dendy, James W.
Dendy, William F.
Deneen, Carl D.
Dennis, Leona J.
Dennis, Mary G.
Depew, Bella D.
Derington, James C.
Derrickson, James H.
Derryberry, Mary W.
Deshield, Annette H.
Detko, George J.
Deuel, Earl C.
Deuel, Glen A.
Devenish, Robert
Devers, James M.
Devine, Paul A.
Dew, Homer H.
Dew, Jimmie L.
Dewberry, Edward P.
Dhom, Friedrich
Diamond, Ralph G.
Dickerson, Lee A.
Dickerson, Preston B.
Dickerson, Sandra O.
Dickerson, Thomas D.
Dickey, Mary R.
Dickinson, David L.
Dickinson, James S.
Dickinson, William E.
Dieteman, Clifford J.
Dieter, Moritz A.
Digesu, Fred E.
Dilbeck, Troy M.
Dill, Charlie C. Jr.
Dill, Francis W.
Dill, Glenn A.
Dillard, Dell D.
Dillard, Felox O.
Dillon, Nelis E.
Dinardi, Vincent J.
Dison, Gordon N.
Ditto, Eugene D.
Dixon, Robert L.
Doane, George B.
Dobbs, Glenn
Dobbs, Stephen J.
Dockham, Robert K.
Dodd, Kenneth A.
Dodd, M. F.
Dodd, William H.
Dodson, Charles B.
Dodson, Gayle W.
Dodson, Paul
Doherty, Robert S.
Dolan, Fred J.
Dolan, Lynda C.
Dolerhie, Bert D. Jr.
Dolin, Irene R.
Dollins, Chalmus B.
Dollman, Thomas S.
Dolve, Nels O.
Domal, Alexander F.
Donahoo, James E.
Donald, Erskine G. III
Donaldson, Ray W.
Donehoo, Larry K.
Doran, Billie J.
Doremus, James R.
Dorman, Martin C.
Dorning, Claude W.
Doss, Neal
Dossett, Louis A.
Doster, Hiram H.
Dothard, Stanley B.

NASA - Marshall Flight Space Center

Dotson, Richard E.
Doty, Patricia M.
Doud, Loxie L.
Doughty, Gerald N.
Douglas, Joseph D.
Douglas, Mosey P.
Douglass, Arville K.
Dow, Carl J.
Dowden, Janet L.
Dowdy, James F.
Downey, James A. III
Downs, Sanford
Downs, Vernon L.
Doyle, Clarence D.
Dozier, James B.
Drachlis, David B.
Drake, Franklin
Drake, Howard W.
Drake, Jean M.
Drake, Juanita H.
Drake, Ruth C.
Drawe, Gerhard P.
Dreher, Percy E.
Drinkard, Rodney
Drinnon, Paul E.
Driscoll, Daniel H.
Driscoll, Michael P.
Driver, Clarence J.
Driver, Orville E.
Droke, Bernice P.
Drost, Edward J.
Drummond, Aubrey S.
Drummond, Charles W.
Drummond, Dorothy C.
Drummonds, Robert M.
Duckwall, Lois R.
Dudley, Carl M.
Dudley, Hugh J.
Dudley, John E.
Duerr, Friedrich
Duffey, Frances S.
Duffey, Paul J.
Duffy, John E.
Duggan, Olene T.
Duke, Alvin S.
Duke, Floyd
Duke, Leonard C.
Duke, Lum
Duke, Thomas R.
Dull, Miriam W.
Dumpis, Talivaldis
Dunaway, William A.
Duncan, Billy J.
Duncan, Christine S.
Duncan, Dyerl W.
Duncan, John R.
Duncan, W. C.
Dunkin, James A.
Dunlap, Joseph V.
Dunlap, Porter
Dunn, Bobby G.
Dunn, Charles L.
Dunn, J. B.
Dunn, Judy S.
Dunnavant, Patricia M.
Dupre, Oscar L. Jr.
Dupree, Donald R.
Dupree, Robert E.
Duren, Paul C.
Durham, James E.
Durham, Robert L.
Durham, Shirley B.
Durkee, Mary Q.
Durrett, Robert H.
Duve, George C.
Dyar, Joseph D.
Dyar, Richard C.
Dyer, Henry C.
Dyer, Morris K.
Eady, James R.
Eagan, Joyce F.
Eakin, Charles A.
Ealy, Doyle W.
Ealy, William H.
Earl, Hazel G.

Earle, James M.
Easley, Charles H.
Eason, Ella B.
Eason, James E.
Eason, Lester L.
Eastep, Doyle E.
Eastwood, Marion F.
Eaton, John W.
Eaton, Peggy A.
Eberly, Esauntuck L.
Eby, Peter B.
Echols, Martha E.
Echols, Sherman D.
Eddleman, Helen P.
Eddleman, Otis N.
Eddleman, Terry M.
Eddy, Francis M.
Eddy, Margaret M.
Edens, William H.
Edge, Roy
Edgemon, James H.
Edmonds, H. C.
Edmonds, Imogene S.
Edmonds, William B.
Edmonson, Nat
Edmonson, Searcy E.
Edwards, Billy E.
Edwards, Catherine S.
Edwards, Clayton
Edwards, Fred G.
Edwards, Georgia A.
Edwards, Homer G.
Edwards, J. B.
Edwards, Katherine S.
Edwards, Odie H.
Edwards, Polly W.
Edwards, Robert C.
Edwards, Thomas E.
Edwards, Willard E.
Eggler, Dolores M.
Eggler, Donald
Ehl, James H.
Eichelberger, Robert P.
Eilerman, Richard N.
Eisenhardt, Otto K.
Elkin, Robert F.
Elkins, Bobby R.
Elledge, Larkin A.
Elledge, Lloyd
Ellet, William S.
Ellis, James M.
Ellis, John R.
Ellis, Lilian A.
Ellis, Roy H.
Ellis, Shirley H.
Ellison, Russell H.
Ellner, Andrew J.
Ellsworth, Charles R.
Ellsworth, Joseph D.
Elrod, Earnest L.
Elrod, Norman W.
Elrod, Robert S.
Emanuel, Garvin R.
Emanuel, William A.
Emens, Frank H.
Emens, Mildred G.
Emery, Luther R.
Emmons, Charlene E.
Enfield, Hugh A.
Enger, Martin D.
England, Elmer P.
England, Gordon W.
England, Harold B.
Engle, Burma G.
Engler, Erich E.
English, Charles E.
English, James B.
Engman, Lloyd J.
Enoch, James B.
Enochs, Jack G.
Eoff, William L.
Erb, Ernest E.
Erickson, Stanley R.
Ernsberger, Gale R.
Erwin, Albert B.

Erwin, Bobby E.
Erwin, Orene H.
Erwin, Robert W.
Escue, William T.
Eslick, Willa M.
Eslick, Willard.
Espy, Patrick N.
Esslinger, James A.
Esslinger, Sherrill H.
Estes, Danny O.
Etheredge, James L.
Etheredge, Ray D.
Ethridge, Orval E.
Ettingoff, Nathaniel V.
Eubanks, Earl W.
Eudy, Robert G.
Euler, Harold C. Jr.
Eulitz, Werner R.
Eutsler, Charles W.
Evans, Alvin V.
Evans, Francis E.
Evans, James D.
Evans, John O.
Evans, Page K.
Evans, Phineas D.
Evans, Ross W.
Evans, William H. Jr.
Everett, Marion A.
Evers, John E.
Fabian, Louis
Facemire, Barbara R.
Fagan, Andre G.
Faile, Gwyn C.
Fallon, Robert A.
Falls, Lee W.
Fann, James T.
Fann, William E.
Fanning, Jonah R.
Fanning, Roy L.
Farabee, Arlie J.
Farley, Benny B.
Farmer, David W.
Farmer, John E.
Farmer, Ralph W.
Farner, Johnny B.
Farr, Wallace B.
Farrell, John F.
Farrer, John A.
Farrow, Hershal M.
Farrow, James H.
Faucett, Harold R.
Faught, Edsel
Faulkner, Billie T.
Faulkner, Charles A.
Faulkner, Willie H.
Fears, Homer R.
Feather, Hugh M.
Fedler, John J.
Fedrowisch, Sytha J.
Fee, John S.
Feemster, Patty D.
Fehlberge, Erwin
Felch, Jerald L.
Felder, James T.
Felix, August R.
Felker, John H.
Fenner, Otis F.
Fenner, William
Fennimore, Emmett L.
Fenton, Robert G.
Ferguson, Foster M.
Ferguson, Robert F.
Ferguson, T. L.
Ferguson, William A.
Ferrario, Martin A.
Ferrell, George A.
Ferrell, Joseph H.
Ferrell, Toon R.
Fersch, Willard C.
Festa, Rudolph K.
Fetzer, Carl H.
Few, Albert G.
Fichtl, Carrie L.
Fichtl, George H.
Fichtner, Hans J.

Field, Elmer L.
Fields, James B.
Fields, Stanley A.
Fife, James P.
Fikes, Eugene H.
Fikes, Joseph E.
Fikes, William K.
Filip, George L.
Finigan, John
Fink, William L.
Finlaw, James M.
Finlaw, Mildred L.
Finley, Gordon W.
Finley, Stewart A.
Finnegan, Daniel V.
Finnell, Woolsey
Finzel, Alfred J.
Firestone, Daryl D.
Fischer, Kenneth C.
Fisher, Doris M.
Fisher, Elmer E.
Fisher, Paul H.
Fisher, Robert R.
Fiske, Donald F.
Fitzgerald, George O.
Fitzjarrald, Daniel E.
Flack, Raymond H.
Flake, Harry M.
Flanagan, Gerald F.
Flanigen, Barrington L.
Fleming, Branch R.
Fleming, Burl R.
Fletcher, James W.
Fletcher, Norman C.
Flora, Clifford L.
Florence, Namon
Flowers, Doris K.
Floyd, Billy R.
Floyd, Henry B.
Fly, James B.
Flynt, Dillard Y.
Flynt, William E.
Fogg, Ella E.
Fogle, Valmore
Folks, Fletcher L.
Folsom, Anne B.
Ford, Jonathan H.
Fordice, Dorinne D.
Forney, James A.
Forney, Margaret P.
Forrest, Earl E.
Forrest, Kenneth E.
Forrester, Robert L.
Forsythe, Douglas J.
Fortenberry, William H.
Fortner, Carl
Fortune, William C.
Foster, Clyde
Foster, George
Foster, James W.
Foster, Jay
Foster, Lee D.
Foster, Louis F.
Foster, Nick D.
Foster, Susan C.
Fountain, James A.
Fountain, Walter F.
Fowler, James L.
Fowler, Randy J.
Fowler, William L.
Fox, Thomas H.
Fox, Vernon J.
Foxworth, Wallace P.
Foxworthy, Davis E.
Foy, Everett Jr.
Foy, John H.
Frady, Florida P.
Fragg, Stanley L.
Francis, Robert C.
Frankel, Adolph
Franklin, Douglas B.
Franklin, Elmer E.
Franklin, Gladys C.
Franklin, Junedale M.
Franklin, Marshall C.

Franklin, Reba M.
Franklin, William D.
Franklin, William J.
Franks, Jerry B.
Franz, Hans W.
Frary, Spencer G.
Fraser, Raymond H.
Frazier, John L.
Frazier, Peggy L.
Frederick, Paul W.
Freedman, Jack A.
Freeman, Dan A.
Freeman, Inellia F.
Freeman, John W.
Freeman, Leonard M.
Freeman, Myrle W.
Freeman, Novelene C.
Freeman, Robert M.
Freeman, Sara B.
Freeman, Thomas B.
Freeze, Daniel Jr.
French, Donald E.
French, Orean K.
French, Price I.
French, Thomas C.
Friday, Ruth B.
Fritz, Carl G.
Frost, Martha C.
Frost, Walter O.
Frye, Albert D.
Fryer, Jo Ann T.
Fryer, Reuben R.
Fryman, Jay G.
Fuchtman, Charles M.
Fuhrmann, Herbert W.
Fuller, Carl A.
Fuller, Foy S.
Fuller, James A.
Fuller, Thomas A.
Fullwood, Travis O.
Fulmer, Alfred H.
Fulmer, Clarence R.
Fults, John L.
Funderburk, Bobby J.
Fuqua, Harold E.
Fuqua, Walter E.
Furman, Jack H.
Futral, Bobby H.
Futral, Nolan G.
Gaffin, Robert D.
Gaines, Bobby J.
Gaines, Charles B.
Galaboff, Zachary J.
Galbreath, James M.
Galey, Joe T.
Gallaher, Buford E.
Gallien, James M.
Galligan, Joseph D.
Galloway, William E.
Gamble, Gerald G.
Gandy, Leroy
Gangl, Hubert R.
Gann, Lee
Gant, Clarence
Gantt, James P.
Garber, Harry W.
Garcia, Francisco F.
Garcia, Raymond P.
Gardiner, David B.
Gardiner, Inez C.
Gardner, Clyde B.
Gardner, Lloyd B.
Garland, Ann R.
Garmon, Betty B.
Garner, Claudia M.
Garner, James P.
Garnett, Raymond E.
Garofalo, Hilda A.
Garrard, Johnny S.
Garrett, Afton H.
Garrett, Harrison
Garrett, Robert E.
Garrett, Robert S.
Garrett, Sidney W.
Garrison, James M.

Garrison, Lawrence
Gary, Gilmer A.
Gassaway, Gilbert G.
Gaston, Grady P.
Gates, Carl D.
Gates, Daniel W.
Gatlin, Maurice
Gattis, John E.
Gattis, Tennie M.
Gattis, W. D.
Gault, Ira
Gause, Raymond L.
Gautreaux, Jean O.
Geiger, James V.
Geiger, Martin H.
Geiger, Ralph E.
Geiger, William A.
Geissler, Ernst D.
Geller, Gerald
Genglebach, Werner K.
Genter, John C.
Gentle, Cecil B.
Gentry, Leroy
George, Henry L.
George, James H.
George, John M.
George, Lawrence J.
George, Ollie B.
Gerry, George B.
Gerry, Mark D.
Gerstlauer, William R.
Getten, Edna V.
Gibbs, Gerald F.
Gibbs, Luther K.
Gibbs, Talmage G.
Gibson, Billy B.
Gibson, Carl E.
Gibson, John C.
Gibson, Nancy J.
Gibson, Walter P.
Gibson, William E.
Gierow, Herman P.
Gilbert, Ward T.
Gilino, Norman R.
Gill, William W.
Gillespie, Charles S.
Gillespie, Reba M.
Gillespie, Victor G.
Gillespie, Walter F.
Gilliam, Thomas A.
Gillies, Donald C.
Gililan, John T.
Gilmore, Herman L.
Gilmore, Robert N.
Gilstrap, Jack T.
Ginn, Nathan L.
Gipson, Fred Z.
Giudici, Robert J.
Given, Jack F.
Gladish, Darrell W.
Glaser, Rudolf F.
Glasgow, Spencer L.
Glass, Charles R.
Glass, James W.
Glazner, Bettye J.
Glazner, John E.
Gleason, Edmund H.
Gleaves, Julian T.
Glen, Clemons T.
Glenn, C. G.
Glover, Frankie D.
Glover, James C.
Goans, Samuel K.
Goans, Vera G.
Godfrey, Roy E.
Godwin, Connie H.
Godwin, Delta I.
Goerner, Erich E.
Goetz, Harold O.
Goetz, Otto K.
Goins, Fred A.
Gold, James R.
Golden, Betty K.
Golden, Harvey
Golden, Jimmie W.

50 Years of Rockets and Spacecraft

Goldsby, William D.
Goldsmith, James H.
Goldstein, Hans
Goldston, Jim E.
Goldston, Robert L.
Golley, Paul T.
Golmon, James H.
Gooch, Jo Ann
Good, Robert L.
Goode, George M.
Goode, Hazel G.
Goodhue, Walter H.
Goodlett, Howard L.
Goodrick, Thomas F.
Goodrum, John C.
Goodwin, David C.
Goodwin, James C.
Goodwin, Larry B.
Goodwin, Ralph F.
Goosby, Marilyn M.
Gore, William R.
Gorum, Ruth E.
Gosdin, Dennis R.
Goss, Robert B.
Gould, John M.
Gover, King C.
Gover, Wilbert M.
Goza, John F.
Grace, Donald H.
Grace, Elizabeth A.
Grace, Velma S.
Gracy, John D.
Grady, Ethel D.
Graff, Charles B.
Grafton, William L.
Graham, Arlie H.
Graham, Claude E.
Graham, Frank E.
Graham, Frank M.
Graham, James J.
Graham, John H.
Graham, Joseph L.
Graham, Lofton
Graham, Marcellus H.
Graham, Ronald J.
Graham, Samuel L.
Graham, Thomas E.
Graham, William C.
Grant, Lynnon F.
Grau, Dieter
Graves, Brenda F.
Graves, James R.
Graves, Robert E.
Gravette, Marvin C.
Gray, Charlotte M.
Gray, Lawson S.
Gray, Lloyd V.
Gray, Olga C.
Gray, Robinson A.
Gray, William A.
Green, Claude E.
Green, Clifford F.
Green, Clifton L.
Green, Erwin E.
Green, Jacob B.
Green, John M.
Green, Marvin R.
Green, Onis C.
Green, Robert L.
Green, William B.
Greene, Hubert S.
Greenwood, Carol L.
Greenwood, Carolyn E.
Greenwood, Lawrence R.
Greenwood, Terry F.
Greenwood, Thomas L.
Greer, Aubrey T.
Greer, Charles L.
Greer, Dewey L.
Greer, Julian A.
Gregg, Cecil C.
Gregory, Hobert
Gresham, Albert E.
Gresham, John W.
Grider, Mary C.

Griffin, Cornelius D.
Griffin, Joyce M.
Griffin, Madge L.
Griffith, Grady
Griffith, J. R. III
Griggs, William F.
Grimes, Robert P.
Grimes, Victor S.
Griner, Carolyn S.
Griner, Roland F.
Grisby, James W.
Grisham, Molly C.
Grissett, Floyd R.
Grissett, Harry E.
Gritzmacher, Thomas G.
Groeneveld, Robert D.
Gross, Jack D.
Gross, Klaus W.
Gross, Loren A.
Groves, Armond S.
Grubb, Ralph L.
Grubbs, Doris M.
Grubbs, James P.
Guerin, Edd
Guerin, Edward G.
Guest, Stanley H.
Guffin, Orville T.
Guice, Harvey E.
Guilian, William E.
Guillebeau, William W.
Guin, Judith E.
Guire, Nancy M.
Gullion, Inez C.
Gullion, Mary S.
Gullion, William H.
Gunn, Howard D.
Gunn, Mildred R.
Gunn, Thomas M.
Gunnde, Pierre D.
Gunner, Robert M.
Gunter, Arnice F.
Gunter, Elmer A.
Gurley, F. Clifton
Gurley, Forshem
Gurley, Marion E.
Gurley, Raymond
Gustin, Robert A.
Guthrie, Roy.
Guthrie, Willie L.
Guttman, Charles H.
Guynes, Barry V.
Guynes, Buddy V.
Guyse, Paul T.
Guyton, Benson
Guzinsky, Leon
Gwathney, Rita S.
Gwin, Hal S.
Haaker, Carl A.
Hackleman, Virginia L.
Haenisch, Hilmar W.
Haeussermann, Walter
Hafner, Alexander
Hagen, William A.
Haggard, Chester
Hagood, Carlos C.
Hagood, Margaret J.
Hagyard, Mona J.
Hahn, Richard L.
Haire, John H.
Haislip, David H.
Halbrooks, William J.
Hale, Daniel P.
Hale, John S.
Hale, Kenneth B.
Hale, Sara B.
Haley, Aubrey D.
Haley, Betty J.
Haley, Foster A.
Haley, Fred A.
Haley, Lawrence B.
Hall, Atlee B.
Hall, Carl E.
Hall, Charles E.
Hall, Charles T.
Hall, Dewey W.

Hall, Gerald E.
Hall, Gloria P.
Hall, Harold C.
Hall, Howard D.
Hall, Jeanette M.
Hall, Jewell M.
Hall, John E.
Hall, Leonard B.
Hall, Orlyn C.
Hall, Raymond
Hall, Richard E.
Hall, Richard L.
Hall, Richard T.
Hall, Robert D.
Hall, Stephen B.
Hall, Vanassa V.
Hall, William J.
Hall, Wilmer E.
Hallisey, Harold W.
Hallmark, A. Clayton
Ham, Ronald F.
Hamby, Herman G.
Hamby, Paul V.
Hamil, Joe A.
Hamilton, Clifford B.
Hamilton, Dale R.
Hamilton, Edward C.
Hamilton, Joe J.
Hamilton, Julian S.
Hamilton, Marvin R.
Hamilton, Richard S.
Hamiter, Leon C.
Hamlet, John F.
Hammac, Houston M.
Hammer, Donald L.
Hammer, Morris W.
Hammer, Wesley F.
Hammers, Fred C.
Hammick, Earl G.
Hammon, William E.
Hammond, Arlice G.
Hammond, Vondaleen S.
Hammonds, Victor E.
Hamner, Elaine W.
Hampton, John P.
Hanby, Cecil B.
Hancock, Fredella
Hancock, Hugh T.
Hancock, Reva H.
Hancock, Sherrel E.
Hancock, Virginia J.
Haney, Beverly K.
Haney, William H.
Hankins, James D.
Hanks, Charles F.
Hannan, Eugene C.
Hansen, Betty J.
Hansen, James F.
Hanson, Margaret J.
Hanvey, Lee G.
Harber, Harry G.
Harbison, John E.
Hardage, James N.
Hardage, Onice M.
Hardee, Marion S.
Hardee, Mikael A.
Hardin, Ernest G. Jr.
Hardin, Mary M.
Hardy, George B.
Hardy, Jimmie N.
Hardy, John P.
Hargett, Walter S.
Hargrove, A. D.
Hargrove, Richard M. Jr.
Harlow, David E.
Harlow, John H.
Harman, Harlan S.
Harmon, Grady D.
Harmon, William H.
Harness, Paul R.
Harnisch, Fred W.
Harp, James A.
Harp, Seldon L.
Harpe, Edith B.
Harpe, Kenneth L.

Harper, Carolyn S.
Harper, Edward M.
Harper, Ernest W.
Harper, Frank M.
Harper, James W.
Harper, Jesse E.
Harper, John D.
Harper, Julia M.
Harrell, William B.
Harrelson, James C.
Harrington, Michael M.
Harrington, Reba E.
Harris, Albert V.
Harris, David P.
Harris, Earle G.
Harris, George A.
Harris, George G.
Harris, Henry C.
Harris, Henry H.
Harris, John D.
Harris, Joseph D.
Harris, Martin G.
Harris, Mary P.
Harris, Ozell F.
Harris, Ransom B.
Harris, Richard S.
Harris, Robert T.
Harris, Ronald P.
Harris, Stanley C.
Harris, Thomas L.
Harrison, Allen H.
Harrison, Bettye L.
Harrison, Calvin V.
Harrison, Charles A.
Harrison, James K.
Harrison, James L.
Harrison, Paul E.
Harrison, William O.
Harsh, Marcellus G.
Harshaw, Oliver A.
Hartlein, Evelyn H.
Hartlein, John
Hartley, Donald L.
Harton, Vance
Hartsfield, Gene A. Jr.
Harwell, Iva D.
Harwell, John E.
Harwell, Lester A.
Harwell, Robert F.
Harwell, Rogers J.
Harvey, George E.
Hassler, Preston L.
Hastings, Leon J.
Hatfield, James E.
Hatfield, Marion R.
Hathcock, Addison S.
Hatton, Leslie O.
Hauff, Christian
Haukohl, Guenther H.
Hauser, John A.
Haussler, Jonathan B.
Hauth, John A.
Haven, Harry C.
Havrilla, Emil J.
Hawk, Grady L.
Hawkins, Billy J.
Hawkins, Charles A.
Hawkins, Franklin D.
Hawkins, Gerald W.
Hawkins, Henry W.
Hawkins, James T.
Hawkins, V. B.
Hay, Frank E.
Hayes, Benita C.
Hayes, Billie F.
Hayes, Carlos C.
Hayes, Carolyn P.
Hayes, Evelyn J.
Hayes, James F.
Hayes, Katie S.
Hayes, Richard B.
Haynes, Carolyn C.
Haynes, Enoch
Haynes, Joe D.
Haynes, John S.

Haynes, Theodore R.
Haynes, Thomas W.
Hays, Harvey O.
Hays, Philip J.
Hazel, Charlotte S.
Hazle, Charles E.
Head, Mildred W.
Head, Robert R.
Headrick, John H.
Heaman, John P.
Heard, Roy E.
Heard, William B.
Heaton, Greta J.
Heck, Arno E.
Heckman, Richard T.
Hefferan, John T.
Hegeman, Richard C.
Heim, Stephen F.
Heimburg, Karl L.
Heintz, Frank J.
Heinze, Sidney A.
Hellebrand, Emil A.
Heller, Clarence
Heller, Raymond G.
Helm, Bruno K.
Helms, Alton D.
Helms, Curtis R.
Helms, James C.
Helton, Clifford M.
Helton, Henry F.
Helton, King P.
Hembree, Ray V.
Henderson, Deward L.
Henderson, Horton E.
Henderson, Janet M.
Henderson, John C.
Henderson, Marietta M.
Henderson, Sandra G.
Henderson, Willie M.
Hendrick, Bedford D.
Hendricks, Bennett
Hendricks, Evan S.
Hendrix, Flemon L.
Hendrix, James L.
Hendrix, Jessie C.
Hendrix, Neal D.
Henke, Charles E.
Henken, Theodore F.
Henning, Eugene S.
Henrie, Frank J.
Henritze, Richard M.
Henry, Alexander
Henry, James D.
Henry, Robert L.
Henry, Robert W.
Henry, William R.
Hensley, Almon
Hensley, Derrel R.
Henson, Bobby B.
Henson, Junior W.
Henson, Orval E.
Henson, Victor Keith
Heptinstall, Carey C.
Herda, Donald R.
Herlong, Dwight L.
Herman, John
Herndon, Earle P.
Herndon, Ralph H.
Herren, Blair J.
Herren, Kenneth A.
Herrin, Charles F.
Herrin, Joe H.
Herrin, Milton T.
Herring, J. W.
Herring, James B.
Herrmann, Adolf L.
Herrmann, Frederick T.
Herrmann, Melody C.
Herron, Vicel A.
Hess, Cecil K.
Hess, James H.
Hess, Mary L.
Hester, Charles E.
Hester, Howard B.
Hester, James F. Jr.

Hethcoat, Jerry P.
Heuser, William W.
Heybey, Willi H.
Heyer, James W.
Hickerson, Alton L.
Hicklen, Willie B.
Hickman, Homer H.
Hicks, Edwin L.
Hicks, Howard L.
Hicks, John I.
Hicks, Maria E.
Hicks, Olan W.
Hicks, William A.
Higgins, Artis L.
Higgins, John B.
Higgins, Nannie R.
Higgins, Ralph H.
Hight, Elizabeth H.
Hight, Hermon H.
Hightower, Clyde M.
Hilbert, Doris A.
Hilchey, John D.
Hildreath, Edward D.
Hiley, Nelda C.
Hill, Betty G.
Hill, Daniel C.
Hill, Douglas E.
Hill, Edna H.
Hill, Ewing K.
Hill, Fred J.
Hill, Harry L.
Hill, Henry C.
Hill, John M.
Hill, John P.
Hill, John W.
Hill, Joseph E.
Hill, Kenneth A.
Hill, Loyd
Hill, Mattie P.
Hill, Olen E.
Hill, Robert W.
Hill, Roy L.
Hill, Thomas E.
Hill, Walter H.
Hill, Walton W.
Hill, Waymon B.
Hill, Wildes M.
Hill, William E.
Hillenbrand, Joseph S.
Hilliard, James W.
Hillis, Linda M.
Hilson, Henry L.
Hilten, Heinz H.
Hilton, Nelda H.
Himes, Ruthie A.
Hinesley, Joseph D.
Hinkle, Adrian J.
Hinkle, Kenneth M.
Hinton, Wylie M.
Hintze, Geoffrey C.
Hintze, Willard P.
Hipp, Doris W.
Hirlston, Whitt H.
Hirsch, Oliver M.
Hirshburg, Ivan E.
Hoag, Phillip C.
Hoard, Earl G.
Hobbs, William I.
Hoberg, Otto A.
Hobson, Joseph D.
Hochberger, Norman L.
Hodge, James E.
Hodge, Millard W.
Hodges, Bobby C.
Hodgins, David
Hoekenschneider, Lee R.
Hoelzer, Helmut
Hoffman, Conrad W.
Hofmann, Helmut R.
Hoffman, Patricia C.
Hofues, John L.
Hogan, William J.
Hoge, Kathryn E.
Hogue, Mildred R.
Holbrook, Clyde E.

214

NASA - Marshall Flight Space Center

Holcomb, Velma	Houston, Shannon D.	Hurford, Robert L.	Jenke, Richard K.	Jones, Charlie T. III	Kelley, Harrison
Holcomb, William H.	Houston, Vance L.	Hurlburt, Mary H.	Jenkins, James E.	Jones, Clara S.	Kelley, Joseph W.
Holden, Billy J.	Houston, William C.	Hurst, Barbara R.	Jenkins, Milton C.	Jones, Clyde E.	Kelley, Nina B.
Holder, Oscar C.	Hovis, Bertie M.	Hurst, Paul D.	Jenkins, Warren H.	Jones, Clyde S.	Kelley, Robert O.
Holder, Ray C.	Hovis, Howard B.	Hurst, Paul R.	Jenkins, William H.	Jones, David V.	Kelley, Timothy O.
Holladay, Alvis M.	Howard, Cherry C.	Hurt, John H.	Jennings, Anne M.	Jones, Dewey E.	Kelley, William J.
Holladay, James C.	Howard, Dorothy C.	Husbands, David L.	Jennings, Charles A.	Jones, Frank	Kelly, Glen H.
Holladay, John W.	Howard, Eschol	Huskey, Ray M.	Jennings, Cornal	Jones, George H.	Kelly, Max A.
Holland, Ben F.	Howard, Everett D.	Huston, Jack E.	Jennings, James B.	Jones, Glenn H.	Kelly, Robert E.
Holland, Elsie T.	Howard, Gladys J.	Hutchens, Cindy F.	Jensen, Marvin L.	Jones, Homer E.	Kelly, Thomas E.
Holland, James G.	Howard, Marvin W.	Huth, Chauncey W.	Jensen, Warren S.	Jones, Homer F.	Kelly, Warren
Holland, Jessie J.	Howard, Nellie E.	Hutson, Edgar D.	Jenson, Earl L.	Jones, Ira P.	Kelm, George P.
Holland, Lou K.	Howard, Paul W.	Hutto, Boyce M.	Jenson, Joseph J.	Jones, Isaac W. Jr.	Kelsey, Lyle J.
Holland, Paul S.	Howard, William D.	Hutto, Ivey C.	Jeralds, William M.	Jones, Jack A.	Kelt, Julius J.
Holland, Robert L.	Howard, William L.	Hyatt, Ernest M.	Jernigan, Robert E.	Jones, James D.	Kemp, Raymond M.
Holland, Sandra S.	Howard, William S.	Hyatt, Sherry M.	Jernigan, Robert T.	Jones, Jess H.	Kendall, Jack M.
Holland, Wayne B.	Howell, Evan S.	Hyde, James L.	Jeurgensen, Klaus	Jones, Joe E.	Kennamer, Lareu P.
Hollander, John E.	Howell, Joe T. Jr.	Hyer, John H.	Jex, David W.	Jones, John P.	Kennedy, Billy G.
Hollich, Stephen E.	Howell, Leonard W. Jr.	Hyfield, Clarence O.	Jex, Leo W.	Jones, John T.	Kennedy, Bobby W.
Hollingsworth, Annie J.	Howell, William F.	Hypes, Marvin C.	Jobe, Grady S.	Jones, Joseph B.	Kennedy, Edith K.
Hollingsworth, Cynthia B.	Howton, Tom M.	Hyter, Delano R.	Jobe, Thomas B.	Jones, Joseph M.	Kennedy, Hubert O.
Hollingsworth, Lula E.	Hubbard, Dorthy S.	Igou, James M.	Jobe, William B.	Jones, Kenneth W.	Kennedy, Jack S.
Hollingsworth, Thomas H.	Hubbard, James B.	Ikard, James E.	Johanson, Walter A.	Jones, Lee W.	Kennedy, John W.
Hollis, Ben R.	Hubble, Barbara K.	Inabinet, Michael J.	Johns, Ila H.	Jones, Leo L.	Kennedy, Joseph D.
Hollis, Ida Sue J.	Hubble, Sant W.	Inman, Chieko O.	Johns, Stanley A.	Jones, Marjorie H.	Kennedy, Joyce J.
Hollis, James C.	Huber, Harold J. Jr.	Inman, Reginald A.	Johnson, Albert A.	Jones, Mckinley	Kennedy, Thesa T.
Hollis, Joe M.	Huber, Hazel M.	Irby, Thomas M.	Johnson, Albert S.	Jones, Merle E.	Kennedy, Thomas E.
Hollis, Rose E.	Huber, William G.	Irick, Clarence L.	Johnson, Alvyn	Jones, Milton S.	Kennedy, Walter L.
Holloway, Henry D.	Huberdeau, Laura C.	Irvine, Charles N.	Johnson, Annie W.	Jones, Paul W.	Kennel, Hans F.
Holloway, Lloyd K.	Huckaby, Stewart W.	Irvine, James F.	Johnson, Arvel	Jones, Prince T.	Kennemer, George J.
Holloway, Melanie J.	Huddleston, Charlie W.	Isbell, Bryant W.	Johnson, Bert E.	Jones, Richard W.	Kennemer, Otis E.
Holly, James Jr.	Hudgins, Arvin Q.	Isbell, Isaac H.	Johnson, Billy E.	Jones, Robert E.	Kenney, William K.
Holmes, Bonnie G.	Hudgins, Jerry L.	Isbell, John T.	Johnson, Bobby G.	Jones, Russell	Kent, Ancil L.
Holmes, Charles W.	Hudgins, Mellina R.	Isbell, Thomas P.	Johnson, Bobby J.	Jones, Stuart H.	Kent, Barbara L.
Holmes, Clyde M.	Hudson, Ernest D.	Ise, Jerry B.	Johnson, Charles W.	Jones, Veronica G.	Kent, Everett W.
Holmes, James R.	Hudson, James E.	Ise, Rein	Johnson, Charlie F.	Jones, Warren G. Jr.	Kent, Marion I.
Holmes, Sam M.	Hudson, John T.	Isom, Larmon B.	Johnson, Clarence H.	Jordan, J. D.	Kephart, Andrew J.
Holt, Grady B.	Hudson, Kermit O.	Ito, James	Johnson, Dale L.	Jordan, John R.	Kerr, Arthur L.
Holt, Katy M.	Hudson, Price W.	Ito, James Nmn	Johnson, Edwin C.	Jordan, Leonard J.	Kerr, Joseph H.
Holweger, Dennis A.	Hudson, William A.	Ives, Robert E.	Johnson, Everett W.	Jordan, Lowell W.	Kerschner, Robert L.
Honea, Naomi B.	Hueter, Uwe	Ivey, Edward W.	Johnson, Frances W.	Jordan, Miron A.	Kersten, Frederick C.
Honea, Nolan L.	Huff, James B.	Ivey, Vickie J.	Johnson, Franklin B.	Jordan, Sandra Q.	Kesler, Herman E.
Hood, Henry L.	Huff, William A.	Ivy, Gladys B.	Johnson, Gary W.	Jordan, Wallace E.	Kessler, Robert H.
Hoodless, Ralph M. Jr.	Huffaker, Charles F.	Jacks, Charles L.	Johnson, Hallman T.	Jordan, Wallace J.	Ketchum, Robert P.
Hooker, William T.	Huffman, Gaylord M.	Jacks, Coleman G.	Johnson, James B.	Joyner, Romely G.	Key, Carol M.
Hoop, James M.	Huffman, Jerry D.	Jacks, Francis M.	Johnson, James H.	Julian, Aaron S.	Key, Carols F.
Hooper, Albert W.	Huffman, Nishman N.	Jacks, Orval D.	Johnson, James R.	Julich, Jule M.	Key, Glenn E.
Hooper, Clyde M.	Huggins, Carl T.	Jackson, Aaron	Johnson, Jane W.	Jupiter, Alvin M.	Key, Naomi G.
Hooper, Thomas M.	Hughes, Alvin E.	Jackson, Archie L.	Johnson, Jerry L.	Justice, David W.	Key, Ollie L.
Hooper, William H.	Hughes, Billy C.	Jackson, Charles H.	Johnson, Lee R.	Justice, Ollie E.	Keyes, Patrick F.
Hoover, Alvie L.	Hughes, Clinton L.	Jackson, Condy L.	Johnson, Lorena C.	Justice, Wilbur C.	Kidd, Jerry W.
Hoover, Charles D.	Hughes, Elbert D. Jr.	Jackson, Earnest C.	Johnson, Melvin	Justin, Gladys W.	Kiefling, Larry A.
Hoover, William L.	Hughes, James C.	Jackson, Guy B.	Johnson, Owen H.	Kaminsky, Joseph A.	Kile, Nell G.
Hope, Harley R.	Hughes, James E.	Jackson, James G.	Johnson, Robert C.	Kampmeier, Heinz W.	Kilgore, Edward W.
Hopkins, James E.	Hughes, Jennie C.	Jackson, Jim M.	Johnson, Robert E.	Kantor, Francis S.	Kilgore, William A.
Hopkins, Miriam S.	Hughes, John J.	Jackson, Junior L.	Johnson, Sandra S.	Kanupp, Jack D.	Killian, Wanda L.
Hopkins, William L.	Hughes, Oliver L.	Jackson, Lathan C.	Johnson, Sharon M.	Kanupp, Phyllis H.	Kilpatrick, Mary R.
Hopper, Edgar H.	Hughes, Pleasant M.	Jackson, Lloyd G.	Johnson, Shields M.	Karigan, William A.	Kilpatrick, William W.
Hopper, Richard L.	Hughes, Robert W.	Jackson, Robert J.	Johnson, Thomas A.	Kaschig, Erich K.	Kimball, Warren E.
Hopper, Volie L.	Hughes, Roland G.	Jackson, Robert L.	Johnson, Warren L.	Kastanakis, John H.	Kimberlin, Marie S.
Hoppers, David H.	Hughey, Vickie A.	Jackson, Robert T.	Johnson, Wiley L.	Katz, Harold K.	Kimery, Donald P.
Hoppes, Robert V.	Hugus, Robert E.	Jackson, Walter W.	Johnson, Willard	Katz, Lester	Kimmons, William L.
Hopson, George D.	Huie, Hubert H.	Jacobi, Walter W.	Johnson, William	Kauffman, Ralph W.	King, Alvis
Horn, Earl F.	Hull, Elsie C.	Jacobs, Edwin P.	Johnson, William A.	Kaufman, John W.	King, Charles A.
Horn, Helmut J.	Hullett, Arthur	Jacobs, Elmer M.	Johnson, William H.	Kaufmann, James J.	King, Charles C.
Horn, Wallace A.	Hullett Smith, Gloria A.	James, Earl C.	Johnston, Danny D.	Kay, Billy W.	King, Curley
Horn, William G.	Hulsey, Norman L.	James, Jean M.	Johnston, Dorothy L.	Kay, Elizabeth R.	King, Edgar
Horne, Charles H.	Humphreys, John T.	James, Lee B.	Johnston, Garland D.	Kay, Linda P.	King, Evangelist W.
Horne, J. D. Jr.	Humphries, Thomas S.	James, Linda K.	Johnston, James D.	Kearney, John D.	King, Hugh J.
Horner, Jack B.	Humphries, William R.	James, Mary T.	Johnston, James E.	Kearns, George B.	King, J. B.
Horst, Max A.	Hundley, Paul L.	James, Solon B.	Johnston, Lawrence M.	Keeffe, Thomas A.	King, James A.
Hortin, Francis H.	Huneidi, Farouk	Jamison, Glenn T.	Johnston, Richard E.	Keel, John M.	King, James E.
Horton, Ovoy Y.	Hunkapiller, Barbara T.	Jandebeur, Frederick A.	Johnston, Thomas S.	Keel, Ralph H.	King, Jimmy
Horton, Robert F.	Hunt, Frances L.	Jandebeur, Sheila C.	Johnstone, Harry M.	Keeney, Joe	King, John A.
Horton, William P.	Hunt, Gaylord C.	Janoski, Joseph W.	Joiner, James B.	Keim, Robert D.	King, Kathrine L.
Hosenthien, Hans H.	Hunt, Robert M.	Jarrell, John K.	Jolley, Horace L.	Keith, Carol H.	King, Norman D.
Hoskins, Robert E.	Hunt, Roy D.	Jarrell, Ralph	Jolley, James O.	Keith, Rita S.	King, Quinton A.
Houchens, Richard B.	Hunter, Harold D.	Jarrett, George T.	Jolliff, Alfonso C.	Keith, William H.	King, Thomas A.
Houk, Nancy C.	Hunter, James R.	Javins, David R.	Jonakin, Lynn	Keller, Billy B.	King, Welby M.
House, Edward G.	Hunter, John O.	Jayroe, Robert R. Jr.	Jones, Alfred M.	Keller, George A.	King, Worthie C.
Householder, Leonard R.	Hunter, Oran E.	Jean, James F.	Jones, Belton	Keller, Vernon W.	Kingsbury, James E.
Houser, William W.	Hunter, Ross W.	Jean, Lawrence R.	Jones, Billy P.	Kellett, Austin C. Jr.	Kinkle, Daniel S.
Housley, John A.	Hunter, Samuel H.	Jean, Otha C.	Jones, Cecil T.	Kelley, Calvin M.	Kinney, Linda H.
Houston, Charles E.	Hunter, Shelby J.	Jefferson-Tisdale, Vernit	Jones, Charles B.	Kelley, David L.	Kinser, Morris
Houston, Richard	Hunter, Warren G.	Jeffreys, Beth R.	Jones, Charles O.	Kelley, Fred W.	Kinser, Thomas E.

215

50 Years of Rockets and Spacecraft

Kirby, Clifton A.
Kirby, Fernander R.
Kirchhuber, Edwin A.
Kirchmyer, Robert H.
Kirk, Boyce L.
Kirk, Marion K.
Kirkindall, Arthur S.
Kirkland, Eugene I.
Kirksey, G. Keith
Kirsopp, Frank L.
Kissel, Ralph R.
Kitchens, Herbert A.
Kite, Burnice E.
Kittredge, Howard B.
Klan, Frederick J.
Klan, Richard L.
Klauss, Ernst K.
Kleminsky, Michael W.
Klima, Jaroslav V.
Klinger, Robert H. Jr.
Klinner, Lavern M.
Klonowski, Floyd A.
Kloock, Mary E.
Knadler, John M.
Knight, Billy L.
Knight, Houston E.
Knight, Joy
Knighton, Maurice H.
Knott, Don M.
Knott, Rex F.
Knowles, William
Knowling, Ted L.
Knox, Ellen H.
Koczor, Ronald J.
Kohr, Rex R.
Koonce, Glennis P.
Koonce, James T.
Kornfeld, Dale M.
Kosis, Alexander
Koslow, Martin
Kozub, George M.
Kramer, Fritz
Kramer, Joseph H.
Kraus, Gerhard W.
Krause, Helmut G.
Krebsbach, Alfred N.
Kreider, William J.
Krempl, William L.
Kress, Mary K.
Kress, Sigmund G.
Kreynus, Leigh A.
Krieg, Gordon C.
Krivutza, John J.
Kroeger, Arthur J.
Kroeger, Frederick H. Jr.
Kroeger, Hermann W.
Kroes, Roger L.
Kroh, Hubert E.
Kroll, Gustav A.
Kroll, Ulrich G.
Krome, Henning O.
Kromis, Andrew G.
Kromis, Theodore
Krone, Robert L.
Kruidenier, Robert D.
Krupnick, Albert C.
Kucks, Louis R.
Kuers, Werner R.
Kurtz, George W.
Kurtz, Herman F.
La Hatte, William F.
La Roche, Lillian B.
Labbe, Robert H.
Lacey, Sarah A.
Lackey, Donald L.
Ladner, Gerald C.
Ladner, James E.
Lafans, Inez D.
Lagarde, Paul M.
Lagrone, Daniel A.
Lahser, Heinz F.
Laiacona, Felix P.
Lake, John F.
Lake, Robert E.
Lakey, Donald W.

Lakin, David R.
Lamb, Carl D.
Lamb, Dale D.
Lamb, Eddie K.
Lamb, Eugenia T.
Lamb, Thyele M.
Lamb, Vincent S.
Lambert, James L.
Lambert, Joe D.
Lambert, Orbra L.
Lambert, William H.
Lambertson, George F.
Lamunyon, Teddy M.
Lance, Brenda C.
Land, Bernice P.
Landau, Morton
Landers, Edward H.
Landers, Joe C.
Landers, Lee
Landers, Margarete K.
Landers, Sherry H.
Landers, Tamara S.
Landers, Vivian B.
Landers, William A.
Lands, John E.
Lane, Billy R.
Lane, Mildred
Lane, Solomon N.
Laney, Edna S.
Laney, Fred A.
Lange, Ernst
Langley, Michael E.
Lanier, Flonzo A.
Lanier, James A.
Lanier, John R.
Lanier, Nellie P.
Lankford, William
Lanter, Murray C.
Lanza, Loretta C.
Larkin, Leo W.
Larry, Curtis N.
Largen, Odis E.
Larsen, Paul A.
Larson, Donald J.
Larson, John B.
Lasseter, James R.
Lassiter, Max C.
Lauderdale, William R.
Laue, Jay H.
Laurine, Donald A.
Laurine, Jerry M.
Laux, Joseph G.
Lavender, Robert E.
Law, Cecil A.
Law, Charles B.
Lawler, Maurice H.
Lawrence, Joe M.
Lawrence, Joel D.
Lawrence, Raymond L.
Lawson, Billy D.
Lawson, Bobby E.
Lawson, Ernest R.
Lawson, Harry
Lawson, Katherine D.
Lawson, Lary F.
Lawter, Charles W.
Layman, Robert E.
Layne, Bennie S.
Layne, Billy A.
Layton, Ferris W.
Lea, William R.
League, Daniel D.
League, Howard T.
League, Woodrow W.
Leak, Betty W.
Leath, Ruby K.
Leatherwood, Robert L.
Leberman, Michael D.
Leberte, Angelo P.
Leblanc, Pearl G.
Lechner, Larry E.
Leckie, Gene A.
Lecroy, Eber W.
Ledbetter, James D.
Ledbetter, Smith E.

Ledford, Harold
Lee, Betty G.
Lee, Charles E.
Lee, Cullen Q.
Lee, Curtis E.
Lee, Don B.
Lee, Elizabeth J.
Lee, Gene
Lee, Henry M.
Lee, Thomas J.
Lee, William E.
Lee, William M.
Lefan, Joseph F.
Legette, Marvin A.
Legg, Mary B.
Legg, Ruby L.
Lehman, Joe J.
Lehner, Jack W.
Leigh, Alfred T.
Leinbaugh, George H.
Leinsetter, Emerson J.
Lemay, Jacqueline R.
Lemay, John W.
Lemley, Delbert C.
Lemley, George W.
Lemley, J. H.
Lenoir, Wroten A.
Lenox, Herbert M.
Lentz, Elbern C.
Lentz, Ronald B.
Leonard, Edward L.
Leonard, Stephen V.
Leslie, James B.
Leslie, Robert
Lessels, Robert Jr.
Lester, Thomas O.
Levan, Charles O.
Levie, James K.
Levine, Jack
Levine, Norman
Levinson, John R.
Lewedag, Robert J.
Lewis, Charles F.
Lewis, Charles M.
Lewis, Elmer H.
Lewis, Jackie W.
Lewis, James L.
Lewis, James L.
Lewis, John E.
Lewis, Malcome M.
Lewis, Robert C.
Lewis, Rosalie A.
Lewis, Theodore W.
Lewis, Thomas A.
Lewis, William P.
Lewter, Billy J.
Lewter, Fred L.
Leytham, Orval L.
Lide, Wilton C.
Lienau, Harry G.
Light, David J.
Light, Susan K.
Lightfoot, Thomas C.
Lincoln, Alonzo W.
Lindberg, James P.
Lindberg, Patricia N.
Lindquist, Robert V.
Lindsay, Jim J.
Lindsey, Andy D.
Lindsey, Roberta E.
Lindsey, Robert O.
Lindsey, William C.
Lindstrom, Robert E.
Linn, Charles C.
Linnan, Thomas A.
Linskey, Andrew F.
Linsley, Edward L.
Linstead, William F.
Linton, Roger C.
Lipsey, Lynda J.
Lirette, Murray J.
Lishman, Sidney H.
Lister, Talbert F.
Little, Cecil S.
Little, Clarence E.

Little, Dewey G.
Little, Eugene C.
Little, Robert P. Jr.
Little, Sally A.
Little, Stephen T.
Little, Thomas T.
Little, Victor C.
Littlefield, Billy W.
Littles, J. Wayne
Lively, Everett A.
Lively, Frank P.
Lively, Mary P.
Livingston, John M.
Lloyd, Howard P.
Lochridge, Robert W.
Locke, Dwight J.
Locke, Juanita C.
Locker, James L.
Loftis, Lewis S.
Loftis, Willis B.
Lofton, Louie R.
Logan, James L.
Logan, Tully L.
Loggins, Robert W.
Loggins, Sam D.
Lomax, James L.
Lombardo, Joseph A.
Lominick, James M.
Lones, Phyllis D.
Long, Barbara A.
Long, Marlin
Long, William O.
Looney, Frank M.
Looney, Juanita H.
Looney, Maxie S.
Looney, Vernon L.
Loose, Jack D.
Lopardo, Arthur M.
Lott, Benny
Loughead, Aaron G.
Love, Asberry D.
Love, Max
Love, Reece J.
Love, Richard A.
Love, Thomas H.
Loveday, Celia A.
Loveless, Cecil R.
Lovell, Charles W.
Lovell, Mable K.
Lovingood, Judson A.
Lovoy, Charles V.
Lowe, Estel G.
Lowe, James T.
Lowery, Brelen A.
Lowery, Harold R.
Lowery, James R.
Lowery, John H.
Lowery, Richard L.
Lowery, Thomas J.
Lowrey, Betty D.
Lowrey, Dane O.
Lowry, Robert L.
Loy, Carl A.
Loynes, Chester R.
Lucas, Jack H.
Lucas, William R.
Luce, William W.
Lucero, William H.
Ludewig, Hermann R.
Luehrsen, Hannes
Luna, Howard D.
Luna, Marianne B.
Luna, Posey E.
Luna, William A.
Lunde, Julian W.
Lundquist, Charles A.
Lusk, Dewey L.
Lusk, Joe L.
Lutz, James P.
Lyle, John C.
Lyle, Johnie F.
Lyles, Coleen
Lyles, Willard A.
Lynch, Thomas J.
Lynch, William W.

Lynn, Charlotte W.
Lynn, Emory E.
Lynn, John C.
Lynn, Richard A.
Lyons, Daniel E.
Lysaght, James O.
Mabie, Clifford N.
Mabie, Edward L.
Mabry, James E.
Mabry, Wilmer C.
MacDonald, James C.
Machen, Jyles L.
Mack, Jerry L.
Mack, Mary H.
Mackay, Charles A.
Maclean, James K.
Macomber, Chester A.
Macpherson, Carol S.
Macpherson, John F.
Maddox, Libby F.
Madewell, Clyde B.
Madison, Charles R.
Madole, John P.
Madole, Sue B.
Madyda, Alexander L.
Maffett, Charles W.
Magee, Anne M.
Magee, Dwight H.
Mahan, James C.
Mahan, W. L.
Mahathey, James T.
Mahone, Morgan
Major, Charles R.
Maley, William E.
Malkmus, Bernard R.
Mallernee, Max E.
Mallory, Edward T.
Mallory, James M.
Mallory, Joyce C.
Malmede, Clovis B. Jr.
Malone, Billy F.
Malone, Deward B.
Malone, James A.
Malone, Johnnie W.
Malone, Lee B.
Malone, Sarah C.
Malone, Terry H.
Maltbie, J. L.
Mance, Fred
Mandel, Carl H.
Mankoski, Steve S.
Manley, Fred
Mann, Charles D.
Mann, Eldon W.
Mann, Era N.
Mann, Franklin D.
Mann, Kathleen K.
Mann, Marvin E.
Mann, Muse E. Jr.
Mann, William H.
Manning, Fred T.
Manning, Harold S.
Manning, Martha K.
Manning, Speairs A.
Manry, Jack H.
Mantooth, Ivis R.
Maples, Donald R.
Maples, Eugene M.
Maples, Gardner S.
Maples, Glenn O.
Maples, Samuel A.
Marberry, Wanda E.
Marcato, Roy C.
Marchant, James R.
Marchese, Michael J.
Marion, Don S.
Marks, Jack A.
Marks, Lloyd B.
Marmann, Richard A.
Maroney, Carl H.
Marrero, Peter J.
Marsh, Gordon R.
Marsh, Martha P.
Marshall, Carlos V.
Marshall, Delbert A.

Marshall, Emerson S.
Marshall, Harry A.
Marshall, Ronald C.
Marshall, William R.
Marthet, Lorraine M.
Martin, Buddie J.
Martin, Charles H.
Martin, Charles R.
Martin, Chester W.
Martin, David A.
Martin, Edward E.
Martin, Elbert W.
Martin, Fonzie O.
Martin, George L.
Martin, Henry L.
Martin, James H.
Martin, James R.
Martin, Judy B.
Martin, Judy D.
Martin, Lee K.
Martin, Lowell D.
Martin, Michael D.
Martin, Nathanial W.
Martin, Patricia L.
Martin, Roy H.
Martineck, Hans G.
Martinez, Eutiquio
Martinez, Nelson G.
Martz, Vergil L.
Massey, Francis H.
Massey, John W.
Massey, Martha D.
Massey, William R.
Masters, Annie F.
Masters, Merlyn M.
Mastin, William C.
Matheney, Jack L.
Mather, Bruce D.
Mathews, Frank M.
Mathews, Thomas H.
Mathews, William H.
Mathis, Estel R.
Mathis, James A.
Matkin, William J.
Mattox, Coy W.
Mattox, Russell M.
Mauldin, Charles R.
Maurer, Paul H.
Maus, Hans H.
Maxcy, Thomas R.
Maxwell, Bly O.
Maxwell, James E.
May, Chester B.
May, Dana F.
May, Ellery B.
May, Frances J.
Mayes, Bill C.
Mayhall, Frankie D.
Mayhall, James C.
Maynard, Jesse A.
Maynard, Marvin C.
Maynard, William E.
Maynor, James L.
McAdam, Michael B.
McAlister, Brenda H.
McAllister, Calvin J.
McAllister, Ella M.
McAllister, Lloyd E.
McAnnally, Robert C.
McAuley, Van A.
McBay, Bernard L.
McBrayer, Rene L.
McBrayer, Robert O.
McBride, Claude H.
McBride, Darmer
McBride, George W.
McBride, Jack S.
McBride, James E.
McBride, Lionel M.
McBride, Mavis E.
McBride, Ruth S.
McCaffrey, Roger C.
McCaig, James C.
McCain, Bernice W.
McCain, Edison H.

NASA - Marshall Flight Space Center

McCaleb, Rebecca C.
McCaleb, Harold K.
McCall, Ernest
McCall, Stanley E.
McCampbell, William M.
McCanless, George F. Jr.
McCann, Dennis D.
McCarrick, Patrick F.
McCarrol, Arthur
McCarter, James W.
McCarty, Estel W.
McCarty, John P.
McCarty, Martha J.
McClain, Paul
McClard, Truman E.
McClearen, Donald L.
McClearen, Richard W.
McClendon, Lemuel F.
McClendon, Mary F.
McClendon, Randy K.
McClung, Paul W.
McClure, Marena M.
McClure, Samuel J.
McCollum, Ralph T.
McCollum, Robert W.
McComas, Robert L.
McCool, Alexander A. Jr.
McCool, James A.
McCord, Andrew J.
McCord, Janet V.
McCord, Robert E.
McCormick, Murphy B.
McCown, Betty B.
McCoy, James T.
McCoy, John C.
McCoy, Margaret I.
McCoy, Richard D.
McCranie, George T. M.
McCrary, Burmah M.
McCrary, Samuel E.
McCrickard, Thomas L.
McCright, James
McCullar, Barney R.
McCullar, Jeanette S.
McCulloch, James C.
McCulloch, James R.
McCullough, Houston M.
McCullough, Hugh S.
McCullough, Thomas J.
McCutcheon, Bernard C.
McCutcheon, Paul B.
McDaniel, Connie S.
McDaniel, Frederick V.
McDaniel, Gerald E.
McDaniel, James D.
McDaris, Harry L.
McDonald, Charles H.
McDonald, D. L.
McDonald, Elton E.
McDonald, James E.
McDonald, Searcy R.
McDonough, George F. Jr.
McDougal, Arnold C.
McDougle, Thomas L.
McDowell, Anna T.
McDowell, Corbin L.
McElrath, Henry P.
McElroy, William P.
McElyea, Carl J.
McElyea, Hollis B.
McElyea, Nannette V.
McElyea, Rupert W.
McEuen, James L.
McFadden, William D.
McFolin, Roy L.
McGahee, Luther M.
McGee, Bobbie H.
McGee, Richard H.
McGee, Robert C.
McGlathery, Dave M.
McGlathery, Laverta M.
McGowan, Edward B.
McGraw, Doris C.
McGraw, James R.
McGuire, Robert M.

McHugh, James L.
McInnis, Erich L.
McIntosh, Charles R.
McIntosh, Walter R.
McIntyre, Stanley D.
McKamie, Hollas B.
McKannan, Eugene C.
McKay, George H.
McKay, Therman W.
McKean, John A.
McKee, Bobby E.
McKee, James W.
McKemie, Robert L.
McKenzie, Edker C.
McKenzie, Maurice W.
McKinley, Jeff W.
McKinney, Howard C.
McKinney, William J.
McLain, Warren C.
McLaughlin, James J.
McLemore, J. C.
McLemore, Joe
McLemore, John D.
McLemore, Sammie Z.
Mclemore, Vicki A.
Mclemore, Virginia C.
McLendon, Samuel P.
McMahen, Cowles M.
McManus, Maurice G.
McMeans, Grady L.
McMeans, Henry P.
McMenemy, James R.
McMillan, Harold F.
McMillan, Laurie N.
McMillan, Orville T.
McMillan, Vernon J.
McMillion, James M.
McMinn, Travis J.
McMullen, Dixie L.
McMullins, Margaret M.
McMurray, Carl P.
McMurray, Gail G.
McMurtrey, Ernest L.
McNabb, Walter D.
McNair, Lewis
McNamara, Jean D.
McNamee, William
McNeely, Ralph D.
McNeese, William L.
McNiel, Daniel O.
McPeak, Billy J.
McPeters, James B.
McPherson, Donald P.
McPherson, William B. Jr.
McQueen, James A.
McQueen, Paul H.
McQuiston, John H.
McRight, Davis M.
McSheehy, Richard K.
McVay, Ann A.
McWhirter, Willadean
McWhorter, Justus K.
McWhorter, Vernon E.
Medal, Edward D.
Medal, Linda L.
Meadows, Charles E.
Meadows, Mahlon L.
Meadows, Malissa B.
Medina, Alexander
Medlen, William F.
Medley, John A.
Meegan, Charles A.
Meek, Opal L.
Meeks, Dewey H.
Meeks, Glynda H.
Meeks, Howard B.
Mefford, John W.
Meinzer, Arthur E.
Melonas, John V.
Melton, Carl E.
Melton, Darrell E.
Melton, Marion S.
Mercer, Hugh A.
Mercier, David R.
Mercier, Joseph C.

Meredith, Oliver D.
Merrell, J. D.
Merrell, Robert L.
Merritt, Ellis L.
Merritt, Richard A.
Messer, Cecil W.
Meyer, Eloise D.
Meyer, Robert E.
Meyer, Warren C.
Meyers, Charles H.
Michael, Calvin L.
Michael, Phylis S.
Mick, Bobbye R.
Mick, Robert W.
Middleton, Robert L.
Mienk, William J.
Milam, Tennis A.
Milde, Hans W.
Miles, Campbell G.
Miles, Elfrida Y.
Miles, Ivon O.
Milewicz, Leo
Miller, Alice G.
Miller, Alton B.
Miller, Arthur A.
Miller, Bura V.
Miller, Edgar R.
Miller, Eugene
Miller, Gerald V.
Miller, James C.
Miller, James W.
Miller, John E.
Miller, John Q.
Miller, Lois S.
Miller, Loren W.
Miller, Marion P.
Miller, Reuben T.
Miller, Robert R.
Miller, Roger I.
Miller, Ruth G.
Miller, Shelvie L.
Miller, Virgil R.
Miller, William J.
Miller, Wilma N.
Milligan, Mary R.
Milly, John J.
Milly, Nancy J.
Milner, Robert W.
Mims, Katherine K.
Mims, Milton D.
Mincher, George W.
Mink, Harold J.
Minney, Harold T.
Mintz, Anna N.
Mintz, Edward S.
Mitchell, Arthur R.
Mitchell, Charles M.
Mitchell, Craig C.
Mitchell, Dan C.
Mitchell, Edna P.
Mitchell, Erby L.
Mitchell, Frank G.
Mitchell, Gilbert D.
Mitchell, James W.
Mitchell, Jerry L.
Mitchell, Joseph R.
Mitchell, Kenneth L.
Mitchell, Margaret P.
Mitchell, Margaret S.
Mitchell, Millard
Mitchell, Royce E.
Mitchell, Thomas W.
Mitchell, Walter T.
Mitchell, William H.
Mitchem, Frances H.
Mitchum, James O.
Mixon, William R.
Mize, Rondal C.
Mizzell, Joye F.
Moak, Rodney L.
Mobley, David H.
Moffitt, Fred L.
Moffitt, James D.
Mohlere, Edward D.
Monguillot, Linus J.

Monk, Jan C.
Monks, Leonard D.
Monks, Thomas C.
Montana, Eugene N.
Montano, Joseph W.
Montenegro, Justino
Montgomery, Brian O.
Montgomery, Clarence C.
Montgomery, Clyde P.
Montgomery, Robert H.
Montgomery, William D.
Montgomery, Yvonne D.
Monty, Bertram E.
Moody, Edwin T.
Moody, Jewell W.
Moon, Margie B.
Moon, Orman E.
Moon, Reedus J.
Moon, Russell D.
Moon, Ruth S.
Moon, William A.
Moon, William M.
Mooney, Charles F.
Mooney, Peggy B.
Moore, Angela L.
Moore, Audrey B.
Moore, Connie E.
Moore, Darrell W.
Moore, Edwin L.
Moore, F. Brooks
Moore, Houston G.
Moore, Jack S.
Moore, James W.
Moore, John C.
Moore, Joyce E.
Moore, Leonard T.
Moore, Lonia R.
Moore, Morgan W.
Moore, Philip
Moore, Robert T.
Moore, Webster W.
Moore, William L.
Moorehead, Tauna W.
Moorhead, John W.
Mordan, Gustav W.
Morea, Saverio F.
Morelan, Jeannie B.
Morelan, Michael R.
Moreland, Donald J.
Morelli, Robert E.
Morelock, James C.
Morgan, Alton O.
Morgan, Eulice
Morgan, Fred H.
Morgan, Irvin T.
Morgan, Jean C.
Morgan, Linda M.
Morgan, Samuel H.
Morgan, Travis A.
Morgan, Wayne R.
Morrell, Presley L.
Morring, Robert E.
Morris, Charles E. K. Jr.
Morris, Charles W.
Morris, Donald E.
Morris, Irma M.
Morris, John K.
Morris, Lena F.
Morris, Paul W.
Morris, Ralph R.
Morris, Robert L.
Morris, Theo T.
Morris, Thomas F.
Morris, Virgil E.
Morris, William E.
Morris, Woodrow W.
Morrison, James A.
Morrison, James S.
Morrison, Robert E.
Morrow, Ben F.
Morrow, Charles W.
Morrow, Earl W.
Morrow, James H.
Morrow, William P.
Morse, Alfrd R.

Morse, Robert J.
Morse, Robert O.
Morton, Edward E.
Morton, Rex R.
Moseley, William A.
Moses, James L.
Moses, O. D.
Moses, Vann C.
Moss, Arlon R.
Moss, Jackson C.
Moss, John D.
Moss, Velma B.
Mossawir, Harvey
Mottar, Roland F.
Mottar, Rose K.
Moultrie, Billy F.
Mounts, Wilfred V.
Mowell, Carl C.
Mowell, Lakie S.
Mowery, David K.
Moyers, Stephen G.
Mrazek, William A.
Mudd, Carl H.
Mudd, Katherine T.
Mueller, Charles W.
Muirhead, Georgia H.
Muirhead, Walter B.
Mulac, Elsworth K.
Mullaley, Leroy E.
Mullen, Mary B.
Mullen, Ray P.
Muller, Gary R.
Muller, Philip L.
Mullin, Daniel F.
Mullins, Alton R.
Mullins, Clemmie L.
Mullins, Hilliard
Mullins, Jerry L.
Mullins, Larry D.
Mullins, Leaman H.
Mullins, Linda W.
Mullins, Vernon B.
Mullins, Wilburn C.
Mullins, William J.
Mulloy, Lawrence B.
Munafo, Paul M.
Mundie, Charles E.
Murphree, Hughlen I.
Murphree, William D.
Murphy, Arthur M.
Murphy, James T.
Murphy, Ralph W.
Murphy, Welcome W.
Murray, James W.
Musgrove, William C.
Myers, Myron L.
Myers, Warren E.
Mynatt, Elaine B.
Naftel, Lee A.
Nagle, Carl W.
Namie, Moses M.
Napier, Estelle J.
Napper, Frank E.
Naramore, Sarah R.
Nash, Emmett C.
Nathan, Ernest B.
Naumann, Robert J.
Naumcheff, Michael
Neal, Billy C.
Neal, William P.
Neece, Ronald L.
Needham, Charles C.
Neeley, Gladys T.
Neighbors, Alice K.
Neighbors, Billy A.
Neighbors, Billy H.
Neiland, Victor R.
Neill, Sidney S.
Nein, Hans J.
Nein, Max E.
Neisler, Lorena G.
Nelson, Doris B.
Nelson, Eli E.
Nelson, Harold F.
Nelson, Howard J.

Nelson, James R.
Nelson, Julian F.
Nelson, Paul E.
Nelson, Raymond H.
Nelson, Richard E.
Nelson, Virginia P.
Nerren, Billy H.
Nesmith, Hubert B.
Nesmith, Malcolm F.
Nesmith, Max L.
Neubert, Erich W.
Neuschaeffer, Robert W.
Neutze, Jefferson C.
Newton, Carol A.
Newton, John A. III
Newton, Lowell C.
Neville, Don C.
Nevins, Clyde D.
Newberry, Jimmie H.
Newberry, Murl H.
Newbill, Howard B.
Newby, Danvis T.
Newby, David H.
Newby, George E.
Newcomb, James I.
Newell, Dorothy A.
Newell, Shepard B.
Newman, Charles D.
Newman, Dewey F.
Newsom, William G.
Newton, Alice K.
Newton, Gloria P.
Newton, James H.
Newton, Patricia G.
Newton, Turner
Nicaise, Placide D.
Nichols, Franklin T.
Nichols, Jack J.
Nichols, Ronald L.
Nichols, Tharel N.
Nicholson, Carolyn S.
Nicholson, Edward T.
Nicholson, Frederick R.
Nicholson, Roger M.
Nicholson, William P.
Nickelson, Del C.
Nicolas, David P.
Nieman, Alexander F.
Nipper, Charles H.
Nisen, Daniel
Nix, Mary J.
Nix, Michael B.
Nix, N. J.
Nixon, James W.
Nixon, Melvin H.
Nixon, Robert F.
Noel, Edward F.
Noel, George W.
Noel, John E.
Nola, Frank J.
Nolley, Royal R.
Noor, Robert A.
Norman, Edwin M.
Norris, Lester M.
Northcutt, Ervin E.
Norton, Carla M.
Norton, Orea A.
Norton, William E.
Nowak, Max E.
Nowakowski, Michael F.
Nuber, Robert J.
Nunley, Anita S.
Nunley, Billy W.
Nunley, Thomas V.
Nunn, Rufus A.
Nunnelley, Benny
Nunnelley, John R.
Nurre, Gerald S.
Nyhan, Stanley L.
Oakes, Henley A.
O'Bryant, Carvel M.
O'Bryant, Thomas O.
Ochoa, Richard H.
O'Connor, Albert A.
Oden, Ralph W.

Oddo, Johnny M.
Odom, James B.
O'Donnell, Marvin I.
O'Donnell, Thomas F.
Offord, Vernon G.
Ogle, Hobert H.
Ogle, Lloyd C.
Ogletree, Thomas W.
Ogozalek, Edmund F.
Ogozalek, Nance Jo
O'Hara, Junior R.
Old, Dodge H.
Oldfield, Leonidas H.
Oldham, Leonard T.
Olive, William D.
Oliver, Annie A.
Oliver, Elwood L.
Oliver, Foster F.
Olivier, Jean R.
Olton, Barry S.
O'Meara, Howard F.
Ondrak, Benedict T.
O'Neal, James E.
O'Neal, Margery K.
O'Neil, Anthony M.
Orem, Ernest L.
Orillion, Alfred G.
Ornburn, William C.
Orr, James P.
Orr, Robert F.
Orswell, Prentice L.
Orton, Bert W.
Osborn, Dorothy E.
Osborne, Robert E.
Osborne, Thomas L.
Osburn, Eli G.
Ousley, Arthur T.
Ousley, Harry K.
Overton, Jesse L.
Owen, Clark M.
Owen, William T.
Owens, Floyd L.
Owens, James M.
Owens, John E.
Owens, John H.
Owens, Joseph D.
Owens, Lawrence W.
Owens, William D.
Ozbolt, Joe D.
Pabon Medina, Norman
Pace, Robert E.
Pack, Homer C.
Pack, Lonnie D.
Pack, Marvorie
Padgett, Billy R.
Paetz, Robert
Page, Billy R.
Page, Milton A.
Page, Rabon P.
Pagenkopf, Benjamin L.
Pajak, John A.
Palaoro, Hans R.
Palladino, Patsy G.
Pallos, Lorant
Palmer, Euple I.
Palmer, Herbert W.
Palmer, Jack R.
Palmer, Lyda S.
Palmer, Stella J.
Paludan, Charles T.
Panciera, Robert L.
Panzer, Walter B.
Park, James F.
Parker, Billie G.
Parker, Dewey L.
Parker, Eva Charlene
Parker, Glenn A.
Parker, Harold C.
Parker, Henry W.
Parker, James W.
Parker, Joe R.
Parker, John B.
Parker, John C.
Parker, Lois C.
Parker, Walter H.

Parkes, Jack C.
Parkes, Martha H.
Parks, Bentley
Parks, Harold E.
Parks, Leonard A. D.
Parks, Paul G.
Parks, Robert W.
Parks, Virginia T.
Parmer, Joseph C.
Parnell, Thomas A.
Parr, Geraldine C.
Parr, Richard A.
Parrish, Edwin L.
Parrish, Marvin L.
Parsley, Raymond A.
Partain, Mary E.
Partin, Marvin E.
Parton, Herman A.
Parton, Sim B.
Parton, Sylvia B.
Paschal, Lloyd E.
Paseur, Barbara M.
Patrick, Donald R.
Patterson, Carlton J.
Patterson, Clyde A.
Patterson, Edgar E.
Patterson, Floyd T.
Patterson, James W.
Patterson, Joe L.
Patterson, Tom
Patterson, Wayne A.
Patterson, Wayne H.
Patterson, William A.
Patterson, William C.
Patterson, William F.
Patterson, William J.
Patton, Burgess
Paul, Hans G.
Paulk, Gerald L.
Pavlick, John F.
Payne, Billy W.
Payne, Charles J.
Payne, Clifton H.
Payne, Clovis J.
Payne, Edris J.
Payne, L. D.
Payne, Molly H.
Payne, William R.
Paysinger, Robert L.
Payton, Faye S.
Peace, Cecil E.
Pearce, James C.
Pearson, Dallias S.
Pearson, James C.
Pearson, Linda M.
Pease, Robert E.
Peck, James B.
Peck, Russell L.
Peck, Stuart M.
Peddycoart, Jerry N.
Pedigo, George P.
Pegg, Estelle F.
Peigler, Frank K.
Pemberton, Lurie H.
Pence, Denton
Pendergraft, J. B.
Pendergrass, Reita
Pennington, Buford
Penny, Robert G.
Penrod, Ned
Pentecost, James C.
Peoples, Jerry A.
Pepper, James E.
Pepper, Mitchell
Perkins, Alvin
Perkins, Harold
Perkinson, Don T.
Perkinson, Linda J.
Perrine, Barton S.
Perry, Charles R.
Perry, Cortes L.
Perry, Deledia P.
Perry, Guy D.
Perry, John L.
Perry, John W.

Perry, Nina L.
Perry, Paul C.
Perry, Thomas L.
Perry, William R.
Pesavento, Victor
Peshek, Edwin J.
Pessin, Myron A.
Peters, Gerald L.
Peters, Palmer N.
Peters, Richard L.
Peters, William L.
Peterson, Cleo J.
Peterson, Willard B.
Pettis, Margaret R.
Pettus, Gracie M.
Petty, Allious L.
Petty, Evelyn S.
Peuler, Bernard L.
Pfaff, Helmuth
Phelps, Benton E.
Phillips, Carl E.
Phillips, Charlie E.
Phillips, Emmett
Phillips, Hardee L.
Phillips, Jack A.
Phillips, James H.
Phillips, Jerry C.
Phillips, Jimmy E.
Phillips, John E.
Phillips, John O.
Phillips, Margie
Phillips, Margie O.
Phillips, Mauree E.
Phillips, Norris W.
Phillips, Robert W.
Phillips, Roy H.
Pickard, Mitchell F.
Pierce, Frank E.
Pierce, Howard J.
Pierce, James A.
Pierce, James C.
Pierce, John P.
Pierce, Ollie F.
Pierce, Raymond E.
Pike, Lenard J.
Pillischer, Evelyn T.
Piner, John R.
Pinkerton, Alfred C.
Pinkerton, John W.
Pinson, Charles F.
Pirtle, Henry M.
Pirtle, Johnnie C.
Pirtle, Ruthie M.
Pisani, Raymond W.
Pitcock, Paul
Pitcock, Robert E.
Pitre, Gaston B.
Pitsenberger, Forrest D.
Pitts, Ellington H.
Pitts, John R.
Pizarro, Juan II
Pizzano, Frank
Platt, Gordon K.
Plattenburg, Stanley M.
Poarch, Nolen F.
Poe, Brenda K.
Poe, James W.
Poe, Katherine J.
Poe, Linda P.
Polites, Michael E.
Politis, Venus K.
Politis, X. Zeny
Polk, Doris J.
Pollock, Joe E.
Polstorff, Walter K.
Pomeroy, William R.
Ponder, Troy W.
Pool, Carl L.
Poole, Alfred D.
Poole, Elbert L.
Poole, James C.
Poorman, Richard M.
Pope, Carter H.
Pope, J. D.
Popejoy, Joyce L.

Porter, Doris J.
Porter, George J.
Porter, Granville D.
Porter, James M.
Porter, Robert L.
Porterfield, Harold D.
Porterfield, Lavera S.
Portwood, Joseph N.
Posey, Emil L.
Posey, Linda B.
Potaczala, Stanley A.
Potate, John S.
Poteet, Joy V.
Potter, Alvin R.
Potter, Jackie H.
Potter, Peggy Y.
Potter, Richard A.
Potter, William R.
Potts, Minnie P.
Pour, Joseph J.
Powell, Edward F.
Powell, Elmer A.
Powell, James T.
Powell, Joseph H.
Powell, Luther E.
Powers, Bobby L.
Powers, Emmett C.
Powers, Homer C.
Powers, James R.
Powers, Luther B.
Powers, Mary G.
Powers, William D.
Powers, William T.
Prager, John W.
Prasthofer, Willibald P.
Pratt, James H.
Pratt, Richard B.
Prendergast, Francis G.
Pressnell, Aubrey M.
Pressnell, Odie L.
Preston, Ores M.
Preston, Owen R.
Prestwood, Sam B.
Price, Clifford E.
Price, David E. III
Price, David M.
Price, Everett D.
Price, Jesse C.
Price, John M.
Price, John P.
Price, Mary L.
Price, William B.
Pride, Robert R.
Priest, Claude E.
Priest, Ruth A.
Prince, Carl
Prince, Warnie B.
Pritchett, David W.
Pritchett, Jack
Pritchett, James C.
Pritchett, William T.
Proffitt, Ray
Pruett, Bumon J.
Pruitt, Jimmy R.
Pruitt, Lawrence L.
Pruitt, Robert W.
Pruitt, Walter H.
Pryor, Donald E.
Pschera, Karl J.
Puckett, Alton K.
Puckett, Coy W.
Puckett, Dock O.
Puckett, Thomas E.
Puett, John A.
Pugh, Elizabeth H.
Pugh, Odell A.
Purcell, Glyndon L.
Purdy, Burma C.
Purinton, Steven C.
Purser, Inez C.
Queen, Donald R.
Queen, Kenneth M.
Quick, Sheron A.
Quick, Walter
Quillen, Lillian L.

Quillin, Gus
Quillin, Mason E.
Quinn, Alberta W.
Quinn, Christopher J.
Quong, Johnny
Rabatin, Michael
Raby, James E.
Raby, Lorraine K.
Ragland, Ollie G.
Ragland, Susan
Rainey, Paula M.
Rainey, Theodore H.
Rains, John C.
Raley, Nathaniel G.
Ralls, Glenda A.
Ralston, James T.
Ralston, William Stephens
Ramage, William E.
Ramey, Walter F.
Ramsey, Lee O.
Ramsey, Paul Elbert
Ramsey, Robert D.
Randall, Joseph L.
Randel, Jack
Randles, Emmett H.
Raney, Bobby H.
Raney, Bunah H.
Rankin, Thomas O.
Rape, Amy P.
Rasberry, Lona L.
Rasco, Walter R.
Rash, Samuel J.
Rasquin, John R.
Ratliff, Clynton T.
Ratliff, John H.
Rawlings, Glen H.
Rawls, Francis C.
Ray, Charles D.
Ray, Jack D.
Ray, John W.
Ray, Julius E.
Reed, Kenneth L. Jr.
Ray, Lena S.
Ray, William L.
Rayfield, Helen N.
Rea, John L.
Read, Preston C.
Reagh, Mary E.
Reasor, Arland
Reaves, John H.
Reaves, Joseph S.
Reavis, James G.
Reavis, Lee W.
Reavis, Norman H.
Recio, Bolivar T.
Record, Walter A.
Rector, James C.
Redd, Johnson E.
Redden, Calvin T.
Redden, Laverne S.
Reddick, Lawrence L.
Redman, Bobby W.
Redmon, John W.
Redus, Jerome R.
Reece, Orvil Y.
Reed, Billy R.
Reed, James H.
Reed, John H.
Reed, Kenneth L. Jr.
Reed, Kenneth W.
Reed, Laura S.
Reed, Leila S.
Reed, Mary O.
Reed, Nelda M.
Reed, Stencil L.
Reed, Thomas G.
Reed, Wanda M.
Reed, William J.
Rees, Eberhard M.
Rees, Jimmy W.
Reese, Robert
Reeves, Elton H.
Reeves, Fredrick A.
Reeves, Jacquelyn W.
Reeves, Robert L.

Reeves, Ruby C.
Rega, John A.
Regnier, O. Patricia
Rehage, Jon R.
Reichmann, Edwin J.
Reid, Harry
Reiker, Paul J.
Reily, Jack C. Jr.
Reinartz, Stanley R.
Reinbolt, Earl J.
Reisig, Gerhard H.
Reitzel, Elva J.
Remer, Ira
Rendall, John B.
Renfrow, Nan A.
Reynolds, James D.
Reynolds, Jewell T.
Reynolds, John M.
Reynolds, Sherry B.
Reynolds, William R.
Rheinfurth, Mario H.
Rhoades, Carl B.
Rhodes, Charles M.
Rhodes, Jacob R.
Rhodes, James E.
Rhodes, Percy H.
Rhods, Alvin L.
Rice, Raymond P.
Rich, Shelby
Richard, Lasalle A.
Richard, Ludie G.
Richard, Victor I. Jr.
Richards, Carol K.
Richards, James S.
Richards, Leo O.
Richardson, Edis C.
Richardson, Euell C. Jr.
Richardson, John R.
Richardson, Robert
Richardson, Syble C.
Richardson, William F.
Richmond, Jeanne D.
Richmond, Ralph
Richmond, Robert J.
Richter, Ellsworth M.
Richter, Frank J.
Richter, Katherine E.
Rickard, Carl E.
Ricketts, Henry
Ricketts, John A.
Rickles, Peggy S.
Ricks, Gordon S.
Riddick, Edwin L.
Riddle, Charlotte J.
Ridgeway, Gerald D.
Rieger, Carl J.
Riehl, Wilbur A.
Riemer, Robert L.
Rieves, Marshal H.
Riggins, Donald B.
Riggs, Howell R.
Riggs, Kenneth E.
Rigsby, Charles E.
Rikard, James O.
Riley, Charles L.
Riley, Donald L.
Riley, John O.
Riley, Joy I.
Riquelmy, James R.
Ritch, Ernest R.
Rithmire, Brenda B.
Ritter, Glen D.
Rivers, Robert H.
Rives, Harold E.
Rives, James M.
Roach, Jack A.
Robbe, Thelma M.
Roberson, Eunice P.
Roberson, James C.
Roberts, David G.
Roberts, Devers C.
Roberts, Eloitte W.
Roberts, George E.
Roberts, Herman B.
Roberts, James B.

NASA - Marshall Flight Space Center

Roberts, Joel A.
Roberts, Joseph M.
Roberts, Lillian O.
Roberts, Lymon L.
Roberts, Marion E.
Roberts, William T.
Roberts, Woodrow W.
Robertson, Billie S.
Robertson, Edward L.
Robertson, Harvey C.
Robertson, John S.
Robertson, Joseph G.
Robertson, Lois M.
Robertson, Loyde L.
Robertson, Ralph E.
Robinette, David R.
Robinson, Billy E.
Robinson, Curtis L.
Robinson, Howard E.
Robinson, James C.
Robinson, James H.
Robinson, Jane M.
Robinson, Johnny F.
Robinson, Maxul F.
Robinson, Michael B.
Robinson, Nancy S.
Robinson, Tracy R.
Robinson, William J.
Rockefeller, Arthur K.
Roden, Donald N.
Roden, Doris E.
Rodgers, Elizabeth B.
Rodgers, Richard N.
Rodman, Martha S.
Rodman, Richard D.
Rodrigue, Freddie A.
Roe, Fred D.
Roe, Fred D. Jr.
Roe, Millard M.
Rogers, James T.
Rogers, Turner L.
Rogers, Virgil H.
Rohr, Stephen V.
Roland, George W.
Roland, Jennie D.
Roland, Odis H.
Roman, Harold A.
Romine, Doris D.
Rood, Robert W.
Rooks, Jahazy
Roper, Albert L.
Roper, James F.
Roper, John P.
Rorex, Beatrice A.
Rorex, James E.
Rose, James B.
Rose, Millard F.
Rose, Stephen D.
Rosenthal, Max E.
Rosinski, Werner K.
Ross, Gordon Mack
Rossman, Kenneth L.
Rosson, Bernard
Roth, Axel
Rothe, Kurte W.
Rougon, Charles A.
Rountree, James W.
Routh, Donald E.
Rowan, Arthur D.
Rowan, James T. Sr.
Rowan, Jesse E.
Rowden, John C.
Rowe, Gordon J.
Rowe, Owen
Rowe, Robert J.
Rowe, Robert R.
Rowell, James G.
Rowell, Thomas F.
Rowzee, Nancy E.
Rozear, Carrol J.
Rubel, Werner H.
Ruck, Elmore F.
Rudd, George B.
Rudder, Elbert
Rudolph, Arthur H.

Rudolphi, Michael U.
Ruff, Rudolph C.
Ruhl, Robert K.
Ruhl, Sylvia E.
Rule, Gladys R.
Runkle, Roy E.
Rupp, Charles C.
Rupp, Henry C.
Rush, Randolph
Rushbrook, Patricia D.
Rushing, Haley W.
Rushing, Robert L.
Rusk, Billy G.
Russ, James W.
Russel, George R.
Russel, Mark L.
Russell, Alfred
Russell, Jacob B.
Russell, James W.
Russell, Jim K.
Russell, Larry W.
Russell, Marion L.
Russell, Willa M.
Rutherford, Perry G.
Rutherford, Robert H. Sr.
Rutland, Cary H.
Rutledge, Charles S.
Rutledge, Henry F.
Rutledge, Mary J.
Rutledge, William H.
Rutledge, William S.
Ryan, Robert S.
Ryan, Thomas F
Ryland, Charles R.
Rylant, Wendell M.
Sabelhaus, Ida L.
Sackheim, Robert L.
Sadler, Benjamin D.
Sadler, Jesse E.
Sadler, Marvin C.
Sage, Richard L.
Saidla, Robert L.
Sainker, Irving S.
Saint, Lester D.
Sallis, Bobby G.
Salmon, Carl V.
Salmon, Ralph D.
Salter, Larry D.
Samaniego, Ramon J.
Sample, Olga S.
Sampson, J. W.
Sampson, Robert D.
Sams, Alice F.
Sams, James C.
Sanders, Curtis T.
Sanders, Fred G. III
Sanders, Grover H.
Sanders, Hazel H.
Sanders, Horace B.
Sanders, James L.
Sanders, Lelan A.
Sanders, Leroy
Sanders, Lois H.
Sanders, Margaret J.
Sanders, Marvis W.
Sanders, Mary S.
Sanders, Richard M.
Sanders, Virgil J.
Sanders, William E.
Sanderson, Arthur E.
Sanderson, Mabel C.
Sanderson, Virginia P.
Sandlin, James R.
Sandlin, Paul D.
Sandlin, Pryer W.
Sandlin, William N.
Sanford, John G.
Sanford, R. L.
Sanford, Virgle R.
Sapp, John W.
Sarver, Laverne A.
Satterfield, James B.
Satterfield, Paul H.
Satterfield, Willard A.
Saucier, Sidney P. III

Saunders, Grady H.
Saunders, Jerry M.
Savage, Michael F.
Savage, Robert S.
Savage, William H.
Sawyer, James T.
Sawyer, William R.
Saxton, Donald R.
Saxton, Floyd M.
Sayer, William C.
Saylor, Woodie M.
Scarbrough, Albert O.
Scarbrough, Charles L.
Scarbrough, Horace S.
Schaefer, Jerome D.
Schaffer, Anthony J.
Schauer, Reyman C.
Scheidt, Jonathan R.
Schell, John T.
Scheuplein, Carl V.
Schilb, Theodore W.
Schlagheck, Ronald A.
Schlemmer, Norman C.
Schlosser, John E.
Schmidt, Dalton M.
Schmidt, Eunice D.
Schmidt, George A.
Schmidt, William F.
Schneider, Ernest L.
Schock, Richard J.
Schock, Richard W.
Schocken, Klaus
Schorsten, Edward S.
Schoulin, Charles V.
Schrick, Byron J.
Schrimsher, Herman E.
Schrimsher, Mary C.
Schuerer, Paul H.
Schuler, Albert E.
Schultz, David N.
Schulz, Malcolm A.
Schulze, Heinrich A.
Schulze, William A.
Schutzenhofer, Luke A.
Schwaniger, Arthur J.
Schwartz, Adger H.
Schwarzwalder, Althea H.
Schwindt, Paul J.
Schwinghamer, Robert J.
Scofield, Harold N.
Scollard, Joseph H.
Scott, Clyde T.
Scott, Clyde W.
Scott, Daniel T.
Scott, Donald R.
Scott, Ewell M.
Scott, Frances E.
Scott, Julie
Scott, Martha F.
Scott, Ramon C.
Scott, Ronald C.
Scott, Royal L.
Scott, Virginia D.
Scott, Wesley W.
Scott, William A.
Scrip, Robert J.
Scruggs, Benjamin I.
Scruggs, Roosevelt
Scruggs, Vermon A.
Seagle, Charles L.
Seal, Charles D.
Seal, Jesse L.
Seale, William L.
Seaver, Cecil A.
Seay, Nell Y.
Seay, William M.
Seborg, Dave W.
Seeley, Donald W.
Seely, Edgar T.
Segewitz, Willi
Seguin, Fernand J.
Seiler, Bernd K.
Seiler, Ernst E.
Seiser, William R.
Seitz, Robert N.

Selens, John K.
Self, Angeliene I.
Self, Deemer O.
Sellers, Dennis A.
Sellers, Robert M.
Sells, Harold R.
Seltzer, Sherman M.
Selvage, George R.
Selvidge, Wilma C.
Semmes, Edmund B.
Sentell, Melvin D.
Serich, Robert D.
Serio, Frank J.
Serpas, Edwin
Serpas, Kathryn B.
Settle, Gray L.
Sever, Thomas L.
Sexton, Frank J.
Sexton, Willie
Seymour, David C.
Seymour, Norman E.
Seymour, Sandra G.
Shackelford, Benjamin W. Jr.
Shafer, James D.
Shafer, Vivian I.
Shahan, Lillian O.
Shamblin, Pat
Shanahan, Montague X.
Shaner, Thomas L.
Shankle, Robert W.
Shariett, Charles A. III
Sharp, Albert
Sharp, Herman H.
Sharp, Shefton O.
Sharpe, Max H.
Sharpe, Mitchell R.
Sharpe, Terry H.
Shaw, Ezell L.
Shaw, Faye H.
Shaw, Judy B.
Shaw, Roy C.
Shaw, Russell C.
Shaw, Thomas C.
Shearer, John B.
Shearl, Fred E.
Sheats, John P.
Shefield, Beatrice W.
Shelton, Billy W.
Shelton, Clyde H.
Shelton, Edward J.
Shelton, Harvey L.
Shelton, Jean
Shelton, Melvin E.
Shelton, Willard
Shepard, Robert P.
Shepherd, James T.
Sheppard, Robert G.
Sherbert, Kermit W.
Sherbert, Yvonne P.
Shields, James M.
Shields, William G.
Shipman, David L.
Shipp, Richard C.
Shippey, Willie L.
Shirey, John E.
Shockley, Wayne J.
Shofner, George E.
Shook, Reab
Short, Henry F.
Short, Joe R.
Showers, Nathan
Shrader, William L.
Shriver, Edward L.
Shulman, Sam
Shultz, Bennie
Shumaker, Wayne B.
Shurden, Walter A.
Shurney, Robert E.
Siddall, Calvin C.
Siddall, Wanda A.
Sidick, James G.
Siebel, Mathias P.
Sieber, Werner H.
Siebigteroth, Maximilian
Sieja, Ralph E.

Siler, Harvey L.
Silvey, Jesse F.
Simmons, Dorothy L.
Simmons, Magalou D.
Simmons, Melvin H.
Simmons, Wayne J.
Simmons, William K.
Simmons, William O.
Simmons, Woodrow W.
Simms, Barbara L.
Simms, George M.
Simms, James K.
Simon, Erwin H.
Simonds, Judy M.
Simpson, John G.
Simpson, Robert M.
Simpson, William
Simpson, William G.
Simpson, William H. Jr.
Sims, Anne D.
Sims, Clifton R.
Sims, Frank W.
Sims, Garner L.
Sims, Joe W.
Sims, John C.
Sims, John H.
Sims, Joseph L.
Sims, Melvin C.
Sinclair, James B.
Singley, Maurice E.
Sinko, Louis
Sinyard, Kenneth E.
Sires, Herbert H.
Sirratt, Cardinal F.
Sisco, Winbern O.
Sisk, Amelia M.
Sisk, David M.
Sisk, Robert C.
Sisson, James M.
Sittason, Ann S.
Sittason, Frederick M.
Sivley, Helon M.
Skeen, Clarence A.
Skinner, James E.
Skinner, Marjorie K.
Skrobiszewski, Edward J.
Slaby, Jerry A.
Slate, George B.
Slaten, Doyle W.
Slattery, Bart J.
Slatton, Lester G.
Slaughter, Albert L.
Slayden, Howard A.
Slayden, Murrell D.
Slayton, J. B.
Sledge, Arthur O.
Sledge, Larry D.
Sloan, James A.
Sloan, Samuel J.
Slobe, Stanley F.
Slone, Bobby M.
Slone, Helen M.
Smartt, Gladys J.
Smartt, Patricia L.
Smelser, Jerry W.
Smith, Allen F.
Smith, Allen T.
Smith, Bessie L.
Smith, Betty L.
Smith, Carl M.
Smith, Carlyle R.
Smith, Charles E.
Smith, Charles F.
Smith, Charles L.
Smith, Charles R.
Smith, Charleston A.
Smith, Charlotte D.
Smith, Clark E.
Smith, Clifford M.
Smith, Cynthia A.
Smith, David N.
Smith, Deon F.
Smith, Earl E.
Smith, Earl W.
Smith, Earnest C.

Smith, Edward E.
Smith, Ellis L.
Smith, Estella M.
Smith, Fletcher B.
Smith, Fred A.
Smith, Fuller G.
Smith, Gene T.
Smith, George E.
Smith, Gilbert
Smith, Grady E.
Smith, Harvey F.
Smith, Henry P.
Smith, Homer M.
Smith, Howard S.
Smith, Hubert E.
Smith, J. B.
Smith, Jack G.
Smith, Jacqulin
Smith, James
Smith, James C.
Smith, James D.
Smith, James M. O.
Smith, James R.
Smith, James R.
Smith, James W.
Smith, Janell M.
Smith, Jeanne S.
Smith, John
Smith, John H.
Smith, John O.
Smith, John V.
Smith, John W.
Smith, Joseph E.
Smith, Kenneth A.
Smith, L. C.
Smith, Lawrence J.
Smith, Leland R.
Smith, Leonard A.
Smith, Leonard B.
Smith, Lindsey B.
Smith, Lois A.
Smith, Lowell C.
Smith, Malcolm H.
Smith, Marvin V.
Smith, Mary F.
Smith, Mary W.
Smith, Mattie L.
Smith, Maxime L.
Smith, Orvel E.
Smith, Orville L.
Smith, Ossie L.
Smith, Paul N.
Smith, Ralph A.
Smith, Ralph F.
Smith, Ralph J.
Smith, Ramon G.
Smith, Raymond
Smith, Richard A.
Smith, Richard B.
Smith, Robert
Smith, Robert A.
Smith, Robert B.
Smith, Robert E.
Smith, Robert L.
Smith, Robert P.
Smith, Ronald B.
Smith, Rufus P.
Smith, Sadenya W.
Smith, Sam E.
Smith, Samuel T.
Smith, Shelby L.
Smith, Spencer E.
Smith, Susan McGuire
Smith, Teresa C.
Smith, Thelmer P.
Smith, Thomas H.
Smith, Thomas H.
Smith, Tommie C.
Smith, Tommy C.
Smith, Verdell
Smith, Virginia W.
Smith, William C.
Smith, William D.
Smith, William E.

50 Years of Rockets and Spacecraft

Smith, William G.
Smitherman, Daniel M.
Smithers, Gweneth A.
Smithers, Martin E.
Smock, Alden W.
Smoot, Charles T.
Smoot, William V.
Smyly, Harold M.
Snead, G. W.
Sneed, Bill H.
Snell, Donald L.
Snell, Zelma A.
Snellgrove, James C.
Snoddy, David E.
Snoddy, James A.
Snoddy, Martha H.
Snoddy, Muriel C.
Snoddy, William C.
Snow, Edward C.
Snow, Oliver
Snyder, Louis E.
Snyder, Robert S.
Sockwell, Ronald W.
Soileau, Luke A.
Solmon, Gordon W.
Solomon, David C.
Solomon, Richard H.
Soltis, Andrew E.
Sommers, Louis H.
Soprano, Quentin C.
Sorensen, Victor C.
Sorter, Dan
Sotherland, Melba R.
Soule, Helen S.
Southard, Judy C.
Southard, Lesly D.
Souther, Charles H.
Southers, Joe W.
Sowell, Kenneth D.
Sowell, William F.
Spangler, Daniel E.
Spann, Judith J.
Spann, Melvin L.
Sparks, Charles H.
Sparks, Mary J.
Sparks, Owen L.
Sparks, Robert O.
Spaulding, Mary D.
Spear, James S.
Spears, Albert P.
Spears, James P.
Spears, Luther T.
Speer, Fridtjof A.
Speer, Spencer W.
Spencer, Bluford K.
Spencer, Clayton M.
Spencer, Harry L.
Spencer, Robert L.
Sperling, Hans J.
Spier, Raymond A.
Spink, Raymond C.
Spivey, Doris P.
Spivey, Phyllis S.
Spivey, William T.
Splawn, James L.
Spradlin, Jim M.
Spradlin, John H.
Spradlin, William G.
Spray, Carolyn T.
Spray, Lillian M.
Springfield, Miles G.
Sprinkle, Charles E.
Sprouse, Cleona N.
Spry, Robert E.
Spurlin, Amon A.
Stacy, Robert B.
Stafford, Walter H.
Standard, Richard M.
Standish, Myles M.
Stanley, J. B.
Stanton, William P.
Stapler, James M.
Stapler, Vachel
Staples, Evelyn M.
Staples, Harry T.

Starkey, Theo T.
Steadman, Jackie D.
Steele, Bobby G.
Steely, Arthur J.
Steely, Jess O.
Steffy, Regina G.
Stein, Arnold B.
Stein, Richard J.
Steinberg, Alvin
Steincamp, James W.
Stem, William M.
Stephens, Donal J.
Stephens, Gordon E.
Stephens, James K.
Stephens, John C.
Stephens, Ralph L.
Stephens, Teddy W.
Stephens, Thomas E.
Stephenson, Henry D.
Stephenson, John U.
Sternaman, Rollo M.
Stevens, John T.
Stevens, Robert L.
Stevens, Winfred G.
Stevenson, Billy M.
Stevenson, Harold H.
Stewart, Betty C.
Stewart, Charles C.
Stewart, Donald L.
Stewart, Francis M.
Stewart, Harvey R.
Stewart, Homer J.
Stewart, Homer R.
Stewart, Milton G.
Stewart, Milton R.
Stewart, Robert W.
Stewart, Rodney D.
Stewart, Sanders R.
Stewart, Thomas E.
Stidger, Buford C.
Stiles, Alfred W.
Stiles, Laura W.
Still, Harry M.
Stimler, Robert S.
Stinnett, Edward J.
Stinnett, William L.
Stinson, Henry P. Jr.
Stluka, Edward F.
Stocks, Charles D.
Stoffregen, Elvin
Stokes, Jack W. Jr.
Stone, Donald
Stone, Edward L.
Stone, John F.
Stone, Lewis M.
Stone, Lyndon T.
Stone, Nobie H.
Stone, Richard N.
Stone, Russell L.
Stone, Walter P.
Stonemetz, Richard E.
Storms, Francis D.
Story, Katie J.
Stovall, John W.
Stover, Calvin C. Jr.
Stover, Herman C.
Stover, Judy H.
Stowe, Elizabeth C.
Strandemo, Herbert C.
Strange, Marion I.
Strange, Tommy D.
Strauss, Emmanuel M.
Street, Harry A.
Streeter, Warren S.
Strickland, Earl M.
Strickland, Homer W.
Strickland, James N.
Strickland, William H.
Stricklin, Carous E.
Strider, Dale K.
Stripling, Ruby D.
Strong, Henry L.
Strong, James S.
Strong, Julia C.
Stroud, Finis B.

Stroud, John D.
Stroud, Mildred W.
Stroud, Robert L.
Struck, Heinrich G.
Stucker, John A.
Stuckey, James M.
Stuhlinger, Ernst
Stull, James T.
Stulting, James B.
Sturdivant, George W.
Sturdivant, James G.
Sturgill, Ralph K.
Styles, Marcella S.
Styles, Paul L.
Suess, Steven T.
Suhr, Lois M.
Sulcer, Clarence
Sullens, Thomas R.
Sullins, Samuel L.
Sullivan, Elbert L.
Summers, Delsie R.
Sumner, Craig E.
Sumner, George A.
Sundstrom, Melvin B.
Suns, Forrest E.
Suns, Joseph R.
Superneau, Walter C.
Susko, Michael
Sutherland, Brenda J.
Sutherland, William C.
Sutton, Carl B.
Sutton, William G.
Swafford, Colonel L.
Swafford, Jesse E.
Swafford, Judith S.
Swaggerty, Thomas B.
Swaim, Leland L.
Swain, Bernice F.
Swain, Robert L.
Swalley, Frank E.
Swann, Allie C.
Swann, Linda R.
Swann, Sarah C.
Swanson, Charles A.
Swearingen, Charles N.
Swearingen, Jack C.
Sweat, Sidney J.
Sweetland, William C.
Swindall, Paul M.
Swinford, Billie K.
Swingley, Andrew B.
Switzer, George D.
Swords, Ben B.
Swords, William A.
Tabor, James W.
Tabor, Opal P.
Tackett, Robert D.
Tackett, Robert R.
Taft, Charles E.
Talbot, James W.
Taliaferro, Lanny R.
Tallent, Leonard F.
Talley, Drayton H.
Talmadge, Gilbert A.
Tandberg-Hanssen, Einar A.
Tankersley, Clarence E.
Tannehill, Bobbie K.
Tanner, Ernest R.
Taormina, Larry K.
Tate, Elmer O.
Taylor, Alfred C.
Taylor, Allan C.
Taylor, Annie R.
Taylor, Arthur S.
Taylor, Billy J.
Taylor, Bobbie N.
Taylor, Christine W.
Taylor, David T.
Taylor, Esten R.
Taylor, Gloria S.
Taylor, James
Taylor, James C.
Taylor, James H.
Taylor, John B.
Taylor, John L.

Taylor, Kenneth R.
Taylor, Lera A.
Taylor, Pauline N.
Taylor, Phillip H.
Taylor, Richard F.
Taylor, Roy A.
Taylor, Shirley L.
Taylor, Summers W.
Taylor, William E.
Tays, Venalou D.
Teague, Cecil E.
Teague, Elijah W.
Teague, Ernest W.
Teague, Everett R.
Teal, Marion L.
Teal, Walker
Teasley, Ralph B.
Tedford, Oscar W.
Teenor, Martha H.
Teir, William
Templeton, Corvin K.
Tennison, Marjorie N.
Tereschuk, Joseph
Terrell, Norma O.
Terry, Jennie F.
Terry, L. C.
Terry, Lois M.
Terry, Mary W.
Terry, Newman B.
Tesney, Joe
Tesney, Joe H.
Tessitore, Frank J.
Tessmann, Bernhard R.
Teuber, Dieter L.
Thacker, Rodell
Thacker, Sarah S.
Thackston, Joe E.
Thaxton, Jimmie B.
Thayer, Harry I.
Theiss, John M.
Thionnet, Charles L.
Thomas, Agnes J.
Thomas, Billy W.
Thomas, Charles E.
Thomas, Charles N.
Thomas, Douglas T.
Thomas, Ethelene F.
Thomas, Fred K.
Thomas, Glynda S.
Thomas, James E.
Thomas, James L.
Thomas, James W.
Thomas, Joe H.
Thomas, John W.
Thomas, Leon W.
Thomas, Leslie J.
Thomas, Linnis G.
Thomas, Orville C.
Thomas, Paul E.
Thomas, Sylvia B.
Thomas, Troy E.
Thomas, Vaude G.
Thomas, Victor
Thomas, Walter H.
Thomas, Willard C.
Thomason, Garlynda D.
Thomason, Herman E.
Thompson, Ara W.
Thompson, Arthur F.
Thompson, Arthur W.
Thompson, Bobby J.
Thompson, Donald D.
Thompson, Earl E.
Thompson, Gayla S.
Thompson, Gerald M.
Thompson, Harold E.
Thompson, Hoyt E.
Thompson, James F.
Thompson, James R.
Thompson, Joseph C.
Thompson, Leldon M.
Thompson, Louis C.
Thompson, Mack
Thompson, Minnie M.
Thompson, Richard A.

Thompson, Richard L.
Thompson, Robert E.
Thompson, Shirley M.
Thompson, Tempie W.
Thompson, Wilbur E.
Thompson, Wyllodene P.
Thompson, Zack
Thomson, Jerry
Thornhill, Claude B.
Thornhill, Gaynell M.
Thornton, Charles F.
Thornton, George E.
Thornton, James E.
Thornton, Oscar
Thornton, Robert T.
Thornton, Rupert B.
Thornton, Shirley F.
Thornton, Stephen E.
Thornton, Wilbur G.
Thrasher, Dennis E.
Thrasher, Harvel N.
Thrasher, James T.
Threlkeld, William B.
Thrower, George R.
Thurman, Barbara V.
Thurman, Donald W.
Tidd, John L.
Tidmore, Bobby N.
Tidmore, Dorothy G.
Tidmore, T. V.
Tidwell, Norman O.
Tidwell, Paul E.
Tiede, Jack E.
Tielking, Sara G.
Tiller, Newton G.
Tiller, Werner G.
Tillery, Clarence W.
Tines, Thomas W.
Tingle, Annette K.
Tingley, Joe J.
Tinius, Richard E.
Tippins, William H.
Tipps, Ruby W.
Tipton, Elva F.
Tipton, Harvey L.
Tjulander, Raymond V.
Tobias, Eleanor S.
Tockley, Arnolf B.
Toelle, Ronald G.
Tolbert, Mitchell
Toles, David O.
Tolson, George B.
Tomlin, Donald D.
Tomlinson, Virginia M.
Tondera, Steve E.
Toney, Fred
Toney, John
Toney, Marshall A.
Torkar, Joseph E.
Torode, William G.
Torruella, Antonio R.
Torstenson, Charles R.
Totty, John C.
Touchstone, Armand A.
Tovar, George
Towery, George C.
Townsend, Elizabeth C.
Townsend, John B.
Towsend, Robert L.
Trafton, Raymond F.
Trapalis, Charles P.
Travis, Woodrow J.
Travis, Woodrow J. Jr.
Traywick, Joe D.
Treece, Ollie J.
Treister, Leon
Trenkle, Charla S.
Trenkle, John J.
Trentham, Norman E.
Trewhitt, Willard L.
Trexler, Harold D.
Tribble, Charlotte S.
Tribble, Ford H.
Tribble, John R.
Tribble, Martha C.

Trimble, Harry H.
Trott, Jack
Troup, Clifford E.
Troupe, Albert F.
Troupe, William L.
Troy, Josephine A.
Trucks, Howard F.
Trucks, Judy B.
Tuck, Peggy H.
Tucker, Billy G.
Tucker, Bobby J.
Tucker, Donald J.
Tucker, Elmo
Tucker, Francis W.
Tucker, Grover C.
Tucker, Joseph T.
Tucker, Nancy B.
Tucker, Thomas E.
Tuell, Lenox P.
Tuggle, Richard H.
Tumlison, James C.
Tunstill, Jimmy L.
Turan, Winston J.
Turgeon, Claude C.
Turner, Anne F.
Turner, Barbara B.
Turner, Gerald L.
Turner, Harold K.
Turner, James R.
Turner, John E.
Turner, Joyce E.
Turner, Ovie M.
Turner, Robert E.
Turner, Thomas B.
Turner, Tom W.
Turner, Tommy B.
Turney, George W.
Turney, Joan I.
Turpen, Donald L.
Tuten, Bedford F.
Tuten, Willis
Tutt, Richard H.
Tyler, Victor A.
Tyson, Ella M.
Tyson, Overton S.
Tyson, Timothy E.
Tyvoll, Frederick M.
Uebele, John J.
Underwood, Jack R.
Underwood, Willard D.
Uptagrafft, Frederick
Upton, Garthell W. T.
Urban, Eugene W.
Urbanski, Arthur
Urlaub, Matthew W.
Vaccaro, Michael L.
Vadasy, Kenneth
Valentine, Richard E.
Vallely, Donald P.
Vallely, Sandra G.
Van Ark, Susan J.
Van Auken, Paul
Van Nimwegen, Lawrence J.
Van Orden, Edmond R.
Van Vlack, William T.
Van Zandt, David J.
Vance, Robert J.
Vandergriff, Sandra C.
Vandersee, Fritz A.
Vanderzyl, Robert J
Vandiver, Ruth H.
Vanhook, Michael E.
Vaniman, Jerold L.
Vann, James W.
Vann, R. E.
Vann, Truett N.
Vardaman, William K.
Varnado, Clinton L.
Varnedoe, William W.
Vasil, John J.
Vaughan, Herbert W.
Vaughan, Lemuel F.
Vaughan, Otha H.
Vaughan, William W.
Vaughn, William C.

NASA - Marshall Flight Space Center

Vedane, Charles R.
Veitch, Robert H.
Velvet, Camille S.
Venus, James D.
Verble, Adas J.
Verderaime, Vincent S.
Verschoore, Charles P.
Vest, Curtis
Vibbart, Charles M.
Vick, Howell G.
Vickers, Charles M.
Vickers, Dallas N.
Villella, Felminio
Vines, Joseph L.
Vines, Marvin W.
Vines, William E.
Vinson, Charlotte M.
Vinson, Mack
Vinson, Tommy L.
Vinz, Frank L.
Vlasse, Marcus
Vogtner, Charles
von Braun, Wernher
Von Pragenau, Geroge L.
Von Saurma, Ruth G.
Von Tiesenhausen, Georg F.
Voss, Werner
Vreuls, Frederick E.
Vucurevich, Milan T.
Vukovich, Milan J.
Waddell, Norman E.
Wade, George H.
Wade, Harry H.
Wade, Stanley H.
Wade, Thomas M.
Wadlington, George L.
Waggoner, Gerald B.
Waggoner, Harold
Wagner, Hermann
Wagner, Ignatius A.
Wagner, Robert J.
Wagnon, Fredrick W.
Wainscott, Harold V.
Waite, Jack H.
Waites, Henry Burton
Waits, William B.
Wakefield, John B.
Walden, Fred
Waldrep, Sherman P.
Waldrop, James A.
Waldrop, John H.
Waldrop, Wilson L.
Wales, William H.
Wales, William W.
Walker, Barbara
Walker, Carl T.
Walker, Cecil C.
Walker, Edward O.
Walker, Edwin C.
Walker, Gordon R.
Walker, Herbert V.
Walker, Hill M.
Walker, Jackie C.
Walker, James D.
Walker, Lewis L.
Walker, Richard
Walker, Robert J.
Walker, Robert W.
Walker, Rubye M.
Walker, Russell D.
Walker, Talmadge L.
Walker, William H.
Wall, Lester C.
Wall, Richard H.
Wall, William A.
Wallace, Gabriel R.
Wallace, George D.
Wallace, James G.
Wallace, John L.
Wallace, L. D.
Wallace, Roy S.
Waller, Walker A.
Walls, Bobby F.
Walls, Foster M.
Walls, Georgia L.

Walls, J. C.
Walls, Jerry C.
Walls, Regnal L.
Walls, Samuel L.
Wally, Billy C.
Walmsley, Carolyn H.
Walsh, J. Russell
Walter, Wanda M.
Walters, Charles L.
Walters, Dwain T.
Walters, Estel F.
Walters, Joseph M.
Walters, William R.
Walton, John B.
Walton, Julia A.
Walton, Timothy P.
Wambeke, Francis G.
Ward, Elmer
Ward, Eugene C.
Ward, Johnnie P.
Ward, Raymond W.
Ward, Samuel R.
Ward, Vurt
Warden, Alta R.
Ware, Robert L.
Warmbrod, John D.
Warren, Alfred P.
Warren, Buford E.
Warren, Charles E.
Warren, Doris S.
Warren, Julius G.
Warren, Marvin
Warren, Nettie P.
Warren, Sarah H.
Warren, William H.
Warrick, Clara M.
Wasserman, Dale J.
Watkins, James T.
Watkins, Jimmy R.
Watkins, Mary T.
Watson, Charles E.
Watson, Gwynneth E.
Watson, Harold C.
Watson, Pauline C.
Watters, Harry H.
Watts, Ethridge B.
Watts, Gaines L.
Watts, John W. Jr.
Watts, Sandra N.
Way, Arthur L.
Weaks, Sallie M.
Wear, Lawrence O.
Weatherbee, James E.
Weathers, Dudley M.
Weathers, Hoyt M.
Weathers, Shelby N.
Weaver, Edwin A.
Weaver, Flora W.
Weaver, James F.
Weaver, John J.
Weaver, Larry A.
Weaver, Orville
Weaver, Richard W.
Weaver, Willie E.
Webb, Charles J.
Webb, Clarence L.
Webb, David D.
Webb, Harding
Webber, Robert E.
Weber, Alfred H.
Weber, Fritz H.
Weber, Sandra G.
Webster, George S.
Webster, Loyce E.
Webster, Sherman T.
Webster, Thomas E.
Weckwarth, Fred
Weece, Woodrow
Weeks, Billie R.
Weeks, Charles R.
Weeks, David J.
Weeks, Gus E.
Weeks, Howard E.
Weeks, Lawrence M.
Weeks, Ozel

Weems, Raymond
Weems, Ruth W.
Weesner, Ronald G.
Wegrich, Richard D.
Weidner, Hermann K.
Weidner, Joe D.
Weil, Jack
Weiler, Jerry D.
Weir, Edward T.
Weir, Marguerite T.
Weir, Robert H.
Weisler, August C. Jr.
Welch, Lee V.
Welch, Ricky A.
Wells, Betty C.
Wells, Edward L.
Wells, Emma S.
Wells, Ernest H.
Wells, Forrest T.
Wells, Hubert B.
Wells, Marie P.
Wells, Valeria G.
Welzyn, John S.
Werner, Judy M.
Wessel, Bertram E.
Wesson, Annie C.
Wesson, Robert L.
West, Arnon E.
West, Eugene
West, George S.
West, John L.
Westbrooks, James A.
Westendorf, Ann H.
Westrope, Dewitt T.
Westrope, Douglas W.
Weyler, George M.
Whaley, Terry J.
Wheeler, Alton L.
Wheeler, Horace R.
Wheeler, John T.
Wheeler, William E.
Whirley, Jere D.
Whisenant, Earl L.
Whisenant, Emmett R.
Whisenant, Judy R.
Whisenant, L. D.
Whitacre, Walter E.
Whitaker, Ann F.
Whitaker, Ervin B.
Whitaker, Olan G.
Whitaker, Peggy S.
White, Alice N.
White, Annie R.
White, Arthur F.
White, Barbara A.
White, Cecelia K.
White, Cecil C.
White, Charles O.
White, Edwin
White, George B.
White, Gilbert G.
White, J. B.
White, James E.
White, Joe E.
White, John K.
White, Lonnie C.
White, Oma B.
White, Ovic M.
White, Paul R.
White, Robert A.
White, Robert S.
White, William
White, William B.
White, William T.
White, Willie J.
Whitehead, Catherine D.
Whitehead, Milton E.
Whiteley, Robert F.
Whitfield, Joyce G.
Whitley, Gerald W.
Whitley, Vivian S.
Whitsitt, Donald E.
Whitson, Isaac
Whitt, Charley R.
Whitt, Lonnie R.

Whitt, Sarah W.
Whitt, William D.
Whittington, Morris J.
Whitworth, Charles B.
Whitworth, Margaret S.
Wicks, Donald C.
Wicks, Thomas G.
Wieland, Paul O.
Wiesenmaier, Bernhard L.
Wiesman, Walter F.
Wiggins, Bobby W.
Wiggins, Herbert H.
Wiggins, James W.
Wiggins, Larry G.
Wilbanks, Burwell L.
Wilbanks, Edd
Wilbourn, Early D.
Wilbourn, James N.
Wild, John D.
Wilder, Archie D.
Wilder, John W.
Wilder, Paul
Wiley, Ivan M.
Wiley, James A.
Wiley, Robert G.
Wilhold, Gilbert A.
Wilkerson, Mildred S.
Wilkes, Nelma L.
Wilkes, Wallace E.
Wilkinson, Clarence O.
Wilkinson, Lilburn T.
Wilkinson, Terry S.
Wilkinson, Tyree H.
Wilkinson, William E.
Wilks, John L.
Willard, James D.
Williamon, Claude F.
Williamon, Morris W.
Williams, Calvot C.
Williams, Charles A.
Williams, Dan C.
Williams, Donald E.
Williams, Elgin
Williams, Ellen M.
Williams, Franklin E.
Williams, Harvell P.
Williams, Henry J.
Williams, Herbert E.
Williams, Imogene H.
Williams, James E.
Williams, James R.
Williams, Jimmy R.
Williams, John A.
Williams, John B.
Williams, John G.
Williams, John H.
Williams, Lenwood A.
Williams, Martha A.
Williams, Mike C.
Williams, Noah G.
Williams, Oliver B.
Williams, Omer E.
Williams, Patricia L.
Williams, Pearl M.
Williams, Robert N.
Williams, Roy
Williams, Thomas J.
Williams, Underwood
Williams, Walter H.
Williamson, Clarence O.
Williamson, Harry B.
Williamson, James G.
Williamson, Kenneth S.
Williamson, Lucian
Willis, Albert B.
Willis, Albert E.
Willis, Bettye P.
Willis, James M.
Willis, Margaret J.
Willis, Maurice C.
Willis, Vicki R.
Willmon, Ann C.
Willmore, Inoma P.
Willoughby, Grover C. Jr.
Wills, Fred D.

Willson, Jean W.
Wilmer, Glenn E.
Wilmer, Glenn E. Jr.
Wilson, David A.
Wilson, Charles
Wilson, George W.
Wilson, Homer
Wilson, Homer B.
Wilson, Hugh H.
Wilson, Leo D.
Wilson, Linda R.
Wilson, Margit C.
Wilson, Moses L.
Wilson, Robert B.
Wilson, Robert T.
Wilson, Vernon
Wilson, Warren G.
Wilson, Wilburn M.
Wilson, William A.
Wilson, Willie S.
Wimbish, Norma M.
Wimmer, James C.
Windham, John M.
Windham, John O.
Windlin, Deane D.
Winford, Garland
Winkler, Carl E.
Winkler, Charles B.
Winn, Charles B.
Winsett, Dorothy J.
Winsett, James T.
Winstead, Thomas W.
Winterstein, William E.
Wise, James H.
Wiser, Cecil W.
Wiser, James N.
Wisner, Lou E.
Wittmann, Albin E.
Witty, Walter H.
Wobrock, Doris W.
Wofford, Leon D.
Wojtalik, Fred S.
Wolf, Robert K.
Wolfe, George W.
Wolfsberger, John W.
Wolk, John P.
Womac, Dorothy B.
Womac, George L.
Wood, Andrew J.
Wood, Carl M.
Wood, Charles C.
Wood, Charles H.
Wood, Charles L.
Wood, Gayle K.
Wood, Gerald
Wood, Gordon A.
Wood, Jack T.
Wood, James E.
Wood, Jewel J.
Wood, Larry W.
Wood, Lelous C.
Wood, Lewis P.
Wood, Patrecia J.
Wood, Thomas E.
Woodham, Robert T.
Woodis, Kenneth W.
Woodman, Muriel K.
Woodruff, Bobby R.
Woodruff, Lee D.
Woodruff, Nancye C.
Woods, Herman O.
Woods, Joseph R.
Woods, Orville H.
Woods, Wilbert R.
Woody, Grady L.
Woody, Wilburn J.
Woolbright, James E.
Wooldridge, James T.
Woolf, Herbert L.
Woosley, Alvan P.
Woosley, James E.
Wooton, E. C.
Word, Dessie J.
Word, James B.

Workman, Floyd
Works, John A.
Worley, Carl L.
Worlund, Armis L.
Wormell, Perley H.
Worrell, Donald O.
Worthy, Wallas L.
Wright, Fletcher
Wright, Geraldine S.
Wright, Herbert N.
Wright, Jerry J.
Wright, Jessie B.
Wright, John J.
Wright, John O.
Wright, Martha J.
Wright, Mary S.
Wright, Peter
Wright, Robert D.
Wright, Thomas D. Jr.
Wyatt, John P.
Wyckoff, James E.
Wyckoff, Kathy J.
Wyman, Charles L.
Wynn, James D.
Wynn, Lonnie
Wynn, Oza
Wynn, Samuel B.
Wynn, William F.
Wynne, Bertha L.
Wysong, Earl M.
Wyss, Gervaise L.
Xenofos, Danny
Yancey, Ronald B.
Yancu, Theodore
Yarbrough, Clyde A.
Yarbrough, Katherine L.
Yarbrough, Nina M.
Yarbrough, Thomas C.
Yates, Gloria F.
Yates, Iva C.
Yates, James E.
Yeager, Bernard D.
Yeager, Charles L.
Yeager, James D.
Yeargin, James C.
Yearwood, Hoyle E.
Yell, Clarence E.
Yell, Maurice K.
Yohannan, Philip H.
Yolton, Virginia A.
Yongue, Monroe L.
York, James A.
York, Van C.
Yost, Vaughn H.
Young, Archie C.
Young, Howard C.
Young, James D.
Young, Janice M.
Young, Leighton E.
Young, Ronald L.
Young, Roy D.
Younger, Johnny E.
Youngkin, Ima B.
Zachary, Edward E.
Zagrodzky, Robert G.
Zahnd, James W.
Zaun, Helen C.
Zeanah, Hugh W.
Zeigler, Billy S.
Zeigler, William H.
Ziak, William J.
Zielenski, Leo
Zielenski, Linda M.
Ziesmer, Erich W.
Zimmerman, Claude R.
Zimmerman, Joe E.
Zoller, Lowell K.
Zur Burg, Frederick W.
Zur Burg, Mildred M.
Zwiener, James Milton

INDEX

The Marshall Retiree Association Members list and the Retiree Roster are not included in the index. They can be found in alphabetical order in their respective sections.

A

Abbott, Bud – 168
Abernathy, "Ab" – 208
Abernathy, Ralph – 65
Able, monkey – 41, 42, 151, 201
Absher, Archie C. – 92
Adair, B. – 165
Adams, Luther – 149
Advanced Research Projects Agency (ARPA) – 39, 42
Advanced Space Transportation Program – 79, 80
Advanced X-ray Astrophysics Facility (AXAF) – 78, 81, 82
Advanced X-ray Astrophysics Facility (AXAF) program – 82
Akridge, Max – 199
Albrecht, Dean – 149
Aldrich, Arnold – 201
Aldrin, Edwin "Buzz" – 5, 10, 161, 209
Altair – 87, 88
Anderson, Dorrance L. – 164
Angele, Wilhelm – 202
Apollo – 42, 43, 47, 51, 55, 56, 59, 62, 63, 64, 65, 66, 69, 74, 83, 84, 94, 97, 99, 144, 146, 148, 158, 159, 160, 168, 171, 172, 190, 197, 199
Apollo 11 – 5, 10, 25, 51, 52, 65, 74, 78, 94, 158, 161, 162, 165, 172, 181, 191, 192, 196, 197, 202, 207, 209
Apollo 12 – 165, 190
Apollo 13 – 15, 165, 166, 195
Apollo 14 – 165
Apollo 15 – 51, 156, 157, 165, 185, 202
Apollo 16 – 51, 165, 202
Apollo 17 – 51, 56, 158, 165, 202
Apollo 4 – 160
Apollo 6 – 160
Apollo 7 – 206, 207
Apollo 8 – 16, 160, 202, 207
Apollo I – 94
Apollo program – 31, 52, 90, 94, 146, 156, 158, 159, 160, 168, 188, 191, 192, 205
Apollo Project – 15
Apollo Telescope Mount – 54, 55, 56, 57, 58, 157, 189, 192
Apollo XI – 160, 161
Apollo-Saturn – 52, 65, 78, 161, 171, 197, 207

Apollo-Saturn program – 42, 50, 95, 197, 206
Apollo-Saturn V – 207
Apollo-Soyuz – 60
Apollo-Soyuz Test Project – 29, 60, 158, 172, 195
Ardrey, Cecil – 149
Ares – 5
Ares 1 – 91
Ares I – 19, 87, 88, 89, 90, 206
Ares I-X – 31, 89, 90, 153
Ares I-Y – 90
Ares program – 24
Ares Project – 87
Ares V – 87, 88, 89, 90, 91, 206
Armstrong, Bill – 199
Armstrong, Neil – 5, 10, 51, 161, 166, 193, 195, 201, 208
Army Ballistic Missile Agency (ABMA) – 10, 37, 38, 39, 40, 41, 42, 47, 51, 62, 95, 96, 147, 148, 149, 152, 154, 156, 171, 173, 174, 201, 202, 203, 206
Arnold, Henry Harley "Hap" – 36
Arsement, Leo – 165
Artis, Paul – 171
Artley, Gordon – 152, 153, 174, 178, 187
Atlantis – 30, 81
Austin, Gene – 188
Avery, John – 165

B

Backer, Ron – 182
Baggett, Tom – 196
Bailey, Ray – 160
Baker, Mike – 6
Baker, monkey – 41, 42, 201
Baker, Robert P. "Bob" – 158
Balch, Jack – 172
Baldwin, Claude – 158
Ball, Edward – 208
Bannister, Tommy – 58
Barclay – 174
Barnard, Christian – 155, 195
Barnes Beasley – 165
Barnes, Bill – 149
Barnes, Tom – 165
Barr, Thomas "Tom" – 164, 165
Barron, Bill – 149
Batts, Wade – 192
Beaman, Robert "Bob" – 165, 166
Bean, Alan – 190
Bean, Don – 179
Beggs, James – 81, 84
Beichel, Rudie – 152

Belew, Leland – 53, 54, 58, 68, 152, 162
Bell, Leon – 165
Bell, Lucian – 145, 165
Beltran, Antonio – 152, 164, 165
Berry, Ron – 160
Bethay, J.A. "Woody" – 6, 38, 40, 49, 50, 52, 195
Bigott, Dick – 149
Bilbro, James W. – 95
Bishop, Don – 188
Black, Paul – 160
Blackstone, John – 160
Blanton, James E. "Jim" – 180
Bleeker, Johan A.M. – 97
Bobo, Archie C. – 149, 150
Bobo, Mildred B. – 149
Bodie, Bill – 162
Bonestell, Chesley – 84, 147
Bordelon, Carroll – 165
Borman, Frank – 172
Boyer, Keith – 178
Bradford, Jack – 160
Bradford, Lyn – 186
Bradley, Jesse N. – 172
Bradshaw, Charles – 208
Brady, Bill – 160, 166
Brady, Hugh – 160, 166
Bramlet, James – 144
Brand, Vance – 29
Bratcher, Andrew – 165
Brazzell, Forrest S. – 179
Brewer, James L. – 179
Bridwell, G.P. (Porter) – 99
Brooks, Charlie – 172
Brothers, Bobby – 160
Broussard, Pete – 184
Brown, Harrison K – 166
Brown, Max – 149
Bryant, Charles – 180
Buchanan, Don – 162, 163
Bucher, George C. – 156
Bucher, Mr. – 123
Buchhold, Theodor – 154
Buckbee, Ed – 5, 9, 92, 96, 122, 175, 180
Bug Nebula – 30
Bugg, Charlie – 76
Buhmann, Eugene J. – 153
Buhmann, Mr. – 123
Buhmann, Renee – 153
Bullman, Jack – 188
Burch, John – 185
Burdine, Bill – 202
Burke, Harlan – 165
Burns, Ralph A. – 152
Burson, George – 165
Busby, Carol – 180
Bush, George – 78, 87
Butterfly Nebula – 30

C

Cagle, Gene – 158
Callisto – 26

Camarda, Charles J. – 73
Cannon, Buck – 149
Canright, Dick – 156
Carey, Ted – 201
Carlin, Robert – 180
Carpenter, Scott – 171
Carruthers, John – 58
Caruso, Vincent P. – 164
Case, Bill – 165
Casey, Chuck – 160
Castenholtz, Paul – 162
Cataldo, Gene – 187
Centaur – 151
Cernan, Eugene – 51
Chaffee, Roger – 95
Challenger – 70, 71, 72, 78, 81, 98, 168, 195, 200, 201
Chambers, Charles – 165
Chandra – 5, 56, 70, 81, 82
Chandra X-Ray Observatory – 26, 82, 83, 94, 188
Chandra X-Ray Telescope – 53, 81
Chandranekhar, Subrahmanyan – 58, 82
Chase, John – 165
Chase, Mr. – 123
Chassay, Roger – 200
Chittam, Ronald – 174, 208
Choate, Earl – 167, 194
Christensen, David – 203
Cimijotti, Lew – 203
Clanton, Price – 174
Clark, Arthur – 192, 193
Clark, Larry – 201
Clark, Rocky – 160
Clarke, Arthur C. – 5
Clinton, Randy – 165
Cobb, Bill – 160
Cochran, Jackie – 155
Coffee, R. – 165
Coldiron, Jack – 182
Collins, Eileen M. – 73
Collins, Michael – 5, 161
Collins, Rufus – 171
Columbia – 72, 81, 87, 161
Comet Kohoutek – 57
Conne, Jack – 182
Conrad, Charles – 56
Constan, George – 208
Cook, Richard – 209
Cook, Steve – 87
Cooper, Charlie – 189
Cooper, Gordon – 171
Cornelius, Don – 182
Corrective Optics Space Telescope Axial Replacement (COSTAR) – 81
Covington, Joe – 179
Cox, John – 165
Craft, James – 166
Craig, Daryl – 165

Cramer, Bud – 11
Cronkite, Walter – 155, 174
Crutcher, A.R. – 179
Cunningham, Walter – 206, 207
Curtis, Ben – 174

D

Dahm, Werner – 204
Daniels, James L. "Jim" – 202, 203
Dannenberg, Konrad – 92, 93, 94, 95, 97, 144, 145, 204
Darwin, Charles – 85, 187, 188
Davis, Alonza – 165
Davis, Ed – 162, 163
Davis, Steve – 90
Dawsey, Harry – 149
Day, Dwayne – 97
De Sanctis, Carmine – 188
Deaton, Terry – 160
deBeek, Gerd – 206, 207
Debus, Kurt – 50, 150, 163, 182
Decher, Rudolph – 165
Denver, John – 155, 195
Derington, James – 165
Deuel, Glen A. – 150
Dick, Steven J. – 97
Digesu, Fred – 188
Discovery – 19, 21, 26, 28, 72, 142
Disney, Walt – 41, 155, 173, 175, 180
Dison, Gordon – 182
Doane, George B. III – 154
Dobbs, Stephen J. – 172
Dodd – 163
Dodd, Kenneth – 195
Dodd, R.P. – 162
Doheny, Bob – 149
Donaruma, L. Guy – 204
Dornberger, Walter – 144
Douglas, Donald – 145
Downey, James – 6, 198
Downey, Jim – 40, 54, 60, 61, 80, 81, 188
Downs, Hugh – 155
Downs, Sanford – 165
Draper, Charles Stark – 154
Driscoll, Dan – 171, 172
Driver, Clarence Jackson – 157
Driver, Orville – 168, 196
Dryden, Hugh – 156
Duggan, O.T. – 165
Duke, Charlie – 51
Duke, Floyd – 206
Duke, Leonard C. – 179
Dulles, John Foster – 151
Dunn, Joe – 163
Dupree – 202

Durant, Frederick C. III – 147
Dynamic Test Stand – 24, 70, 90, 177

E

Eagle – 161
Edens, Bill – 165
Edens, R. – 165
Edison, Thomas – 122
Ehricke, Krafft – 145, 147, 178
Eichelberger, Robert – 165
Eisenhower, Dwight D. – 39, 40, 148, 151, 155, 157, 175
Ellis, Tom – 149
Ely, Olin – 165
Emens, Frank – 165
Emme, Eugene – 123
Endeavour – 23, 85
England, Anthony – 51
Enterprise – 24, 25, 69, 70
Erwin, Bobby – 163
Escue, William Tom – 151, 165
Eslick, Willa – 165
Ethridge, Ethel R. – 179
Eubanks, Earl – 182
Europa – 26
European Space Agency – 38
Evans, Ross – 165
Expedition 11 – 73
Explorer – 38, 39, 55, 93, 151, 157, 189
Explorer 1 – 97, 154, 181
Explorer I – 5, 10, 39, 40, 41, 59, 92, 148, 154, 157, 164, 171, 181
Explorer VII – 165
External Tank – 64, 67, 68, 69, 70, 71, 72, 73, 78, 99, 151, 200, 206
External Tank program – 66

F

Facemire, Barbara – 58
Faget, Max – 54
Farish, Preston – 161, 207
Farley, Clare – 193
Feast, Charles – 145
Felch, Jerry – 158
Fellenz, Deiter – 199
Ferrell, George A. – 202, 203
Ferringno, Frank – 149
Feustel, Andrew – 30
Fichtner, Hans – 158, 204
Fiedler, Arthur – 155
Field, Spike – 156
Fikes, Ken – 188
Fischel, Edward – 164
Fletcher, James – 58, 65, 71, 158, 186

Flynn, Al – 188
Ford, Gerald – 155, 175
Foster, J.N. "Jay" – 6, 9, 38, 40, 48, 50, 52, 65, 122, 168, 169, 192, 193
Foxworth, Wally – 182
Foxworthy, Grace – 150
Frank Emens – 165
Freedom – 85
Freedom 7 – 10, 20
Freeman, Buford – 165
Freeman, Fred – 147
French, Donald – 164
Frost, W.O. – 165
Fullerton, Charles – 51

G

Gagarin, Yuri – 42
Galileo – 122
Gallaher, Gene – 158
Gamma Ray Observatory – 78
Ganymede – 26
Garcia, Ray – 160
Gardner, Benny – 149
Garriott, Owen – 57, 75
Gassaway, Charles "Chuck" – 162
Geiger, Hans – 97
Geissler, Dr – 191
Gemini – 66, 97, 189
Gemini 8 – 166
Gemini program – 50, 54, 154, 191
Giacconi, Riccardo – 100
Gibson, Ed – 189
Gierow, Herman – 188
Gillespie, Charlie – 170, 171
Gilruth, Robert – 45, 193, 208
Glass, Bill – 158
Gleason, Ed – 165
Glen, Batrice – 9
Glenn, John – 171
Glennan, T. Keith – 40
Goddard, Robert – 93
Godfrey, Roy – 196
Goerner, Erich – 188
Goldberg, Leo – 57
Golden, Harvey – 165
Goldin, Dan – 79
Goldsmith, Jim – 182
Goodhue, Walt – 158
Goodrich, Bill – 182
Gorman, Harry – 209
Gould, John M. – 164
Grace, Gus H. – 179
Graff, Charles – 158
Grafton, Bill – 170, 174
Grau, Dieter – 204
Gravity Probe A – 157
Gravity Probe B – 83
Green, Jay – 160
Greenstein, Jessie – 169
Greenwood, Frances – 179

222

Greenwood, L. Ron – 204, 205
Greer, Dewey – 158
Greever, Bill – 153
Gregg, Cecil – 188
Gregg, Jack – 158
Gregory, John – 165
Griessen, John III – 149
Griffin, Michael – 87
Grimes, Victor – 161, 162, 206
Grissom, Gus – 95, 144, 171
Grunsfeld, John – 30
Guerin, Ed – 158
Guire, Nancy – 6, 122

H

Haase, Maria – 95
Haeussermann, Walter – 154, 165, 172
Haise, Fred – 51
Hale, Lee – 149
Hall, G.E. – 160
Hallisey, Harold "Bill" – 9, 196
Hamaker, Joe – 188
Hamill, James P. – 149
Hamilton, Julian – 175
Hamlet, John – 165
Hammack, Houston – 194
Hanks, Tom – 195
Hanley, Jeff – 87
Hardage, Monroe – 160
Hardy, George – 9, 67, 196
Harmon, Harlan – 187
Harness, Paul – 192
Harper, Warren – 164, 165
Harris, Dave – 165
Harris, George – 165
Harris, Gordon – 181
Harris, Jesse – 149
Harris, Ron – 188
Harrison, Colin – 196
Harrison, Jim – 182
Harsh, George – 194
Hart, Johnny – 155
Hauessermann, Walter – 195, 208
HEAO-1 – 61
HEAO-2 – 61, 81
HEAO-3 – 61
HEAO-A – 61
HEAO-B – 61
HEAO-C – 61
Heim, Mr. – 123
Heimburg, Karl – 38, 153, 172, 173, 175, 178, 180, 182, 183, 187, 191
Heinisch, Kurt – 93
Heisenberg, Werner – 97
Heiser, Mr. – 123
Heller, Gerhard – 145, 147, 178, 182, 202
Henderson, Mack – 160
Hermann, Rudolph – 204
Hess, Jim – 149
Hester, Howard – 172
Hiers, Jim – 179
High Energy Astronomical Observatory – 94
High Energy Astronomy Observatory (HEAO) – 53, 60, 61, 62, 97, 98
High Energy Astronomy Observatory (HEAO) project – 80
High Energy Astrophysical Observatory (HEAO) – 186, 187, 188, 189
Hight, A. Hill Sr. – 180
Hight, Anita – 181
Hight, Hermon H. – 180
Hight, Pearl – 181
Hight, Vicky – 181
Hildreth, E.D. – 179
Hill, Jim – 179
Hines, Jerome – 155
Hirsch, Ollie – 182
Hirschler, Otto – 204
Hoberg, Otto – 92, 164, 165
Hoffman, Jeffrey – 23
Holder, Ray – 165
Hollingsworth, Bill – 149
Holmes, Bonnie – 6, 122, 155
Hooper, Howell – 160
Hopkins Ultraviolet Telescope – 77
Horeff, Tom – 149
Horiuchi, Kats – 149
Houston, Betty Foxworthy – 150
Houston, Cecil – 197
Howard, Ron – 195
Howell, James – 165
Hubble – 5, 56, 70, 82, 98, 180
Hubble Space Telescope – 23, 30, 53, 76, 78, 79, 80, 81, 83, 94, 183, 188, 200
Hubble Telescope – 156, 189, 199
Hubble, Edwin P. – 81
Hube, Mr. – 123
Huggins, Carl – 165
Hughes, Kaylene – 6
Humphrey, Hubert H. – 155, 175
Hunt, Roger – 149
Hunter, Ross – 182
Huntsville Arsenal – 36, 37
Huss, Carl – 160
Hutchins, Jim – 166
Huth, Chauncey – 180
Hyer, Jack – 182

I

Igou, Jim – 160
Inman, Chieko – 6
International Space Station – 5, 26, 28, 60, 70, 73, 74, 77, 78, 79, 81, 84, 85, 86, 87, 88, 91, 93, 99, 142, 144, 172, 188, 189, 199, 205, 209
International Space Station program – 99
Irwin, Bentley – 171
Isbell, B. Spencer – 202, 203
Ise, Rein – 58, 169

J

Jackson, Jim – 188
Jacobi, Walter – 204
James, Lee – 206
Jaques, Bob – 6
Jastrow, Robert – 193
Jean, O.C. – 188
Jennings, Ralph E. – 203
Jessick, Don – 149
Jobe, Grady Sherman – 202
Johnson, Dale – 191
Johnson, Jim – 179
Johnson, Lady Bird – 155, 175
Johnson, Lyndon B. – 155, 175
Johnson, Robert E. "Bob" – 195
Johnson, Warren – 171
Johnson, William – 59
Johnston, G.D. – 162
Johnston, Garland – 162, 163
Johnstone, Harry – 153, 178, 196
Jolliff, Al – 182
Jones, Alfred M. – 179
Jones, Bob – 36
Jones, Jack – 182
Jones, Joe – 181
Jones, Stu – 182
Joseph, Sandra – 21
Jucunda, Sister Mary – 14
Juno – 93, 154, 157, 201
Juno I – 40, 97
Juno II – 41
Juno V – 42
Jupiter – 5, 37, 41, 42, 47, 92, 97, 148, 151, 152, 154, 157, 162, 169, 174, 177, 178, 187, 201
Jupiter (AM-18 flight) – 42
Jupiter (planet) – 26
Jupiter C – 10, 39, 40, 41, 93, 157, 170, 171, 201
Jupiter C program – 37
Jupiter C-Explorer I – 201
Jupiter H-1 – 42
Jupiter program – 44

K

Kampmeier, Heinz – 165
Karabinos, Andy – 149
Kaschig, Erich – 174, 208
Kassell, Jerry – 149
Kastanakis, John H. Sr. – 152, 153
Kefauver, Estes – 36
Kelly, James M. – 73
Kennedy, Jackie – 176
Kennedy, John F. – 5, 42, 50, 148, 155, 168, 171, 172, 174, 175, 176, 178, 180, 185, 190, 196
Kennedy, Paul – 174, 179
Kent, Ancil – 182
Kerr, Joe – 165
Kerwin, Joe – 56
Kesti, Dick – 149
Keyes, Patrick F. – 148
King, David – 73, 99
King, Mrs. – 123
King, Olin – 165
Kingsbury, James "Jim" – 37, 62, 65, 66, 71, 149, 158, 194
Kittrell, Don – 149
Klan – 202
Klep, Rolf – 147
Kline, Ray – 193
Koelle, Heinz H. – 122
Koelle, Hermann – 147
Korolev, Sergei – 97
Koroleva, Natalya – 97
Kosis, Al – 165
Kraft, Christopher C. – 161
Kramer, Fritz – 187
Kreiger, Robert L. – 193
Kriklev, Sergei K. – 73
Kroll, Mr. – 162, 163
Kubasov – 195
Kummersdorf Arsenal – 95
Kutyna, Don – 201

L

LAGEOS I – 62
Laika, dog – 41
Lakey, Pat – 165
Lambertson, George – 165
Landau, Walter – 149
Landsat-1 – 183
Lange, Ernil – 204
Lange, Oswald – 144
Large Space Telescope – 80, 81
Large Space Telescope project – 80
Larsen, Mr. – 123
Laser Geodynamic Satellite (LAGEOS) – 62
Latham, Rayford H. – 179
Laue, Jay – 198
Launius, Roger D. – 97
Law, Cecil – 182
Lawrence, Wendy B. – 73
Lawson, T. – 165
Layke, Pat – 165
Ledford, Harold – 158
Lee, Leland – 149
Lee, T.J. "Jack" – 6, 38, 67, 74, 75, 78
Lee, Thomas J. – 98, 99
Lemley, Clark – 149
Leonov – 195
Lewedag, Bob – 158
Lewter, Bill – 165
Ley, Willy – 147
Lichtenberg, Byron – 75
Light, Sue – 9
Lightfoot, Robert M. – 7, 99
Lincoln, Evelyn – 155
Lindberg, Jim – 160
Lindbergh, Ann Morrow – 100
Lindbergh, Charles – 161, 207
Lindsey, Bill – 196
Lindstrom, Robert "Bob" – 6, 37, 38, 49, 64, 66, 67, 123, 149, 152, 153, 196
Littles, J. Wayne – 79, 99
Lively, Bud – 182
Lo – 26
Locke, Jim – 182
Long, Jesse E. Jr. – 179
Looney, Frank – 187
Love, Quincy – 202
Love, Richard – 186
Low, George M. – 193
Lowe, George – 183
Lowery, Dave – 165
Lowery, Ray – 165
Lucas, William R. – 52, 53, 59, 60, 63, 83, 84, 85, 98, 151, 159, 179, 188
Lunar Rover – 50, 157
Lunar Roving Vehicle – 51, 183, 184, 185, 186
Lundin, Bruce – 193
Lundquist, Charles A. – 14, 37, 59, 60, 148, 149, 203, 204, 205
Lusk, Dick – 149
Lybrand, Yewell – 203
Lydon, Tom – 196

M

MacFadden, Don – 166
Machnikowski, Frank – 149
Mack, Jerry – 146, 160
Macomb, Tom – 149
Main Propulsion Test Article (MPTA) – 64
Mallernee, Max – 182
Malone, Lee – 165
Mandel, Karl – 154
Mandy, George – 149
Maples, Gardner – 174
Mariner I – 166
Mariner II – 166, 167
Mark, Hans – 84
Marmon, Richard – 188
Marshall, Bob – 188
Marshall, George C. – 6, 40
Martin, Ed – 165
Mathews, Charles W. – 193
Matthews, Frank – 158
Mauldin, Charles – 158, 200
Maus, Hans – 48, 168, 169
Max Planck Institute – 82
Maxwell, Jim – 182
May, Chester B. – 97, 196
May, Ellery – 158, 171
Mayer, John – 160
McBrayer, Robert – 205
McCall, Jerry C. – 122, 123, 204, 208
McClendon, Toddy – 174
McCool, Alex – 37, 38, 45, 47, 56, 70, 201
McCutcheon, Paul – 182
McDonald, Al – 201
McDonough, George – 72, 183
McKay, George – 160
McLemore, Gene – 174
McMahan, Bill – 174
McMillan, Harold – 182
McMillan, Jim – 160
McMillion, Jim – 188
McNabb, Walter D. – 173, 174
McNair, Ann – 160
McPeak, Bill – 156
Meadows, Charles E. – 195
Medaris, John B. – 38, 39, 40, 41, 95, 153, 154, 174, 181, 203
Meister, Vernon – 149
Merbold, Ulf – 75
Mercury – 66, 97, 144, 155, 158, 171, 189, 197
Mercury 7 – 171
Mercury program – 50, 75
Mercury-Redstone – 5, 10, 20, 41, 170, 171
Mick, Bob – 182
Microgravity Glove Box – 63
Microgravity Science Glovebox – 85
Microgravity Science Program – 189
Milton, Tom – 159
Milwitski, Ben – 184
Minney, Hal – 182
Mitchell, Craig – 182
Mitchell, Ferdinand H. – 204
Mitchell, Morton – 149
Mitchell, Royce – 158
Mixon, Bob – 165
Mohlere, Ed – 122
Moll, Andy – 149
Moore, Brooks – 6, 9, 46, 94, 146, 154, 180
Moore, Morgan – 182
Morea, Saverio "Sonny" – 51, 183
Morgan, Leonard – 149
Morris, Charles – 165
Morris, Mr. – 150
Morris, Owen – 196
Morrison, Jim – 182
Mrazek, William – 42, 45, 152
Mueller, Fritz – 93, 204
Mueller, George – 49, 53, 65, 74, 170, 172, 183, 199, 208, 209
Mueller, Ursula – 93
Murphy, James T. – 9, 159, 188
Murphy, Jim – 9
Musgrave, Story – 23
Myers, Dale D. – 193

N

Naugle, John E. – 193
Naumann, Robert "Bob" – 6, 59, 76
Neal, Billy C. – 161
Near Infrared Camera and Multi-Object Spectrometer (NICMOS) – 81
Nebel, Rudolf – 93, 94
Nein, Max – 199
Neubert, Erich – 153, 180
Neutral Buoyancy Simulator (NBS) – 54, 55, 56, 76, 79, 189
Neutral Buoyancy Tank – 56
Newby, Dave – 176
Newell, Homer – 193
Nicase, P.D. – 165
Nichols, Wayne – 182
Nicholson, Rosco – 196
Nicks, Oran W. – 193
Nixon, Richard – 65, 74, 192[o]
Noblitt, Bob G. – 9
Noguchi, Soichi – 73
Noyes, George – 149

O

O'Brian, Hugh – 155
O'Connell, Kevin – 21
O'Conner, Edmund – 48
O'Connor, Edward – 172, 180, 182

O'Keefe, Sean – 81
Oberlies, Norma – 165
Oberth, Hermann – 64, 93, 94, 145, 147, 203
Odom, Jim – 81, 85, 196
Olivier, Jean – 199
Olsen, Ronald K. – 193
Operation Highwater – 38
Operation Paperclip – 36
Ordnance Missile Laboratory – 37, 38
Ordway, Frederick I. III – 6, 96, 97, 147, 156
Orion – 87, 88, 89, 90, 192, 194, 206
Owen, Clark – 166
Owens, Alvin – 149

P

Pace, Robert E. Jr. – 198
Pack, Pam – 160
Paine, Thomas – 65, 74, 192, 193, 209
Palaora, Hans – 178
Palaors – 145
Paludan, Ted – 92, 150, 164, 165, 183
Pao, Dr. – 195
Parker, Robert – 75
Parks, Wayne – 188
Parsons, Ralph M. – 152
Pascal, Ethridge – 158
Pass, Joe – 152
Patterson, Floyd (Pat) – 174
Paul, Hans – 152, 201
Pearson, Jim – 170, 171, 187
Pegasus – 42, 55, 155
Pegasus program – 38, 59
Peoples, Jerry – 186
Perry, Ace – 208
Perry, Guy – 174
Pershing – 47, 97, 144
Peters, Elbert – 165
Peterson, Donald – 51
Petroff, Ralph – 97
Petrone, Rocco – 53, 58, 60, 67, 98, 162, 163, 194, 209
Phillips, John L. – 73
Phillips, Sam – 159
Piccard, Jacques – 155, 196
Pickering, James – 39
Pickering, William H. – 193
Pierce, Ernie – 149
Pinkerton – 202
Pioneer 4 – 97
Pioneer-Jupiter – 99
Pittman, William – 165
Pitts, Ellington – 165
Plowden, John – 196
Polis, Gran – 160
Ponder, Troy – 165
Poppel – 145
Porter, George – 164
Porter, Richard W. – 147
Posey, Linda – 9
Pottenger, John – 92
Powell, J.T. – 165
Powell, Luther – 74, 75, 85
Power, James – 165
Powers, Emmitt – 208
Presto, Al – 149
Price, Harold – 203

R

Price, John – 165
Priddy, Erwin – 202
Priest, Pete – 188
Prince, Mr. – 123
Proffitt, Ray – 152, 173
Project Highwater – 42, 59
Project Mercury – 40
Project Paperclip – 149
Pylant, William F. – 179

R

Raketenflugplatz (Rocket Airfield) – 93
Randall, Joseph L. – 94
Ranger Program – 166
Reagan, President – 70, 71, 84, 201
Redstone – 5, 18, 38, 39, 40, 42, 47, 92, 93, 94, 95, 97, 144, 146, 148, 150, 151, 152, 153, 154, 157, 165, 169, 170, 174, 187, 189, 194, 201
Redstone Airfield – 25, 176
Redstone Arsenal – 36, 37, 39, 40, 41, 92, 93, 94, 95, 96, 97, 144, 146, 147, 148, 150, 153, 154, 164, 170, 174, 176, 179, 201, 202, 203
Redstone program – 44
Redstone Project – 38
Redstone Test Stand – 18
Reed, Billy – 165
Rees, Eberhard – 48, 60, 70, 94, 95, 96, 98, 154, 159, 169, 172, 180, 184, 186, 192, 209
Rees, Gerlinde – 95
Reeser, Al – 196
Reeves, Bob – 208
Rehm, Gil – 149
Reinartz, Stan – 199
Reisig, Gerhard – 204
Reusable Launch Vehicle (RLV) program – 79
Rhoades, Richard G. – 205
Richards, Ludie – 158
Richardson, Jerry – 75
Ricks, Gene – 160
Ridgeway, Matthew – 37
Riedel, Klaus – 93
Rieger, Mr. – 123
Riehl, Bill – 149, 151
Riley, Charles L. – 158
Rivers, Mr. – 123
Roberts, Ray – 165
Robinson, Stephen K. – 73
Rocket-based Combined Cycle (RBCC) program – 79
Rockwell, Norman – 207
Rogers, William P. – 71
Roman, Nancy – 208
Rorex, James – 164, 165
Rose, President – 203, 204
Rosinski, Erika – 95
Rosinski, Kurt – 95
Rosinski, Werner – 147, 152
Rossman, Ken L. – 146, 147
Roth, Axel – 6, 95, 188

Roth, Ludwig – 95, 145
Rouse, Carroll – 159
Rudolph, Arthur – 144, 159, 162
Rutland, Cary H. – 156
Rutledge, Bill – 188
Rutledge, Frank – 171, 173
Ryan, Cornelius – 181, 182
Ryan, Robert – 69

S

Sabatino, Ray – 149
Saidla, Robert "Bob" – 172, 174, 196, 208
Sainker, Irv – 180
Sallo, Bud – 173
Saturn – 5, 40, 41, 50, 51, 52, 53, 59, 66, 69, 78, 84, 92, 97, 98, 99, 148, 150, 153, 155, 156, 157, 158, 159, 164, 168, 172, 175, 176, 178, 180, 186, 189, 190, 199, 202, 208
Saturn 1 – 43
Saturn 1B – 43, 172, 185, 205, 206
Saturn I – 22, 40, 41, 42, 44, 45, 46, 47, 49, 50, 68, 95, 148, 151, 153, 158, 162, 164, 169, 174, 175, 176, 178, 187, 190, 191, 192, 209
Saturn I program – 47, 53
Saturn IB – 42, 45, 46, 47, 49, 60, 95, 148, 151, 157, 158, 172, 187, 190, 191, 200, 201, 208, 209
Saturn IB program – 47
Saturn IB-Apollo – 158
Saturn program – 37, 41, 45, 47, 50, 55, 61, 65, 67, 68, 69, 88, 89, 144, 206
Saturn V – 21, 27, 31, 32, 43, 44, 45, 46, 47, 48, 49, 52, 54, 56, 68, 69, 88, 93, 94, 95, 144, 146, 148, 151, 152, 156, 157, 158, 159, 160, 161, 162, 163, 164, 165, 168, 169, 170, 171, 172, 176, 177, 181, 182, 185, 190, 191, 192, 193, 197, 204, 205, 206, 207, 208, 209
Saturn V program – 47
Saturn V-Apollo – 162, 182, 191
Saturn V-Apollo program – 193
Saturn-Apollo – 44, 51, 65, 162, 181, 206, 208
Saturn-Apollo program – 72, 96, 98, 122, 162, 208
Saturn-Skylab – 55
Saunders, Grady – 165
Scarbrough, Lou – 171
Schafer, Herb – 186
Scharnowski, Heinz – 152
Scheer, Julian W. – 193
Schilling, Martin – 152, 202
Schirra, Walter "Wally" – 171, 196

Schlidt, Rudi – 204
Schmidt, Edward – 193
Schneider, E.L. – 186
Schultz, Charles – 160
Schultz, Harold – 165
Schulze, W.A. – 204
Schweitzer, Albert – 16
Schwidetzky, Walter – 164
Schwinghamer, R.J. "Bob" – 47, 67, 160, 189
Scofield, Harold – 160
Scoggins, Jim – 178
Scollard, Joe – 149
Scott, David R. – 166
Scott, Donald R. "Don" – 180
Seal, Jesse – 165
Seamans, Dr. – 169
Searcy, R.B. "Spec" – 182
Seibert, Gunther – 76
Seick, Bob – 159
Sendler, Karl – 152
Sessions, Jeff – 13
Shapley, Willis – 193
Sharpe, Max – 155, 167
Sharpe, Mitchell R. – 161
Shelby, Richard – 12
Shepard, Alan – 5, 10, 18, 20, 42, 95, 97, 144, 171
Shepherd, J.T. – 122
Shepherd, Jim – 198
Shields, Bill – 158
Shields, William – 165
Shippey, Willie – 177
Shirey, John – 177
Shockley, Wayne J. – 148
Shoun, Charles W. – 148, 149
Shriver, Sergeant – 181
Shuskus, Al – 149
Shuttle – 70, 74, 75, 76, 77, 78, 80, 81, 83, 84, 98, 99, 178, 180, 191, 192, 194, 198, 200, 207
Shuttle External Tank program – 85
Shuttle program – 60, 64, 68, 70, 80, 84, 88, 155, 159, 168
Shuttle-Spacelab – 76, 77
Shymkus, Bob – 149
Siebel, Mathias – 57, 62
Sieber, Werner – 152
Silverstein, Abe – 45
Singer, Chris – 80
Singer, Fred – 147
Skylab – 5, 50, 52, 53, 54, 55, 56, 57, 58, 59, 60, 62, 63, 74, 75, 84, 97, 99, 148, 151, 153, 154, 155, 157, 158, 180, 183, 186, 190, 192, 193, 194, 197, 202, 207, 208, 209
Skylab 2 – 56, 57
Skylab 3 – 58
Skylab 4 – 52, 57
Skylab B – 58
Skylab I – 157, 209
Skylab II – 209
Skylab III – 209
Skylab program – 37, 64, 65, 67, 68, 95, 98, 157, 158, 202
Skylab Project – 208, 209
Slattery, Bart – 96, 181
Slayton, Donald "Deke" – 29, 171

Sloan, Joe – 191
Smith, Dick – 158
Smith, Edward – 193
Smith, Gerald – 67
Smith, J.W. – 191
Smith, Martha W. – 179
Smith, Mr. – 123
Smith, O.E. – 191
Smith, Orville – 165
Smith, Richard – 68, 194
Smithsonian Astrophysical Observatory – 82
Sneed, Bill – 159, 188
Snoddy, Bill – 38, 53, 59, 60, 188
Solar Array Passive Long Duration Exposure Facility (LDEF) – 63
Solid Rocket Booster – 64, 67, 70, 71, 72, 73, 78, 194, 195, 200
Solid Rocket Booster program – 66
Solid Rocket Development Motor-2 (DM-2) – 66
Sorensen, Victor – 96, 150, 180
Sortie Can – 74, 153
Sortie Lab – 74, 75
Souther, Charles – 147
Sowell, Ken – 182
Space Processing Applications Rockets (SPAR) – 62
Space Shuttle – 5, 53, 58, 60, 62, 68, 69, 70, 72, 73, 74, 77, 78, 79, 80, 81, 84, 87, 88, 89, 91, 92, 93, 98, 99, 142, 148, 151, 157, 158, 178, 186, 188, 189, 194, 195, 197, 198, 199, 200, 201, 202, 205
Space Shuttle External Tank – 19, 64, 157
Space Shuttle Main Engine (SSME) – 66, 67, 69, 72, 157, 164, 191
Space Shuttle Main Engine (SSME) program – 66
Space Shuttle program – 59, 64, 88, 91, 99, 151, 191
Space Shuttle Project – 159
Space Shuttle Solid Rocket Booster – 148, 157, 195
Space Station – 43, 53, 77, 144, 188, 198, 199, 200
Space Station program – 99
Space Telescope Imaging Spectrograph (STIS) – 81
SpaceHab – 76
Spacelab – 5, 53, 60, 63, 70, 74, 75, 76, 77, 78, 80, 98, 99, 153, 157, 188, 189, 198
Spacelab 1 – 76, 77
SpaceLab 3 – 202
Spacelab I – 74, 75
SpaceLab J – 202

Spacelab One – 153
Spacelab program – 38, 74, 77, 95, 99
Sparkman, John – 36, 192
Sparks, Robert O. – 179
Spaulding, Mary – 9
Speer, Fred – 186
Spilman, James C. – 161
Spinak, Abraham – 193
Splawn, Jim – 8, 9, 10
Sputnik – 39, 41, 97, 145, 147, 151, 154, 189
Sputnik I – 147, 148, 181
Sputnik II – 39, 181
Stafford, Thomas – 29, 195
Stamey, Jim – 208
Stephens, K. – 165
Stephenson, Arthur G. – 78, 79, 99
Stewart, Frank M. – 162
Stocks, Charles D. – 189
Stone, Donald – 165
Stone, Mr. – 123
Stone, Richard (Dick) N. – 177
Story, Milt – 182
Stowe, Mary – 165
Stroud, John – 158
Stuhlinger, Ernst – 14, 16, 37, 38, 52, 55, 59, 60, 61, 78, 80, 93, 96, 97, 100, 147, 156, 202, 204, 208
Stulting, Jim – 158
Super Guppy – 48, 176
Susko, Michael – 197
Sutherland, W. – 165
Swanson, Conrad – 202
Swanson, Gloria – 155
Swindall, Paul – 165
Swindell, Calvin L. – 161

T

Talley, Drayton – 165
Tanner, Ray – 160
Taylor, James C. "Tex" – 164
Teal, M. – 165
Technology Test Bed – 25, 69
Tepool, Ron – 173
Tessman, Bernard – 187, 191
Teuber, Dieter – 154
Thiel, Adolf – 37
Thiel, Walter – 94
Thomas, Andrew S.W. – 73
Thomas, James W. Jr. – 190
Thomas, Leslie – 165
Thomas, Sylvia B. – 178, 179
Thomason, Herman – 188
Thompson, J.R. – 66, 71, 98, 196, 201
Thompson, Robert (Bob) – 196, 197
Thompson, Wilbur E. – 197
Thompson, Zack – 164
Threlkeld, Bill – 165
Threlkeld, William – 165
Todd, Ashford – 148
Toftoy, Holger N. – 36, 37, 39, 146, 154

Tondura, Steve – 160
Touchstone, A. – 165
Townsen, Don – 166
Trapalis, Catherine – 155
Trapalis, Charles P. Jr. – 155
Trexler, Harold – 186
Troxell, Bob – 149
Truman, President – 147
Truszynsky, Jerry – 193
Tschinkel, Johann – 204
Tsiolkovsky, Constantine – 93
Tucker, Francis – 163
Tucker, Grover – 165
Turner, Gerry – 202

U

Unity – 85

V

van Allen, James – 39
Vandersee, Fritz – 171, 174, 187
Vanguard – 148
Vanguard missile – 39, 40
Vann, Ursula Mrazek – 165
VanRensselaer, Frank – 166
Vaughan, Bill – 191
Vaughan, Otha H. Jr. – 201
Vaughan, William W. – 178
Vaughn, Farley – 175
Vedane, Chuck – 46
Verbal, Jim – 187
von Braun, Magnus – 145
von Braun, Maria – 147, 176, 192
von Braun, Wernher – 5, 10, 17, 36, 38, 39, 40, 41, 43, 44, 45, 46, 47, 49, 50, 52, 53, 55, 63, 64, 65, 70, 80, 83, 84, 88, 93, 94, 95, 96, 97, 98, 100, 122, 123, 144, 145, 146, 147, 148, 149, 150, 152, 153, 154, 155, 156, 159, 160, 161, 162, 169, 170, 171, 172, 173, 174, 175, 176, 178, 179, 180, 181, 182, 183, 185, 188, 189, 192, 193, 195, 197, 198, 202, 203, 204, 205, 207, 208, 209
von Saurma, Friedrich Graf – 96
von Saurma, Ruth – 6, 96, 181
Voyager – 52, 183

W

Wade, Thomas M. – 150
Waite, Jack – 187
Wales, William W. Jr. – 185, 186
Wallace, Gabriel – 165
Ward, Elmer – 178
Ward, George – 180
Warren, Sylvia B. – 179
Weaver, Willie E. – 148
Weidner, Hermann – 47, 152
Weiler, Jerry – 160, 166
Weisskopf, Martin – 81, 82

Weitz, Paul – 56
West, George – 191
Whitaker, Ann – 63
Whitbeck, Phil – 149, 150
White Sands Proving Ground – 36, 95, 97, 144, 147, 152
White, Ed – 95
Whitesides, George – 97
Whittenstein, Gerry – 166
Wible, Mr. – 123
Wicks, Gary – 9, 153
Wiesenmaier, Bernie – 149
Wilkins, Stanley C. – 149
Willard, James D. – 97
Willhite, Gordon D. – 203
Williams, Frank – 198
Williams, Frazier – 165
Williams, Jim – 178, 197
Williams, John – 165
Williams, Manley – 182
Williams, R. – 165
Williams, Robert (Bob) – 195
Willis, Albert E. – 164
Wilson – 202
Wilson, Hamp – 182
Wilson, Jerry – 196
Wilson, Peggy – 182
Winkler, Carl – 149, 163, 164
Wise, Jim – 177
Wittenstein, Gerald – 159, 160
Woerdemann, Hugo – 164
Wojtalik, Fred – 82
Wolf, Robert – 160
Wolfsburger, John – 160
Wood, Charles – 149
Wood, Charles C. – 192
Woodham, Bob – 182
Woosley, Al – 158
Worlund, Leonard – 69
Wright, Mike – 6, 122
Wright, Nancy R. – 179
Wright, Orville – 123
Wright, Wilbur – 123
Wuenscher, Hans – 58
Wyatt, D.D. – 193
Wyne, Bob – 149
Wysong, Earl M. – 179

X

X-33 program – 89, 99
X-34 program – 99

Y

Yardley, John – 159
Yarkin, Col. – 172
Young, John – 51
Young, Ronald L. – 199

Z

Zarya – 85
Zeanah, Hugh – 164
Zeman, Sam – 149
Zimmerman, Joe – 165
Zoike, Helmut – 93, 164

Astronaut Jim Irwin sa
of the Moon, April 1, 19
are the lunar module
(LRV). This was the firs